Complex Analysis and Applications

Second Edition

Alan Jeffrey

University of Newcastle upon Tyne
United Kingdom

Published in 2006 by
Chapman & Hall/CRC
Taylor & Francis Group
6000 Broken Sound Parkway NW, Suite 300
Boca Raton, FL 33487-2742

Printed in the United States of America on acid-free paper
10 9 8 7 6 5 4 3 2 1

International Standard Book Number-10: 1-58488-553-X (Hardcover)
International Standard Book Number-13: 978-1-58488-553-5 (Hardcover)

Library of Congress Cataloging-in-Publication Data

Catalog record is available from the Library of Congress

Taylor & Francis Group
is the Academic Division of Informa plc.

Visit the Taylor & Francis Web site at
http://www.taylorandfrancis.com

and the CRC Press Web site at
http://www.crcpress.com

Preface

This volume develops complex analysis for students of applied mathematics and engineering, with special attention directed toward its applications. The second edition of this book has been restructured and completely revised. By rearranging the order of the chapters, the general development of complex analysis now forms the first three chapters of the book, while conformal mapping and its application to boundary value problems for the two-dimensional Laplace equation form the subject matter of the last two chapters. This separation of material makes the book more convenient for readers whose interests lie mainly in complex analysis and also for those who wish to gain an understanding of the important geometrical interpretation of complex analysis, and its applications to Dirichlet and Neumann boundary value problems.

This new edition, with very few exceptions, retains all of the topics in the first edition of the book and the revisions of the text have been made to emphasize important points in both the theory and application of complex analysis. As in the first edition, the introduction of each new idea is followed by examples illustrating its application. Sections are supported by large sets of exercises, the working of a selection of which is essential to develop an understanding of complex analysis. The purpose of the more routine exercises is to develop a familiarity with the manipulation that is essential when working with complex analysis. However, many other exercises are more demanding and they may involve small extensions of ideas found in the text. In the main, the more challenging exercises are collected at the end of an exercise set.

Although computers play no part in the analytical development of complex analysis, the use of a sophisticated computer algebra system is invaluable when making applications. The more complicated graphical plots in the text have all been produced using such a system and both the need for, and the benefits of the use of such systems is reflected in the small number of straightforward exercises that require a computer algebra system for their solution.

As in the first edition, the application of complex analysis to two-dimensional boundary value problems has been confined to the consideration of temperature distribution, fluid flow, and electrostatic problems. To limit the size of the book, other applications of complex analysis, for example elasticity, have been omitted. In each case, in order to show the relevance of complex analysis, each application is preceded by a concise introduction to the mathematical background of the subject, designed to show how a real valued potential function and its related complex potential can be derived from the mathematics that describe the physical situation.

No book can be free from external influences, and this one is no exception because it reflects the many books that have influenced the author, discussions with colleagues over the years, and the responses of students to the courses on which this book is based. The suggested reading and bibliography list at the end of the book is not intended to be comprehensive. The first group of books has been chosen because they are in some ways similar to this one, and so can serve to present different accounts of much of the material found here, while the second more advanced group lists some of the books the author has found useful.

Acknowledgments

It has been a pleasure to have collaborated with the many individuals who worked on this project. I wish to express my thanks to Lisa Van Horn, whose copyediting determined the final style of the book; and to the staff of Macmillan India, Ltd.: John Sollami, production manager, and the production team in Bangalore for converting rough files and diagrams into elegant pages. Thanks are also due to the staff of Taylor & Francis: Julie Spadaro, project editor, whose efficiency and guidance have expedited the various stages of production; Helena Redshaw, manager, Editorial Project Development, for her help during pre-production; and, finally, Sunil Nair, publisher, for agreeing to publish the second edition of my book.

Complex Analysis and Applications

Second Edition

Contents

Chapter 1 Analytic Functions
1.1 Review of Complex Numbers 1
1.2 Curves, Domains, and Regions 27
1.3 Analytic Functions 34
1.4 The Cauchy–Riemann Equations: Proof and Consequences 53
1.5 Elementary Functions 63

Chapter 2 Complex Integration
2.1 Contours and Complex Integrals 89
2.2 The Cauchy Integral Theorem 107
2.3 Antiderivatives and Definite Integrals 120
2.4 The Cauchy Integral Formula 128
2.5 The Cauchy Integral Formula for Derivatives 135
2.6 Useful Results Deducible from the Cauchy Integral Formulas 145
2.7 Evaluation of Improper Definite Integrals by Contour Integration 158
2.8 Proof of the Cauchy–Goursat Theorem (Optional) 198

Chapter 3 Taylor and Laurent Series: Residue Theorem and Applications
3.1 Sequences, Series, and Convergence 203
3.2 Uniform Convergence 219
3.3 Power Series 229
3.4 Taylor Series 242
3.5 Laurent Series 255
3.6 Classification of Singularities and Zeros 280
3.7 Residues and the Residue Theorem 288
3.8 Applications of the Residue Theorem 303
3.9 The Laplace Inversion Integral 322

Chapter 4 Conformal Mapping
4.1 Geometrical Aspects of Analytic Functions: Mapping 333
4.2 Conformal Mapping 348
4.3 The Linear Fractional Transformation 360
4.4 Mappings by Elementary Functions 378
4.5 The Schwarz–Christoffel Transformation 392

Chapter 5 Boundary Value Problems, Potential Theory, and Conformal Mapping
5.1 Laplace's Equation and Conformal Mapping: Boundary Value Problems 409
5.2 Standard Solutions of the Laplace Equation 421
5.3 Steady-State Temperature Distribution 446
5.4 Steady Two-Dimensional Fluid Flow 466
5.5 Two-Dimensional Electrostatics 499

Solutions to Selected Odd-Numbered Exercises 521

Bibliography and Suggested Reading List 555

Index 557

To the memory of my dear wife Lisl
and to our children and grandchildren
who have given us so much happiness

1

Analytic Functions

1.1 Review of Complex Numbers

This section reviews the elementary properties of complex numbers that are encountered in any first course on the subject. Its purpose is to collect, in a concise form, the basic concepts and algebraic operations on complex numbers that will be needed in what is to follow.

In the Cartesian representation of a complex number, also called the real and imaginary form of a complex number, the general complex number z is written

$$z = x + iy, \tag{1.1}$$

where x and y are real numbers, and i is the imaginary unit in the complex number system with the property that

$$i^2 = -1. \tag{1.2}$$

The quantity iy in Equation (1.1) is to be interpreted as the imaginary unit i scaled (multiplied) by the real number y; the symbol i is placed before the number y in Equation (1.1) to emphasize the role played by the imaginary unit. Hereafter when y has a specific value, such as 3, it will be more natural to write $3i$ instead of $i3$, although $i3$ and $3i$ have the same meaning.

In Equation (1.1), the real number x is called the real part of z, which is shown by writing

$$x = \mathrm{Re}\{z\}. \tag{1.3}$$

Correspondingly, the real number y in Equation (1.1) is called the imaginary part of z, which is shown by writing

$$y = \mathrm{Im}\{z\}. \tag{1.4}$$

As usual in analysis, the set of all real numbers will be denoted by $\mathbb{R}$. When a is a real number, it will be shown symbolically by writing $a \in \mathbb{R}$. Here we have used the symbol $\in$ to denote membership in a set, which is read either as "is a member of the set" or, more simply, as "belongs to the set." Using this notation in Equation (1.1), we can write $x, y \in \mathbb{R}$. The negated symbol $\notin$ is to be read "is not a member of the set," so in Equation (1.1) we can write $i \notin \mathbb{R}$ because the square of every real number is non-negative, so the imaginary unit i cannot be a real number because $i^2 = -1$.

A complex number z in which $y = 0$ is said to be purely real, whereas one in which $x = 0$ is said to be purely imaginary. For conciseness, the set of all complex numbers will be denoted by $\mathbb{C}$, so in Equation (1.1) we have $z \in \mathbb{C}$. As the set of all real numbers is obtained from $\mathbb{C}$ by excluding all numbers z that are purely imaginary, it follows that $\mathbb{R} \subset \mathbb{C}$, where the set theoretic symbol $\subset$ is to be read as "is a proper subset of." Here, by a *proper subset*, we mean that all numbers in $\mathbb{R}$ belong in $\mathbb{C}$ as special cases, but some numbers in $\mathbb{C}$ do not belong to $\mathbb{R}$.

The equality of two complex numbers is defined as the equality of their respective real and imaginary parts, so if $z_1 = a_1 + ib_1$ and $z_2 = a_2 + ib_2$, writing $z_1 = z_2$ implies the *two* real results $a_1 = a_2$ and $b_1 = b_2$.

Associated with every complex number $z = x + iy$ is another complex number called its complex conjugate, denoted by $\bar{Z}$, to be read "z bar," and defined as

$$\bar{z} = x - iy. \tag{1.5}$$

A comparison of Equations (1.1) and (1.5) shows that the operation of forming the complex conjugate of a complex number, called the *operation of conjugation*, simply involves reversing the sign of the imaginary part of z, leading to the obvious result that

$$\overline{(\bar{z})} = z. \tag{1.6}$$

An immediate consequence of the definition of a complex conjugate is that when z is purely real, $\bar{z} = z$; but when z is purely imaginary, $\bar{z} = -z$.

The algebra of complex numbers becomes clearer if they are represented geometrically. This is accomplished by representing a complex number $z = a + ib$ as a point (a, b) with respect to the ordinary rectangular Cartesian axes $O(x, y)$, with a the x-coordinate of z and b its y-coordinate. Thus the complex numbers represented by points on the x-axis are purely real numbers, while those represented by points on the y-axis are purely imaginary numbers. With this representation in mind, a point with coordinate b on the y-axis is understood to be the complex number ib. For obvious reasons the x-axis is called the *real axis* and the y-axis is called the *imaginary axis*. When using this geometrical approach, the (x, y)-plane is called the *complex plane*, which for convenience is often denoted by $\mathbb{C}$, although originally the graphical representation of a complex number was called an argand diagram. Another name for the complex plane, which will be used later, is the z-plane because complex numbers are usually represented by the symbol z. It will be seen later that the z-plane plays an important role throughout complex analysis, and that it is particularly important when *conformal transformations* are used to solve boundary value problems for a partial differential equation called the *Laplace equation.*

The zero or null complex number 0 is defined as the complex number with zero real and imaginary parts. So $z = 0$ if and only if, $\text{Re}\{z\} = 0$ and $\text{Im}\{z\} = 0$, so when written out in full the complex number $z = 0$ has the form $z = 0 + 0i$.

The sum of two complex numbers $z_1 = a_1 + ib_1$ and $z_2 = a_2 + ib_2$, written $z_1 + z_2$, is defined as the complex number whose real part is the sum of the real parts of z_1 and z_2, and whose imaginary part is the sum of the imaginary parts of z_1 and z_2, so that

$$z_1 + z_2 = (a_1 + a_2) + i(b_1 + b_2). \tag{1.7}$$

To proceed further, it will be convenient to anticipate two special cases of the multiplication of complex numbers, the general definition of which will be given shortly. Let us agree that $1 \times z = z$; so if $z = a + ib$ then $1 \times z = a + ib$, while the negative of z, denoted by $-z$, is given by $-z = -a - ib$.

The difference of two complex numbers z_1 and z_2 now follows from Equation (1.7) by replacing the + sign on the left of the equality sign by a − sign, and the + signs inside each of the two brackets on the right by − signs, so that

$$z_1 - z_2 = (a_1 + ib_1) + (-a_2 - ib_2) = (a_1 - a_2) + i(b_1 - b_2). \tag{1.8}$$

It follows directly from Equation (1.8) that $z + (-z) = z - z = 0$.

The Cartesian representation of complex numbers in the z-plane allows them to be interpreted as two-dimensional vectors (directed quantities) which can be seen by examination of Figure 1.1(a), where the origin represents the initial point, or base of the vector $z = a + ib$, and the point with the Cartesian coordinates (a, b) represents the terminal point, or tip, of the vector. The vector z itself is the straight line segment directed from the base of the vector to its tip. Figure 1.1(a) also shows that geometrically, the vector $\overline{z} = a - ib$ is obtained from vector $z = a + ib$ by reflecting z in the real axis, and conversely that $\overline{(\overline{z})} = z$.

The parallelogram rule for the addition of vectors is shown in Figure 1.1(b) for vectors $z_1 = a_1 + ib_1$ and $z_2 = a_2 + ib_2$. There, vector z_2 is added to vector z_1 by translating vector z_2 parallel to itself, without change of length, until the base of z_2 coincides with the tip of z_1. The vector $z_1 + z_2$ then becomes the straight line segment with its base at the base of z_1(the origin) and its tip at the tip of the translated vector z_2, the coordinates of which are $(a_1 + a_2, b_1 + b_2)$. Figure 1.1(b) also shows that the addition of complex numbers is commutative because $z_1 + z_2 = z_2 + z_1$. Figure 1.1(c) shows how the difference of the

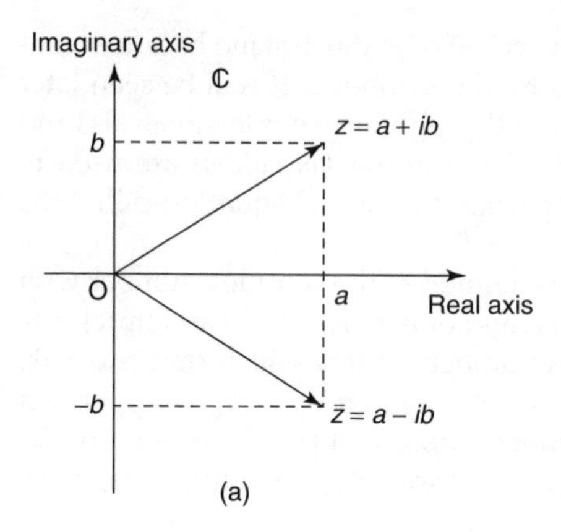

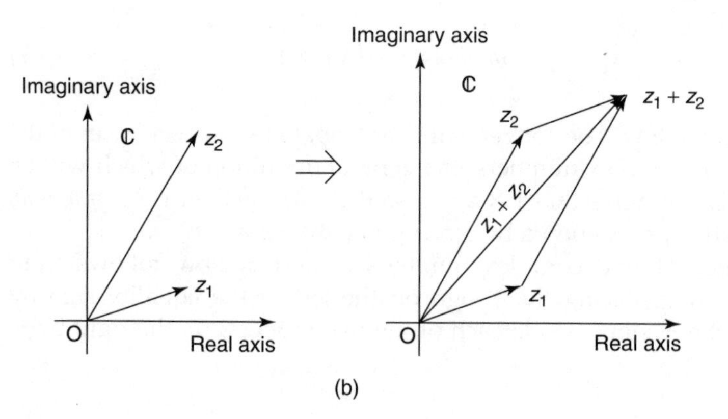

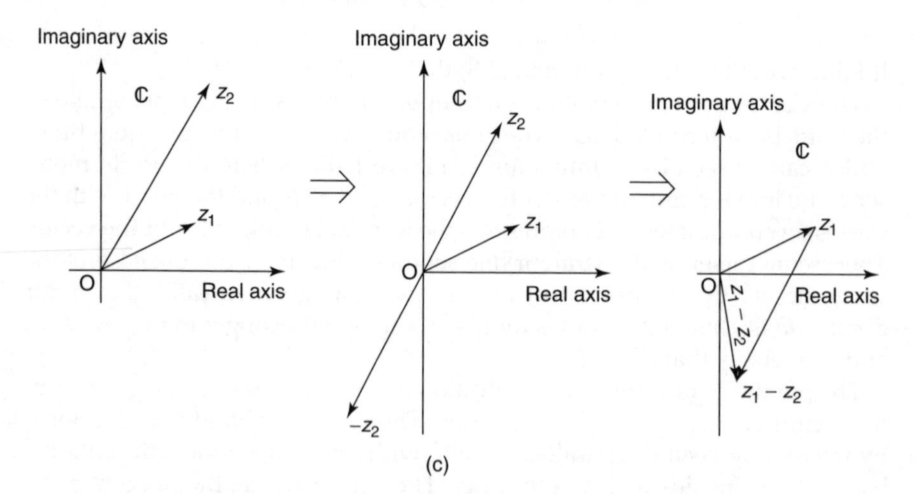

FIGURE 1.1
(a) z and $\overline{z}$ as vectors, (b) the sum $z_1 + z_2$, (c) the difference $z_1 - z_2$.

complex numbers $z_1 - z_2$ is represented geometrically as the sum $z_1 - z_2 = z_1 + (-z_2)$.

Unlike the real numbers in $\mathbb{R}$, complex numbers have no natural order, so if z_1 and z_2 are complex, it is meaningless to write $z_1 < z_2$. However, associated with any complex number $z = a + ib$ is a real number $|z|$ called its modulus, and defined as

$$|z| = \sqrt{a^2 + b^2}, \tag{1.9}$$

where the *positive* square root is always taken, so that $|z| \geqslant 0$. We remark here that $|z|$ is the analogue of the absolute value of a real number. Inspection of Figure 1.1(a) shows $|z| = \sqrt{a^2 + b^2}$ to be the length of the vector z drawn from its base (the origin) to its tip; and because $\bar{z} = a - ib$, it follows that $|z| = |\bar{z}|$. As $|z|$ is a real number, the moduli of complex numbers *can* be ordered, though this does not impose any natural order on the complex numbers themselves because all complex numbers with the same modulus, such as $|z| = r$, lie on a circle of radius r in the z-plane centered on the origin.

Given that $z = a + ib$, it is a routine matter to verify that

$$\begin{aligned} \mathrm{Re}\{z\} &= \tfrac{1}{2}(z + \bar{z}), \\ \mathrm{Im}\{z\} &= \tfrac{1}{2i}(z - \bar{z}) = \tfrac{i}{2}(\bar{z} - z), \\ \overline{z_1 + z_2} &= \bar{z}_1 + \bar{z}_2, \\ \overline{z_1 z_2} &= \bar{z}_1 \bar{z}_2. \end{aligned} \tag{1.10}$$

We will define the multiplication of the complex numbers $z_1 = a + ib$ and $z_2 = c + id$ in Cartesian form as the result of expanding the product $z_1 z_2 = (a + ib)(c + id)$ in the usual way, using the result $i^2 = -1$, and then collecting the real and imaginary parts of the product, so that

$$\begin{aligned} z_1 z_2 &= (a + ib)(c + id) \\ &= ac + iad + ibc + i^2 bd \\ &= (ac - bd) + i(ad + bc). \end{aligned} \tag{1.11}$$

In actual computations, necessitating the multiplication of complex numbers, rather than using the formal result $z_1 z_2 = (ac - bd) + i(ad + bc)$ in Equation (1.11), it is simpler to arrive at the product using the steps leading to result Equation (1.11).

Example 1.1.1

Given $z_1 = 2 + 3i$ and $z_2 = -1 + 4i$, find $z_1 z_2$, $z_1 \bar{z}_2$ and $|z_1 z_2|$.

SOLUTION

$$\begin{aligned} z_1 z_2 &= (2 + 3i)(-1 + 4i) \\ &= (2)(-1) + (2)(4i) + (3i)(-1) + (3i)(4i) \\ &= -2 + 8i - 3i - 12 = -14 + 5i. \end{aligned}$$

A similar calculation gives $z_1\bar{z}_2 = 10 - 11i$, while $|z_1 z_2| = \sqrt{10^2 + (-11)^2} = \sqrt{221}$. ◇

If z_1 is purely real, so that $b = 0$ in Equation (1.11), it follows that $az_2 = ac + iad$, while if z_1 is purely imaginary, so $a = 0$ in Equation (1.11), then $ibz_2 = -bd + ibc$. Thus, for real λ and μ,

$$\lambda z = \lambda(a + ib) = \lambda a + i\lambda b \quad \text{and} \quad i\mu z = i\mu(a + ib) = -\mu b + i\mu a, \tag{1.12}$$

justifying our earlier use of the results $1 \times z = 1 \times (a + ib) = (a + ib)$, and $-z = -a - ib$.

An important consequence of multiplication is that if $z = a + ib$, then

$$z\bar{z} = (a + ib)(a - ib) = a^2 + b^2 = |z|^2 = |\bar{z}|^2, \tag{1.13}$$

showing that the product $z\bar{z}$ is always real and such that $z\bar{z} \geqslant 0$. This simple result finds many applications, one of which occurs when complex numbers are divided.

When defining the quotient (division) z_1/z_2 of the complex numbers $z_1 = a + ib$ and $z_2 = c + id$ with $z_2 \neq 0$, we will again use the approach similar to the one used when deriving the product in Equation (1.11). The method involves first multiplying both the numerator and denominator of the quotient by $\bar{z}_2 = c - id$, when because of Equation (1.13) the denominator becomes the real number $c^2 + d^2$. The product in the numerator is then evaluated, after which the result of the quotient follows by dividing the real and imaginary parts in the numerator by $c^2 + d^2$. This is, of course, equivalent to *multiplying* the complex number in the numerator by $1/(a^2 + b^2)$. Carrying out these steps we gives

$$\begin{aligned}\frac{z_1}{z_2} &= \frac{a + ib}{c + id} = \frac{(a + ib)(c - id)}{(c + id)(c - id)} \\ &= \frac{(a + ib)(c - id)}{c^2 + d^2} \\ &= \frac{ac + bd}{|z_2|^2} + i\frac{bc - ad}{|z_2|^2}.\end{aligned}$$

So the formal definition of the quotient z_1/z_2 is

$$\frac{z_1}{z_2} = \frac{ac + bd}{|z_2|^2} + i\frac{bc - ad}{|z_2|^2}, \quad \text{when } z_2 \neq 0, \tag{1.14}$$

where the quotient z_1/z_2 is not defined when $z_2 = 0$.

This definition of a quotient is difficult to remember, so in practice when calculating z_1/z_2 it is usual to arrive at the result step by step, as in the derivation of Equation (1.14).

Example 1.1.2
Find the quotient $(2 + 3i)/(1 - 4i)$.

SOLUTION

$$\frac{2+3i}{1-4i} = \frac{(2+3i)(1+4i)}{(1-4i)(1+4i)} = \frac{(2+3i)(1+4i)}{17}$$
$$= \frac{-10+11i}{17} = -\tfrac{10}{17} + \tfrac{11}{17}i. \qquad \diamond$$

Notice that in the Examples 1.1.1 and 1.1.2 we have used of the convention introduced earlier by writing $2 + 3i$ and $1 + 4i$, in place of the notation $2 + i3$ and $1 + i4$ used in Equation (1.1), as this seems more natural when working with specific numbers.

The following important and very useful result involving the modulus of complex numbers is called the *triangle inequality*. If z_1 and z_2 are any two complex numbers, the *triangle inequality* asserts that

$$|z_1 + z_2| \leqslant |z_1| + |z_2|. \tag{1.15}$$

To prove this result notice that

$$|z_1 + z_2|^2 = (z_1 + z_2)\overline{(z_1 + z_2)} = (z_1 + z_2)(\bar{z}_1 + \bar{z}_2)$$
$$= z_1\bar{z}_1 + z_1\bar{z}_2 + z_2\bar{z}_1 + z_2\bar{z}_2$$
$$= |z_1|^2 + (z_1\bar{z}_2 + \overline{z_1\bar{z}_2}) + |z_2|^2.$$

Now

$$z_1\bar{z}_2 + \overline{z_1\bar{z}_2} = 2\mathrm{Re}\{z_1\bar{z}_2\} \leqslant 2|z_1\bar{z}_2| = 2|z_1|\,|z_2|,$$

so substituting this inequality into the previous result we obtain

$$|z_1 + z_2|^2 \leqslant |z_1|^2 + 2|z_1|\,|z_2| + |z_2|^2 = \left(|z_1| + |z_2|\right)^2.$$

The expressions on each side of the inequality are non-negative, so taking the square root of both sides of this inequality we arrive at the triangle inequality

$$|z_1 + z_2| \leqslant |z_1| + |z_2|.$$

Replacing z_1 by $z_1 - z_2$ and z_2 by $z_2 - z_3$ in the triangle inequality gives another form of the triangle inequality that is often useful

$$|z_1 - z_3| \leqslant |z_1 - z_2| + |z_2 - z_3|. \tag{1.16}$$

The triangle inequality can be used to obtain another useful inequality. Applying the inequality in Equation (1.15) to the identity $z_1 = z_2 + (z_1 - z_2)$ gives

$$|z_1| \leqslant |z_2| + |z_1 - z_2|, \quad \text{so } |z_1| - |z_2| \leqslant |z_1 - z_2|.$$

Repeating the argument, but this time starting from the identity $z_2 = z_1 + (z_2 - z_1)$, we find that

$$|z_2| - |z_1| \leqslant |z_2 - z_1|.$$

Combining these two inequalities, and using the result $|z_1 - z_2| = |z_2 - z_1|$, shows that

$$-(|z_1| - |z_2|) \leqslant |z_1 - z_2| \leqslant |z_1| - |z_2|,$$

which is equivalent to the inequality

$$|z_1 - z_2| \geqslant \big||z_1| - |z_2|\big|, \tag{1.17}$$

that was to be established. Because Equation (1.17) has been derived from Equation (1.15) the two inequalities are equivalent, though the triangle inequality is used more frequently than Equation (1.17).

Example 1.1.3

Given $z_1 = 1 - \sqrt{3}i$ and $z_2 = 3 + 4i$, verify the inequalities in Equations (1.15) and (1.17).

SOLUTION

$$|z_1| = 2, |z_2| = 5, |z_1 + z_2| = \sqrt{9 - (3 - \sqrt{3})^2} = 3.2569,$$

and

$$|z_1 - z_2| = \sqrt{1 + (3 + \sqrt{3})^2} = 4.8366.$$

Equation (1.15) is satisfied because $3.2569 < 2 + 5 = 7$, while Equation (1.17) is satisfied because $4.8366 > |2 - 5| = 3$. ◇

The triangle inequality in Equation (1.15) illustrated in Figure 1.2(a) states that complex numbers can be interpreted as vectors, which is equivalent to the familiar result due to Euclid that the length of a side of a triangle is less than or equal to the sum of the lengths of the other two sides (hence the name *triangle inequality*). Clearly, equality can only occur in result Equation (1.15) when the origin, z_1 and z_2 are all collinear, as in Figure 1.2(b).

The inequality in Equation (1.15) can be generalized by mathematical induction to the case of n complex numbers $z_1, z_2, \ldots, z_n$, when it takes the form

$$|z_1 + z_2 + \cdots + z_n| \leqslant |z_1| + |z_2| + \cdots + |z_n|. \tag{1.18}$$

Complex numbers have a different representation called the *polar form*, arising from the use of the plane polar coordinates r and θ to describe z. This

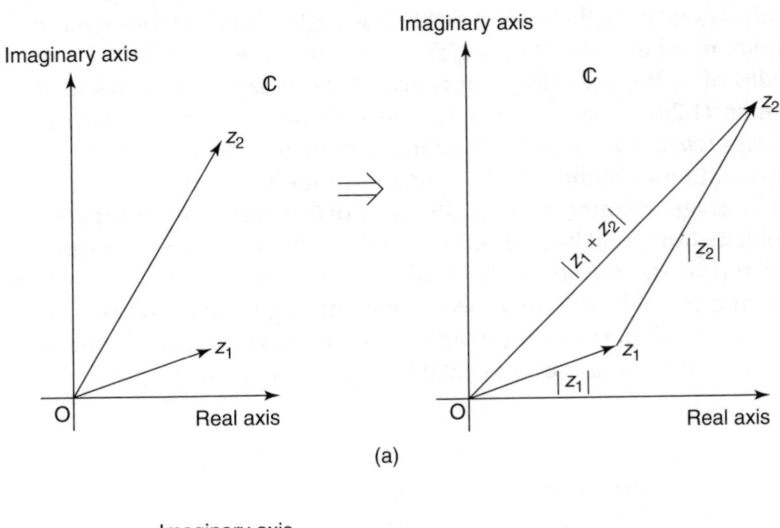

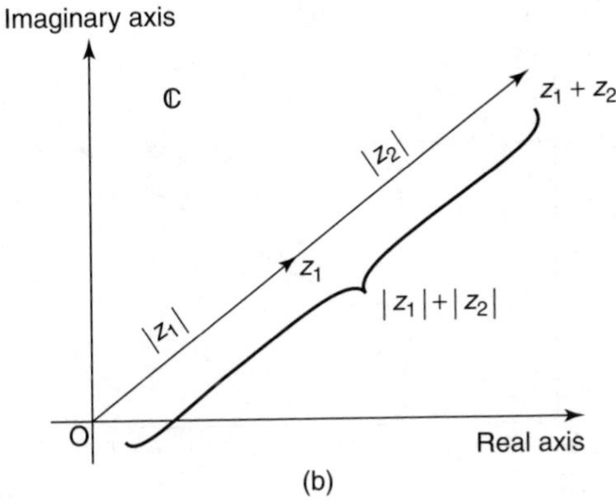

FIGURE 1.2
(a) $|z_1 + z_2| \leqslant |z_1| + |z_2|$, (b) $|z_1 + z_2| = |z_1| + |z_2|$.

form is illustrated in Figure 1.3, where point P in its real and imaginary form $z = a + ib$ is identified by its radial polar coordinate $r = |z| \geqslant 0$, measured from the origin, and the polar angle θ, that is always measured counterclockwise from the positive real axis.

We see from Figure 1.3 that

$$x = r\cos\theta \quad \text{and} \quad y = r\sin\theta, \tag{1.19}$$

so $z = x + iy$ becomes the polar form

$$z = r(\cos\theta + i\sin\theta). \tag{1.20}$$

The radial distance $r = |z|$ from the origin to a complex number z represented by a point P in the complex plane is unique, but the polar angle θ is not because replacing θ by $\theta \pm 2k\pi$, with $k = 0, 1, 2, \ldots$, while keeping r unchanged, will always identify the same point P. The angle θ is called the *argument* of the complex number z, written $\arg(z)$, so $\theta = \arg(z)$, while $r = |z|$ is called the *modulus* of z. Representing z in terms of the polar coordinates (r, θ), as in Equation (1.20), is said to specify z in its *polar form* or, equivalently, in its *modulus-argument form*. Notice that the argument of the polar form of the zero complex number is undefined, though its modulus $r = 0$.

At first sight the ambiguity in the value of $\theta = \arg(z)$ might appear to cause a problem, but it will be seen shortly that it is in fact the many-valued nature of θ that provides the key to the resolution of situations where more than one solution is possible, as will be seen when finding roots of complex numbers. From among all possible arguments θ associated with a given complex number z, we identify one that is called the *principal argument*, or the *principal*

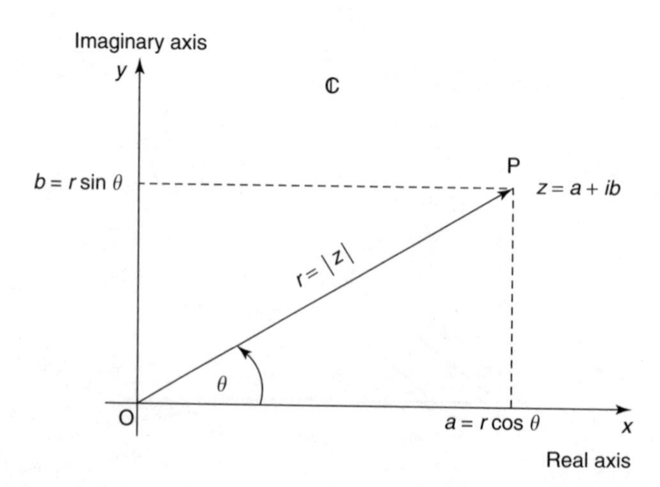

FIGURE 1.3
The polar representation of z.

value of z, and denoted by $\text{Arg}(z)$. This is the value of $\theta = \text{Arg}(z)$ chosen such that $-\pi < \theta \leqslant \pi$, corresponding to

$$-\pi < \text{Arg}(z) \leqslant \pi. \tag{1.21}$$

Notice the convention used here, where $\arg(z)$ represents the set of all possible values of θ which differ one from the other by the addition of a multiple of 2π, whereas $\text{Arg}(z)$ denotes the *unique* value of θ such that $-\pi < \text{Arg}(z) \leqslant \pi$. The connection between $\arg(z)$ and $\text{Arg}(z)$ is given by

$$\arg(z) = \text{Arg}(z) \pm 2k\pi, \quad k = 0, 1, 2, \ldots. \tag{1.22}$$

The *equality* of two complex numbers z_1 and z_2 in *polar form* is defined as the requirement that their moduli are equal, so that $|z_1| = |z_2|$, while their arguments are such that

$$\arg(z_1) = \arg(z_2) \pm 2k\pi,$$

for k zero or some positive integer. For example, if $|z_1| = |z_2| = 3$, $\arg(z_1) = 13\pi/6$ and $\arg(z_2) = -23\pi/6$, then $z_1 = z_2$, because the moduli of z_1 and z_2 are equal while $\arg(z_1) = \arg(z_2) + 6\pi$. This definition of the *equality* of the complex numbers z_1 and z_2 in polar form simply amounts to the obvious requirement that $|z_1| = |z_2|$, and $\text{Arg}(z_1) = \text{Arg}(z_2)$. Thus, in the preceding example, it is easily seen that $\text{Arg}(z_1) = \text{Arg}(z_2) = \pi/6$.

It follows directly from Equation (1.19) that

$$\theta = \tan^{-1}(y/x), \tag{1.23}$$

but θ is not unique because the inverse tangent function is many-valued, with each value differing from neighboring values by π. In addition, the interpretation of $\tan^{-1}(y/x)$ will depend on the interval chosen for the principal value of the inverse tangent function.

The standard convention used in analysis, and by calculators and computers when working with the inverse tangent function, is $-\pi/2 < \tan^{-1}(y/x) < \pi/2$. However this interval confines $\tan^{-1}(y/x)$ to the first and fourth quadrants, whereas z may also lie in either the second or the third quadrant, in which case $\tan^{-1}(y/x)$ will not satisfy the requirement that the principal value $\text{Arg}(z)$ is such that $-\pi < \text{Arg}(z) \leqslant \pi$.

This difficulty in determining the correct value for $\text{Arg}(z)$ has arisen because when forming the quotient y/x, no account was taken of the quadrant in which z was located. For example, if z lies in the *second* quadrant, then $x < 0$ and $y > 0$, so $y/x < 0$, in which case the convention $-\pi/2 < \tan^{-1}(y/x) < \pi/2$ will give a value for $\tan^{-1}(y/x)$ in the *fourth* quadrant. So to find the value of $\text{Arg}(z)$, it will be necessary to *add* π to this value of $\tan^{-1}(y/x)$. Similarly, if z lies in the *third* quadrant $x < 0$ and $y < 0$, so $y/x > 0$, in which case the convention $-\pi/2 < \tan^{-1}(y/x) < \pi/2$ will give a value of $\tan^{-1}(y/x)$ in the *first* quadrant. So to find the value of $\text{Arg}(z)$, it will be necessary to *subtract* π from the value of $\tan^{-1}(y/x)$. Given an arbitrary complex number $z = x + iy$, when using the standard convention $-\pi/2 < \tan^{-1}(y/x) < \pi/2$, the above arguments lead to the following rules for finding $\text{Arg}(z)$:

Rules for Computing Arg(*z*)

The convention $-\pi/2 < \tan^{-1}(y/x) < \pi/2$ will give the correct value $\text{Arg}(z) = \tan^{-1}(y/x)$ when $z = x + iy$ lies in the first quadrant, corresponding to $x > 0, y > 0$.

The convention $-\pi/2 < \tan^{-1}(y/x) < \pi/2$ will give the correct value $\text{Arg}(z) = \tan^{-1}(y/x)$ when $z = x + iy$ lies in the fourth quadrant, corresponding to $x > 0, y < 0$.

If z lies in the second quadrant, corresponding to $x < 0, y > 0$, then to find $\text{Arg}(z)$ it is necessary to add π to the value of $\tan^{-1}(y/x)$ that is found when using the convention that $-\pi/2 < \tan^{-1}(y/x) < \pi/2$.

If z lies in the third quadrant, corresponding to $x < 0, y < 0$, then to find $\text{Arg}(z)$ it is necessary to subtract π from the value of $\tan^{-1}(y/x)$ that is found when using the convention that $-\pi/2 < \tan^{-1}(y/x) < \pi/2$.

Example 1.1.4

Find $r = |z|$ and $\text{Arg}(z)$ when (a) $z = 1 + \sqrt{3}i$, (b) $z = 1 - \sqrt{3}i$, (c) $z = -\sqrt{3} - i$, (d) $z = -1 + i$.

SOLUTION

(a) $r = 2$, and as z lies in the first quadrant the convention will give the correct value $\text{Arg}(z) = \tan^{-1}(\sqrt{3}/1) = \frac{1}{3}\pi$.

(b) $r = 2$, and as z lies in the fourth quadrant the convention will give the correct value $\text{Arg}(z) = \tan^{-1}(-\sqrt{3}/1) = -\frac{1}{3}\pi$.

(c) $r = 2$, but here z lies in the third quadrant, so to find the correct value of $\text{Arg}(z)$ it will be necessary to subtract π from the value of θ found using the convention, so that $\text{Arg}(z) = \tan^{-1}((-1)/(-\sqrt{3})) - \pi = \frac{1}{6}\pi - \pi = -\frac{5}{6}\pi$.

(d) $r = \sqrt{2}$, but here z lies in the second quadrant, so to find the correct value of $\text{Arg}(z)$ it will be necessary to add π to the value of θ found using the convention, so that $\text{Arg}(z) = \tan^{-1}(1/(-1)) + \pi = -\frac{1}{4}\pi + \pi = \frac{3}{4}\pi$. ◇

The polar form of complex numbers makes the operations of multiplication and division very simple. To see why this is, consider the two arbitrary complex numbers in polar form

$$z_1 = r_1(\cos\theta_1 + i\sin\theta_1) \quad \text{and} \quad z_2 = r_2(\cos\theta_2 + i\sin\theta_2),$$

where r_1 and r_2 are the moduli of z_1 and z_2, and $\theta_1 = \arg(z_1)$ and $\theta_2 = \arg(z_2)$ are not necessarily the principal arguments of z_1 and z_2. Then

$$\begin{aligned} z_1 z_2 &= r_1 r_2(\cos\theta_1 + i\sin\theta_1)(\cos\theta_2 + i\sin\theta_2) \\ &= r_1 r_2[\cos\theta_1\cos\theta_2 - \sin\theta_1\sin\theta_2 + i(\sin\theta_1\cos\theta_2 + \cos\theta_1\sin\theta_2)] \\ &= r_1 r_2[\cos(\theta_1 + \theta_2) + i\sin(\theta_1 + \theta_2)], \end{aligned}$$

where the trigonometric identities $\cos(A + B) = \cos A\cos B - \sin A\sin B$ and $\sin(A + B) = \sin A\cos B + \cos A\sin B$ have been used.

A similar argument using the identities $\cos(A - B) = \cos A\cos B + \sin A\sin B$ and $\sin(A - B) = \sin A\cos B - \cos A\sin B$ shows that when $z_2 \neq 0$,

$$\frac{z_1}{z_2} = \frac{r_1}{r_2}[\cos(\theta_1 - \theta_2) + i\sin(\theta_1 + \theta_2)].$$

1.1.1 Products and Quotients in Polar Form

We have the general results that when

$$z_1 = r_1(\cos\theta_1 + i\sin\theta_1) \quad \text{and} \quad z_2 = r_2(\cos\theta_2 + i\sin\theta_2), \tag{1.24}$$

the product

$$z_1 z_2 = r_1 r_2[\cos(\theta_1 + \theta_2) + i\sin(\theta_1 + \theta_2)], \tag{1.25}$$

and when $z_2 \neq 0$ (that is $r_2 \neq 0$) the quotient

$$\frac{z_1}{z_2} = \frac{r_1}{r_2}[\cos(\theta_1 - \theta_2) + i\sin(\theta_1 + \theta_2)]. \tag{1.26}$$

When expressed in words, results in Equations (1.25) and (1.26) show that when complex numbers in polar form are multiplied, their respective moduli are *multiplied* and their respective arguments are *added*; while when they are *divided* their respective moduli are *divided* and their respective arguments are *subtracted*. The relationships between $\arg(z_1)$ and $\arg(z_2)$ are thus

$$\arg(z_1 z_2) = \arg(z_1) + \arg(z_2) \quad \text{and} \quad \arg(z_1/z_2) = \arg(z_1) - \arg(z_2), \tag{1.27}$$

but because of the definition of arg, these results will not necessarily be true if arg is replaced by Arg.

For future use we record the following results:

(a) The number 1 is represented by the point $(1, 0)$ on the real axis, so its modulus is $|1| = 1$ and $\mathrm{Arg}(1) = 0$.
(b) The number -1 is represented by the point $(-1, 0)$ on the real axis, so its modulus is $|-1| = 1$ and $\mathrm{Arg}(-1) = \pi$.
(c) The imaginary unit i is represented by the point $(0, 1)$ on the imaginary axis, so its modulus is $|i| = 1$ and $\mathrm{Arg}(i) = \pi/2$. The *geometrical effect* of multiplying a vector z (a complex number) by i is to leave its length unchanged but to rotate the vector *counterclockwise* through an angle $\pi/2$, so $\mathrm{Arg}(z)$ is increased by $\pi/2$.
(d) The negative imaginary unit $-i$ is represented by the point $(0, -1)$ on the imaginary axis, so its modulus is $|-i| = 1$ and $\mathrm{Arg}(i) = -\pi/2$.

Setting $z_1 = z_2 = z$ in Equation (1.25), with $|z| = r$ and $\arg(z) = \theta$ shows that

$$z^2 = r^2(\cos 2\theta + i \sin 2\theta),$$

while setting $3\theta = \theta + 2\theta$ and repeating this reasoning shows that

$$z^3 = r^3(\cos 3\theta + i \sin 3\theta).$$

Routine mathematical induction then establishes the general result for integral n that

$$z^n = r^n(\cos n\theta + i \sin n\theta) \quad \text{for } n = 0, 1, 2, \ldots . \tag{1.28}$$

This result remains true when n is a negative integer because setting $z_1 = 1$ in Equation (1.26), using the fact that $\mathrm{Arg}(1) = 0$ and writing z in place of z_2 and r in place of r_2, leads to the result

$$1/z = (1/r)(\cos \theta - i \sin \theta),$$

from which it follows that

$$1/z^n = (1/r^n)(\cos n\theta - i \sin n\theta), \quad \text{for } n = 1, 2, \ldots . \tag{1.29}$$

Thus result from Equation (1.29) is contained in Equation (1.28) if we allow $n = \pm 1, \pm 2, \ldots$. The special case of Equation (1.28) with $r = 1$ gives *de Moivre's theorem*

$$(\cos \theta + i \sin \theta)^n = \cos n\theta + i \sin n\theta. \tag{1.30}$$

So to find the solutions z it will be necessary to find ρ and θ in terms of the known values of R, ϕ, and n.

Recalling that the equivalence of complex numbers expressed in polar form requires their moduli to be equal, and their arguments to be equal to within an additive multiple of 2π, we see that

$$\rho^n = R \quad \text{and} \quad n\theta = \phi + 2k\pi \quad \text{for } k = 0, \pm 1, \pm 2, \ldots. \tag{1.35}$$

Thus

$$\rho = R^{1/n} \quad \text{and} \quad \theta = \frac{\phi + 2k\pi}{n} \quad \text{for } k = 0, \pm 1, \pm 2, \ldots. \tag{1.36}$$

where ρ is the positive nth root of R.

The trigonometric functions $\cos\theta$ and $\sin\theta$ are periodic with period 2π, so as the integer k increases it follows that the n permissible values of θ will be determined from the second result in Equation (1.36) by allowing k to run through any set of n consecutive integers. For convenience it is usual to take this set of integers to be $k = 0, 1, 2, \ldots, n-1$. Allowing k to increase beyond the value $n - 1$, or to decrease below zero, will simply generate the same set of n roots $z_0, z_1, \ldots, z_{n-1}$. So the roots $z_0, z_1, \ldots, z_{n-1}$ of $z = \zeta^{1/n}$ are given by

$$z_k = R^{1/p}\left(\cos\left(\frac{\phi + 2k\pi}{n}\right) + i\sin\left(\frac{\phi + 2k\pi}{n}\right)\right), \quad k = 0, 1, 2, \ldots, n-1, \tag{1.37}$$

where $R = |\zeta|$ and $\theta = \mathrm{Arg}(\zeta)$, or equivalently by $\theta = \arg(\zeta)$.

Example 1.1.7

For any fixed n, the n roots of $z^n = 1$, denoted sequentially by $\omega_0, \omega_1, \ldots, \omega_{n-1}$, are called the *nth roots of unity*. Find the form of these roots and use the result to find the cube roots of unity.

SOLUTION

In terms Equation (1.34), $\zeta = e^0 = 1$, so $R = |1| = 1$, while $\phi = \mathrm{Arg}(1) = 0$. So from Equation (1.37) the nth roots of unity are given by

$$\omega_k = \cos\left(\frac{2k\pi}{n}\right) + i\sin\left(\frac{2k\pi}{n}\right), \quad \text{with } k = 0, 1, 2, \ldots, n-1.$$

When the nth roots of unity are plotted in the complex plane they are seen to lie on a circle of radius 1 with its center at the origin, called a *unit circle* (see Figure 1.4). The roots are uniformly distributed around the unit circle, and the vector from the origin to each root is inclined at an angle $2\pi/n$ radians to the vectors drawn to the adjacent roots, with the root ω_0 located at the point $z = 1$. Thus the polygon formed by chords joining adjacent roots on the unit circle is a regular n-sided polygon. Because all points on the unit circle are at a constant unit distance from the origin, the equation of this unit circle is $|z| = 1$.

To find the cube roots of unity we set $n = 3$ in the expression for ω_k, when we find that

$$\omega_0 = 1, \quad \omega_1 = -\tfrac{1}{2} + i\tfrac{\sqrt{3}}{2}, \quad \omega_2 = -\tfrac{1}{2} - i\tfrac{\sqrt{3}}{2}. \qquad \diamond$$

Notice that because the nth roots of unity are symmetrically spaced around the unit circle with $\omega_k = \cos(2k\pi/n) + i\sin(2k\pi/n)$, it follows from the polar representation of complex numbers that $\omega_k^2 = \omega_{k+1}$, $\omega_k^3 = \omega_{k+2}$, $\omega_k^4 = \omega_{k+3}, \ldots,$ $\omega_k^n = \omega_{k+n-1}$. Thus, knowing any one of the nth roots of unity other than $\omega_0 = 1$, ω_k, enables all of the other roots to be generated by raising it to suitable powers.

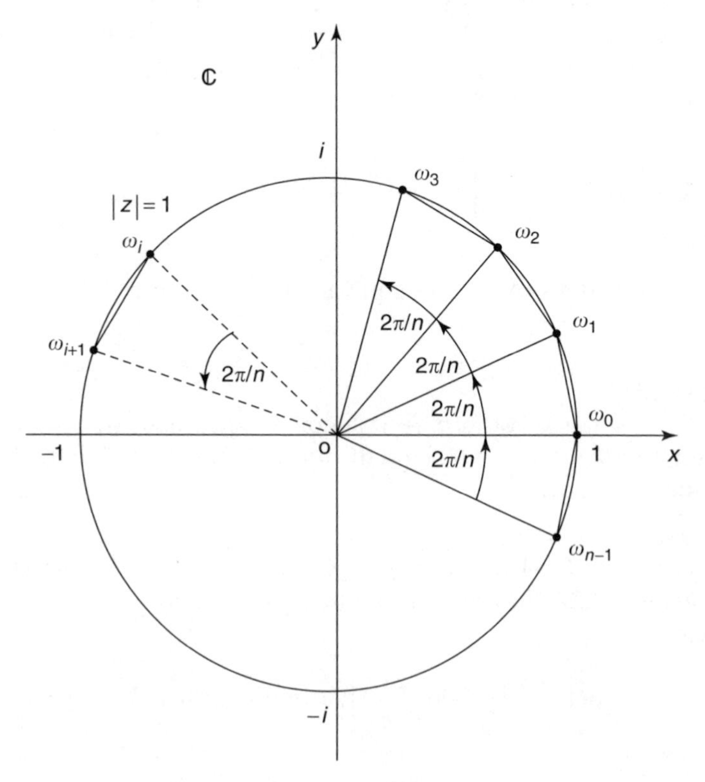

FIGURE 1.4
A plot of the nth roots of unity.

This property can be illustrated by considering $\omega_1 = -\frac{1}{2} + i\frac{\sqrt{3}}{2}$, that is one of the cube roots of 1 found in Example 1.1.7 because a simple calculation shows that

$$\omega_1^2 = -\tfrac{1}{2} - i\tfrac{\sqrt{3}}{2} = \omega_2,\ \omega_1^3 = 1, \quad \text{while } \omega_1^4 = -\tfrac{1}{2} + i\tfrac{\sqrt{3}}{2} = \omega_1.$$

Example 1.1.8

Find the values of z such that $z^3 = -4 - 4\sqrt{3}i$, and plot them in the complex plane.

SOLUTION

Using the notation of Equation (1.37), we have $n = 3$ and $\zeta = -4 - 4\sqrt{3}i$, so $R = |\zeta| = ((-4)^2 + (-4\sqrt{3})^2)^{1/2} = 8$, and $\phi = \text{Arg}(\zeta) = -\frac{2}{3}\pi$. Thus $\rho = R^{1/3} = 8^{1/3} = 2$, and $\theta = (-\frac{2}{3}\pi + 2k\pi)/3 = \frac{1}{3}(-2 + 6k)\pi$, with $k = 0, 1$ and 2. Thus the three cube roots of $\zeta = -4 - 4\sqrt{3}i$ are

$$\begin{aligned} z_0 &= 2(\cos(2\pi/9) - i\sin(2\pi/9)) \approx 1.5321 - 1.2856i, \\ z_1 &= 2(\cos(4\pi/9) + i\sin(4\pi/9)) \approx 0.3473 + 1.9696i, \\ z_2 &= 2(\cos(10\pi/9) + i\sin(10\pi/9)) \approx -1.8794 - 0.6840i. \end{aligned}$$

All three roots lie on the circle of radius 2 centered on the origin shown in Figure 1.5, with the equation $|z| = 2$. The roots are symmetrically spaced

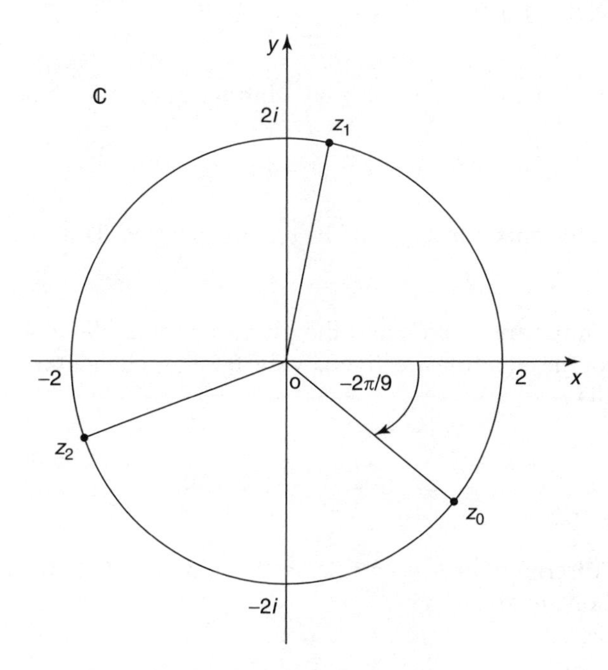

FIGURE 1.5
The cube roots of $\zeta = -4 - 4\sqrt{3}i$ spaced around the circle $|z| = 2$.

around the circle with their respective vectors inclined to one another at an angle of $2\pi/3$ radians. ◇

It is sometimes necessary to compute an expression such as $w^{q/p}$, where w is a complex number and p and q are integers. To do this we set $z = w^{q/p}$, and use the result that this is equivalent to $z^p = w^q$. As w and q are known, it is a simple matter to compute w^q, after which results in Equations (1.34) and (1.37) can be used with $\zeta = w^q$.

Example 1.1.9
Find the fourth roots of $(1 + i)^{3/4}$.

SOLUTION
Setting $z = (1 + i)^{3/4}$, it follows that $z^4 = (1 + i)^3$. Now $(1 + i)^3 = -2 + 2i$, so we need to find the fourth roots of $\zeta = -2 + 2i$. We have $R = |\zeta| = 2\sqrt{2}$, and $\phi = \mathrm{Arg}(\zeta) = 3\pi/4$, so the roots z_k given by Equation (1.38) are

$$z_k = 2^{3/8}\left[\cos\frac{(3+8k)\pi}{16} + i\sin\frac{(3+8k)\pi}{16}\right], \qquad k = 0, 1, 2, 3.$$

Thus $z_0 \approx 1.0783 + 0.7205i$, $z_1 \approx -0.7205 + 1.0783i$, $z_2 \approx 1.0783 - 0.7205i$ and $z_3 = 0.7205 - 1.0783i$. ◇

We are now in a position to solve an arbitrary quadratic equation

$$az^2 + bz + c = 0, \tag{1.38}$$

where the coefficients a, b and c can be real or complex. Three cases are to be considered:

(i) If a, b and c are all real, and the discriminant $\Delta = b^2 - 4ac \geqslant 0$, the roots of the equation are given by the familiar elementary quadratic formula

$$z_{\pm} = \frac{-b \pm \sqrt{\Delta}}{2a}, \quad \Delta \geqslant 0. \tag{1.39}$$

(ii) If the discriminant $\Delta = b^2 - 4ac < 0$, the two roots of the quadratic equation are given by

$$z_1 = \frac{b^2 - i\sqrt{-\Delta}}{2a} \quad \text{and} \quad z_2 = \frac{b^2 + i\sqrt{-\Delta}}{2a}, \quad -\Delta = 4ac - b^2 > 0. \tag{1.40}$$

(iii) If, however, the discriminant $\Delta = b^2 - 4ac$ is complex with $\Delta = \alpha + i\beta$, the $\pm$ sign in the numerator of Equation (1.39) must be replaced by a + sign, when the two roots of the quadratic equation are then given by

$$z = \frac{-b + \sqrt{\alpha + i\beta}}{2a}. \tag{1.41}$$

This is because finding the square root of a complex number automatically generates two values.

Example 1.1.10

Solve the quadratic equations:
(a) $z^3 - 3z + 4 = 0$ and (b) $4z^2 + 4z + 1 - i = 0$.

SOLUTION

(a) The coefficients are real, but the discriminant $\Delta = b^2 - 4ac = -7 < 0$, so by (ii) the two roots are $z_0 = \frac{3}{2} - \frac{1}{2}\sqrt{7}i$ and $z_1 = \frac{3}{2} + \frac{1}{2}\sqrt{7}i$.

(b) The coefficients a and b in the quadratic equation are real, but the coefficient $c = 1 - i$, and the discriminant $\Delta = i$, so by (iii) the roots are given by

$$z = \frac{-4 + \sqrt{16i}}{8}, \quad \text{which is equivalent to } z = -\tfrac{1}{2} + \tfrac{1}{2}\sqrt{i}.$$

Reasoning as in Example 1.1.8 is easily shown that the two square roots of i are given by

$$\sqrt{i} = \cos\left(\frac{(1+4k)\pi}{4}\right) + i\sin\left(\frac{(1+4k)\pi}{4}\right), \quad k = 0, 1$$

so $\sqrt{i} = -\frac{1}{\sqrt{2}}(1+i)$ and $\sqrt{i} = \frac{1}{\sqrt{2}}(1+i)$. Thus the two roots of the quadratic equation become

$$z_0 = -\tfrac{1}{2} - \tfrac{1}{2\sqrt{2}}(1+i) \quad \text{and} \quad z_1 = -\tfrac{1}{2} + \tfrac{1}{2\sqrt{2}}(1+i). \qquad \diamond$$

It is no coincidence that the roots of equation (a) in Example 1.1.10 are complex conjugates, while those of equation (b) are not. We now show why this is so by establishing a very useful property of polynomial equations in general.

The need to perform algebraic operations on complex numbers has been illustrated when solving the most general form of the quadratic equation in Equation (1.38), and the necessity will become even clearer when we come to work with *functions of a complex variable*. In anticipation of the forthcoming discussion of general functions of a complex variable, and to explain the relationships among the roots in Example 1.1.10, we will use an elementary argument to establish an important property of polynomial equations. We define a *polynomial of degree* n to be an expression of the form

$$P(z) = a_0 + a_1 z + a_2 z^2 + \cdots + a_n z^n, \tag{1.42}$$

where the term *degree* refers to the highest power of z that occurs in $P(z)$. The numbers $a_0, a_1, \ldots, a_n$ in Equation (1.42) are called the *coefficients* of the polynomial. A *root* of the polynomial equation $P(z) = 0$ refers to a specific number $z = \zeta$, say, that is a solution of the equation. The term *zero* when applied to the polynomial $P(z)$ refers to a value of z, say $z = \zeta$, with the property that $P(\zeta) = 0$. Thus a *root* is a solution of an *equation*, while a *zero* is a property of a *function*. To illustrate matters, the quadratic polynomial (function) $P(z) = z^2 + 4z + 3$ becomes zero when $z = -1$ and $z = -3$, so these are its zeros, whereas the quadratic equation $P(z) = 0$ has the roots $z = -1$ and $z = -3$. It is a fundamental algebraic result that a polynomial equation $P(z) = 0$ of degree n has n roots, where a root repeated p times is counted as p roots and called a *degenerate root*. This is a consequence of a result known as the *fundamental theorem of algebra*, and the result is true irrespective of whether the coefficients $a_0, a_1, \ldots, a_n$ of Equation (1.42) are real or complex. This result will be proved later; but in what follows we prove a useful theorem concerning polynomials because its proof only requires some of the elementary results concerning complex conjugates that have already been established.

THEOREM 1.1.1 A Property of Polynomials with Real Coefficients
If the coefficients $a_0, a_1, \ldots, a_n$ of the polynomial

$$P(z) = a_0 + a_1 z + a_2 z^2 + \cdots + a_n z^n, \tag{1.43}$$

are all real, then either all of the roots of $P(z) = 0$ are real or, if complex, they must occur in complex conjugate pairs, while if the degree of $P(z)$ is odd it must have at least one real root.

PROOF

We start from the equation determining the roots of $P(z) = 0$, namely

$$a_0 + a_1 z + a_2 z^2 + \cdots + a_n z^n = 0.$$

Taking the complex conjugate of this equation, and using the fact that all a_r are real $\overline{(a_r z^r)} = a_r \overline{z}^r$, it follows that

$$a_0 + a_1 \overline{z} + a_2 \overline{z}^2 + \cdots + a_n \overline{z}^n = 0,$$

showing that if z is a root, then so is its complex conjugate $\overline{z}$. The first part of the theorem is proved.

If $z = \zeta$ is a complex root of $P(z) = 0$, then $(z - \zeta)$ and $(z - \overline{\zeta})$ must be factors of $P(z)$, so the product

$$(z - \zeta)(z - \overline{\zeta}) = z^2 - (\zeta + \overline{\zeta})z + \zeta\overline{\zeta}$$

must also be a factor of $P(z)$. However, $\zeta + \overline{\zeta}$ and $\zeta\overline{\zeta} = |\zeta|^2$ are always real so the pair of complex conjugate roots corresponds to this quadratic factor with real coefficients. Thus to produce a polynomial with real coefficients, the roots must either all be real leading to real factors, or some may be real while the remaining pairs of complex conjugate roots will correspond to real quadratic factors.

If the degree of $P(z)$ is odd and equal to $2m + 1$, then $P(z)$ can have at most m real quadratic factors; so if the product of factors is to yield a polynomial with real coefficients, the remaining factor or factors must be real and $P(z)$ will have at least one real root. ◆

This theorem explains the behavior of the roots in Example 1.1.10 because in case (a), the quadratic polynomial had real coefficients so the roots of $P(z) = 0$ could either both be real or if complex, they must occur as a complex conjugate pair, as indeed they did. In case (b), the coefficients of the quadratic polynomial were *not* all real, so Theorem 1.1.1 did not apply, and the two complex roots found were *not* complex conjugates.

Example 1.1.11

Given the cubic polynomial $P(z) = z^3 - 3z^2 + 7z - 5$, find its roots and hence factor the polynomial.

SOLUTION

The polynomial has real coefficients so Theorem 1.1.1 applies. Because the degree of the polynomial is odd, it must have at least one real root, and inspection (trial and error) shows this to be $z = 1$, so $z - 1$ must be a factor. Dividing $P(z)$ by $z - 1$ gives

$$\frac{z^3 - 3z^2 + 7z - 5}{z - 1} = z^2 - 2z + 5,$$

so the remaining roots of $P(z) = 0$ must be the roots of $z^2 - 2z + 5 = 0$. The quadratic formula shows these to be $1 + 2i$ and $1 - 2i$, so when factored we find that

$$\begin{aligned} P(z) &= z^3 - 3z^2 + 7z - 5 \\ &= (z - 1)(z^2 - 2z + 5) \\ &= (z = 1)(z = 1 - 2i)(z = 1 + 2i). \end{aligned}$$

◇

A final example of the way an elementary argument can provide information about the roots of a polynomial is shown by Theorem 1.1.2.

THEOREM 1.1.2 The Eneström–Kakeya Theorem
Let the polynomial

$$P(z) = a_0 + a_1 z + \cdots + a_n z^n$$

have real coefficients such that $a_0 > a_1 > \cdots > a_n > 0$. *Then the polynomial* $P(z)$ *has no zeros inside the unit circle* $|z| = 1$.

PROOF

We start from the identity

$$(1 - z)P(z) = a_0 - \sum_{k=0}^{n-1} (a_k - a_{k+1})z^{k+1} - a_n z^{n+1}.$$

Then as $a_k - a_{k+1} > 0$, we have $|a_k - a_{k+1}| = a_k - a_{k-1}$ for $k = 0, 1, \ldots, n - 1$, so an application of Equation (1.17) gives

$$|(1 - z)P(z)| \geqslant a_0 - \sum_{k=0}^{n-1} (a_k - a_{k+1})|z|^k - a_n|z|^{n+1}.$$

When $|z| < 1$ we have $|z|^k < 1$, so this inequality can be strengthened to

$$|(1 - z)P(z)| > a_0 - \sum_{k=0}^{n-1} (a_k - a_{k+1}) - a_n,$$

but after expanding the right-hand side, all terms are found to cancel so we have shown that $|(1 - z)P(z)| > 0$, when $|z| < 1$. Thus $P(z)$ has no zeros when $|z| < 1$ and consequently the equation $P(z) = 0$ has no roots strictly inside the unit circle $|z| = 1$. The theorem offers no information about the behavior of the function $P(z)$ or the roots of $P(z) = 0$ on the unit circle. ◆

Example 1.1.12
Given that $P(z) = z^3 + 3z^2 + 8z + 12$, find its zeros. Check that Theorem 1.1.2 (the Eneström–Kakeya Theorem) applies, and verify that the roots of $P(z) = 0$ all lie outside the unit circle $|z| = 1$.

SOLUTION
Inspection shows that $z = -2$ is a zero of $P(z)$, so $(z + 2)$ must be a factor. As $P(z)/(z + 2) = z^2 + z + 6$, the other two zeros of $P(z) = 0$ must be the zeros

of $z^2 + z + 6$, that is, the roots of $z^2 + z + 6 = 0$, and from the quadratic formula these are found to be $-\frac{1}{2} + \frac{\sqrt{23}}{2} i$ and $-\frac{1}{2} - \frac{\sqrt{23}}{2} i$. Thus the three roots of $P(z) = 0$ are $z_1 = -2$, $z_2 = -\frac{1}{2} - \frac{\sqrt{23}}{2} i$ and $z_3 = -\frac{1}{2} + \frac{\sqrt{23}}{2} i$. The coefficients of $P(z)$ satisfy the conditions of Theorem 1.1.2, so $P(z)$ can have no zeros inside the unit circle, which is confirmed by the fact that $|z_1|$, $|z_2|$ and $|z_3|$ are all greater than 1. Equivalently, $P(z) = 0$ has no roots inside the unit circle. ◇

Exercises 1.1

1. Given $z_1 = 2 - 3i$ and $z_2 = 1 + 4i$, find: (a) $z_1 + 2z_2$ (b) $3z_1 - 4z_2$ (c) $2z_1 - 3\bar{z}_2$ (d) $4\bar{z}_1 - 2\bar{z}_2$.
2. Given $z_1 = 2 + i$ and $z_2 = 1 - 2i$, find: (a) z_1/z_2 (b) $\bar{z}_2/z_1$ (c) $z_2/\bar{z}_1$ (d) $\bar{z}_1/\bar{z}_2$.
3. Given $z_1 = 2 - 2\sqrt{3}i$ and $z_2 = 1 + i$, find: (a) $|z_1/z_2|$ (b) $z_2/|z_1|$ (c) $|z_1|/|z_2|$.
4. Find the Cartesian form of $(1 + 3i)/(1 + 2i) + (1 + 2i)/(1 - 3i)$.
5. Find the Cartesian form of $(1 - 2i)/(2 + i) + 3i/(1 - 2i)$.
6. Given $z_1 = 2 + 3i$, $z_2 = 1 - i$, $z_3 = -3 - 2i$, find (a) $z_1^2 z_3$ (b) $z_1 \bar{z}_2 z_3$ (c) $(z_2 - z_3)/z_1$.
7. Find z, given that $(z - 1)/(2z + 3i) = 2 - 3i$.
8. Given $z_1 = 3 + 2i$ and $z_2 = 2 - 4i$, verify the triangle inequality in Equations (1.15) and (1.17).
9. Give an example of complex numbers z_1 and z_2 for which $|z_1 + z_2| = |z_1| + |z_2|$.
10. Using three complex numbers of your own choice, verify the generalization of the triangle inequality in Equation (1.18) when $n = 3$. When can the inequality sign be replaced by an equality sign?
11. Find the form of z if $z^2 = (\bar{z})^2$.
12. Is it true that $|z^2| = |\bar{z}|^2$?
13. When is it true that $|(a + ib)z| = |a||z| + |b||z|$?
14. Given that $z_1 = -\frac{1}{2}(1 + i)$ and $z_2 = 1 - 2i$, find the modulus and principal argument of (a) z_1 and z_2 (b) $z_1 z_2$ (c) z_1/z_2 (d) $z_2 z_3/z_1$.
15. Convert $w = (-2\sqrt{3} + 2i)/(1 - i)$ to polar form.
16. Convert $w = (2 + 4i)/(-1 + i)$ to polar form.
17. Find an expression for $\sin 5\theta$ in terms of powers of $\sin\theta$ and $\cos\theta$.
18. Find an expression for $\cos 5\theta$ in terms of powers of $\sin\theta$ and $\cos\theta$.
19. Find an expression for $\sin^4\theta$ in terms of cosines of multiples of θ.
20. Find an expression for $\cos^4\theta$ in terms of cosines of multiples of θ.
21. Is it true that the six roots of the equation $z^6 + z^3 + 1 = 0$ are of the form $z = \cos(2k\theta/\pi) + i\sin(2k\theta/\pi)$ with $k = 0, 1, 2, 3, 4, 5$? Justify your answer.
22. If for any fixed n the number ω is any nth root of unity, show that $1 + \omega + \omega^2 + \omega^3 + \cdots + \omega^{n-1} = 0$.

Use the polar representation of the nth roots of unity to explain this result.

23. Using cross multiplication, or otherwise, verify the identity

$$1 + z + z^2 + \cdots + z^n = \frac{1 - z^{n+1}}{1 - z}.$$

Multiply the numerator and denominator of the expression on the right by $z^{-1/2}$, set $z = \cos\theta + i\sin\theta$, use de Moivre's theorem and equate the real parts on each side of the result to obtain the *Lagrange identity*

$$1 + \cos\theta + \cos 2\theta + \cdots + \cos n\theta = \frac{1}{2} + \frac{\sin\left(n + \frac{1}{2}\right)\theta}{2\sin\left(\frac{1}{2}\theta\right)}.$$

24. Using the method of Exercise 23, equate the imaginary parts on each side of the transformed identity to show that

$$\sin\theta + \sin 2\theta + \cdots + \sin n\theta = \tfrac{1}{2}\cot\left(\tfrac{1}{2}\theta\right) - \frac{\cos\left(n + \frac{1}{2}\right)\theta}{2\sin\left(\frac{1}{2}\theta\right)}.$$

25. Find the values of z such that $z^4 = -\frac{1}{2} + \frac{1}{2}i$.
26. Find the values of z such that $z^5 = 16 - 16\sqrt{3}i$.
27. Find the fourth roots of $-i^{3/4}$.
28. Find the two square roots of $1 - 2i$.
29. Show that if $z = x + iy$ and $w^2 = z$, that the square roots $w_\pm$ of z are given by

$$w_\pm = \pm\left(\sqrt{\frac{|z| + x}{2}} + i\,\mathrm{sgn}(y)\sqrt{\frac{|z| - x}{2}}\right), \quad \text{where } \mathrm{sgn}(y) = \begin{cases} 1 \text{ if } y \geqslant 0 \\ -1 \text{ if } y < 0 \end{cases}.$$

In this result the upper and lower + and − signs on the left are to be taken with the corresponding upper and lower + and − signs on the right.

30. Use the result of Exercise 29 to find the two square roots $w_\pm$ of $-\frac{1}{2} + \frac{\sqrt{3}}{2}i$, and check the result using multiplication to show that $(w_\pm)^2 = -\frac{1}{2} + \frac{\sqrt{3}}{2}i$.
31. Use the result of Exercise 29 to find the square roots of i.
32. Use the result of Exercise 29 to find the fourth roots of i.
33. Use Theorem 1.1.1 to find the roots of $z^3 + 5z^2 + 9z + 5 = 0$.
34. Use Theorem 1.1.1 to find the roots of $z^3 + 3z^2 + 3z + 2 = 0$.

35. Use Theorem 1.1.1 to find the roots of $z^3 + 5z^2 + 10z + 12 = 0$, and hence confirm the result of Theorem 1.1.2.
36. Use Theorem 1.1.1 to find the roots of $z^3 + 5z^2 + 14z + 24 = 0$, and hence confirm the result of Theorem 1.1.2.

1.2 Curves, Domains, and Regions

The need to consider curves and areas in the complex plane occurs throughout the study of complex analysis, so the nature of curves and the way they are defined must be described. An unbroken curve, or path, of finite length comprising a continuous set of points in the complex plane with distinct end points is called an *arc*. A closed curve formed by joining end to end a number of arcs, with the initial point of the first arc joined to the end point of the last arc, is called a *closed curve* or *contour* in the complex plane.

Of special importance are *simple arcs* that do not intersect themselves and contain no loops. A *simple closed curve* is a closed curve that forms a single loop such as a circle, an ellipse, or a triangular path; whereas a curve such as a figure eight, although a closed curve, is *not* a simple closed curve. We take it as axiomatic that a simple closed curve in the complex plane has points (complex numbers) that lie *inside* and points that lie *outside* **the curve**; by convention when a point in the complex plane is moved around a simple closed curve in the *positive sense,* it does so in a *counterclockwise* direction. When considering contour integrals in subsequent chapters, it will be important to distinguish between a point moving around a simple closed curve in the positive sense, and one moving around the curve in the *negative sense,* which is in a *clockwise* direction.

Figure 1.6 shows some typical simple closed curves which are formed by connecting end-to-end circular arcs and straight line segments, with the

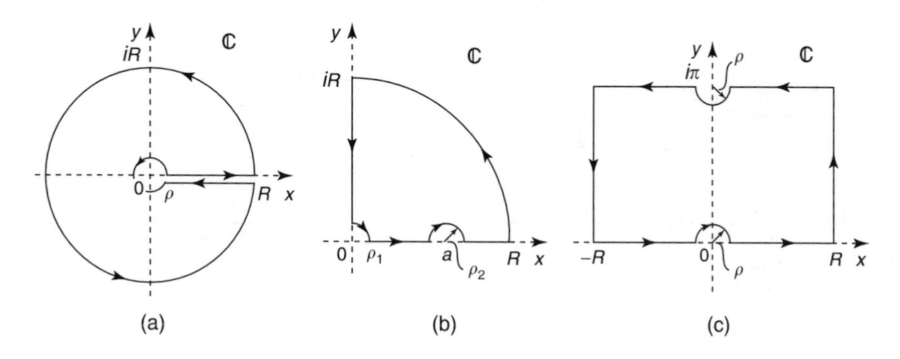

FIGURE 1.6
Examples of simple closed contours formed by joining up end-to-end circular arcs and straight line segments.

positive sense around each curve shown by arrows. Each of the small circular arcs in Figure 1.6(b, c) is called an *indentation* in the contour and when considering contour integration, indentations are necessary to avoid the contour passing through a singularity of a complex function.

When developing the calculus of complex functions, it is necessary to consider complex numbers as *complex variables*, and this happens in a different context when examining the geometrical properties of complex functions in connection with *conformal mappings*. In a conformal mapping, an arc (path) traced out as a point z moves in the z-plane is transformed by a complex function $w = f(z)$ into a related arc traced out by the variable w. So when displaying the geometrical consequences of such a transformation, called a *mapping* from z to w, it is necessary to introduce two different complex planes, one the z-plane, and the other the w-plane, though each will be contain the same set of points $\mathbb{C}$. Figure 1.7 shows how points on a circle with its center slightly displaced from the origin in the z-plane are mapped to points in the w-plane by the complex function $w = z + i/z$. The airfoil like closed simple curve in the w-plane is called a *Joukowski profile*, and is studied in Chapter 4 on conformal mapping.

A significant difference between real and complex numbers, already mentioned in Section 1.1, is that complex numbers have no natural order. So when working with the real number system $\mathbb{R}$, it makes sense when referring to limiting operations involving infinity to write $+\infty$ or $-\infty$, but no such distinction can be made with the complex number system $\mathbb{C}$, although some concept of infinity must still be retained. This is achieved by defining an idealized *"point at infinity,"* written ∞ without a $\pm$ sign, to be the set of all numbers z that lie outside a circle of an arbitrarily large radius in the z-plane with its center at the origin. A more precise definition of the point at infinity is: the set of all complex numbers z that lie within an ε-neighborhood of infinity, defined as those z such that $|z| > 1/\varepsilon$, where ε is an arbitrarily small positive number. As each point in the complex plane represents a specific complex number, from now on the terms point and complex number will be used interchangeably because they are synonymous.

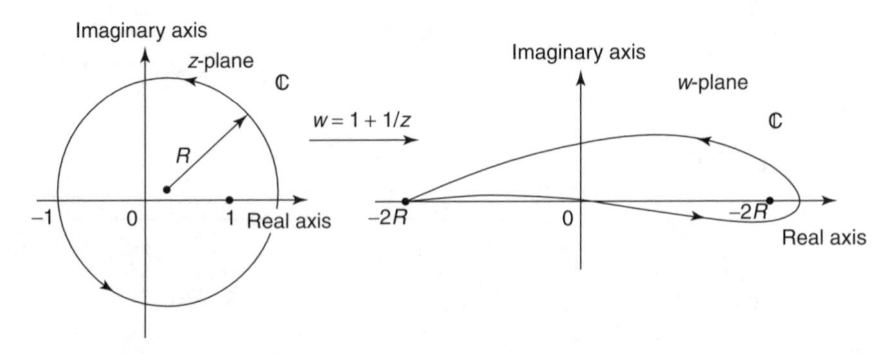

FIGURE 1.7
Two simple closed curves related by the Joukowski transformation.

The points in the complex plane $\mathbb{C}$ represent every complex number with the exception of the point at infinity. When the point at infinity is added to $\mathbb{C}$ the result is called ***the extended complex plane***, and it will always be this plane that will be used whenever limiting operations involving infinity arise.

A convenient finite geometrical representation of the extended complex plane involves using a stereographic projection in which the points in the extended complex plane are brought into one-to-one correspondence with points on a sphere of unit diameter. The idea is illustrated in Figure 1.8, where the sphere of unit diameter stands tangent to the extended complex plane at its origin O, with the point on the sphere vertically above O denoted by N for the *north pole* of the sphere, though the point O is not called the south pole.

A set of three-dimensional axes O(ξ, η, ζ) is located at O with the ξ- and η-axes coinciding, respectively, with the x- and y-axes in the complex plane, while the positive ζ-axis is drawn vertically above O through N. A point representing an arbitrary complex number $z = x + iy$ in the extended complex plane is then joined to the point N on the sphere by a straight line that intercepts the sphere at a point P(ξ, η, ζ). The coordinates (ξ, η, ζ) corresponding to $z = x + iy$ are the *stereographic coordinates* of z, and it can be seen that every finite point in the complex plane corresponds to a unique point on the sphere. The point at infinity in the extended complex plane, that is where the points outside an arbitrarily large circle in the complex plane centered on O correspond,

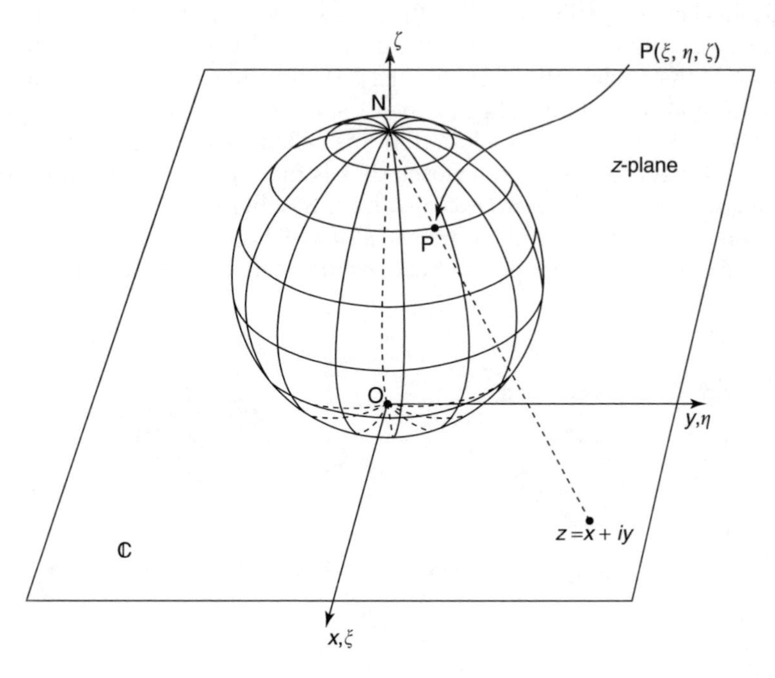

FIGURE 1.8
Stereographic projection of points on the Riemann sphere.

in the limit as the radius of the circle becomes arbitrarily large, to the point N on the sphere. This sphere is called the *Riemann sphere*, so with the sole exception of point N, every point on the Riemann sphere corresponds to a unique complex number z in the extended complex plane.

A detailed discussion of arcs and curves involves using the branch of mathematics called *topology*, but it will be sufficient to outline a few of its most important ideas. One property of any simple closed curve Γ which has already been mentioned is that it separates points in the complex plane into those that lie *inside* the curve and those that lie *outside* it. That is a simple closed curve has an *interior* and an *exterior*. This seemingly obvious result that we take as axiomatic is remarkably difficult to prove. The first attempt at a proof was given by the French mathematician Camille Jordan (1838–1922), though his proof turned out to be incomplete in some respects, and only later in 1905, was a rigorous proof given by the American topologist Oswald Veblen (1880–1960). To honor the work of Camille Jordan, the fact that a simple closed curve has both an inside and an outside is now called the *Jordan curve theorem*, and simple closed curves are also called *Jordan curves*.

Two important simple closed curves that often arise in practice are a rectangle and a circle. Figure 1.9(a) shows a set of points S in the complex plane lying strictly inside the shaded rectangular area defined by $a < \text{Re}\{z\} < b$ and $c < \text{Im}\{z\} < d$, where a, b, c and d are real numbers. Figure 1.9(b) shows a set of points S lying strictly inside a circle of radius ρ centered on the point z_0, that can be described analytically by writing $|z - z_0| < \rho$. When expressed in words this last inequality says that the points z in S are all such that the modulus of $z - z_0$ (that is the distance of z from z_0) is always strictly less than the radius ρ of the circle with its center at z_0. This circular area is called an *open disk* in the complex plane, while an area S such that $|z - z_0| \leqslant \rho$ where points on the bounding circle are included in S is called a *closed disk*. Thus the difference between an open and a closed disk is that in an open disk the points on the circular boundary are *excluded* from S, whereas in a closed disk the points on the circular boundary are *included* in S. Figure 1.9 uses the standard graphical convention that points to be excluded from S are shown as points

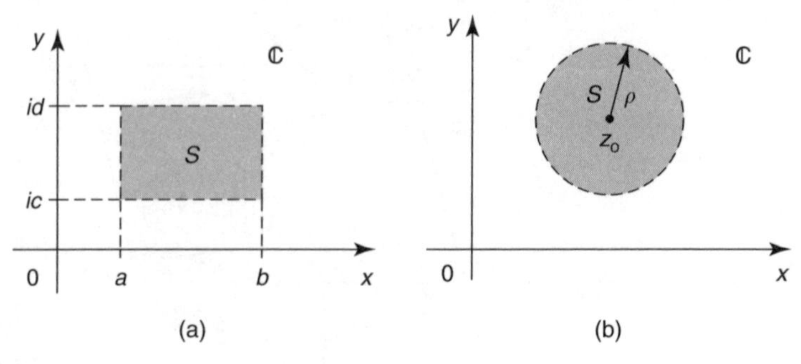

FIGURE 1.9
(a) An open rectangle. (b) A closed circular disk.

on a *dashed line*, whereas points to be *included* in S are shown as points on a *solid* line.

Often the radius ρ of an open disk is small and equal to a positive number δ, in which case the set of points $|z - z_0| < \delta$ inside the disk is called a *δ-neighborhood* of point z_0, sometimes denoted symbolically by writing $N(z_0, \delta)$. On occasion, when a singularity of a complex function arises at the point z_0 in $N(z_0, \delta)$, it is necessary to exclude the single point z_0 itself. When this occurs, we write $N'(z_0, \delta)$ and define this by the inequality $0 < |z - z_0| < \delta$. This set of points from which only the point z_0 is excluded, called a *deleted neighborhood* of z_0, and the deleted disk is then called a *punctured disk*.

A *boundary point* of a set S is a point where *every* neighborhood of the point contains points belonging to S and points not belonging to S. A set of points S with a boundary comprising a simple closed curve Γ is said to be *open* if every neighborhood of a point in S only contains points of S, irrespective of how close the point is to Γ. Correspondingly, a set of points S with a boundary comprising a simple closed curve Γ is said to be *closed* if it contains all of its boundary points. Figure 1.10(a) illustrates an example of an open set S, which shows the open sector

$$\alpha < \mathrm{Arg}\{z\} < \beta \quad \text{and} \quad |z| < R, \quad \text{with } 0 < \alpha < \beta < \pi/2, R > 0,$$

while Figure 1.10(b) shows the closed sector

$$\alpha \leqslant \mathrm{Arg}\{z\} \leqslant \beta \quad \text{and} \quad |z| \leqslant R, \quad \text{with } 0 < \alpha < \beta < \pi/2, R > 0.$$

Figure 1.10(b) also shows a typical boundary point P on the sector and an associated δ-neighborhood of P.

Often the shape of an arc or closed curve Γ in the complex plane is sufficiently complicated that it needs to be represented in *parametric form*. Such a

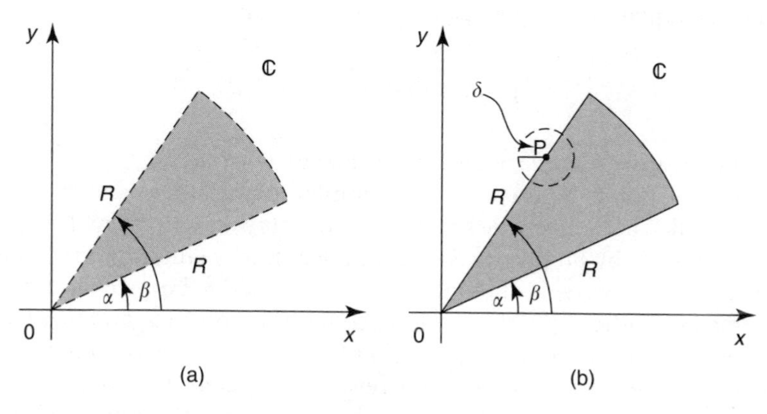

FIGURE 1.10
(a) An open sector. (b) A closed sector and a δ-neighborhood of P.

representation of Γ is accomplished in the z-plane by defining the x- and y-coordinates of points on Γ in the parametric form

$$x = x(s), \quad y = y(s) \qquad \text{for } \alpha \leqslant s \leqslant \beta, \tag{1.44}$$

where $x(s)$ and $y(s)$ are monotonic functions and s is a parameter. In terms of the complex variable z the curve Γ has the representation

$$z(s) = x(s) + iy(s), \quad \text{for } \alpha \leqslant s \leqslant \beta, \tag{1.45}$$

so as s increases the point $z(s)$ moves in a particular direction along the curve Γ. A typical example of a parametric representation involves a circle of radius r centered on the point $z = a + ib$, which can be written as

$$x = a + r\cos\theta, \quad y = b + r\sin\theta, \quad \text{with } 0 \leqslant \theta \leqslant 2\pi, \tag{1.46}$$

where θ is the parameter. It should be recognized that parametric representations of curves are not unique and an equally good parametric representation of the circle described by Equation (1.46) is

$$x = a + \cos 2\theta + i\sin 2\theta, \quad \text{with } 0 \leqslant \theta \leqslant \pi, \tag{1.47}$$

Others are possible. In practice, the parametrization chosen is always the one that is easiest to use in subsequent calculations. The fact that different parametrizations are possible is unimportant when used to describe curves (*contours*) around which complex integration is to be performed because there the nature of the parametrization is taken into account automatically.

A curve represented by Equation (1.45) is said to be *smooth* if $x(s)$ and $y(s)$ are continuously differentiable and to be *piecewise smooth*, or *sectionally smooth*, if the curve is continuous (unbroken) but formed by joining end to end a finite number of piecewise smooth curves. It follows immediately from elementary calculus that when the length l of the smooth arc or curve Γ represented by Equation (1.45) is required, it is given by

$$l = \int_\alpha^\beta \sqrt{[dx/ds]^2 + [dy/ds]^2}\, ds, \tag{1.48}$$

where the positive square root is always taken.

The sets of points S (areas of the complex plane) that concern us here are those which are *connected*. This means for sets S in which any two points can be connected by an unbroken arc (not necessarily smooth), the points of which all lie in S. Two sets of points S_1 and S_2 are said to be *disconnected* if any arc joining an arbitrary point in each set contains points in S_1 and S_2, and also points *outside* both of these sets. The points in an open or closed disk and in a rectangular area are connected, whereas the points in two nonintersecting disks are not connected. A set of points within some neighborhood of the origin (which may be large or small) is said to be *bounded*, otherwise the set of points is said to be *unbounded*. The points in the sectors shown in Figures 1.10(a) and (b) are bounded, while the sets of points belonging to any parallel strip of finite width in the extended complex plane, or in the first quadrant of the plane, are unbounded.

Two terms that will often be used are *domain* and *region*. A *domain* is an open connected set of points, and a *region* is a domain together with all of its boundary points. Figure 1.10(a) shows a typical domain and Figure 1.10(b) a typical region.

A connected set of points S will be said to be *simply connected* if the interior of every simple closed curve drawn in S only contains points of S, otherwise the set will be said to be *nonsimply connected*. Thus a set S will be nonsimply connected if inside it simple closed curves are drawn entirely in S, with the property that the interior of some of the curves *do not* contain points of S. Figure 1.11(a) shows an example of a simply connected set of points S, and Figure 1.11(b) shows a nonsimply connected set of points S, where although the closed curve γ_1 can be contracted to a single arc joining the points P_1 and P_2, the curve γ_2 cannot be contracted to a single arc joining the points Q_1 and Q_2 because it encloses the domain D that does not contain points of S.

Exercises 1.2

In Exercises 1 through 11, shade the required domain or region and indicate a boundary belonging to it by a solid line and one that is excluded from it by a dashed line.

1. $|z - 3| \geqslant 1$.
2. $|\text{Arg}\{z\}| \leqslant \pi/6, \text{Re}\{z\} < 1$.
3. $\text{Im}\{z\} > 0, R_1 \leqslant |z| < R_2\ (0 < R_1 < R_2)$.
4. $R_1 < |z| \leqslant R_2, \text{Re}\{z\} \geqslant \alpha, \text{Re}\{z\} \geqslant \alpha, \text{Re}\{z\} \leqslant \beta\ (0 < \alpha < R_1 < \beta < R_2)$.
5. $\alpha < \text{Arg}(z) < \beta, a < \text{Im}\{z\} < b\ (0 < \alpha < \beta < \pi/2, 0 < a < b)$.

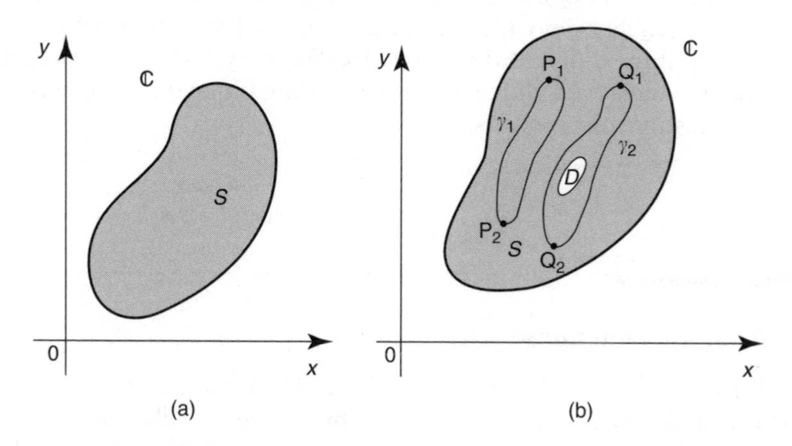

FIGURE 1.11
(a) A simply connected set of points S. (b) A nonsimply connected set of points S.

6. $\dfrac{x^2}{a^2} + \dfrac{y^2}{b^2} \geqslant 1,\ y > x, y > -x\ (0 < b < a)$.
7. $|z| \geqslant 1, -2 \leqslant \text{Re}\{z\} \leqslant 3, -1 \leqslant \text{Im}\{z\} \leqslant 2$.
8. $|z-2| + |z+2| \geqslant 6, |z-2| + |z+2| \leqslant 8$.
9. $\text{Re}\{z\} < \cos\xi, \text{Im}\{z\} < b\sin\xi, a > b > 0\ (-\pi < \xi \leqslant \pi)$.
10. $|z| \leqslant 1 + \cos\theta, \theta = \text{Arg}(z)\ (-\pi < \theta \leqslant \pi)$.
11. $\text{Re}\{z\} > a\cosh\xi, 0 < \text{Im}\{z\} \leqslant b\sinh\xi, a > b > 0\ (0 \leqslant \xi < \infty)$.

In Exercises 12 through 16, sketch the curves in $\mathbb{C}$ defined by the given equation.

12. $|z-1| = |z+2|$.
13. $\left|\dfrac{z+1+i}{z-1-i}\right| = 1$.
14. $4|z+2| = |z-2|$.
15. $\text{Re}\{z + 1/z\} = \alpha x\ (\alpha > 1)$.
16. $\text{Im}\{z + 1/z\} = \alpha y\ (0 < \alpha < 1)$.
17. For what range of values of a are there points of $\mathbb{C}$ common to both $|z-1| \leqslant 1$ and $|z-a| \leqslant 1$.
18. Give inequalities specifying the set of points S in $\mathbb{C}$ that lies inside but not on the boundary of an annulus with inner radius 1 and outer radius 2 with its center at $z = i$, from which points z have been removed such that $0 < \text{Arg}(z) \leqslant \pi/3$. Sketch the boundaries of S, shade its interior points and state if it is a region or a domain.
19. Give inequalities specifying the set of points S in $\mathbb{C}$ that are exterior to an ellipse with semimajor axis 3 and semiminor axis 2 centered on the point $z = 2 - i$, with the semimajor axis parallel to the imaginary axis. Sketch the boundaries of S, shade its interior points, and state if it is a region or a domain.
20. Let the set S of points in $\mathbb{C}$ be those in an open unit circle centered on the origin, from which have been removed the points belonging to any three different diameters. Explain why S is not connected. Give two ways in which S may be modified to make it connected, and justify your answers.

1.3 Analytic Functions

We call z a complex variable if $z \in \mathbb{C}$ is allowed to assume values in some set D in the complex plane. Let us now denote by f a rule (usually a mathematical expression) that assigns to each $z \in D$ a unique complex number, w. Then f is called a *function of the complex variable* z defined on set D, that is usually a region, and we will show this by writing

$$w = f(z), \quad z \in D. \tag{1.49}$$

Expressed more precisely, $f(z)$ in Equation (1.49), is the *value* of the function f at z: that is the complex number w assigned by the function f to the number $z \in D$. By analogy with the real variable case, the set D is called the *domain of definition* of f, while the set R of all the complex numbers w is called the *range* of the function f. Although f is the *function* and $f(z)$ is its *value* for $z \in D$, it is an accepted convention to sometimes misuse this notation by referring to $f(z)$ as the function (rather than f) because this has the advantage of making explicit the independent variable z involved.

The set D on which function f is defined forms part of the definition of f and if D is not specified, it is taken to be the largest part of the complex plane in which f has meaning. Thus $f(z) = z^2$ for $|z| < R$ is a function defined for all points of the open disk of radius R centered on the origin, whereas writing $f(z) = z^2$ implies that z is any point in the complex plane.

Let a complex variable w depending on z be expressed in the real and imaginary form $w = u + iv$. Then if z is expressed in the Cartesian form $z = x + iy$, it follows that in general u and v will depend on x and y, so that

$$w = f(z) = u(x, y) + iv(x, y). \tag{1.50}$$

Here the functions $u(x, y)$ and $v(x, y)$ are called the *real* and *imaginary parts* of $f(z)$, respectively and that Equation (1.50) is the *Cartesian representation* of $f(z)$ which displays the explicit dependence of $f(z)$ on x and y.

Alternatively, if z is expressed in the polar form $z = r(\cos\theta + i\sin\theta)$, the analogous result is

$$w = f(z) = u(r, \theta) + iv(r, \theta), \tag{1.51}$$

which is called the *polar representation* of $f(z)$ where, of course, the functions u and v in Equation (1.51) are not the same as those in Equation (1.50).

Example 1.3.1 A Function Expressed in Cartesian and Polar Form
Let

$$w = f(z) = z^2 + 4z + 3, \quad z \in \mathbb{C},$$

then in this case set D is the set of all complex numbers. The Cartesian representation $f(z) = u(x, y) + iv(x, y)$ is found by setting $z = x + iy$ in $f(z)$ to obtain

$$\begin{aligned} w = f(z) = u(x, y) + iv(x, y) &= (x + iy)^2 + 4(x + iy) + 3 \\ &= x^2 + 2ixy - y^2 + 4x + 4iy + 3. \end{aligned}$$

Equating the respective real and imaginary parts on each side of this result gives

$$u(x,y) = \text{Re}\{f(z)\} = x^2 - y^2 + 4x + 3$$

and

$$v(x,y) = \text{Im}\{f(z)\} = 2xy + 4y.$$

Similarly, the polar representation $f(z) = u(r, \theta) + iv(r, \theta)$ follows by setting $z = r(\cos\theta + i\sin\theta)$ in $f(z)$ to obtain

$$\begin{aligned} f(z) &= r^2(\cos\theta + i\sin\theta)^2 + 4r(\cos\theta + i\sin\theta) + 3 \\ &= r^2(\cos 2\theta + i\sin 2\theta) + 4r(\cos\theta + i\sin\theta) + 3. \end{aligned}$$

Expanding the expression on the right, collecting its real and imaginary parts and equating the real and imaginary parts on each side of the equation gives

$$u(r,\theta) = \text{Re}\{f(z)\} = r^2\cos 2\theta + 4r\cos\theta + 3 \tag{1.52}$$

and

$$v(r,\theta) = \text{Im}\{f(z)\} = r^2\sin 2\theta + 4r\sin\theta. \tag{1.53}$$

If, for example, $z = 1 - i$ for which $r = |z| = \sqrt{2}$ and $\text{Arg}\{z\} = -i\pi/4$, the Cartesian representation becomes $f(1 - i) = 7 - 6i$ and, of course, the polar representation with $r = \sqrt{2}$ and $\theta = -\pi/4$ also gives the same result. ◇

Example 1.3.2 A Function Defined on a Disk

Let

$$f(z) = \begin{cases} z \text{ for } |z| < 1 \\ 1/\bar{z} \text{ for } 1 < |z| < 3. \end{cases}$$

In this example the set D on which $f(z)$ is defined is an open disk of radius 3 centered on the origin. Notice that $f(z)$ is defined differently within the open disk centered on the origin, from its definition in the annular domain $1 < |z| < 3$. Setting $z = x + iy$ in $f(z)$ gives

$$f(z) = \begin{cases} x + iy, & \text{for } |z| < 1 \\ \dfrac{x}{x^2 + y^2} + i\dfrac{y}{x^2 + y^2}, & \text{for } 1 < |z| < 3. \end{cases}$$

This shows that

$$u(x, y) = \text{Re}\{f(z)\} = \begin{cases} x, & \text{for } |z| < 1 \\ \dfrac{x}{x^2 + y^2}, & \text{for } 1 < |z| < 3, \end{cases}$$

and

$$v(x, y) = \text{Im}\{f(z)\} = \begin{cases} y, & \text{for } |z| < 1 \\ \dfrac{y}{x^2 + y^2}, & \text{for } 1 < |z| < 3. \end{cases}$$

Inspection of the different forms of $u(x, y)$ and $v(x, y)$ inside and outside the unit circle shows that they coincide on $|z| = 1$, so instead of omitting to define $f(z)$ on the unit circle, the function could have been defined as

$$f(z) = \begin{cases} z & \text{for } |z| \leqslant 1 \\ 1/\bar{z} & \text{for } 1 < |z| < 3. \end{cases}$$ ◇

1.3.1 Limits and Continuity

The real variable definition of a limit of a function of two variables can be extended in a natural manner to the case of a complex function. Starting with an intuitive definition, suppose that the complex function $f(z)$ defined in a domain D is such that by taking z sufficiently close to a point z_0 in D, it is possible to make $f(z)$ as close as we wish to some complex number L. Then, provided this approach to L is *independent* of the way in which z tends to z_0, written $z \to z_0$, the function will be said to have the *limit* L as z tends to z_0, and we will write

$$\lim_{z \to z_0} f(z) = L. \tag{1.54}$$

Notice that like the real variable case, this definition of a limit assumes nothing about the behavior of the function $f(z)$ at $z = z_0$, where it may or may not be defined, and when it is its value is not necessarily such that $f(z_0) = L$. Notice also the requirement that when taking the limit, it is necessary that the result is *independent* of the way in which z tends to z_0. When we define the derivative of a complex function, based as would be expected on a limit, this condition will be seen to play a fundamental role when arriving at a condition that ensures a complex function has a unique derivative.

The weakness of this intuitive definition is due to its failure to say in what sense two complex numbers are *close*, and this in turn is due to the fact that complex numbers have no natural order. The difficulty is overcome by defining

z to be close to z_0 if $z \neq z_0$ lies within a δ-neighborhood $N(z_0, \delta)$, where $\delta > 0$ is an arbitrarily small real number. The number δ then provides a direct measure of the closeness of z to z_0.

We now formulate the rigorous definition of a limit.

1.3.2 Definition of a Limit

The function $f(z)$ has the **limit** L as $z \to z_0$, denoted by

$$\lim_{z \to z_0} f(z) = L,$$

if for any real number $\varepsilon > 0$, however small, it is possible to find a real number $\delta > 0$, depending on $\varepsilon > 0$ such that for every $z \neq z_0$ in the punctured disk $0 < |z - z_0| < \delta$, $|f(z) - L| < \varepsilon$.

A punctured neighborhood has been used because the value of $f(z)$ at z_0 does not enter into the definition of a limit. It is this use of a punctured neighborhood of z_0 that imposes the condition that z must tend to z_0 *independently* of the way in which this occurs because it means that z may follow *any* path in $N(z_0, \delta)$ as it tends to z_0.

Let us be quite clear about what is meant when we say that z tends to z_0 independently of the way this occurs because this condition must hold in any neighborhood of z_0, and not only in a circular disk centered on the point. Consider Figure 1.12 in which an arbitrary neighborhood D of a point z_0 is shown, together with two arbitrary simple arcs Γ_1 and Γ_2 drawn from points P_1 and P_2 to z_0. Then to say z tends to z_0 independently of the way this happens means that z may move along any simple arc like Γ_1 or Γ_2, with an arrow showing the way z must move.

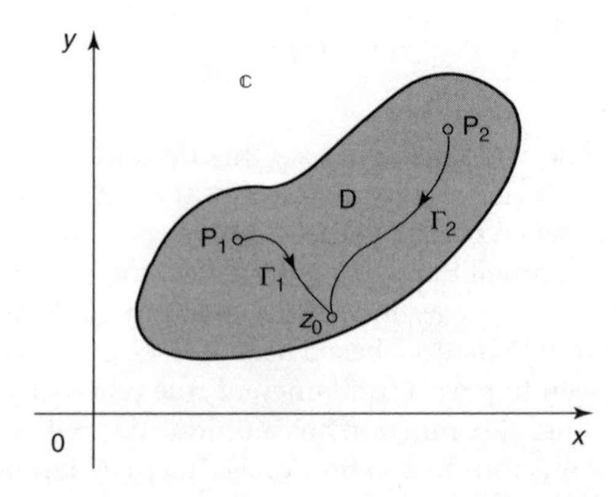

FIGURE 1.12
Two paths Γ_1 and Γ_2 along which $z \to z_0$.

Let us show that this definition of a limit implies that when a limit exists it is unique. Proving this is straight forward will provide an example of how the rigorous definition of a limit can be used. Suppose, if possible, that $f(z)$ has two different limits L_1 and L_2 as $z \to z_0$. Then by definition we can find an $\varepsilon > 0$ such that

$$|f(z) - L_1| < \varepsilon \quad \text{and} \quad |f(z) - L_2| < \varepsilon,$$

for any $z \neq z_0$ in $|z - z_0| < \delta$. Then

$$|L_1 - f(z)| + |L_2 - f(z)| < 2\varepsilon \quad \text{for } |z - z_0| < \delta \quad \text{and} \quad z \neq z_0.$$

To proceed further we will apply the version of the triangle inequality given in Equation (1.16) with $z_1 = L_1$, $z_2 = f(z)$ and $z_3 = L_2$ to the above result. This strengthens and simplifies the inequality to

$$|L_1 - L_2| < 2\varepsilon \quad \text{for all } z \neq z_0 \text{ in } |z - z_0| < \delta.$$

Because ε is arbitrary, so also is 2ε and we conclude that $L_1 = L_2$, thereby establishing the uniqueness of the limit.

The Theorem 1.3.1 summarizes the most important consequences of combining two functions $f(z)$ and $g(z)$ when each has a limit as $z \to z_0$. Only result (i) will be proved, as the other proofs follow in similar fashion and so are left as exercises.

THEOREM 1.3.1 Limit Theorems
If

$$\lim_{z \to z_0} f(z) = \alpha \quad \text{and} \quad \lim_{z \to z_0} g(z) = \beta, \text{ then}$$

$$\lim_{z \to z_0} [f(z) + g(z)] = \alpha + \beta,$$

$$\lim_{z \to z_0} [f(z)g(z)] = \alpha\beta,$$

$$\lim_{z \to z_0} [f(z)/g(z)] = \alpha/\beta, \text{ provided } \beta \neq 0,$$

Let

$$\lim_{w \to \beta} f(w) = k, \text{ then } \lim_{z \to z_0} f(g(z)) = k.$$

PROOF

To prove (i), by definition we can find an $\varepsilon > 0$ and $\delta_1, \delta_2 > 0$ such that

$$|f(z) - \alpha| < \tfrac{1}{2}\varepsilon \quad \text{for all } z \neq z_0 \text{ in } |z - z_0| < \delta_1,$$

and

$$|g(z) - \beta| < \tfrac{1}{2}\varepsilon \quad \text{for all } z \neq z_0 \text{ in } |z - z_0| < \delta_2.$$

Taking $\delta = \min(\delta_1, \delta_2)$, both inequalities are true for all $z \neq z_0$ in $|z - z_0| < \delta$. Adding these two inequalities gives

$$|f(z) - \alpha| + |f(z) - \beta| < \varepsilon.$$

Using triangle Inequality [Equation (1.15)] of Section 1.1 with $z_1 = f(z) - \alpha$ and $z_2 = g(z)$ gives

$$|[f(z) + g(z)] - (\alpha + \beta)| < \varepsilon \quad \text{for all } z \neq z_0 \text{ in } |z - z_0| < \delta.$$

As $\varepsilon > 0$ is arbitrary, this shows that

$$\lim_{z \to z_0} |[f(z) + g(z)] - (\alpha + \beta)| = 0,$$

from which result (i) follows immediately. ◆

When using the results of this theorem to evaluate specific limits, it is usually easiest to work directly with the complex functions $f(z)$ and $g(z)$, though sometimes it is convenient to use the related limits of real functions that are implied by the theorem. To see how this works, starting from $\lim_{z \to z_0} f(z) = L$, by writing $z_0 = x_0 + iy_0$, $L = a + ib$ and $f(z) = u + iv$ we have

$$\lim_{z \to z_0} f(z) = \lim_{\substack{x \to x_0, \\ y \to y_0}} u(x, y) + i \lim_{\substack{x \to x_0, \\ y \to y_0}} v(x, y), \tag{1.55}$$

so equating the respective real and imaginary parts on each side of the equation gives

$$\lim_{\substack{x \to x_0, \\ y \to y_0}} u(x, y) = a \quad \text{and} \quad \lim_{\substack{x \to x_0, \\ y \to y_0}} v(x, y) = b. \tag{1.56}$$

Example 1.3.3 Limits of Complex Functions
Find the following limits when they exist:

$$\text{(i)} \lim_{z\to 1+i}\left(\frac{z^2+4z-1}{z+2}\right), \quad \text{(ii)} \lim_{z\to i}\left(\frac{z^3+2z^2+z+2}{z-i}\right), \quad \text{(iii)} \lim_{z\to 0}\left[(1+z)\sin\left(\frac{1}{|z|}\right)\right].$$

SOLUTION

(i) The polynomials in the quotient have the nonzero values

$$\lim_{z\to 1+i}(z^2+4z-1) = 3+6i \quad \text{and} \quad \lim_{z\to 1+i}(z+2) = 3+i,$$

so

$$\lim_{z\to 1+i}\left(\frac{z^2+4z-1}{z+2}\right) = \frac{3+6i}{3+i} = \tfrac{3}{2}(1+i).$$

(ii) In this case both polynomials in the quotient vanish as $z \to i$, so we cannot use the result of Theorem 1.3.1 directly. However, the fact that each polynomial vanishes when $z = i$ means that $(z - i)$ is a factor of both polynomials. So dividing both numerator and denominator by $(z - i)$ gives

$$\lim_{z\to i}\left(\frac{z^3+2z^2+z+2}{z-i}\right) = \lim_{z\to i}\left[z^2+(2+i)z+2i\right] = -2+4i.$$

(iii) No limit exists in this case, because although the factor $(1 + z)$ becomes 1 in the limit as $z \to 0$, the bounded factor $\sin 1/|z|$ has no limit as it merely oscillates between ± 1 with increasing rapidity as $z \to 0$.

Example 1.3.4 A Limiting Process that Depends on the Direction of Approach
The function of two real variables

$$h(x, y) = \frac{xy}{2x^2+3y^2}$$

is a typical example of a function of two real variables that might form the real or imaginary part of a complex function. Let us show that this function has no limit at the origin. To demonstrate this we set $y = mx$, and let x and y tend to zero along this line by letting x tend to zero. The value of m will determine the direction of approach to the origin. So if the result of this operation depends on m, the function can have no limit because when a limit exists it must be independent of the way the general point $(x, y) \to (0, 0)$. Setting $y = mx$ in $h(x, y)$ gives

$$h(x, mx) = \frac{mx^2}{2x^2 + 3m^2x^2} = \frac{m}{2 + 3m^2},$$

which is independent of x, so in fact $h(x, y)$ has the constant value $m/(2 + 3m^2)$ along the line $y = mx$. Thus

$$\lim_{\substack{x\to 0,\\ y=mx}} \left(\frac{xy}{2x^2 + 3y^2} \right) = \lim_{x\to 0} h(x, mx) = \frac{m}{2 + 3m^2}.$$

As this result depends of m the function can have no limit at the origin. ◇

To develop the study of complex functions, it is now necessary to introduce the important concept of *continuity*. Simply stated, a complex function $f(z)$ is said to be *continuous* at $z = z_0$ if it possesses a limit L as $z \to z_0$ and, furthermore, $f(z_0)$ exists and is equal to L. Thus continuity is a property of a function $f(z)$ in a neighborhood of any point where it is continuous. The formal definition now follows.

1.3.4 Definition of Continuity

The complex function $f(z)$ is said to be *continuous* at z_0 if $f(z_0)$ is defined and

$$\lim_{z\to z_0} f(z) = f(z_0).$$

If the limit exists but does not equal $f(z_0)$, or if it is not defined, the function $f(z)$ is said to be *discontinuous* at z_0.

The following text addresses functions that are continuous throughout some domain of the complex plane. We see from this that the function

$$f(z) = \frac{z^2 - 4z - 1}{z + 2}$$

in (i) of Example 1.3.3 is continuous away from the point $z = -2$ because that is the only point where the quotient is undefined. Both the numerator and denominator of the function

$$f(z) = \frac{z^3 + 2z^2 + z + 2}{z - i}$$

in (ii) of Example 1.3.3 vanish when $z = i$, so the function is not properly defined there. However, there the difficulty was resolved by canceling the factor $(z - i)$ that was common to both the numerator and the denominator. The point $z = i$ is called a *singularity* of the function and when the limit at a singularity such as this can be removed by cancellation of a common factor in the numerator and denominator it is called a ***removable singularity***.

THEOREM 1.3.2 Properties of Continuous Functions
Let $f(z)$ and $g(z)$ be continuous at each point of a domain D. Then

$f(z) \pm g(z)$ is continuous throughout D;
$f(z)g(z)$ is continuous throughout D;
$f(z)/g(z)$ is continuous throughout D except at points where $g(z)$ vanishes but $f(z)$ remains finite and nonvanishing.

PROOF

The proofs of these properties are almost the same as the proofs of the results of Theorem 1.3.1, so as one of those proofs has been given in detail the proofs here will be left as exercises. ◆

As a typical illustration of a consequence of continuity we will prove that $f_n(z) = z^n$ is continuous for $n = 1, 2, \ldots,$ and all $z \in \mathbb{C}$. Taking $\delta = \varepsilon$ in the definition of a limit it follows trivially that $f_1(z)$ is continuous for all $z \in \mathbb{C}$. Now suppose for some positive integer m that $f_m(z_0)$ is continuous for all $z \in \mathbb{C}$. Then by (ii) of Theorem 1.3.2 the function $f_{m+1}(z) = z^{m+1} = z^m z = f_m(z)f_1(z)$ must be continuous for all $z \in \mathbb{C}$.

That $f_n(z) = z^n$ is continuous now follows by mathematical induction, because $f_1(z)$ is continuous, and the continuity of $f_{m+1}(z)$ follows from the continuity of $f_m(z)$, so the result must be true for $n = 1, 2, \ldots$. Clearly, from this result and Theorem 1.3.2(i), polynomials $P_n(z)$ must also be continuous for all $z \in \mathbb{C}$.

1.3.5 Differentiability and Derivatives: Analytic Functions

A function $f(z)$ of a complex variable that is defined at $z = z_0$, and in some neighborhood of z_0, will be said to be *differentiable* at z_0 with *derivative* $f'(z_0)$ at that point, if the limit

$$f'(z_0) = \lim_{z \to z_0} \left(\frac{f(z) - f(z_0)}{z - z_0} \right) \tag{1.57}$$

exists.

It is clear from Equation (1.57) that the first derivative of $f(z)$ with respect to z, denoted as in the real variable calculus by $f'(z_0)$, is a complex number specific to the point $z = z_0$. When $f(z)$ has a derivative at every point $z = z_0$ of some domain D, the suffix zero is omitted from z_0 and $f(z)$ is said to be differentiable throughout D with the *derivative* $f'(z)$ for $z \in D$. A function $f(z)$ that possesses a derivative at every point of a domain D is said to be *analytic* in D, although other names used in place of analytic are *regular* and *holomorphic*. A function $f(z)$ that is analytic throughout the whole of the finite complex plane is said to be an *entire function*. Example 1.3.5 shows that a simple example of an entire function is the polynomial

$$P_n(z) = a_0 + a_1 z + a_2 z^2 + \cdots + a_n z^n,$$

for $n = 1, 2, \ldots$, where the coefficients a_i are complex numbers.

When the nature of the domain D need not be specified it is usual to omit all reference to it, and simply to refer to $f(z)$ as an analytic function. It often happens that a function $f(z)$ is analytic everywhere in some domain D with the exception of a finite number of isolated points $z_1, z_2, \ldots, z_n$ where $f'(z)$ does not exist. Such points are called *isolated singular points*, or *isolated singularities* of $f(z)$. A typical example of a function with isolated singularities is

$$f(z) = \frac{z+4}{(z+1)(z-3)},$$

that Example 1.3.7(ii) shows to be analytic for all z with the exception of the two isolated singularities at $z = -1$ and $z = 3$ where differentiability fails because at each point the numerator is finite and nonzero, while the denominator vanishes.

An important consequence of the differentiability of $f(z)$ at a point z_0 is that it implies the continuity of $f(z)$ in a neighborhood of z_0, which can be seen by writing the difference $f(z) - f(z_0)$, with $z \neq z_0$, as

$$f(z) - f(z_0) = \left(\frac{f(z) - f(z_0)}{z - z_0}\right)(z - z_0),$$

because from Equation (1.57) and (ii) in Theorem 1.3.1(ii) we have

$$\lim_{z \to z_0} [f(z) - f(z_0)] = \lim_{z \to z_0} \left(\frac{f(z) - (z_0)}{z - z_0}\right) \lim_{z \to z_0} (z - z_0) = f'(z_0) \cdot 0 = 0.$$

This shows that

$$\lim_{z \to z_0} f(z) = f(z_0),$$

that is just the definition of continuity given earlier.

An equivalent and often more useful form of the definition of the derivative given in Equation (1.57) is

$$f'(z_0) = \lim_{h \to z_0} \left(\frac{f(z_0 + h) - f(z_0)}{h} \right) \tag{1.58}$$

where, of course, h is a complex variable.

The next example shows that this is a working definition.

Example 1.3.5 Some Elementary Entire Functions

Show that:

(i) The function $f(z) = k$ (a complex constant) is everywhere differentiable and such that

$$f'(z) = \frac{dk}{dz} = 0 \quad \text{for all } z \in \mathbb{C}.$$

(ii) The function $f(z) = z^n$ with $n = 1, 2, \ldots,$ is everywhere differentiable and such that

$$f'(z) = \frac{d[z^n]}{dz} = nz^{n-1} \quad \text{for all } z \in \mathbb{C}.$$

(iii) The complex polynomial of degree n

$$P_n(z) = a_0 + a_1 z + a_2 z^2 + a_3 z^3 + \cdots + a_n z^n,$$

in which $a_0, a_1, \ldots, a_n$ are arbitrary complex constants with $a_n \neq 0$, is everywhere differentiable and such that

$$P'(z) = \frac{d[P_n(z)]}{dz} = a_1 + 2a_2 z + 3a_3 z^2 + \cdots + na_n z^{n-1} \quad \text{for all } z \in \mathbb{C}.$$

SOLUTION

Result (i) follows trivially from definition Equation (1.58) because for any given $z_0 \in \mathbb{C}$,

$$f'(z_0) = \left(\frac{d[k]}{dz} \right)_{z=z_0} = \lim_{h \to 0} \left(\frac{k(z_0 + h) - k(z_0)}{h} \right) = \lim_{h \to 0} \left(\frac{k - k}{h} \right) = 0.$$

The arbitrary nature of z_0 allows the suffix zero to be dropped so that z becomes a complex variable instead of a complex number. As a result the derivative of $f(z)$ as a function of z becomes

$$f'(z) = \frac{dk}{dz} = 0 \quad \text{for all } z \in \mathbb{C}.$$

Result (ii) follows in similar fashion after use of the binomial theorem. For any given $z = z_0$ and $n = 1, 2, \ldots$, we have

$$f'(z_0) = \left(\frac{d[z^n]}{dz}\right) = \lim_{h\to 0}\left(\frac{(z_0+h)^n - z_0^n}{h}\right)$$
$$= \lim_{h\to 0}\left(\frac{z_0^n + nz_0^{n-1}h + \frac{1}{2!}n(n-1)z_0^{n-2}h^2 + \cdots + h^n - z_0^n}{h}\right)$$
$$= \lim_{h\to 0}\left(nz_0^{n-1} + \tfrac{1}{2!}n(n-1)z_0^{n-2}h + \cdots + h^{n-1}\right) = nz_0^{n-1}.$$

Again dropping the suffix zero and allowing z to become an arbitrary complex variable shows that when $f(z)$ is regarded as a function of z it follows that

$$f'(z) = \frac{d[z^n]}{dz} = nz^{n-1} \qquad \text{for all } z \in \mathbb{C}.$$

This result is true for arbitrary n, and not simply for integral values of n, but to establish this it is necessary to modify this proof, although a description of this process is omitted.

To establish result (iii) it is necessary to show that $d[kz^n]/dz = knz^{n-1}$, where k is a constant, and that $d[Kz^m + kz^n]/dz = mKz^{m-1} + nkz^{n-1}$ for m, $n = 1, 2, \ldots$, for all $z \in \mathbb{C}$. The first result follows trivially from result (ii) by replacing z^n by kz^n and noticing that the constant factor k can then be removed from the subsequent calculations and replaced by a multiplication factor k. The second result also follows from result (ii) by replacing z^n by $Kz^m + kz^n$, separating out the two terms and using the result that $d[kz^n]/dz = knz^{n-1}$. Dropping the suffix zero to make $P_n(z)$ a function of z, and making repeated use of the derivative of a sum, we find that

$$P_n'(z) = \frac{d}{dz}[P_n(z)] = a_1 + 2a_2z + 3a_3z^2 + \cdots + na_nz^{n-1} \qquad \text{for all } z \in \mathbb{C}. \quad \diamond$$

Example 1.3.6 A Complex Function Only Differentiable at the Origin

This example provides an illustration of a function of a complex variable whose real and imaginary parts are both continuously differentiable for all x and y, yet the function is only differentiable at the origin. Consider the function $f(z) = z\bar{z}$. Taking an arbitrary point $z_0 \in \mathbb{C}$ and using Equation (1.58) gives

$$f'(z_0) = \lim_{h\to 0}\left(\frac{(z_0+h)(\bar{z}_0+\bar{h}) - z\bar{z}}{h}\right) = \lim_{h\to 0}\left[\bar{z}_0 + \bar{h} + z_0(\bar{h}/h)\right].$$

If $h \to 0$ through purely real values $\bar{h} = h$, s in the limit as $h \to 0$ we find $f'(z_0) = z_0 + \bar{z}_0$ which is a *real* number. However, if $h \to 0$ through purely imaginary values $\overline{h} = -h$, and now in the limit as $h \to 0$ we find $f'(z_0) = z_0 - \overline{z}_0$ which is a *purely imaginary* number. These two limits are equal only when $z_0 = 0$, so elsewhere the value of the limits depends on the way $h \to 0$ and this function of z is only differentiable at the origin.

A closer look at this function shows that the result should not be surprising because $z\overline{z}$ is always a real function, and if $f(z) = u + iv$, it follows that

$$u(x, y) = x^2 + y^2 \quad \text{and} \quad v(x, y) \equiv 0.$$

Although the functions u *and* v are continuous and differentiable everywhere, the function $f(z) = z\overline{z}$ is not differentiable in the complex sense. This example illustrates the need for the test for complex differentiability given in the next theorem.

THEOREM 1.3.3 The Cauchy–Riemann Equations and Analyticity
Let the functions $u(x, y)$ and $v(x, y)$ in $f(z) = u(x, y) + iv(x, y)$ together with their first order partial derivatives be defined and continuous throughout some domain D, and let $u(x, y)$ and $v(x, y)$ satisfy the Cauchy–Riemann equations

$$\frac{\partial u}{\partial x} = \frac{\partial v}{\partial y} \quad \text{and} \quad \frac{\partial u}{\partial y} = -\frac{\partial v}{\partial x}$$

at every point of D. Then $f(z)$ is analytic in D, and its derivative in D when expressed in Cartesian form is given by either of the expressions

$$f'(z) = \frac{\partial u}{\partial x} + i\frac{\partial v}{\partial x} \quad \text{or} \quad f'(z) = \frac{\partial v}{\partial y} - i\frac{\partial u}{\partial y}.$$

To avoid interrupting the development of this section, the proof of this important theorem is deferred until Section 1.4 where after proving the theorem, another consequence of the Cauchy–Riemann equations is established for later use, and it is shown how, when either of the functions u or v is known, it is possible to find the other related function, and hence to construct an analytic function $f(z) = u(x, y) + iv(x, y)$.

Example 1.3.7 An Application of Theorem 1.3.3 to Some Complex Functions

Apply Theorem 1.3.3 to the following complex functions to determine if they are analytic, and in case (i) find $f'(z)$ as a function of x and y:

(i) $f(z) = 2z^3 + z^2 - z + 4$;

(ii) $f(z) = \dfrac{z + 4}{(z + 1)(z - 3)}$;

(iii) $f(z) = \dfrac{x^4 - y^4}{x^3 + y^3} + i\dfrac{x^4 + y^4}{x^3 + y^3}$ with $f(0) = 0$,

(iv) $f(z) = z\bar{z}$.

SOLUTION

(i) Setting $z = x + iy$ in $f(z) = u + iv$, and separating the real and imaginary parts gives

$$u(x, y) = 2x^3 - 6xy^2 + x^2 - y^2 - x + 4$$

and

$$v(x, y) = 6x^2y - 2y^3 + 2xy - y.$$

Routine differentiation then shows that

$$\frac{\partial u}{\partial x} = 6x^2 - 6y^2 + 2x - 1, \quad \frac{\partial u}{\partial y} = -12xy - 2y,$$
$$\frac{\partial v}{\partial x} = 12xy + 2y, \quad \frac{\partial v}{\partial y} = 6x^2 - 6y^2 + 2x - 1.$$

Clearly $u(x, y)$ and $v(x, y)$ are continuous for all x and y and their derivatives satisfy the Cauchy–Riemann equations, so by Theorem 1.3.3 the function $f(z)$ is analytic for all z (it is, of course, an entire function).

Using either of the last results of Theorem 1.3.3 it follows that in Cartesian form the derivative of the function $f(z)$ is

$$f'(z) = 6x^2 - 6y^2 + 2x - 1 + i(12xy + 2y).$$

A simple way of arriving at this result in terms of z is provided by Result (i) of the next theorem, while in Section 1.4 it will be shown how the above result can be converted rapidly into the equivalent expression involving z.

(ii) The function $f(z)$ is the quotient of two polynomials, so it is continuous away from the points $z = -1$ and $z = 3$ where the denominator vanishes. Thus its real and imaginary parts u and v share these same properties where $z = -1$ corresponds to $x = -1$ and $z = 3$ corresponds to $x = 3$. Some routine though tedious differentiation, the details of which are left as an exercise, confirm that the Cauchy–Riemann equations are satisfied, so the function $f(z)$ is analytic everywhere away from the two isolated singularities at $z = -1$ and $z = 3$.

(iii) This example shows the necessity of the requirements in Theorem 1.3.3 that u, v and their first order partial derivatives are continuous in D.

From the definition of differentiation, using the standard suffix notation,

$$u_x = \partial u/\partial x,\ u_y = \partial u/\partial y,\ v_x = \partial v/\partial x \quad \text{and} \quad v_y = \partial v/\partial y,$$

we have

$$\begin{aligned} u_x(0,0) &= \lim_{x\to 0}\left(\frac{u(x,0)-u(0,0)}{x}\right) = \lim_{x\to 0}\left(\frac{u(x,0)}{x}\right) \\ &= \lim_{x\to 0}\left[\left(\frac{x^4-0}{x^3+0}\right)\Big/ x\right] = 1. \end{aligned}$$

Similar arguments show that $u_y(0,0) = -1, v_x(0,0) = 1$ and $v_y(0,0) = 1$. Thus the function satisfies the Cauchy–Riemann equations at the origin, so assuming it is permissible to apply the last result of Theorem 1.3.3 it would seem that $f'(0) = 1 + i$.

However, if the limit is taken along the line $y = x$, that can be parametrized as $h = (1 + i)x$ with $f(h) = ix$ we have

$$\lim_{h\to 0}\left(\frac{f(h)-f(0)}{h}\right) = \lim_{x\to 0}\left(\frac{ix-0}{(1+i)x}\right) = \tfrac{1}{2}(1+i).$$

This is not equal to the value $1 + i$ determined previously, so the function is not differentiable at the origin, which is a singular point, though it is differentiable for all $z \neq 0$. The failure of differentiability at the origin is because this is the only point where u and v and their partial derivatives are not continuous, thereby violating the conditions of Theorem 1.3.3 at the origin and invalidating the application of the last result in the theorem.

(iv) The function $f(z) = \bar{z} = x - iy$ fails to satisfy the Cauchy–Riemann equation at any point in the complex plane. To see this set $f(z) = u + iv$, when $u = x$ and $v = -y$, so that $u_x = 1$, $u_y = 0$, $v_x = 0$ and $v_y = -1$. This shows that throughout the complex plane $u_x \neq v_y$. A function such as this which is said to be *nonanalytic*. Whereas Example (iii) failed to be analytic at one point, this one is nonanalytic for all z. ◇

Theorem 1.3.4 simplifies the task of finding $f'(z)$ as a function of z, thereby rendering it unnecessary to use either of the last results of Theorem 1.3.3. The definition of a complex derivative has the same form as the corresponding definition of the derivative of a real function of a real variable, aside from the fact that the result must be independent of the way $h \to 0$. Thus, formally, the rules for differentiating complex functions are the same as those for differentiating real functions, the difference being that in the complex case, before applying the rules, it is first necessary to ensure the functions involved are differentiable in the complex sense. We state these rules without further justification.

THEOREM 1.3.4 Differentiation Rules for Analytic Functions
If $f(z)$ and $g(z)$ are analytic functions in some domain D, then

(i) $\dfrac{d}{dz}[f(z)+g(z)] = f'(z)+g'(z)$ is analytic in D,

(ii) $\dfrac{d}{dz}[f(z)g(z)] = f'(z)g(z)+f(z)g'(z)$ is analytic in D,

(iii) $\dfrac{d}{dz}\left(\dfrac{f(z)}{g(z)}\right) = \dfrac{f'(z)g(z)-f(z)g'(z)}{[g(z)]^2}$ is analytic for all z in D such that $g(z)\neq 0$.

(iv) $\dfrac{d}{dz}(f[g(z)]) = f'(g(z))g'(z)$ is analytic for all z for which $f(g(z))$ is defined. ◆

Example 1.3.8 An Application of Theorem 1.3.4
Use Theorem 1.3.4 to find $f'(z)$ for the functions (i) and (ii) in Example 1.3.7.

SOLUTION

(i) From Theorem 1.3.4(i) we have $\dfrac{d}{dz}[2z^3+z^2-z+4] = 6z^2+2z-1$;

(ii) From Theorem 1.3.4(iii) we have $\dfrac{d}{dz}\left[\dfrac{z+4}{(z+1)(z-3)}\right] = \dfrac{5-8z-z^2}{(z+1)^2(z-3)^2}$.

◇

THEOREM 1.3.5 L'Hospital's Rule
Let $f(z)$ and $g(z)$ be analytic functions in some domain D such that $z_0\in D, f(z_0)=0$ and $g(z_0)=0$, while $\lim_{z\to z_0} g'(z)\neq 0$ and $\lim_{z\to z_0}[f'(z)/g'(z)]$ exists. Then

$$\lim_{z\to z_0}\left(\frac{f(z)}{g(z)}\right) = \lim_{z\to z_0}\left(\frac{f'(z)}{g'(z)}\right).$$

PROOF

The result is almost immediate, because as $f(z_0)=g(z_0)=0$, using the definition of a derivative, the limit can be rewritten as

$$\lim_{z\to z_0}\left(\frac{f(z)}{g(z)}\right) = \lim_{z\to z_0}\left(\frac{[f(z)-f(z_0)]/(z-z_0)}{[g(z)-g(z_0)]/(z-z_0)}\right) = \lim_{z\to z_0}\left(\frac{f'(z)}{g'(z)}\right).$$

The last limit is well defined and unique because $f(z)$ and $g(z)$ are analytic functions in some neighborhood of z_0, and by hypothesis $g'(z_0) \neq 0$. ◆

Example 1.3.9 An Application of L'Hospital's Rule

Evaluate $\lim_{z\to 1}\left(\frac{1-z^{n+1}}{z^2+3z-4}\right)$ when n is a positive integer.

SOLUTION

This is an indeterminate form to which Theorem 1.3.5 applies, with $f(z) = 1 - z^{n+1}$ and $g(z) = z^2 + 3z - 4$, with $z_0 = 1$. A direct application of Theorem 1.3.5 gives

$$\lim_{z\to 1}\left(\frac{1-z^{n+1}}{z^2+3z-4}\right) = \lim_{z\to 1}\left(\frac{-(n+1)z^n}{2z+3}\right) = -\tfrac{1}{5}(n+1). \qquad \diamond$$

Example 1.3.10 A Combination of L'Hospital's Rule with Other Reasoning

Evaluate $\lim_{z\to i}\left(\frac{z^2+2z+1-2i}{2z^2+z+2-i}\right)^2$.

SOLUTION

To determine the limit we apply L'Hospital's rule to the expression in parentheses which is an indeterminate form of the type to which Theorem 1.3.5 applies, and then raise the result to the power 2.

$$\lim_{z\to i}\left(\frac{z^2+2z+1-2i}{2z^2+z+2-i}\right) = \lim_{z\to i}\left(\frac{2z+2}{4z+1}\right) = \frac{2+2i}{1+4i} = \tfrac{1}{17}(10-6i),$$

so

$$\lim_{z\to i}\left(\frac{z^2+2z+1-2i}{2z^2+z+2-i}\right)^2 = \left[\tfrac{1}{17}(10-6i)\right]^2 = \tfrac{1}{289}(64-120i). \qquad \diamond$$

Exercises 1.3

1. Find $f(1+i)$ and $f(i\sqrt{3})$ if $f(z) = z^2 - 3$.
2. Find $f(-i)$ and $f(-\frac{1}{2}+i)$ if $f(z) = 1/(1+2z)$.
3. Find $f(2-i)$ and $f(3+i)$ if $f(z) = z|z|$.

Find the real and imaginary parts u and v of $f(z) = u + iv$ (a) in the Cartesian representation and (b) in the polar representation given that:

4. $f(z) = z^2 - z + 1$.
5. $f(z) = 1/z^2$.
6. $f(z) = z|z|^2$.

In each of the following exercises, by considering both the z-plane and the w-plane where $w = f(z)$, show graphically the domain in the w-plane that corresponds to the given domain in the z-plane.

7. $f(z) = 2z + 1, 0 \leqslant \text{Re}\{z\} \leqslant 2$.
8. $f(z) = z^2, |\text{Arg}\{z\}| \leqslant \pi/4$.
9. $f(z) = 1/z, \frac{1}{4} \leqslant |z| \leqslant \frac{1}{2}$ and $|z| \geqslant 1$.
10. Use the vector property of complex numbers together with geometrical arguments to sketch the path in the w-plane followed by $w = 2z + i$ as z moves around the unit circle $|z| = 1$ in the z-plane.

    When they exist, find the limits of the following functions $f(z)$, and state whether the functions are continuous at the points where the limits are to be determined.

11. $\lim_{z \to 1-i} f(z)$, with $f(z) = \frac{1}{2}(z^2 + 2z + 4)$.
12. $\lim_{z \to i} f(z)$, with $f(z) = \dfrac{z^2 - 2(1+i)z - 1 + 2i}{z - i}$, for $z \neq i$ and $f(i) = 1 + 2i$.
13. $\lim_{z \to 1} f(z)$, with $f(z) = (1 - z)/(1 - \bar{z})$.
14. $\lim_{z \to 0} f(z)$, with $f(z) = (\bar{z})^2/z$ for $z \neq 0$ and $f(0) = 1$.
15. Show from first principles that if $f(z) = (a + bz)^{-1}$, then

$$f'(z) = \frac{d}{dz}[(a+bz)^{-1}] = \frac{-b}{(a+bz)^2}, \quad z \neq -a/b.$$

16. Show from first principles that if $f(z)$ is analytic and $k =$ const. that

$$\frac{d}{dz}[k + f(z)] = f'(z) \quad \text{and} \quad \frac{d}{dz}[kf(z)] = kf'(z).$$

17. Given that $f(z)$ and $g(z)$ are entire functions, state with reasons which of the following combinations is an entire function:
    (a) $f(z) + g(z)$
    (b) $f(z)g(z)$
    (c) $[f(z)]^2 + [g(z)]^2 + 1$

(d) $f(z)/g(z)$
(e) $f(z+i)g(z+2)$
(f) $f[g(z)]$
(g) $g[f(z)-2i+1)]/[g(z)+f(z)]$.

18. Differentiate the following functions once with respect to z:
$f(z) = 4z^6 - 3z^2 + iz - 2$;
$f(z) = (z^3 + 2z^2 - 1)(z^2 - 2 + i)$;
$f(z) = (z+2)/(3z-1)^2$;
$f(z) = (z^2 + 7z - i)^6$.

Use L'Hospital's rule, together with other arguments where necessary, to determine the following limits:

19. $\lim_{z \to -i}\left(\dfrac{z^{10} + z^2 + 2iz}{z+i}\right).$

20. $\lim_{z \to 1-i}\left(\dfrac{z^2 - 2(1-i)z - 2i}{z^2 - 2z + 2}\right).$

21. $\lim_{z \to -i}\left(\dfrac{z^6 + 1}{z^2 + 1}\right)^2.$

22. $\lim_{z \to i}\left(\dfrac{z^4 - 1}{z^2 + 1}\right)^2 \Big/ \lim_{z \to 3}\left(\dfrac{z^2 - 4z + 3}{2z^2 - 13z + 21}\right).$

1.4 The Cauchy–Riemann Equations: Proof and Consequences

It is a straightforward matter to establish that for a complex function $f(z) = u + iv$ to be analytic is *necessary* that u and v satisfy the Cauchy–Riemann equations $u_x = v_y$ and $u_y = -v_x$. However, to show that *sufficient* conditions for analyticity involve the additional requirements that u, v and all of their first order partial derivatives must be continuous is a little harder, so readers may wish to delay studying that part of the proof until later.

We recall first that when a function $f(z)$ is defined for z in some domain D, the derivative $f'(z)$ is

$$f'(z) = \lim_{s \to 0}\left(\frac{f(z+s) - f(z)}{s}\right), \quad \text{for all } z \in D, \tag{1.59}$$

provided the limit exists and is independent of the way $s \to 0$. Before proceeding to the details of the proof of Theorem 1.3.3 we will outline the basic steps used in the first part to show the necessity of the Cauchy–Riemann equations. These involve setting

$$f(z) = u(x, y) + iv(x, y), \tag{1.60}$$

using Equation (1.60) to express Equation (1.59) in terms of u and v, and then finding the limit in Equation (1.59) by letting $s \to 0$ in two different ways. This will lead to two different looking expressions for $f'(z)$ that must be equal if the derivative is to be independent of the way $s \to 0$. The Cauchy–Riemann equations follow by equating the respective real and imaginary parts of these two expressions, as do the expressions for $f'(z)$ given in Theorem 1.3.3.

1.4.1 Proof of Theorem 1.3.3

Necessity of the Cauchy–Riemann Equations

To derive the Cauchy–Riemann equations, we first proceed to the limit in Equation (1.59) using a purely real value of s by setting $s = h + 0i$, where h is real. Then, after using Equation (1.60), the expression for the derivative in Equation (1.59) becomes

$$\begin{aligned} f'(z) &= \lim_{h \to 0}\left(\frac{u(x+h, y) + iv(x+h, y) - [u(x, y) + iv(x, y)]}{h}\right) \\ &= \lim_{h \to 0}\left(\frac{u(x+h, y) - u(x, y)}{h}\right) + i\lim_{h \to 0}\left(\frac{v(x+h, y) - v(x, y)}{h}\right). \end{aligned}$$

The existence of the derivative $f'(z)$ implies that each of the two limits on the right must exist, so recalling the definition of a partial derivative in real analysis, in the limit this last result is seen to reduce to

$$f'(z) = \frac{\partial u}{\partial x} + i\frac{\partial v}{\partial x}. \tag{1.61}$$

We now repeat this process, although this time proceeding to the limit in Equation (1.59) using purely imaginary values of s by setting $s = 0 + ik$, where k is real. After grouping terms, the derivative in Equation (1.59) becomes

$$f'(z) = \lim_{k \to 0}\left(\frac{u(x, y+k) - u(x, y)}{ik}\right) + i\lim_{k \to 0}\left(\frac{v(x, y+k) - v(x, y)}{ik}\right).$$

Proceeding to the limit this becomes

$$f'(z) = -i\frac{\partial u}{\partial y} + \frac{\partial v}{\partial y}. \tag{1.62}$$

If the derivative in Equation (1.59) is independent of the way s tends to zero, results Equations (1.61) and (1.62) must be identical, which can only occur when their respective real and imaginary parts are equal. So equating the respective real and imaginary parts of Equations (1.61) and (1.62) shows that for $f(z)$ to be analytic, a *necessary* condition is that the Cauchy–Riemann equations

$$\frac{\partial u}{\partial x} = \frac{\partial v}{\partial y} \quad \text{and} \quad \frac{\partial u}{\partial y} = -\frac{\partial v}{\partial x} \tag{1.63}$$

are satisfied. The two expressions for the derivative $f'(z)$ given in Theorem 1.3.3 are simply results Equations (1.61) and (1.62).

Sufficiency of the Cauchy–Riemann Equations

The Cauchy–Riemann equations in Theorem 1.3.3 are *necessary* for the analyticity of $f(z)$ and it remains for us to establish the *sufficiency* conditions. Before proceeding, recall the simplest form of Taylor's theorem for a real function $F(x, y)$ of the two real variables x and y. This is that if $F(x, y)$ together with its first-order partial derivatives exist and are continuous in a domain D containing the region $x_0 \leqslant x \leqslant x_0 + h, y_0 \leqslant y \leqslant y_0 + k$, with h, k arbitrary, then

$$\begin{aligned} F(x_0 + h, y_0 + k) = F(x_0, y_0) &+ hF_x(x_0 + \xi h, y_0 + \eta k) \\ &+ kF_y(x_0 + h\xi, y_0 + \eta k), \end{aligned}$$

where although unknown, the real numbers ξ, η are such that $0 < \xi < 1$, $0 < \eta < 1$. It will be recognized that the last two terms on the right represent the remainder term in this form of Taylor's theorem. For conciseness, in what follows we will use the abbreviations

$$\langle F_x \rangle_{\xi,\eta} = F_x(x_0 + \xi h, y_0 + \eta k) \quad \text{and} \quad \langle F_y \rangle_{\xi,\eta} = F_y(x_0 + \xi h, y_0 + \eta k).$$

If $f(z) = u + iv$ *is analytic in a domain* D containing the points (x_0, y_0) and $(x_0 + h, y_0 + k)$ we write

$$u(x_0 + h, y_0 + k) = h\langle u_x \rangle_{\xi_1,\eta_1} + k\langle u_y \rangle_{\xi_1,\eta_1},$$

and

$$v(x_0 + h, y_0 + k) = h\langle v_x \rangle_{\xi_2,\eta_2} + k\langle v_y \rangle_{\xi_2,\eta_2},$$

for some $0 < \xi_1 < 1, 0 < \eta_1 < 1, 0 < \xi_2 < 1$ and $0 < \eta_2 < 1$.

Setting $z_0 = x_0 + iy_0$, and $s = h + ik$, we have

$$f(z_0 + s) - f(z_0) = h\langle u_x \rangle_{\xi_1,\eta_1} + k\langle u_y \rangle_{\xi_1,\eta_1} + i\left[h\langle v_x \rangle_{\xi_2,\eta_2} + k\langle v_y \rangle_{\xi_2,\eta_2}\right].$$

After using the Cauchy–Riemann equations which are *necessary* for analyticity and adding and subtracting terms where appropriate, this result can be rearranged to give

$$\frac{f(z_0+s)-f(z_0)}{s} = \langle u_x\rangle_{\xi_1,\eta_1} + i\langle v_x\rangle_{\xi_2,\eta_2} + i\left(\frac{\varepsilon_1 h}{s} + \frac{\varepsilon_2 k}{s}\right),$$

where

$$\varepsilon_1 = \langle v_x\rangle_{\xi_2,\eta_2} - \langle v_x\rangle_{\xi_1,\eta_1} \quad \text{and} \quad \varepsilon_2 = \langle u_x\rangle_{\xi_2,\eta_2} - \langle u_x\rangle_{\xi_1,\eta_1}.$$

As $h, k \to 0$, so $x_0 + \xi_1 h \to x_0$, $x_0 + \xi_2 h \to x_0$, $y_0 + \xi_1 h \to y_0$ and $y_0 + \xi_2 h \to y_0$, thereby causing ε_1, $\varepsilon_2 \to 0$, because the *continuity* of the partial derivatives ensures that all of the functions involved tend to their respective limiting values at (x_0, y_0). And because $s = h + ik$ it follows that $|h/s| < 1$, $|k/s| < 1$, showing that h/s and k/s remain bounded as $s \to 0$. So proceeding to the limit we have

$$\begin{aligned} f'(z_0) &= \lim_{s\to 0}\left(\frac{f(z_0+s)-f(z_0)}{s}\right) \\ &= \lim_{s\to 0}\left(\langle u_x\rangle_{\xi_1,\eta_1} + i\langle v_x\rangle_{\xi_2,\eta_2}\right) + i\lim_{s\to 0}\left(\frac{\varepsilon_1 h}{s} + \frac{\varepsilon_2 k}{s}\right) \\ &= u_x(x_0, y_0) + i v_x(x_0, y_0). \end{aligned}$$

The limit is independent of the way $s \to 0$, and z_0 was any point in D, so the result is true throughout D, and the proof of the sufficiency conditions is complete. ◆

Example 1.3.7(iii) involved a function that although satisfying the Cauchy–Riemann equations at the origin was *not* differentiable at that point because u, v and their first-order partial derivatives were not continuous at the origin and therefore did not satisfy the additional conditions of continuity necessary to guarantee analyticity. In fact, with the exception of the point $z = 0$, the function in Example 1.3.7(iii) satisfies the Cauchy–Riemann equations throughout the finite complex plane, and u, v and their first-order partial derivatives are continuous everywhere except for that one point; in fact the origin was the *only* point where the function is not differentiable.

1.4.2 The Cauchy–Riemann Equations in Polar Form

A routine change of variables from the Cartesian variables x and y to the polar coordinates r and θ through the transformation

$$x = r\cos\theta, \; y = r\sin\theta \tag{1.64}$$

shows that in terms of the polar representation, if $f(z) = u(r, \theta) + iv(r, \theta)$, the Cauchy–Riemann equations take the form

$$\frac{\partial u}{\partial r} = \frac{1}{r}\frac{\partial v}{\partial \theta} \quad \text{and} \quad \frac{\partial v}{\partial r} = -\frac{\partial u}{\partial \theta}, \tag{1.65}$$

with

$$f'(z) = \left(\frac{\partial u}{\partial r}\cos\theta - \frac{1}{r}\frac{\partial u}{\partial \theta}\sin\theta\right) + i\left(\frac{\partial v}{\partial r}\cos\theta - \frac{1}{r}\frac{\partial v}{\partial \theta}\sin\theta\right) \tag{1.66}$$

or, equivalently (after using the Cauchy–Riemann equations),

$$f'(z) = \left(\frac{\partial v}{\partial r}\cos\theta - \frac{1}{r}\frac{\partial v}{\partial \theta}\sin\theta\right) - i\left(\frac{\partial u}{\partial r}\sin\theta - \frac{1}{r}\frac{\partial u}{\partial \theta}\cos\theta\right). \tag{1.67}$$

The derivation of these last results is left as an exercise.

Example 1.4.1 An Application of the Cauchy–Riemann Equations in Polar Form

Use the polar form of the Cauchy–Riemann equations to show that $f(z) = z^2 + 3z + 1$ is analytic.

SOLUTION

The analyticity of $f(z)$ for all z has already been established in Section 1.3, but to establish the result using the polar form of the Cauchy–Riemann equations we set $z = r(\cos\theta + i\sin\theta)$ in $f(z)$ and use De Moivre's theorem to obtain

$$f(z) = (u + iv) = r^2\cos 2\theta + 3r\cos\theta + 1 + i(r^2\sin 2\theta + 3r\sin\theta).$$

This shows that $u(r, \theta) = r^2\cos 2\theta + 3r\cos\theta + 1$ and $v(r, \theta) = r^2\cos 2\theta + 3r\sin\theta$, and routine differentiation shows these functions satisfy Equations (1.65) for all $z \neq 0$, while they and their partial derivatives are all continuous away from the origin. So from Theorem 1.3.3, using the polar form of the Cauchy–Riemann equations, the function $f(z)$ has been shown to be analytic everywhere except at $z = 0$.

As already mentioned in Section 1.3, $f(z)$ has been shown to be an entire function (analytic for all z), so a discrepancy appears to be between that result at $z = 0$ and the result obtained using the polar representation. This apparent discrepancy is easily resolved once it is recalled that in the polar form $\theta = \text{Arg}\{f(z)\}$ is *undefined* at the origin, so the polar form of the Cauchy–Riemann equations can tell us nothing about the analyticity of $f(z)$ at the origin. ◇

Differentiating the first of the Cartesian form of the Cauchy–Riemann equations partially with respect to x and the second partially with respect to y gives

$$\frac{\partial^2 u}{\partial x^2} = \frac{\partial}{\partial x}\left(\frac{\partial v}{\partial y}\right) \quad \text{and} \quad \frac{\partial^2 u}{\partial y^2} = -\frac{\partial}{\partial y}\left(\frac{\partial v}{\partial x}\right).$$

However, when a function $v(x, y)$ of the two real variables x and y together with its first and second order partial derivatives are all continuous, its second order mixed partial derivatives are equal. Thus it follows directly by equating the results in these equations that $u(x, y)$ is a solution of the *partial differential equation*

$$\frac{\partial^2 u}{\partial x^2} + \frac{\partial^2 u}{\partial y^2} = 0, \tag{1.68}$$

called *the* ***Laplace equation***, and this result is often written in the abbreviated form $\Delta u = 0$. Here the symbol Δ represents the *partial differential operator* $\Delta \equiv \partial^2/\partial x^2 + \partial^2/\partial y^2$, and when Δ acts on u it shows the differentiation operations that are to be performed in order to arrive at the expression on the left of Equation (1.68).

The symbol Δ itself is called the *Laplacian operator* or, more simply, the *Laplacian*, and when written in terms of Cartesian coordinates x and y, it is called the *Cartesian form* of the Laplacian. Notice that by itself, the Laplacian is simply an instruction to perform certain differentiation operations on a suitably differentiable function, but it is *not* itself a function. Only when Δ acts on a suitably differentiable function $\phi(x, y)$ of two variables x and y does $\Delta\phi$ become a function.

If the differentiation of the Cauchy–Riemann equations had been performed in the reverse order, that is differentiation the first equation with respect to y and then the second with respect to x, it would have shown that in addition to Equation (1.68) being true, it also follows that

$$\frac{\partial^2 v}{\partial x^2} + \frac{\partial^2 v}{\partial y^2} = 0, \tag{1.69}$$

so when $f(z) = u + iv$ is analytic, both $u(x, y)$ and $v(x, y)$ are solutions of Laplace's equation. A solution of Laplace's equation is called an *harmonic function*, and the functions u and v in an analytic function $f(z) = u + iv$ are called *conjugate harmonic functions*. In many parts of mathematics, and especially in applications to physical problems, the Laplace equation and its solutions are of considerable importance. For example, solutions of Laplace's equation describe the steady-state distribution of heat in a heat conducting solid, certain aspects of steady incompressible fluid flow, the electrostatic potential in a cavity, and many other physical situations.

Notice that u and v are only conjugate harmonic functions if they are respectively the real and imaginary parts of the *same* analytic function $f(z) = u + iv$. So the real part u_1 of an analytic function $f_1(z) = u_1 + iv_1$, and the imaginary part v_2 of a *different* analytic function $f_2(z) = u_2 + iv_2$ are *not* conjugate harmonic functions. The significance of Laplace's equation and the role played by conjugate harmonic functions become clear when boundary value problems have been defined and related to conformal mapping. It will suffice to remark here that for the specific analytic function $f(z) = z^2 = u + iv$, if we set $z = x + iy$, the conjugate harmonic functions are found to be

$$u = x^2 - y^2 \quad \text{and} \quad v = 2xy.$$

When the families of curves $u = const.$ and $v = const.$ are superimposed, they intersect each other at right angles, and families of curves possesing this property are called *mutually orthogonal trajectories*, or more simply, *orthogonal trajectories*. Figure 1.13 shows plots of these trajectories in the first quadrant of the (x, y)-plane. It will be shown later that the mutual orthogonality of plots of families of curves corresponding to conjugate harmonic functions is a general property of conjugate harmonic functions and not just a property of this particular choice of analytic function.

It is appropriate at this stage we formulate two simple rules by which to convert an analytic function $f(z) = u(x, y) + iv(x, y)$ into a function of z.

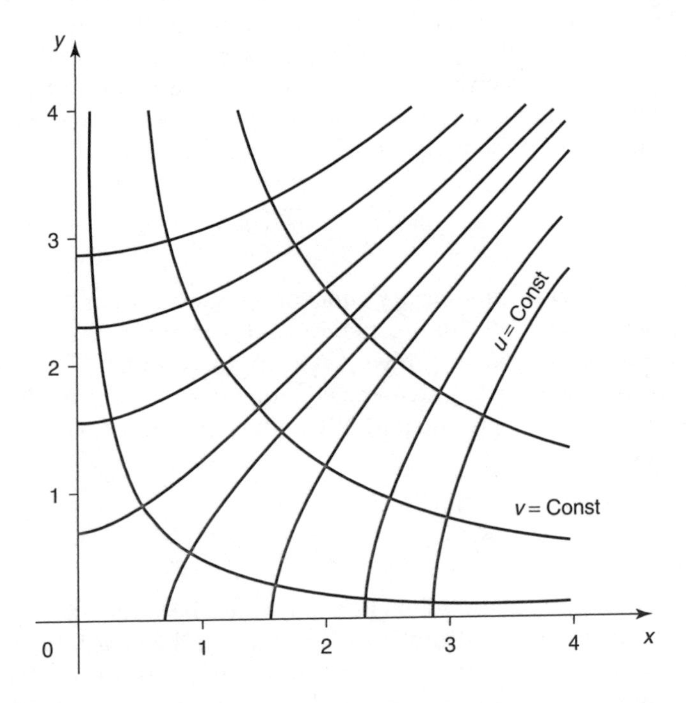

FIGURE 1.13
Typical orthogonal trajectories.

A little thought shows that if $f(z)$ is analytic, the functional form of $f(z)$ as a function of z must, in the Cartesian representation, be the form obtained from $f(z) = u(x, y) + iv(x, y)$ by setting $y = 0$ and replacing x by z in both u and v. Similarly, in the polar representation, the functional form of $f(z)$ as a function of z follows from $f(z) = u(r, \theta) + iv(r, \theta)$ by setting $\theta = 0$ and replacing r by z in both u and v.

Rules for Expressing f(z) = u + iv as a Function of z

If, and only if, $f(z) = u + iv$ is analytic, it can be expressed as a function of z either by

(i) Setting $y = 0$ and replacing x by z in both $u(x, y)$ and $v(x, y)$ in the Cartesian case; or by
(ii) Setting $\theta = 0$ and replacing r by z in both $u(r, \theta)$ and $v(r, \theta)$ in the polar case.

By way of example, applying Rule (i) to the analytic function $f(z) = x^2 - y^2 + 2ixy$ just used to illustrate orthogonal trajectories we find, as expected, that $f(z) = z^2$.

Later, given one harmonic function (either u or v), it will be necessary to find its harmonic conjugate (v or u) in order to construct the analytic function $f(z) = u + iv$, so we now show how this can be done. Suppose $u(x, y)$ is known, then we can find u_x and relate it to v_y through the Cauchy–Riemann equation $v_y = u_x$. Recalling the definition of partial differentiation with respect to y, during which x is regarded as a constant, after integrating the Cauchy–Riemann equations partially with respect to y (i.e., regarding x as a constant) we fine the general result

$$v(x, y) = \int \frac{\partial u}{\partial x} dy + h(x). \tag{1.70}$$

At this stage $h(x)$ is an arbitrary function of x, and it has been included because when Equation (1.70) is differentiated partially with respect to y its derivative will vanish, leaving only u_x on the right. A similar argument, starting from the second Cauchy–Riemann equation $v_x = -u_y$, using the fact that u_y can be found and integrating partially with respect to x (that is by regarding y as a constant) gives

$$v(x, y) = -\int \frac{\partial u}{\partial y} dx + k(y), \tag{1.71}$$

where this time an arbitrary function $k(y)$ must be added to the general result. Both Equations (1.70) and (1.71) are forms of the *same* function $v(x, y)$, so they must be identical for all x and y. The unknown function $h(x)$ is found by identifying it with terms in Equation (1.71) containing *only* functions of x, while the unknown function $k(y)$ is found by identifying it with terms in

Equation (1.70) containing *only* functions of y. A similar approach leads to the determination of $u(x, y)$ when $v(x, y)$ is known and, similarly, to $v(r, \theta)$ when $u(r, \theta)$ is known, or to $u(r, \theta)$ when $v(r, \theta)$ is known.

Example 1.4.2 Finding an Harmonic Conjugate
Given $u = 3x^2 + x - 3y^2 + 4$, find its harmonic conjugate, and hence find $f(z) = u + iv$ in terms of z.

SOLUTION
First it is necessary to check that u is harmonic because $u_{xx} = 6$ and $u_{yy} = -6$, showing that $u_{xx} + u_{yy} = 0$. Then as $u_x - 6x + 1$ and $u_y = -6y$, result Equation (1.70) becomes

$$v(x, y) = \int (6x + 1)dy = 6xy + y + h(x) + c,$$

while result Equation (1.71) becomes

$$v(x, y) = -\int (-6y)dx + k(y) = 6xy + k(y) + d,$$

where c and d are arbitrary real constants of integration. Equating these two forms of $v(x, y)$, that must be identical for all x and y, we find that as the second result contains no terms containing only functions of x, we must set $h(x) \equiv 0$. Then, as the only term in the first result containing only a function of y *is* y itself, we must set $k(y) = y$ and, finally, the arbitrary constants must be equal, so $d = c$. Thus the required harmonic conjugate function is

$$v(x, y) = 6xy + y + c.$$

Combining results to find the analytic function $f(z) = u + iv$ we find that

$$f(z) = 3x^2 + x - 3y^2 + 4 + c + i(6xy + y + c).$$

Rule (i) can be applied to express $f(z)$ in terms of z because $f(z)$ is analytic. Setting $y = 0$ and replacing x by z gives

$$f(z) = 3z^2 + z + 4 + c(1 + i).$$

It should come as no surprise that $f(z)$ contains an arbitrary additive constant because when either u + const. or v + const. is substituted into the Laplace equation, the constant will vanish. The role played by this arbitrary constant will be better understood when boundary value problems are considered in Chapter 5.

Exercises 1.4

In Exercises 1 through 10, use Theorem 1.3.3 to find if the following functions are analytic, and if so for what z this is true.

1. $f(z) = x^3 - 3xy^2 + x - 4 + i(3x^2y - y^3 + y)$.
2. $f(z) = 1/(z-2)$.
3. $f(z) = z + 1/z$.
4. $f(z) = 1/(z^2 - 1)$.
5. $f(z) = z^2|z|^2$.
6. $f(z) = \text{Re}\{z\} + i$.
7. $f(z) = x^2 - y^2 + x + i(2xy + |y|)$.
8. $f(z) = (\cos\theta - i\sin\theta)/r$.
9. $f(z) = r(\cos\theta - i\sin\theta)$.
10. $f(z) = 2r\cos\theta + (3/r^2)\cos 2\theta + i(2r\sin\theta - (3/r^2)\sin 2\theta)$.

In Exercises 11 through 13, use Theorem 1.3.3 to show the following functions are analytic and find $f'(z)$. Express $f(z)$ and $f'(z)$ in terms of z.

11. $f(z) = e^{2x}(\cos 2y + i\sin 2y)$.
12. $f(z) = \sin x\cosh y + i\cos x\sinh y$.
13. $f(z) = \cosh x\cos y + i\sinh x\sin y$.
14. Derive the polar form of the Cauchy–Riemann equations from the Cartesian form by using elementary calculus methods to make the change of variables $x = r\cos\theta$ and $y = r\sin\theta$.
15. Show that the derivatives $f'(z)$ in Equations (1.66) and (1.67) can be written in the form

$$f'(z) = (\cos\theta - i\sin\theta)\left(\frac{\partial u}{\partial r} + i\frac{\partial v}{\partial r}\right)$$

and

$$f'(z) = \frac{1}{r}(\cos\theta - i\sin\theta)\left(\frac{\partial v}{\partial \theta} - i\frac{\partial u}{\partial \theta}\right).$$

16. Use the polar form of the Cauchy–Riemann equations to show that u and v must both satisfy the ***polar form of Laplace's equation***

$$\frac{\partial^2\phi}{\partial r^2} + \frac{1}{r}\frac{\partial\phi}{\partial r} + \frac{1}{r^2}\frac{\partial^2\phi}{\partial\theta^2} = 0.$$

17. Show that although $u = \sin x\cosh y$ and $v = \sinh x\sin y$ are both harmonic functions, they are not conjugate harmonic functions.

In Exercises 18 through 24, show that each of the functions is harmonic and find its harmonic conjugate. Use the result to write down the analytic function $f(z) = u + iv$, and hence determine $f(z)$ in terms of z.

18. $u = 3xy$.
19. $v = \cosh x \sin y$.
20. $v = e^x(y \cos y + x \sin y)$.
21. $u = x \sin x \cosh y - y \cos x \sinh y$.
22. $v = -\sin x \sinh y$ with $f(0) = 3$.
23. $v = -3\theta$ with $f(1) = 4$.
24. $v = k\left(r - \dfrac{a^2}{r}\right)\cos\theta$ $(a, k$ real$)$.

Exercises of Greater Difficulty

25. Prove that if $f(z)$ is analytic in D, and either $\text{Re}\{f(z)\} =$ const. in D or $\text{Im}\{f(z)\} =$ const. in D, then $f(z) \equiv$ const. in D.
26. Prove that if $f(z) = u(x, y) + iv(x, y)$ is analytic in D, then $\Delta\{|f(z)|\} = |f'(z)|^2/|f(z)|$ throughout D.
27. Prove that if $f(z) = u(x, y) + iv(x, y)$ is analytic in D, then $\Delta\{[u(x, y)]^n\} = n(n-1)[u(x, y)]^{n-2}\,|f'(z)|^2$, for $n = 1, 2, \ldots$.
28. Prove that if v is analytic in D then $\phi = uv$ is analytic in D, where $f(z) = u + iv$.
29. Prove that if $f(z) = u + iv$ is analytic in D and $|f(z)|$ is harmonic in D, then $f(z) =$ const. in D.
30. Let $f(z) = u + iv$ be analytic in a bounded domain D, with $|f(z)| = M =$ const. on the boundary of D. By considering the behavior of $\Delta|f(z)|$ prove that $f(z) \equiv M$ in D.

When establishing this result, make use of the standard results from elementary calculus: that if $\phi(x, y)$ is a twice differentiable function in D, then at a local extremum of ϕ inside D:

(i) ϕ has a maximum if $\partial\phi/\partial x = \partial\phi/\partial y = 0$, $\phi_{xx}\phi_{yy} - \phi_{xy}^2 > 0$ and $\phi_{xx} < 0$, and
(ii) ϕ has a minimum if $\partial\phi/\partial x = \partial\phi/\partial y = 0$, $\phi_{xx}\phi_{yy} - \phi_{xy}^2 > 0$ and $\phi_{xx} > 0$.

1.5 Elementary Functions

Theorem 1.3.3 has established necessary and sufficient conditions for a function $f(z)$ to be analytic, so we now examine the basic properties of the most frequently occurring elementary analytic functions. Each is defined as an extension of its real variable counterpart in such a way that the two become identical when z is real. When examining these complex analytic functions we will be concerned with both their similarities and their dissimilarities, while also identifying the domain in which the complex function is analytic. Nonanalytic functions like $f(z) = \overline{z}$ that fail to satisfy the Cauchy–Riemann equations at any point of the complex plane are not used in this section.

1.5.1 Polynomials

The general complex polynomial $P_n(z)$ of degree n has the form

$$P_n(z) = a_0 + a_1 z + a_2 z^2 + \cdots + a_n z^n, \tag{1.72}$$

where the coefficients $a_0, a_1, \ldots, a_n$ are arbitrary complex numbers with $a_n \neq 0$. It has already been established that $P_n(z)$ is a continuous analytic function for all $z \in \mathbb{C}$, and so it is an entire function. Furthermore, its derivative

$$P_n'(z) = \frac{d}{dz}[P_n(z)] = a_1 + 2a_2 z + 3a_3 z^2 + \cdots + na_n z^{n-1} \quad \text{for all } z \in \mathbb{C}. \tag{1.73}$$

1.5.2 Rational Functions

By analogy with real functions, the quotient of two complex polynomials $P_m(z)$ and $Q_m(z)$ of respective degrees m and n, is called a *rational function*, and it will be assumed that any factors common to $P_m(z)$ and $Q_n(z)$ have been removed. The resulting function

$$f(z) = P_m(z)/Q_n(z) \tag{1.74}$$

is, by Theorems 1.3.2(iii) and 1.3.3, analytic for every z that is not a zero of $Q_n(z)$. If $Q_n(z)$ has n distinct zeros $z_1, z_2, \ldots, z_n$ the rational function in Equation (1.74) will only cease to be analytic at these n points. These n points are called *poles* of the function, where a function $f(z)$ will be said to have a *pole* at $z = z_0$ if

$$\lim_{z \to z_0} |f(z)| = \infty. \tag{1.75}$$

Functions other than rational functions can have poles but in applications of complex analysis, the poles of rational functions occur frequently and are important.

If one of the zeros of $Q_n(z)$ is repeated r times, the zero will be said to have *multiplicity* r, in which case the number of points where Equation (1.74) has a pole will be reduced to $n - r + 1$. A rational function like Equation (1.74) is an example of a *meromorphic function*, which is defined as a function that is analytic (holomorphic) throughout a domain D, except for a finite number of points at each of which it has a pole.

It is often helpful to simplify a rational function such as Equation (1.74) by expressing it as a sum of simpler rational functions known as its *partial fraction* expansion. In general, if $P_m(z)$ is of degree $m \leqslant n - 1$, and none of the zeros $z_1, z_2, \dots, z_n$ of $P_n(z)$ is a multiple zero, the *partial fraction* expansion of Equation (1.74) takes the form

$$\frac{P_m(z)}{Q_n(z)} = \frac{A_1}{z - z_1} + \frac{A_2}{z - z_2} + \dots + \frac{A_n}{z - z_n}, \tag{1.76}$$

where the numbers $A_1, A_2, \dots, A_n$ are called *undetermined coefficients*. The undetermined coefficients are found precisely in the way they are found when using a partial fraction expansion in elementary calculus, with the exception that now, in general, the coefficients can be complex numbers. This partial fraction expansion has to be modified if a zero z_s of $Q_n(z)$ has multiplicity p, because then Equation (1.76) will have p terms, each with the same denominator $z = z_s$, in which case these p terms must be replaced by the set of p terms

$$\frac{B_1}{z - z_s} + \frac{B_2}{(z - z_s)^2} + \dots + \frac{B_p}{(z - z_s)^p}. \tag{1.77}$$

Given a rational function $f(z)$ there are various ways of finding its partial fraction expansion, but the classical way, as used in elementary calculus, is illustrated in the next example.

Example 1.5.1 A Partial Fraction Expansion Involving a Repeated Zero

Use partial fractions to simplify

$$f(z) = \frac{3z^4 + 13z^3 + 19z^2 + 13z + 2 + i}{(z + 2)(z + 1)^2}.$$

SOLUTION

The difficult part of any partial fraction expansion involves factoring the denominator $Q_n(z)$, but this does not occur here because the denominator is already factored as $(z + 2)(z + 1)^2$, showing that zero $z = -1$ has multiplicity 1, while zero $z = -2$ has multiplicity 2. However, another difficulty with this example exists because the degree of the numerator is 4 but the degree of the denominator is 3, unlike the case discussed previously in which the degree of the numerator had to be *strictly less* than the degree of the denominator. The way around this difficulty adopted here is to divide the denominator into the numerator longhand, and then to apply partial fractions to the remaining rational function. The result of the division is easily seen to be

$$f(z) = 3z + 1 + \frac{2z + i}{(z + 2)(z + 1)^2},$$

so now it is only necessary to find the partial fraction expansion of the last term on the right. The appropriate form of the partial fraction expansion is

$$\frac{2z+i}{(z+2)(z+1)^2} = \frac{A_1}{z+2} + \frac{B_1}{z+1} + \frac{B_2}{(z+1)^2}.$$

Multiplying this by $(z+2)(z+1)^2$ it becomes

$$2z + i = A_1(z+1)^2 + B_1(z+1)(z+2) + B_2(z+2).$$

This must be an identity, and so it must be true for all z. Two obvious choices for z that lead quickly to undetermined coefficients are $z = -1$ and $x = -2$. Setting $z = -1$ gives $B_2 = -2 + i$, while $z = -2$ gives $A_1 = -4 + i$. To find B_1 we must make a different choice for z, so setting $z = 0$ gives $i = A_1 + 2B_1 + B_2$, from which it follows that $B_1 = 4 - i$. So the required partial fraction expansion is

$$\frac{3z^4 + 13z^3 + 19z^2 + 13z + 2 + i}{(z+2)(z+1)^2} = 3z + 1 + \frac{-4+i}{z+2} + \frac{4-i}{z+1} + \frac{-2+i}{(z+1)^2},$$

with a pole at $z = -1$ and another at $z = -2$. ◇

1.5.3 The Exponential Function

We define the *complex exponential function*, denoted either by e^z or by $\exp z$, in terms of $z = x + iy$, as

$$e^z = e^{x+iy} = e^x(\cos y + i \sin y). \tag{1.78}$$

Setting $e^z = u + iv$, we see that $u = e^x \cos y$ and $v = e^x \sin y$ and these u and v satisfy the Cauchy–Riemann equations (check this) and together with their first-order partial derivatives are continuous, showing that e^z is analytic for all z, and so is an entire function. From Theorem 1.3.3 we also have that

$$\frac{d}{dz}[e^z] = \frac{\partial u}{\partial x} + i\frac{\partial v}{\partial x} = e^x(\cos y + i \sin y) = e^z.$$

This same form of argument establishes the more general result that

$$\frac{d}{dz}[e^{kz}] = ke^{kz}, \tag{1.79}$$

for any complex constant k.

It is left as an exercise to show that an immediate consequence of definition in Equation (1.78) is that

$$e^{z_1+z_2} = e^{z_1}e^{z_2}, \tag{1.80}$$

a special case of which is

$$e^{x+iy} = e^x(\cos y + i \sin y). \tag{1.81}$$

Thus the multiplication of complex exponential functions obeys the same rule as in the real variable case. When the result in Equation (1.80) is written in the equivalent form

$$\exp(z_1 + z_2) = \exp(z_1)\exp(z_2)$$

it expresses more clearly the way in which $\exp(z_1)$ and $\exp(z_2)$ combine when multiplied.

It follows from Equation (1.78) that

$$|e^z| = e^x, \tag{1.81}$$

and that

$$\arg(e^z) = y + 2n\pi, \quad \text{for } n = 0, \pm 1, \pm 2, \ldots . \tag{1.82}$$

Thus the modulus of e^z never vanishes, while the result in Equation (1.82) shows that e^z is periodic in y with period 2π. As a result, if the behavior of e^z is known in any infinite strip in the complex plane of width 2π drawn parallel to the real axis, the periodicity with respect to y will determine the behavior of e^z for all z. The semi-infinite strip in the z-plane

$$-\pi < y \leqslant \pi \tag{1.83}$$

is called the *fundamental strip* for e^z. By restricting y in this manner, one-to-one relationship exists between points z in the fundamental strip and the points $w = e^z$ in the w-plane. Here, a *one-to-one* relationship means that to one point in the fundamental strip in the z-plane there corresponds precisely one point in the w-plane, and conversely. It is important to remember that e^∞ has no meaning in the complex plane.

1.5.4 The Logarithmic Function

The function $\ln z$ is called *the* ***natural logarithm*** of the *complex variable* $z = x + iy$. So to avoid confusion, the natural logarithm of a *real* number r is denoted by $\ln_e r$. The function ln is defined as the function inverse to the exponential function so that

$$w = \ln z \quad \text{when } e^w = z. \tag{1.84}$$

Setting $w = u + iv$ and $z = re^{i\theta}$ with $r = |z|$, the relationship $e^w = z$ becomes $e^{u+iv} = e^u e^{iv} = re^{i\theta}$, so equating the modulus on each side of this equation gives

$$e^u = r = |z|, \quad \text{showing that } u = \ln_e|z|, \tag{1.85}$$

where $\ln_e|z|$ is the natural logarithm of the real number $|z|$, defined for all $z \neq 0$.

Similarly, equating the arguments on each side of the equation gives

$$v = \theta = \arg(z) = \operatorname{Arg}(z) + 2n\pi, \quad \text{for } n = 0, \pm 1, \pm 2, \ldots, \tag{1.86}$$

where

$$-\pi < \operatorname{Arg}(z) \leqslant \pi. \tag{1.87}$$

Combining Equations (1.85) and (1.86) gives

$$\ln z = \ln_e|z| + i(\operatorname{Arg}(z) + 2n\pi), \quad \text{for } z \neq 0,\ n = 0, \pm 1, \pm 2, \ldots. \tag{1.88}$$

This last result shows that $\ln z$ is *infinitely many-valued* for any given z, with all real parts of $\ln z$ the same, but with the imaginary parts differing from one another by integral multiples of 2π.

The *principal part* of $\ln z$, denoted here by $\operatorname{Ln} z$, is chosen to correspond to the situation where $\arg(z)$ and $\operatorname{Arg}(z)$ coincide, and so $\operatorname{Ln} z$ is defined as

$$\operatorname{Ln} z = \ln_e|z| + i\operatorname{Arg}(z) \quad \text{for } z \neq 0. \tag{1.89}$$

Consequently, if $w = \operatorname{Ln} z = u + iv$, then

$$u = \ln_e|z| \quad \text{and} \quad -\pi < v \leqslant \pi \quad \text{for } z \neq 0. \tag{1.90}$$

We now prove that $\operatorname{Ln} z$ is a continuous function of z except when z lies on the negative real axis, across which it is discontinuous. Write $z = re^{i\theta}$, so that

$$\operatorname{Ln} z = \ln_e r + i\theta, \tag{1.91}$$

and let $\theta \downarrow \alpha$ signify that θ decreases to α, and let $\theta \uparrow \alpha$ signify that θ increases to α. Now consider $\lim_{z \to z_0} \operatorname{Ln} z$ at an arbitrary point $z_0 = r_0 e^{i\alpha}$ with

$-\pi < \alpha < \pi$ and $r_0 \neq 0$, so that z_0 is arbitrary, but not on the negative real axis. Then

$$\lim_{\substack{r \to r_0 \\ \theta \uparrow \alpha}} \mathrm{Ln}(re^{i\theta}) = \lim_{r \to r_0} \ln_e r + \lim_{\theta \uparrow \alpha}(i\theta) = \ln r_0 + \alpha i,$$

and, similarly,

$$\lim_{\substack{r \to r_0 \\ \theta \downarrow \alpha}} \mathrm{Ln}(re^{i\theta}) = \lim_{r \to r_0} \ln_e r + \lim_{\theta \downarrow \alpha}(i\theta) = \ln r_0 + \alpha i.$$

As z_0 was arbitrary, this has established the continuity of $\mathrm{Ln}\, z$ except when z lies on the negative real axis.

Finally, let us consider the limiting behavior of $\mathrm{Ln}\, z$ across any point $x = -r_0$ on the negative real axis with $r_0 > 0$. The limits then show that

$$\lim_{\substack{r \to r_0 \\ \theta \uparrow \pi}} \mathrm{Ln}(re^{i\theta}) = \lim_{r \to r_0} \ln_e r + \lim_{\theta \uparrow \pi}(i\theta) = \ln_e r_0 + \pi i,$$

and

$$\lim_{\substack{r \to r_0 \\ \theta \uparrow -\pi}} \mathrm{Ln}(re^{i\theta}) = \lim_{r \to r_0} \ln_e r + \lim_{\theta \uparrow -\pi}(i\theta) = \ln_e r_0 - \pi i.$$

Thus $\mathrm{Ln}\, z$ has a jump of $2\pi i$ across each point on the negative real axis. Thus, we have proved the assertion that $\mathrm{Ln}\, z$ is continuous everywhere except at the origin and across points on the negative real axis where it is discontinuous.

To examine the relationship between $w = \mathrm{Ln}\, z$ and its inverse function $z = e^w$, it is necessary to consider the behavior of these functions in the complex plane. If the point $z = 0$ and all of the points on the negative real axis are removed from the z-plane, a one-to-one correspondence exists between the remaining points in the z-plane and points in the strip $-\pi < v \leqslant \pi$ in the w-plane. Thus for any z in this modified z-plane, the function $w = \mathrm{Ln}\, z$ is determined uniquely. Conversely, for any point w in the strip $-\pi < v \leqslant \pi$ in the w-plane, the function $z = e^w$ is also determined uniquely for all w in this modified w-plane.

When a complex plane is modified in this manner by having removed from it all points on a line to make the correspondence between two functions one-to-one, the plane is said to be **cut**. In this case the z-plane was cut along the negative real axis. The fundamental strip in the w-plane and the cut z-plane are shown in Figure 1.14. The cut is to be regarded as a barrier that may not be crossed by z. In Figure 1.14, a full line shows points that belong to a region and a dashed line points that are excluded from the region.

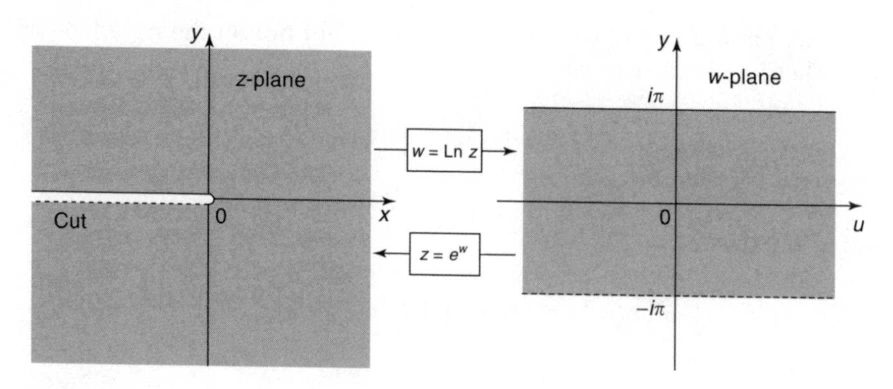

FIGURE 1.14
The cut z-plane for $w = \text{Ln}\, z$ and the fundamental strip in the w-plane.

The individual functions

$$w = \ln_e |z| + i(\text{Arg}(z) + 2n\pi),$$

corresponding to different choices of $n = 0, \pm 1, \pm 2, \ldots$, are called *branches* of $\ln z$, and the logarithmic function is seen to have infinitely many branches. In this context, the cut in the z-plane along the negative real axis is called a *branch cut*.

It is a straightforward matter to show that $\text{Ln}\, z = \ln_e r + i\theta$ satisfies the conditions of Theorem 1.1.3 provided θ is restricted so that $-\pi < \theta \leqslant \pi$, in which case $\text{Ln}\, z$ is an analytic function in the cut z-plane. The same form of argument establishes the fact that each branch of $\ln z$ is also analytic in the cut z-plane.

Writing

$$\begin{aligned}\ln z &= \ln_e[(x^2 + y^2)^{1/2}] + i \arg(z)\\ &= \tfrac{1}{2}\ln_e(x^2 + y^2) + i \tan^{-1}(y/x),\end{aligned}$$

it follows from Theorem 1.3.3 that for any z in the cut z-plane

$$\begin{aligned}\frac{d}{dz}[\ln z] &= \tfrac{1}{2}\frac{\partial}{\partial x}\ln_e(x^2 + y^2) + i\frac{\partial}{\partial x}\tan^{-1}(y/x)\\ &= \frac{x}{(x^2 + y^2)^{1/2}} + i\frac{(-y/x^2)}{1 + (y/x)^2} = \frac{x - iy}{x^2 + y^2} = \frac{1}{z}.\end{aligned}$$

So we have proved that in the cut z-plane the derivative of the logarithmic function $\ln z$ is

$$\frac{d}{dz}[\ln z] = \frac{1}{z}. \tag{1.92}$$

The following familiar properties of $\ln z$ can also be proved in similar fashion, though this is left as an exercise:

$$\ln(z_1 z_2) = \ln z_1 + \ln z_2, \tag{1.93}$$

$$\ln(z_1/z_2) = \ln z_1 - \ln z_2, \tag{1.94}$$

$$\ln(z^{p/q}) = (p/q)\ln z, \quad p, q \text{ integers.} \tag{1.95}$$

These last three results require interpretation because of the many-valued nature of $\ln z$.

They are to be taken to mean that the value of the left-side of each result is included among the set of values of the right side. Results in Equations (1.93) and (1.94) are not necessarily true if $\ln z$ is replaced by the principal value $\operatorname{Ln} z$. This is because of the constraint laced on the arguments involved that must satisfy the principal value condition.

As a generalization of z^n we have z^c where $z = x + iy$ and c is an arbitrary complex constant. This is defined as

$$z^c = \exp(c \ln z), \quad \text{for } z \neq 0. \tag{1.96}$$

The infinitely many-valued nature of $\ln z$ makes the complex power z^c many-valued, and its principal value is defined as

$$z^c = \exp(c \operatorname{Ln} z), \quad \text{for } z \neq 0. \tag{1.97}$$

An immediate consequence of Equation (1.96) is that if α and β are complex number, then when $z \neq 0$,

$$\begin{aligned} z^\alpha z^\beta &= z^{\alpha+\beta}, \quad z^\alpha / z^\beta = z^{\alpha-\beta}, \\ 1/z^\alpha &= z^{-\alpha} \quad \text{and} \quad (z^\alpha)^n = z^{n\alpha} \text{ for integral } n. \end{aligned} \tag{1.98}$$

If the z-plane is cut along the negative real axis up to and including the origin, an application of Theorem 1.3.4(iv) to the principal value of $f(z) = z^c$ shows that in the cut z-plane

$$\frac{d}{dz}[\exp(c \operatorname{Ln} z)] = \frac{c}{z}\exp(c \operatorname{Ln} z). \tag{1.99}$$

This may be re-expressed in the more familiar, though less precise form

$$\frac{d}{dz}[z^c] = c z^{c-1}, \tag{1.100}$$

as long as it is understood that the principal value of z^c is to be used on the right.

For any complex number $a \neq 0$, we define

$$a^z = \exp(z \ln a). \tag{1.101}$$

Once the value to be assigned to the many-valued function $\ln a$ has been chosen, a branch of Equation (1.101) is identified that is analytic in the cut z-plane. It then follows as before that

$$\frac{d}{dz}[a^z] = a^z \ln a, \tag{1.102}$$

where here again it is to be understood that the principal value of a^z is to be used in the expression on the right. The next example illustrates these properties of the complex exponential and logarithmic functions.

Example 1.5.2 Special Values of ln z and z^c

(i) Find $\ln(-1)$ and $\mathrm{Ln}(-1)$. As $e^{i\pi} = -1$, it follows that $|e^{i\pi}| = 1$ and $\mathrm{Arg}(e^{i\pi}) = \pi$. So from Equation (1.88)

$$\begin{aligned}\ln(-1) &= \ln_e|e^{\pi i}| + i(\mathrm{Arg}(e^{i\pi}) + 2n\pi), \quad \text{for } n = 0, \pm 1, \pm 2, \ldots, \\ &= \ln_e 1 + (\pi + 2n\pi)i, \\ &= (1 + 2n)\pi i, \quad \text{for } n = 0, \pm 1, \pm 2, \ldots \quad (\text{because } \ln 1 = 0),\end{aligned}$$

and so

$$\ln(-1) = \pm\pi i, \pm 3\pi i, \pm 5\pi i, \ldots.$$

In particular, setting $n = 0$ the principal value is found to be

$$\mathrm{Ln}(-1) = \pi i.$$

(ii) Find $\ln z$ and $\mathrm{Ln}\, z$, when $z = \frac{5}{2}(1 + \mathrm{i}\sqrt{3})$. Because $z = 5e^{i\pi/3}$, we see that $|z| = 5$ and $\mathrm{Arg}(z) = \pi/3$, so from Equation (1.88)

$$\ln\left(\tfrac{5}{2}(1 + i\sqrt{3})\right) = \ln_e 5 + \tfrac{1}{3}\pi i + 2n\pi i, \quad \text{for } n = 0, \pm 1, \pm 2, \ldots.$$

Thus

$$\ln\left(\tfrac{5}{2}(1 + i\sqrt{3})\right) = \ln_e 5 + \tfrac{1}{3}\pi i, \ln 5 - \tfrac{5}{3}\pi i, \ln_e 5 + \tfrac{7}{3}\pi i, \ldots.$$

In particular, setting $n = 0$, the principal value is found to be

$$\ln\left(\tfrac{5}{2}(1 + i\sqrt{3})\right) = \ln_e 5 + \tfrac{1}{3}\pi i.$$

(iii) Find z^i, when $z = \frac{3}{\sqrt{2}}(1+i)$. Because $z = 3e^{i\pi/4}$, we see that $|z| = 3$ and $\mathrm{Arg}(z) = \pi/4$. From Equation (1.96) we have

$$\begin{aligned} z^i &= \exp(i \ln z) = \exp\left[i\left(\ln_e 3 + \tfrac{1}{4}\pi i + 2n\pi i\right)\right], \qquad \text{for } n = 0, \pm1, \pm2, \ldots, \\ &= \exp\left[-\tfrac{1}{4}(1+8n)\pi + i\ln 3\right], \\ &= \exp\left[-\tfrac{1}{4}(1+8n)\pi\right][\cos(\ln_e 3) + i\sin(\ln_e 3)], \qquad \text{for } n = 0, \pm1, \pm2, \ldots. \end{aligned}$$

This is infinitely many-valued and its principal value, corresponding to $n = 0$, is

$$z^i = e^{-\pi/4}[\cos(\ln_e 3) + i\sin(\ln_e 3)].$$

◇

1.5.5 Trigonometric Functions

For arbitrary $z = x + iy$, we define *the complex sine* and *cosine* functions as

$$\sin z = \frac{e^{iz} - e^{-iz}}{2i}, \qquad \cos z = \frac{e^{iz} + e^{-iz}}{2}. \tag{1.103}$$

These results provide a direct extension of the Euler formula $e^{i\theta} = \cos\theta + i\sin\theta$ which is only defined for real values of θ.

Following the pattern of real-valued trigonometric functions, the other complex trigonometric functions are defined in terms of $\sin z$ and $\cos z$ as:

$$\tan z = \frac{\sin z}{\cos z}, \qquad \csc z = \frac{1}{\sin z} \quad (\text{also written } \operatorname{cosec} z)$$

$$\sec z = \frac{1}{\cos z}, \qquad \cot z = \frac{\cos z}{\sin z}. \tag{1.104}$$

Using the results $e^{2\pi ni} = e^{-2\pi ni} = 1$, for $n = 0, \pm1, \pm2, \ldots$, allows us to write

$$\begin{aligned} \sin z &= \frac{e^{iz} - e^{-iz}}{2i} = \frac{e^{iz}e^{2\pi ni} - e^{-iz}e^{-2\pi ni}}{2i} \\ &= \frac{e^{i(z+2n\pi)} - e^{-i(z+2n\pi)}}{2i} = \sin(z + 2n\pi). \end{aligned}$$

A similar argument applies to the cosine function, so we have shown that, as in the real variable case, these complex functions are periodic with period 2π, because

$$\sin(z + 2n\pi) = \sin z \quad \text{and} \quad \cos(z + 2n\pi) = \cos z, \tag{1.105}$$

for $n = 0, \pm1, \pm2, \ldots$.

The fact that $\tan z$ and $\cot z$ are periodic with period π can be established in the same way, so that

$$\tan(z + n\pi) = \tan z \quad \text{and} \quad \cot(z + n\pi) = \cot z, \tag{1.106}$$

for $n = 0, \pm 1, \pm 2, \ldots$.

The following identities are direct consequences of the definitions in Equations (1.103) and (1.104):

$$\sin^2 z + \cos^2 z = 1, \tag{1.107}$$

$$\sin(z_1 \pm z_2) = \sin z_1 \cos z_2 \pm \cos z_1 \sin z_2, \tag{1.108}$$

$$\cos(z_1 \pm z_2) = \sin z_1 \cos z_2 \mp \cos z_1 \sin z_2, \tag{1.109}$$

$$\sinh(ix) = i \sinh x, \quad \cosh(ix) = \cosh x \ (x \text{ real}). \tag{1.110}$$

Euler's formula also holds for complex z when it becomes

$$e^{iz} = \cos z + i \sin z, \tag{1.111}$$

and it also follows that

$$\sin(-z) = -\sin z, \quad \cos(-z) = \cos z, \quad \tan(-z) = -\tan z. \tag{1.112}$$

The fact that $\sin z$ and $\cos z$ are defined as linear combinations of e^{iz} and e^{-iz} taken together with the fact that the exponential functions are entire functions, means that $\sin z$ and $\cos z$ are also entire functions. Furthermore, the definitions of other complex trigonometric functions in terms of $\sin z$ and $\cos z$ mean that they are analytic except at the zeros of the denominators of their defining relations.

Differentiation of the complex trigonometric functions using the above definitions and the properties of e^z shows that

$$\frac{d}{dz}[\sin z] = \cos z, \quad \frac{d}{dz}[\cos z] = -\sin z \quad \text{for all } z, \text{ Equation text is deleted.}$$

$$\frac{d}{dz}[\tan z] = \sec^2 z, \quad \frac{d}{dz}[\sec z] = \sec z \tan z, \quad \text{for } z \neq \tfrac{1}{2}\pi + n\pi, \ n = 0, \pm 1, \pm 2, \ldots, \tag{1.113}$$

$$\frac{d}{dz}[\csc z] = -\csc z \cot z, \quad \frac{d}{dz}[\cot z] = -\csc^2 z, \quad \text{for } z \neq n\pi, \ n = 0, \pm 1, \pm 2, \ldots .$$

Combining Equations (1.108) to (1.110) and using the real variable definitions of the hyperbolic sine and cosine functions shows that $\sin(x+iy) = \sin x \cosh y + i\cos x \sinh y$, and

$$\cos(x+iy) = \cos x \cosh y - i \sin x \sinh y. \tag{1.114}$$

The nonvanishing of $\sinh y$ for $y \neq 0$ coupled with the results from Equations (1.114) implies that $\sin z$ and $\cos z$ can only vanish on the real axis. Thus the zeros of $\sin z$ occur at the points

$$z = n\pi, \quad \text{for } n = 0, \pm 1, \pm 2, \dots, \tag{1.115}$$

while the zeros of $\cos z$ can only occur at the points

$$z = \tfrac{1}{2}(2n+1)\pi, \quad \text{for } n = 0, \pm 1, \pm 2, \dots. \tag{1.116}$$

Thus the zeros of the complex functions $\sin z$ and $\cos z$ occur at the same points as those of the corresponding real functions.

Example 1.5.3 Roots Involving a Complex Sine Function
Find all of the Roots of $\sin z = \cosh 3$.

SOLUTION
Combining the first result in Equation (1.114) with the equation $\sin z = \cosh 3$, and equating its real and imaginary parts gives

$$\sin x \cosh y = \cosh 3 \quad \text{and} \quad \cos x \sinh y = 0.$$

As $\cosh y \geqslant 1$ for all y, for the first equation to be true it is necessary that $\sin x > 0$ and $y \neq 0$. Using $y > 0$ in the second equation then shows that it can only be satisfied when x is a zero of $\cos x$. The only zeros of $\cos x = 0$ for which $\sin x > 0$ are seen to occur when $x = \frac{1}{2}(4n+1)\pi$, for $n = 0, \pm 1, \pm 2, \dots$. Thus, for these values of x, the first equation becomes $\cosh y = \cosh 3$, which has two solutions, $y = \pm 3$. Consequently the roots of $\sin z = \cosh 3$ are infinite in number and are given by

$$z = \tfrac{1}{2}(4n+1)\pi \pm 3i, \quad \text{for } n = 0, \pm 1, \pm 2, \dots. \qquad \diamond$$

1.5.6 Hyperbolic Functions

For arbitrary $z = x + iy$ the *complex hyperbolic sine* and *cosine* functions are defined as

$$\sinh z = \frac{e^z - e^{-z}}{2}, \quad \cosh z = \frac{e^z + e^{-z}}{2}. \tag{1.117}$$

These are direct generalizations of the corresponding real variable functions and they are obviously entire functions. The following identities follow directly from these definitions:

$$\cosh^2 z + \sinh^2 z = 1, \tag{1.118}$$

$$\sinh(z_1 \pm z_2) = \sinh z_1 \cosh z_2 \pm \cosh z_1 \sinh z_2, \tag{1.119}$$

$$\cosh(z_1 \pm z_2) = \cosh z_1 \cosh z_2 \pm \sinh z_1 \sinh z_2, \tag{1.120}$$

$$\sinh(ix) = i \sin x, \quad \cosh(ix) = \cos x \ (x \text{ real}) \tag{1.121}$$

$$i \sinh z = \sin(iz), \quad \cosh z = \cos(iz) \ (z \text{ complex}), \tag{1.122}$$

and also

$$\sinh(-z) = -\sinh z, \quad \cosh(-z) = \cosh z, \quad \tanh(-z) = -\tanh z, \tag{1.123}$$

where $\tanh z$ and the other complex hyperbolic functions are defined as:

$$\tanh z = \frac{\sinh z}{\cosh z}, \quad \operatorname{csch} z = \frac{1}{\sinh z} \ (\text{also written cosech } z),$$

$$\operatorname{sech} z = \frac{1}{\cosh z}, \quad \coth z = \frac{\cosh z}{\sinh z}. \tag{1.124}$$

The derivatives of these functions found in the usual manner are formally the same as those of the corresponding real variable functions, namely:

$$\frac{d}{dz}[\sinh z] = \cosh z, \quad \frac{d}{dz}[\cosh z] = \sinh z,$$

$$\frac{d}{dz}[\tanh z] = \operatorname{sech}^2 z, \quad \frac{d}{dz}[\operatorname{csch} z] = -\operatorname{csch} z \coth z, \tag{1.125}$$

$$\frac{d}{dz}[\operatorname{sech} z] = -\operatorname{sech} z \tanh z, \quad \frac{d}{dz}[\coth z] = -\operatorname{csch}^2 z.$$

Combining Equations (1.119) to (1.121) leads to the following useful results

$$\sinh(x \pm iy) = \sinh x \cos y \pm i \cosh x \sin y$$

and

$$\cosh(x \pm iy) = \cosh x \cos y \pm i \sinh x \sin y. \tag{1.126}$$

Inspection of these results shows that the zeros of $\sinh z$ occur at the points

$$z = n\pi i, \quad n = 0, \pm 1, \pm 2, \dots, \tag{1.127}$$

while the zeros of $\cosh z$ occur at the points

$$z = \tfrac{1}{2}(2n+1)\pi i, \quad n = 0, \pm 1, \pm 2, \dots . \tag{1.128}$$

Notice that unlike the corresponding real variable case, the hyperbolic functions $\sinh z$ and $\cosh z$ are periodic functions with period $2\pi i$. This can be seen by using the result $e^{2n\pi i} = e^{-2n\pi i} = 1$ for $n = 0, \pm 1, \pm 2, \dots$, in the definitions of $\cosh z$ and $\sinh z$ because, for example,

$$\begin{aligned}\cosh z &= \tfrac{1}{2}(e^{z} + e^{-z}) = \tfrac{1}{2}(e^{z}e^{2\pi ni} + e^{-z}e^{-2\pi ni}) \\ &= \tfrac{1}{2}(e^{(z+2n\pi i)} + e^{-(z+2n\pi i)}) = \cosh(z + 2n\pi i),\end{aligned} \tag{1.129}$$

with the corresponding result

$$\sinh z = \sinh(z + 2n\pi i). \tag{1.130}$$

Example 1.5.4 Roots of the Hyperbolic Cosine Function
Find all of the roots of $\cosh z = -1$.

SOLUTION
As $\cosh z = -1$, from the second result in Equation (1.126) we have

$$\cosh x \cos y + i \sinh x \sin y = -1,$$

so equating the real and imaginary parts gives

$$\cosh x \cos y = -1 \quad \text{and} \quad \sinh x \sin y = 0.$$

The first result is only possible if $x = 0$ and $\cos y = -1$, so that $y = \frac{1}{2}(2n + 1)\pi$, in which case the second result is satisfied automatically, so the roots of the equation are

$$z = \tfrac{1}{2}(2n + 1)\pi i, \quad \text{for } n = 0, \pm 1, \pm 2, \dots. \qquad \diamond$$

1.5.7 Inverse Trigonometric and Hyperbolic Functions

The following text, for the sake of uniformity, denotes functions inverse to the trigonometric and hyperbolic functions just discussed by adding the prefix *arc* to the corresponding function. So, for example, we write

$$w = \arcsin z \text{ when } z = \sin w \quad \text{and} \quad w = \operatorname{arccosh} z \text{ when } z = \cosh w.$$

An equivalent notation also in use involves adding a superscript -1 to a function to denote the inverse function, so in this notation

$$w = \sin^{-1} z \text{ when } z = \sin w \quad \text{and} \quad w = \cosh^{-1} z \text{ when } z = \cosh w.$$

These inverse functions were first encountered in Examples 1.5.3 and 1.5.4, where they were many-valued, though at the time they were not identified as inverse functions. It was possible to evaluate these particular inverse functions by considering their real and imaginary parts separately because only an inverse function with a purely real argument was involved. When inverse trigonometric and hyperbolic functions involve arbitrary complex numbers a different approach becomes necessary. To show the approach that is required we now find all of the complex numbers $w = \arcsin z$ for an arbitrary fixed complex number z.

By definition,

$$z = \sin w = \frac{e^{iw} - e^{-iw}}{2i},$$

so after multiplication by $2ie^{iw}$ this becomes

$$(e^{iw})^2 - 2iz(e^{iw}) - 1 = 0,$$

which is a quadratic equation for e^{iw}. Solving this gives

$$e^{iw} = iz + \sqrt{1 - z^2},$$

where it will be recalled that the $\pm$ usually associated with the square root sign is omitted since in complex analysis the square root operation is understood to be two-valued. Taking the natural logarithm shows that the inverse sine function $w = \arcsin z$ can be written in the form

$$w = -i \ln\left(\sqrt{1 - z^2} + iz\right).$$

Thus in this result the logarithmic function and the square root function both contribute to the infinitely many-valued behavior of this inverse function.

The same form of reasoning applied to the inverse cosine and tangent functions lead to the results

$$\arcsin z = -i \ln\left(iz + \sqrt{1 - z^2}\right), \quad z \in \mathbb{C}, \tag{1.131}$$

$$\arccos z = -i \ln\left(z + i\sqrt{1 - z^2}\right), \quad z \in \mathbb{C}, \tag{1.132}$$

$$\arctan z = \frac{i}{2} \ln\left(\frac{i + z}{i - z}\right), \quad z \neq \pm i. \tag{1.133}$$

If these results are differentiated using implicit differentiation, for any fixed branch the become analytic functions with the derivatives

$$\frac{d}{dz}[\arcsin z] = \frac{1}{\sqrt{1-z^2}}, \quad z \in \mathbb{C}_2, \tag{1.134}$$

$$\frac{d}{dz}[\arccos z] = \frac{-1}{\sqrt{1-z^2}}, \quad z \in \mathbb{C}_2, \tag{1.135}$$

$$\frac{d}{dz}[\arctan z] = \frac{1}{1+z^2}, \quad z \in \mathbb{C}_1, \tag{1.136}$$

where $\mathbb{C}_1$ is the complex plane cut along the imaginary axis from $y = 1$ to $+\infty$ and from $y = -1$ to $-\infty$, and $\mathbb{C}_2$ is the complex plane cut along the real axis from $x = 1$ to $+\infty$ and from $x = -1$ to $-\infty$.

The functions in Equations (1.131) to (1.136) are infinitely many-valued, but they can be made analytic by restricting z in such a way that a particular branch of the logarithmic function is used together with a specific branch of the square root function. The branch of the square root function used in Equations (1.131) or (1.132) must, of course, be the one used in Equations (1.134) or (1.135) for the function and its derivative to be compatible.

Similar arguments applied to the inverse hyperbolic functions show that

$$\operatorname{arcsinh} z = \ln\left(z + \sqrt{z^2+1}\right), \quad \text{for all } z, \tag{1.137}$$

$$\operatorname{arccosh} z = \ln\left(z + \sqrt{z^2-1}\right), \quad \text{for all } z, \tag{1.138}$$

$$\operatorname{arctanh} z = \frac{1}{2}\ln\left(\frac{1+z}{1-z}\right), \quad \text{for } z \neq \pm 1. \tag{1.139}$$

Implicit differentiation of these results for any fixed branch yield analytic functions with the derivatives

$$\frac{d}{dz}[\operatorname{arcsinh} z] = \frac{1}{\sqrt{z^2+1}}, \quad z \in \mathbb{C}_1, \tag{1.140}$$

$$\frac{d}{dz}[\operatorname{arccosh} z] = \frac{1}{\sqrt{z^2-1}}, \quad z \in \mathbb{C}_2, \tag{1.141}$$

$$\frac{d}{dz}[\operatorname{arctanh} z] = \frac{1}{1-z^2}, \quad z \in \mathbb{C}_2, \tag{1.142}$$

Here also, the infinitely many-valued functions in Equations (1.137) to (1.139) can be made analytic by restricting z to a particular branch of the logarithmic function and using a specific branch of the square root function where it occurs in Equations (1.140) and (1.141).

Example 1.5.5 The Inverse Hyperbolic Sine and Its Derivative
Find all of the values of $\text{arcsinh}\, z$ and its derivative when $z = i\sqrt{5}$.

SOLUTION
From Equation (1.137) we have

$$\text{arcsinh}(i\sqrt{5}) = \ln(i\sqrt{5} + \sqrt{-4}) = \ln\left[i(\sqrt{5} \pm 2)\right].$$

However,

$$\left|i(\sqrt{5} \pm 2)\right| = \sqrt{5} \pm 2, \quad \text{and} \quad \text{Arg}\left[i(\sqrt{5} \pm 2)\right] = \tfrac{1}{2}\pi,$$

so

$$\arg[i(\sqrt{5} \pm 2)] = \tfrac{1}{2}\pi + 2n\pi = \tfrac{1}{2}(1 + 4n\pi), \quad \text{for } n = 0, \pm 1, \pm 2, \ldots.$$

Thus

$$\text{arcsinh}(\sqrt{5}) = \ln(\sqrt{5} \pm 2) + i\tfrac{1}{2}(1 + 4n)\pi, \quad \text{for } n = 0, \pm 1, \pm 2, \ldots.$$

Taking the positive sign arising from the square root function gives the infinitely many-valued result

$$\text{arcsinh}(\sqrt{5}) = \ln(\sqrt{5} + 2) + i\tfrac{1}{2}(1 + 4n)\pi, \quad \text{for } n = 0, \pm 1, \pm 2, \ldots,$$

while taking the negative sign gives

$$\text{arcsinh}(\sqrt{5}) = \ln(\sqrt{5} - 2) + i\tfrac{1}{2}(1 + 4n)\pi, \quad \text{for } n = 0, \pm 1, \pm 2, \ldots.$$

If we now take the result corresponding to the positive sign together with the principal branch of the logarithmic function (corresponding to $n = 0$), we find that

$$\text{arcsinh}(\sqrt{5}) = \ln(\sqrt{5} + 2) + \frac{\pi i}{2}, \quad \text{for } n = 0, \pm 1, \pm 2, \ldots.$$

The compatible derivative corresponding to the positive sign follows from Equation (1.140), from which it is seen to be

$$\left[\frac{d}{dz}(\operatorname{arcsinh} z)\right]_{z=i\sqrt{5}} = \frac{1}{2i} = -\tfrac{1}{2}i.$$

If the negative branch of the square root had been taken with, say, the first branch of the logarithmic function (corresponding to $n = 1$) we would have obtained

$$\operatorname{arcsinh}(\sqrt{5}) = \ln(\sqrt{5} - 2) + \frac{5\pi i}{2},$$

and for the compatible derivative the result

$$\left[\frac{d}{dz}(\operatorname{arcsinh} z)\right]_{z=i\sqrt{5}} = \frac{1}{-2i} = \tfrac{1}{2}i. \qquad \diamond$$

The following text is a summary of the properties of analytic functions.

1.5.8 An Analytic Function and Its Derivatives

The function $f(z) = u + iv$ is analytic in a domain D with the derivative

$$f'(z) = \frac{\partial u}{\partial x} + i\frac{\partial v}{\partial x} = \frac{\partial v}{\partial y} - i\frac{\partial u}{\partial y} \quad \text{(Cartesian representation)}$$

or, equivalently,

$$\begin{aligned} f'(z) &= \left(\frac{\partial u}{\partial r}\cos\theta - \frac{1}{r}\frac{\partial u}{\partial \theta}\sin\theta\right) + i\left(\frac{\partial v}{\partial r}\cos\theta - \frac{1}{r}\frac{\partial v}{\partial \theta}\sin\theta\right) \\ &= \left(\frac{\partial v}{\partial r}\sin\theta + \frac{1}{r}\frac{\partial v}{\partial \theta}\cos\theta\right) - i\left(\frac{\partial u}{\partial r}\sin\theta + \frac{1}{r}\frac{\partial u}{\partial \theta}\cos\theta\right) \end{aligned} \quad \text{(Polar representation)}$$

if

(i) the first order partial derivatives of u and v are continuous in D, and
(ii) the Cauchy–Riemann equations are satisfied in D so that

$$\frac{\partial u}{\partial x} = \frac{\partial v}{\partial y}, \quad \frac{\partial u}{\partial y} = -\frac{\partial v}{\partial x} \quad \text{(Cartesian representation)}$$

$$\frac{\partial u}{\partial r} = \frac{1}{r}\frac{\partial v}{\partial \theta}, \quad \frac{\partial v}{\partial r} = -\frac{1}{r}\frac{\partial u}{\partial \theta}. \quad \text{(Polar representation)}$$

Harmonic Functions

The function ϕ is said to be a harmonic function if

$$\Delta\phi = 0,$$

where in two dimensions $\Delta\phi$ is the Laplacian

$$\Delta\phi = \frac{\partial^2\phi}{\partial x^2} + \frac{\partial^2\phi}{\partial y^2} \quad \text{(Cartesian representation)}$$

or, equivalently,

$$\Delta\phi = \frac{\partial^2\phi}{\partial r^2} + \frac{1}{r}\frac{\partial\phi}{\partial r} + \frac{1}{r^2}\frac{\partial^2\phi}{\partial\theta^2}. \quad \text{(Polar representation)}$$

If $f(z) = u + iv$ is an analytic function in a domain D, then both u and v are harmonic in D, so that

$$\Delta u = 0 \quad \text{and} \quad \Delta v = 0.$$

The functions u and v belonging to the analytic function $f(z) = u + iv$ are called conjugate harmonic functions.

Rules for Differentiation

If D is a domain where $f(z)$ and $g(z)$ are analytic, then

1. $\dfrac{d}{dz}[kf(z)] = kf'(z)$ (k is a complex constant);

2. $\dfrac{d}{dz}[f(z) \pm g(z)] = f'(z) \pm g'(z);$

3. $\dfrac{d}{dz}[f(z)g(z)] = f'(z)g(z) + f(z)g'(z);$

4. $\dfrac{d}{dz}\{f[g(z)]\} = \dfrac{d}{dz}\{f[g(z)]\}\dfrac{d}{dz}\{g(z)\};$

5. $\dfrac{d}{dz}\left[\dfrac{f(z)}{g(z)}\right] = \dfrac{g(z)f'(z) - f(z)g'(z)}{[g(z)]^2}, \quad (g(z) \neq 0).$

Derivatives of Elementary Functions

6. $\dfrac{d}{dz}[k] = 0,$ (k a complex constant);

7. $\dfrac{d}{dz}[z^n] = nz^{n-1}$, ($n$ an integer);

8. $\dfrac{d}{dz}[z^c] = cz^{c-1}$, ($c$ an arbitrary constant, and the principal value of z^c used on the right);

9. $\dfrac{d}{dz}[a_0 + a_1 z + \cdots + a_n z^n] = a_1 + a_2 z + \cdots + a_n z^{n-1}$;

10. $\dfrac{d}{dz}[e^{kz}] = ke^{kz}$, ($k$ a complex constant);

11. $\dfrac{d}{dz}[a^z] = a^z \ln a$, ($a$ an arbitrary complex constant and the principal value of a^z used on the right);

12. $\dfrac{d}{dz}[\ln z] = \dfrac{1}{z}$, $(z \neq 0)$.

Definitions of Trigonometric Functions

$$\sin z = \frac{e^{iz} - e^{-iz}}{2i}, \quad \cos z = \frac{e^{iz} + e^{-iz}}{2}, \quad \tan z = \frac{\sin z}{\cos z},$$

$$\csc z = \frac{1}{\sin z}, \quad \sec z = \frac{1}{\cos z}, \quad \cot z = \frac{\cos z}{\sin z}.$$

$\sin z$, $\cos z$, $\sec z$ and $\csc z$ have a fundamental period of 2π, while $\tan z$ and $\cot z$ have a fundamental period of π.

The following results are true for $n = 0, \pm 1, \pm 2, \ldots$

13. $\dfrac{d}{dz}[\sin z] = \cos z$, all z.

14. $\dfrac{d}{dz}[\cos z] = -\sin z$, all z.

15. $\dfrac{d}{dz}[\tan z] = \sec^2 z$, $z \neq \left(n + \frac{1}{2}\right)\pi$.

16. $\dfrac{d}{dz}[\csc z] = -\csc z \cot z$, $z \neq n\pi$.

17. $\dfrac{d}{dz}[\sec z] = \sec z \tan z$, $z \neq (n + \frac{1}{2})\pi$.

18. $\dfrac{d}{dz}[\cot z] = -\csc^2 z$, $z \neq n\pi$.

Definitions of Hyperbolic Functions

$$\sinh z = \frac{e^z - e^{-z}}{2}, \quad \cosh z = \frac{e^z + e^{-z}}{2}, \quad \tanh z = \frac{\sinh z}{\cosh z},$$
$$\operatorname{csch} z = \frac{1}{\sinh z}, \quad \operatorname{sech} z = \frac{1}{\cosh z}, \quad \coth z = \frac{1}{\tanh z}.$$

$\sinh z$, $\cosh z$, $\operatorname{sech} z$ and $\operatorname{csch} z$ have a fundamental period of $2\pi i$, while $\tanh z$ and $\coth z$ have a fundamental period of πi.

The following results are true when $n = 0, \pm 1, \pm 2 \ldots$.

19. $\dfrac{d}{dz}[\sinh z] = \cosh z.$

20. $\dfrac{d}{dz}[\cosh z] = \sinh z.$

21. $\dfrac{d}{dz}[\tanh z] = \operatorname{sech}^2 z, \quad z \neq (n + \tfrac{1}{2})\pi.$

22. $\dfrac{d}{dz}[\operatorname{csch} z] = -\operatorname{csch} z \coth z, \quad z \neq n\pi i.$

23. $\dfrac{d}{dz}[\operatorname{csch} z] = -\operatorname{csch} z \coth z, \quad z \neq n\pi i.$

24. $\dfrac{d}{dz}[\operatorname{sech} z] = -\operatorname{sech} z \coth z, \quad z \neq (n + \tfrac{1}{2})\pi i.$

25. $\dfrac{d}{dz}[\coth z] = -\operatorname{csch}^2 z, \quad z \neq n\pi i.$

Definitions of Inverse Trigonometric Functions

$$\arcsin z = -i \ln\left(iz + \sqrt{1 - z^2}\right), \quad \arccos z = -i \ln\left(z + i\sqrt{1 - z^2}\right),$$

$$\arctan z = \frac{i}{2} \ln\left(\frac{i + z}{i - z}\right).$$

26. $\dfrac{d}{dz}[\arcsin z] = \dfrac{1}{\sqrt{1 - z^2}},$ (for a fixed branch).

27. $\dfrac{d}{dz}[\arccos z] = \dfrac{-1}{\sqrt{1 - z^2}},$ (for a fixed branch).

28. $\dfrac{d}{dz}[\arctan z] = \dfrac{1}{1 + z^2},$ (for a fixed branch).

Definitions of Inverse Hyperbolic Functions

$$\operatorname{arcsinh} z = \ln\left(z + \sqrt{z^2+1}\right), \qquad \operatorname{arccosh} z = \ln\left(z + \sqrt{z^2-1}\right),$$

$$\operatorname{arctanh} z = \frac{1}{2}\ln\left(\frac{1+z}{1-z}\right).$$

29. $\dfrac{d}{dz}[\operatorname{arcsinh} z] = \dfrac{1}{\sqrt{z^2+1}}$, (for a fixed branch).

30. $\dfrac{d}{dz}[\operatorname{arccosh} z] = \dfrac{1}{\sqrt{z^2-1}}$, (for a fixed branch).

31. $\dfrac{d}{dz}[\operatorname{arctanh} z] = \dfrac{1}{1-z^2}$, (for a fixed branch).

Useful Identities and Properties of Complex Functions

32. $e^{iz} = \cos z + i \sin z$.
33. $\exp(z_1 + z_2) = \exp(z_1)\exp(z_2)$.
34. If $z = x + iy$, then

$$|e^z| = e^x, \quad \arg(e^z) = y + 2\pi n, \quad n = 0, \pm 1, \pm 2, \ldots.$$

35. If $z = x + iy$ and $w = \ln z = u + iv$, then

$$u = \ln|z|, \quad v = \operatorname{Arg}(z) + 2\pi n, \quad n = 0, \pm 1, \pm 2, \ldots.$$

36. $\ln(z_1 z_2) = \ln z_1 + \ln z_2$.
37. $\ln(z_1/z_2) = \ln z_1 - \ln z_2$.
38. $\ln(z^{p/q}) = (p/q)\ln z$, ($p, q$ integers).
39. $\sin^2 z + \cos^2 z = 1$.
40. $\sin(z_1 \pm z_2) = \sin z_1 \cos z_2 \pm \cos z_1 \sin z_2$.
41. $\sin 2z = 2\sin z \cos z$.
42. $\cos(z_1 \pm z_2) = \cos z_1 \cos z_2 \mp \sin z_1 \sin z_2$.
43. $\cos 2z = \cos^2 z - \sin^2 z = 1 - 2\sin^2 z = 2\cos^2 z - 1$.
44. $\sin(iz) = i \sinh z$, $\cos(iz) = \cosh z$.
45. $\sin(-z) = -\sin z$, $\cos(-z) = \cos z$, $\tan(-z) = -\tan z$
46. The zeros of $\sin z$ occur at $z = n\pi, n = 0, \pm 1$
47. The zeros of $\cos z$ occur at $z = (2n+1)\pi/2,$
48. The zeros of $\tan z$ occur at $z = n\pi, n = 0, \pm 1$
49. If $z = x + iy$, then $|\sin z|^2 = \sin^2 x + \sinh^2 y, |$
50. $\cosh^2 z - \sinh^2 z = 1$.
51. $\sinh(z_1 \pm z_2) = \sinh z_1 \cosh z_2 \pm \cosh z_1 \sinh$
52. $\sinh 2z = 2\sinh z \cosh z$.
53. $\cosh(z_1 \pm z_2) = \cosh z_1 \cosh z_2 \pm \sinh z_1 \sinh z$
54. $\cosh 2z = \cosh^2 z + \sinh^2 z = 1 + 2\sinh^2 z =$
55. $\sinh(iz) = i \sin z$, $\cosh(iz) = \cos z$.

56. The zeros of $\sinh z$ occur at $z = n\pi, n = 0, \pm 1, \pm 2, \ldots$.
57. The zeros of $\cosh z$ occur at $z = \frac{1}{2}(2n + 1)\pi, n = 0, \pm 1, \pm 2, \ldots$.
58. The zeros of $\tanh z$ occur at $z = n\pi i, n = 0, \pm 1, \pm 2, \ldots$.
59. If $z = x + iy$, then $|\sinh z|^2 = \sin^2 y + \sinh^2 x$, $|\cosh z|^2 = \cos^2 y + \sinh^2 x$

Exercises 1.5

Simplify the following rational functions by means of partial fractions.

1. $\dfrac{3z}{(z-1)(z+1)^2}$.

2. $\dfrac{1}{z^3 - 2z^2 + z}$.

3. $\dfrac{z^4 - 3}{z^2 + 2z + 1}$.

4. $\dfrac{z^2 + 2}{(z+1)^3(z-2)}$.

In each of the following exercises use the given value of z to find e^z, e^{-z}, and $\exp(z^2)$.

5. $z = \frac{3}{2}\pi i$.
6. $z = 2 + \frac{1}{2}\pi i$.
7. $z = 1 + 2i$.
8. $z = -\frac{1}{4}\pi i$.

In each of the following exercises use the given value of z to find $\ln z$ and $\operatorname{Ln} z$.

9. $z = i$.
10. $z = -i$.
11. $z = e^{-3}$.
12. $z = e^{5i}$.
13. $z = 4$.
14. $z = -5i$.

Find the values and the principal value of the following expressions.

$(-4)^i$.
[illegible]$)^{-3i}$.

20. 1^{2i}.
21. $3^{(1-i)}$.
22. $(i)^i$.
23. Find all of the roots of $\sin z = 5$.
24. Find all of the roots of $\cos z = 2$.
25. Find all of the roots of $\cos z = 3i$.
26. Find all of the roots of $\sinh z = i$.
27. Find all of the values of arctan $\sqrt{7}$.
28. Find all of the values of arccos $4i$.
29. Find all of the values of arccosh$(\sqrt{3}/2)$.
30. Find all of the values of arctanh i.
31. By considering $|\sinh z|^2$ with $z = x + iy$, prove that

$$\sinh|y| \leqslant |\sin z| \leqslant \cosh y.$$

32. By considering $|\sinh z|^2$ with $z = x + iy$, prove that

$$\sinh|x| \leqslant |\sinh z| \leqslant \cosh x.$$

Exercises of Higher Difficulty

33. Derive the result that if $P(z)$ is a polynomial of degree m and $Q(z)$ is a polynomial of degree n, with $m > n$, and $Q(z)$ has the n simple zeros $z_1, z_2, \ldots, z_n$ (each with multiplicity 1), that the unknown coefficients A_r, $r = 1, 2, \ldots, n$ in the partial fraction expansion

$$\frac{P(z)}{Q(z)} = \frac{A_1}{z - z_1} + \frac{A_2}{z - z_2} + \cdots + \frac{A_n}{z - z_n},$$

are given by

$$A_r = \frac{P(z_r)}{Q'(z_r)}, \quad r = 1, 2, \ldots, n.$$

Apply the result to the rational function

$$\frac{z^2 + z - 3i}{(z-2)(z-3)(z+i)}.$$

34. Given that $f(z) = z^2 = u + iv$, with $z = x + iy$, prove that the families of curves u = const. and v = const. form orthogonal trajectories (See Figure 1.13). Does the mutual orthogonality of these two families of curves hold everywhere in the (x, y)-plane, and if not where does it fail and why?

2

Complex Integration

The theory of complex integration is founded on the fundamental Cauchy integral theorem and its extension to the Cauchy integral formulas. The first form of the integral theorem was proved by the French mathematician Augustin-Louis Cauchy in 1825 under the hypothesis that the derivative of a function $f(z)$, analytic in a domain D, is *continuous* on the boundary of D. This proof of the theorem restricts its application to domains D with smooth boundaries, and so would seem to imply that the result of the theorem does not apply to domains with piecewise smooth boundaries, such as, semicircles and rectangles. Later in 1883, this restriction was lifted by Edouard Goursat who, using a different method of proof, showed the theorem to be applicable to domains with *piecewise smooth* boundaries. This extension of the validity of the theorem is extremely important because it enables it to be applied in a great variety of situations, most of which arise in applications. Because of Goursat's contribution, the theorem is now called the *Cauchy–Goursat theorem*. Both versions of the theorem will be proved, as well as the Cauchy integral formulas and many of their consequences, but although this generalization of the Cauchy formula will be used throughout this section, its proof will be postponed until the end of the section. The proof can be omitted at a first reading by those mainly interested in applications of the theorem, many of which were first made by Cauchy.

2.1 Contours and Complex Integrals

The integral of a complex function $f(z)$ with respect to the complex variable z involves integrating $f(z)$ along a curve C in the complex plane. The curve will always be either a *simple arc*, which although continuous may be only piecewise smooth, or a *simple closed curve*. The path of integration C followed by z in the course of the integration is called a *contour*, and the integral itself a *contour integral*.

As already mentioned, it is usual to specify C in parametric form by writing the representative point z on C as

$$z(s) = x(s) + iy(s), \tag{2.1}$$

where the parameter s is in an interval $\alpha \leqslant s \leqslant \beta$, and $x(s)$ and $y(s)$ are continuous functions of s while α and β may be either finite or infinite. It follows from Equation (2.1) that as s increases, so $z(s)$ moves in a specified direction along the contour C in the z-plane, thereby inducing a natural *direction* along C. By this we mean that the direction of motion of z along C is the one associated with the direction in which $z(s)$ moves as s increases. Reversing the sense of s, so that it *decreases* from β to α, will reverse the direction of z along C. In diagrams, the direction along C is indicated by the addition of arrows to C.

A typical example of a *simple arc* that often arises in complex integration is a semicircle of radius R centered on the origin and confined to the upper half z-plane, as shown in Figure 2.1(a). This can be parameterized in many ways, one of which involves setting $z(s) = x(s) + iy(s)$, with

$$x(s) = R\cos s, \quad y(s) = R\sin s, \quad 0 \leqslant s \leqslant \pi.$$

Then, as s increases from 0 to π, so $z(s)$ starts from D ($s = 0$), passes E ($s = \pi/2$), and finishes at F ($s = \pi$), with the direction along the semicircle indicated by the arrows in Figure 2.1(a).

Also of interest is the *simple closed curve* shown in Figure 2.1(b) formed by open the semicircular arc in Figure 2.1(a), closed by the addition of the diameter FD.

A typical parameterization of this closed curve involves again setting $z(s) = x(s) + iy(s)$ but this time with

$$x(s) = \begin{cases} R\cos s, & 0 \leqslant s \leqslant \pi \\ R(s - 2\pi)/\pi, & \pi < s \leqslant 3\pi \end{cases} \qquad y(s) = \begin{cases} R\sin s, & 0 \leqslant s \leqslant \pi \\ 0, & \pi < s \leqslant 3\pi. \end{cases}$$

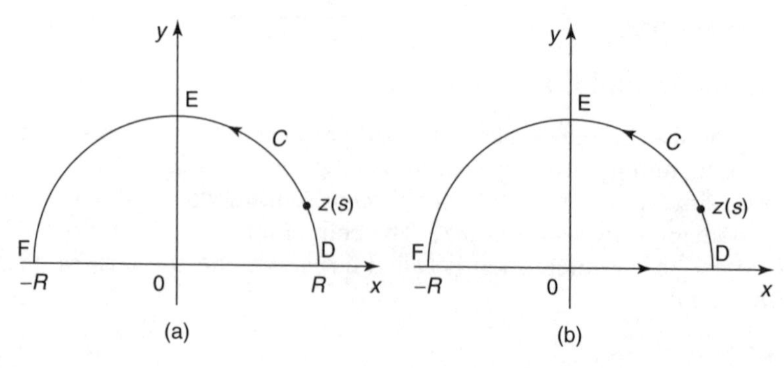

FIGURE 2.1
Typical contours with a parametric representation.

The fundamental idea underlying the complex integration of $f(z)$ along a contour C is as follows. Once $f(z)$ and C have been specified, the arc C is divided into n contiguous simple arcs $C_1, C_2, \ldots, C_n$ as shown in Figure 2.2, the end points $z_0, z_1, \ldots, z_n$ of which are chosen by selecting $s_0, s_1, \ldots, s_n$ arbitrarily, subject only to the condition that

$$\alpha = s_0 < s_1 < s_2 < \cdots < s_n = \beta.$$

Then the end points of the ith arc C_i are $z_{i-1} = z(s_{i-1})$ and $z_i = z(s_i)$, for $i = 0, 1, \ldots, n$.

Setting $\Delta z_i = z_i - z_{i-1}$, and taking z_i^* to be any point on the simple arc C_i, the sum

$$S_n = \sum_{i=1}^{n} f(z_i^*)\Delta z_i \tag{2.2}$$

is formed. The complex integral of $f(z)$ along C, called a *contour integral*, follows from Equation (2.2) in the limit as $n \to \infty$ by letting n become arbitrarily large while simultaneously $\max|\Delta z_i| \to 0$. This contour integral is denoted by

$$\int_C f(z)dz, \tag{2.3}$$

where z moves along the contour C in a specific direction. When C is a simple closed curve, the *positive* direction, or *orientation* around C will always be taken to be such that the *interior* of C lies to the *left* as z moves around the contour, as illustrated in Figure 2.1(b).

To understand the meaning of Equation (2.3), return to the approximating sum S_n in Equation (2.2) whose limit provided the definition of the contour

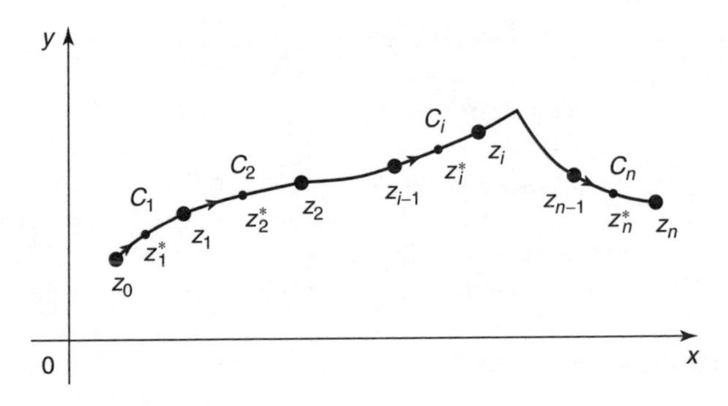

FIGURE 2.2
Contiguous segments along C.

integral. Setting $f(z) = u(x, y) + iv(x, y)$, $z_i^* = x_i^* + iy_i^*$, and $\Delta z_i = \Delta x_i + i\Delta y_i$, the approximating sum S_n in Equation (2.2) becomes

$$S_n = \sum_{i=1}^{n} [u(x_i^*, y_i^*)\Delta x_i - v(x_i^*, y_i^*)\Delta y_i] + \sum_{i=1}^{n} [u(x_i^*, y_i^*)\Delta y_i + v(x_i^*, y_i^*)\Delta x_i]. \tag{2.4}$$

Then, if $S = A + iB$ and $\delta_i = \max\{|\Delta x_i|, |\Delta y_i|\}$,

$$A = \mathrm{Re}\{S\} = \lim_{\substack{n\to\infty,\\ \delta_i\to 0}} \sum_{i=1}^{n} [u(x_i^*, y_i^*)\Delta x_i - v(x_i^*, y_i^*)\Delta y_i] \tag{2.5}$$

and

$$B = \mathrm{Im}\{S\} = \lim_{\substack{n\to\infty,\\ \delta_i\to 0}} \sum_{i=1}^{n} [u(x_i^*, y_i^*)\Delta y_i + v(x_i^*, y_i^*)\Delta x_i]. \tag{2.6}$$

The expressions on the right of Equations (2.5) and (2.6), whose existence ensures the existence of the contour integral in Equation (2.3), are now seen as ordinary line integrals involving the real functions $u(x, y)$ and $v(x, y)$ of the real variables x and y. In the usual notation for line integrals we write Equations (2.5) and (2.6) as

$$A = \int_C u\,dx - v\,dy, \quad B = \int_C u\,dy + v\,dx. \tag{2.7}$$

Thus the contour integral in Equation (2.3) is defined in terms of these two real integrals as

$$\int_C f(z)dz = \int_C u\,dx - v\,dy + i\left(\int_C u\,dy + v\,dx\right). \tag{2.8}$$

As $z(s)$ is constrained to lie on C, which is parameterized by

$$z(s) = x(s) + iy(s), \quad \alpha \leqslant s \leqslant \beta,$$

it follows directly that $u(x, y)$ and $v(x, y)$ in Equation (2.8) are the functions of s given by $u[x(s), y(s)]$ and $v[x(s), y(s)]$, respectively. Similarly on C, $dx = x'(s)ds$ and $dy = y'(s)ds$, where the prime indicates differentiation with respect to s, while the integral over C becomes a definite integral with respect to s from $s = \alpha$ to $s = \beta$.

Combining these results gives

$$\int_C u\,dx - v\,dy = \int_\alpha^\beta \{u[x(s), y(s)]x'(s) - v[x(s), y(s)]y'(s)\}ds, \tag{2.9}$$

and

$$\int_C u\,dy + v\,dx = \int_\alpha^\beta \{u[x(s), y(s)]y'(s) + v[x(s), y(s)]x'(s)\}ds. \tag{2.10}$$

Results Equations (2.8), (2.9), and (2.10) combine to allow us to write Equation (2.8) in the concise form

$$\int_C f(z)dz = \int_\alpha^\beta f[z(s)]z'(s)ds. \tag{2.11}$$

The real integrals Equations (2.9) and (2.10) always exist when C is piecewise smooth, which is the case that will be of interest here.

An immediate consequence of these integrals is that interchanging the limits reverses the direction of integration, and so reverses the sign of the contour integral. A convenient notation to use in this context is to denote by C_- the contour obtained from C by reversing the direction of integration around C. Thus we have the result:

2.1.1 Reversal of the Direction of Integration

$$\int_C f(z)dz = -\int_{C_-} f(z)dz. \tag{2.12}$$

Other important properties of contour integrals that follow directly from Equations (2.8), (2.9), and (2.10) when combined with the familiar properties of definite integrals of real functions, are:

2.1.2 Scaling Property

$$\int_C kf(z)dz = k\int_C f(z)dz, \quad (k = \text{constant}), \tag{2.13}$$

2.1.3 Linearity Property

$$\int_C [\alpha f(z) + \beta g(z)]dz = \alpha\int_C f(z)dz + \beta\int_C g(z)dz, \quad (\alpha, \beta \text{ constants}), \tag{2.14}$$

2.1.4 Partitioning of a Contour

If C comprises n smooth arcs $C_1, C_2, \ldots, C_n$ joined end to end, not necessarily forming a closed curve, then

$$\int_C f(z)dz = \int_{C_1} f(z)dz + \int_{C_2} f(z)dz + \cdots + \int_{C_n} f(z)dz, \tag{2.15}$$

2.1.5 An Integral Inequality

If, on a contour C of length L, it is known that $|f(z)| \leqslant \mu$, then

$$\left|\int_C f(z)dz\right| \leqslant \int_C |f(z)|\,d\tau \leqslant \mu L, \tag{2.16}$$

where $d\tau$ is the element of arc length along C.

Only Equation (2.16) requires proof because the other results are self-evident. Starting from Equation (2.4) and using inequalities familiar from real analysis we can write

$$\left|\int_C f(z)dz\right| = \left|\lim_{\substack{n\to\infty,\\ \delta_i\to 0}} \sum_{i=1}^{n} |f(z_i^*)\Delta z_i|\right| \leqslant \lim_{\substack{n\to\infty,\\ \delta_i\to 0}} \sum_{i=1}^{n} |f(z_i^*)||\Delta z_i|. \tag{2.17}$$

Proceeding to the limit, and recognizing that $|\Delta z_i|$ approaches the element of arc length $d\tau$ along C as $n \to \infty$, we arrive at the result

$$\left|\int_C f(z)dz\right| \leqslant \int_C |f(z)|\,d\tau,$$

which is the first part of Equation (2.16). Using the sharp upper-bound $|f(z)| \leqslant \mu$ when z is on C, this becomes

$$\left|\int_C f(z)dz\right| \leqslant \int_C |f(z)|dz \leqslant \int_C \mu\, d\tau = \mu \int_C d\tau = \mu L,$$

and the result is proved.

Example 2.1.1 An Integral along a Simple Arc
Evaluate

$$\int_C z^2\, dz,$$

given that C is the directed contour shown in Figure 2.3 formed by the straight line paths C_1 and C_2 joining the points P at $z = i$, Q at $z = 1 + i$ and R at $z = 3 + 3i$. Verify that integral inequality in Equation (2.16) is satisfied.

SOLUTION
We parameterize C_1 and C_2 as follows:

$$C_1\colon\ x(s) = s, \quad y(s) = 1, \quad 0 \leqslant s \leqslant 1,$$

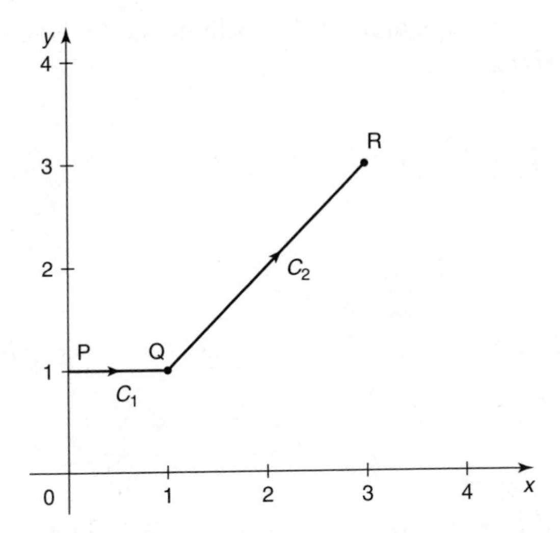

FIGURE 2.3
A simple piecewise smooth arc.

and

$$C_2\colon\ x(s) = s, \quad y(s) = s, \quad 1 \leqslant s \leqslant 3.$$

From Equation (2.15) the contour integral can be written as the sum

$$\int_C z^2\,dz = \int_{C_1} z^2\,dz + \int_{C_2} z^2\,dz,$$

and in this form the parameterizations of C_1 and C_2 can be used.

Setting $z = x + iy$, it follows that $f(z) = z^2 = x^2 - y^2 + 2ixy$, when if $f(z) = u + iv$, we find that

$$u(x, y) = x^2 - y^2 \quad \text{and} \quad v(x, y) = 2xy.$$

On C_1, for $0 \leqslant s \leqslant 1$,

$$u[x(s), y(s)] = s^2 - 1 \quad \text{and} \quad v[x(s), y(s)] = 2s,\ x'(s) = 1 \quad \text{and} \quad y'(s) = 0.$$

On C_2, for $1 \leqslant s \leqslant 3$,

$$u[x(s), y(s)] = 0 \quad \text{and} \quad v[x(s), y(s)] = 2s^2,\ x'(s) = 1 \quad \text{and} \quad y'(s) = 1.$$

Using these result in Equations (2.9) and (2.10), and then combining them with Equation (2.8), gives

$$\int_{C_1} z^2\,dz = \int_0^1 (s^2-1)ds + i\int_0^1 2s\,ds = -\tfrac{2}{3} + i,$$

and

$$\int_{C_2} z^2\,dz = \int_1^3 (-2s^2)ds + i\int_1^3 2s^2\,ds = -\tfrac{52}{3} + \tfrac{52}{3}i,$$

so

$$\int_C z^2\,dz = \int_{C_1} z^2\,dz + \int_{C_2} z^2\,dz = -\tfrac{2}{3} + i - \tfrac{52}{3} + \tfrac{52}{3}i = -18 + \tfrac{55}{3}i.$$

To verify inequality in Equation (2.16), it is necessary to find sharp upper bounds for $|z^2|$ on C_1 and C_2.

On C_1,

$$|z^2| = [(s^2-1)^2 + 4s^2]^{1/2} = s^2 + 1, \quad \text{for } 0 \leqslant s \leqslant 1, \text{ so } \max|z^2|_{C_1} = 2.$$

On C_2,

$$|z^2| = (4s^4)^{1/2} = 2s^2, \quad \text{for } 1 \leqslant s \leqslant 3, \text{ so } \max|z^2|_{C_2} = 18.$$

So the required upper bound on C is $\mu = \max_C|z^2| = \max_{C_1+C_2}|z^2| = 18$. The length L of lines C_1 and C_2 is $1 + 2\sqrt{2}$, so $\mu L = 18(1 + 2\sqrt{2}) = 68.912$. However

$$\int_C z^2\,dz = -18 + \tfrac{55}{3}i, \quad \text{so} \quad \left|\int_C z^2\,dz\right| = [18^2 + (55/3)^2]^{1/2} = 25.693,$$

and $25.693 < \mu L = 68.912$, confirming that the inequality in Equation (2.16) is satisfied. ◇

Estimates of this type will be required later, together with much sharper ones of a different kind—when contour integrals are evaluated by means of the residue theorem.

Example 2.1.2 An Integral around a Closed Rectangular Contour
Show that

$$\int_C \sin z\, dz = 0,$$

where C the directed path around the rectangle in Figure 2.4 with its corners at

$$\text{A}\ (z = a + ic), \text{B}\ (z = b + ic), \text{C}\ (z = b + id), \text{ and D}\ (z = a + id).$$

The contours C_1 to C_4 in Figure 2.4 can be parameterized as follows:

$$\begin{aligned}
&C_1:\ x(s) = s, \quad y(s) = c, \quad a \leqslant s \leqslant b, \quad \text{with} \quad dz = dx = ds\ (dy = 0),\\
&C_2:\ x(s) = b, \quad y(s) = s, \quad c \leqslant s \leqslant d, \quad \text{with} \quad dz = idy = ids\ (dx = 0),\\
&C_3:\ x(s) = s, \quad y(s) = d, \quad a \leqslant s \leqslant b, \quad \text{with} \quad dz = dx = ds\ (dy = 0),\\
&C_3:\ x(s) = a, \quad y(s) = s, \quad c \leqslant s \leqslant d, \quad \text{with} \quad dz = idy = ids\ (dx = 0).
\end{aligned}$$

Setting $z = x + iy$ and $f(z) = \sin z$ and writing $f(z) = u + iv$, we have

$$u = \sin x \cosh y \quad \text{and} \quad v = \cos x \sinh y.$$

Then from Equations (2.8), (2.9), and (2.10) we have

$$\begin{aligned}
\int_{C_1} \sin z\, dz &= \int_a^b (\sin s \cosh c + i \cos s \sinh c) ds\\
&= (\cos a \cosh c - \cos b \cosh c) + i(\sin b \sinh c - \sin a \sinh c).
\end{aligned}$$

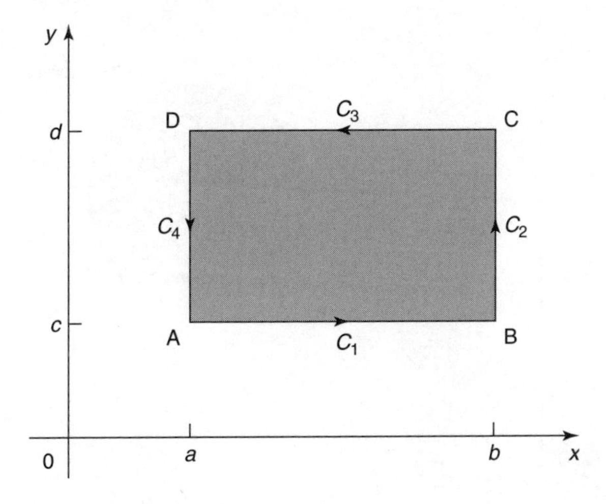

FIGURE 2.4
A closed-rectangular contour.

$$\int_{C_2} \sin z\,dz = \int_c^d (-\cos b \sinh s + i \sin b \cosh s)ds$$
$$= (\cos b \cosh c - \cos b \cosh d) + i(\sin b \sinh d - \sin b \sinh c),$$

$$\int_{C_3} \sin z\,dz = \int_{CD} \sin z\,dz = -\int_{DC} \sin z\,dz = -\int_a^b (\sin s \cosh d + i \cos s \sinh d)ds$$
$$= (\cos b \cosh d - \cos a \cosh d) + i(\sin a \sinh d - \sin b \sinh d),$$

$$\int_{C_4} \sin z\,dz = \int_{DA} \sin z\,dz = -\int_{AD} \sin z\,dz = -\int_c^b (-\cos a \sinh s + i \sin a \cosh s)ds$$
$$= (\cos a \cosh d - \cos a \cosh c) + i(\sin a \sinh c - \sin a \sinh d).$$

Adding these four integrals gives

$$\int_C \sin z\,dz = 0,$$

as was to be shown. ◇

In the next example the same type of approach is used to evaluate integrals around a circular contour. If a contour is in the form of a circle of radius R centered on the origin, then in the polar representation a point on the circle can always be written $z = Re^{i\theta}$, in which case when z is on the circle $dz = iRe^{i\theta}\,d\theta$. The polar representation also shows that the circle can be represented as $|z| = R$ which is essentially the Cartesian representation, because if $z = x + iy$, then $|z|^2 = R^2$ becomes the familiar equation of a circle $x^2 + y^2 = R^2$. It may also happen that a contour is in the form of a circle of radius R centered on a general point $z = z_0$. To find its polar representation, it is necessary to consider Figure 2.5 and to use the vector properties of complex numbers to write

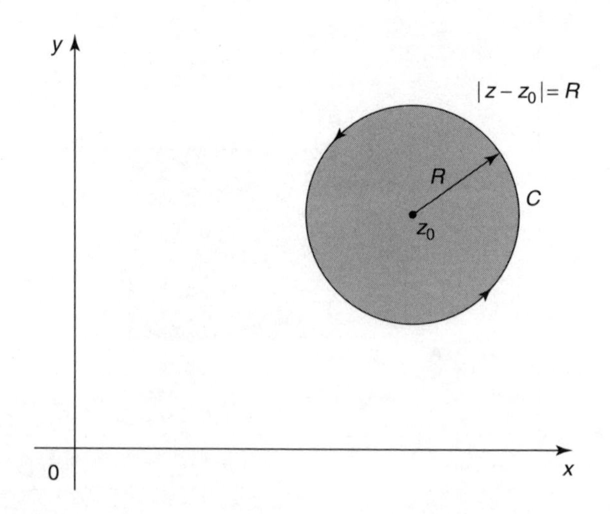

FIGURE 2.5
The polar representation of a circle of radius R centered on z_0.

a general point z on the circle as $z = z_0 + Re^{i\theta}$ or, equivalently, as $z - z_0 = Re^{i\theta}$. When z is on the circle it again follows that $dz = iRe^{i\theta}\,d\theta$. Taking the modulus of $z - z_0 = Re^{i\theta}$ shows that this circle can be described by $|z - z_0| = R$, and when $z = x + iy$ and $z_0 = x_0 + iy_0$, after squaring this last result it is seen to be equivalent to the familiar Cartesian equation of a circle $(x - x_0)^2 + (y - y_0)^2 = R^2$.

Example 2.1.3 Two Integrals around a Circular Contour

Show that

(i) $\displaystyle\int_C \frac{dz}{z - z_0} = 2\pi i$ and

(ii) $\displaystyle\int_C (z - z_0)^n\,dz = iR\int_0^{2\pi} e^{i(n+1)\theta}d\theta = 0, \ n = 0, \pm1, \pm2, \ldots .$

where C is the circular contour of radius R shown in Figure 2.5 in which the arrows show the positive direction of integration.

SOLUTION

(i) Writing the circle in the parametric form $z = z_0 + Re^{i\theta}$ with $0 \leqslant \theta \leqslant 2\pi$, and using the fact that on C we have $dz = iRe^{i\theta}\,d\theta$, the integral becomes

$$\int_C \frac{dz}{z - z_0} = \int_0^{2\pi} \frac{Rie^{i\theta}}{Re^{i\theta}}\,d\theta = 2\pi i.$$

(ii) This same parameterization yields

$$1\ n = 0, \pm1, \pm2, \ldots .$$

Notice that in (i) the integrand is analytic throughout the interior of C with the exception of the point $z = z_0$ where the singularity in the integrand at $z = z_0$ is called a *pole*. Notice also that in (ii), even when $n = -2, -3, \ldots$ the integral around C is still zero. Simple though they are, these two integrals will prove to be of fundamental importance in what is to follow. ◇

Domains around which a complex function $f(z)$ is to be integrated are not always simply connected, and typically it may be necessary to integrate around the boundary of a domain D with one or more internal boundaries, as shown in Figure 2.6(a). An integral of $f(z)$ taken around the entire boundary of a nonsimply connected domain, corresponding to an integral taken around both the *external* boundary and all *internal boundaries*, is performed by making a cut from the external boundary to each of the internal boundaries, and then integrating around each boundary and along each side of a cut in the manner indicated by the arrows in Figure 2.6(b). It is to be understood that no two cuts intersect, and that each cut is chosen in such a way that $f(z)$ is continuous across the cut.

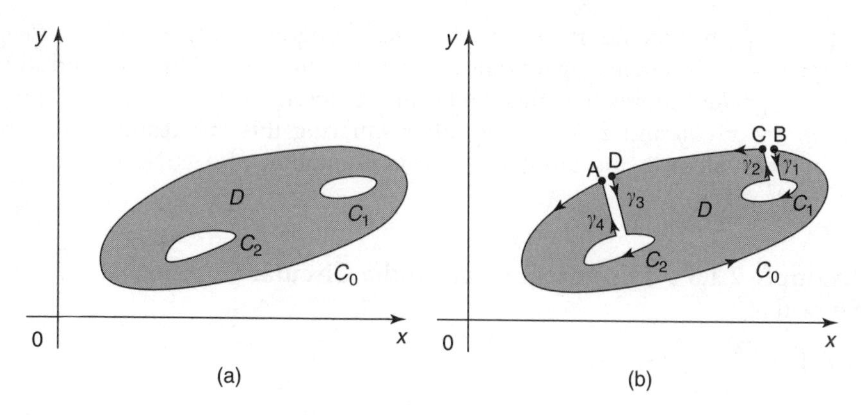

FIGURE 2.6
Integral around a nonsimply connected domain.

Thus if C denotes the entire boundary of the domain D shown in Figure 2.6(b), where C_0^{AB} is the part of C_0 from A to B, and C_0^{CD} is the part of C_0 from C to D, with both directed paths being in the counterclockwise (positive) direction, it follows that

$$\int_C f(z)dz = \int_{C_0^{AB}} f(z)dz + \int_{\gamma_1} f(z)dz + \int_{C_1} f(z)dz + \int_{\gamma_2} f(z)dz \\ + \int_{C_0^{CD}} f(z)dz + \int_{\lambda_3} f(z)dz + \int_{C_2} f(z)dz + \int_{\gamma_4} f(z)dz. \quad (2.18)$$

The location of the cuts is unimportant provided $f(z)$ is continuous across the cuts because the continuity of $f(z)$ across each cut ensures that the integral along one side of the cut is canceled by the integral along the other side of the cut, which is taken in the opposite direction. Thus, for example, in Figure 2.6(b),

$$\int_{\gamma_1} f(z)dz = -\int_{\gamma_2} f(z)dz \quad \text{and} \quad \int_{\gamma_3} f(z)dz = -\int_{\gamma_4} f(z)dz,$$

so that

$$\int_C f(z)dz = \int_{C_0} f(z)dz + \int_{C_1} f(z)dz + \int_{C_3} f(z)dz, \quad (2.19)$$

with the direction of integration taken *counterclockwise* around the *external boundary* and *clockwise* around each *internal boundary*. That is, the direction of integration always proceeds in such a way that the domain D lies to the left as the contours are traversed.

This result extends in an obvious manner to an integral around a multiply connected domain D bounded externally by a simple closed curve C_0, and internally by n simple closed nonintersecting closed curves $C_1, C_2, \dots, C_n$, to give

$$\int_C f(z)dz = \int_{C_0} f(z)dz + \int_{C_1} f(z)dz + \cdots + \int_{C_n} f(z)dz \qquad (2.20)$$

where, as before, the direction of integration is taken counterclockwise around the external boundary C_0, and clockwise around each internal boundary $C_1, C_2, \dots, C_n$.

Example 2.1.4 An Integral around a Doubly Connected Domain
Evaluate

$$\int_C \frac{dz}{z - z_0},$$

where C is the boundary of the annular domain $R_1 < |z - z_0| < R_2$.

SOLUTION
The domain D is doubly connected, and $1/(z - z_0)$ is analytic inside and on the annulus. We make a cut from the outer boundary C_1 to the inner boundary C_2 as shown in Figure 2.7 and denote the lower edge of the cut by γ_1 and the upper edge by γ_2, with the direction of integration shown by the arrows in Figuer 2.7. Then as $1/(z - z_0)$ is continuous across

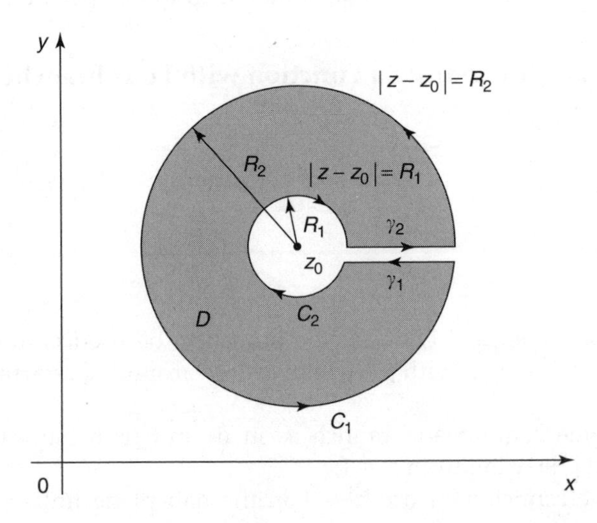

FIGURE 2.7
Integral around a doubly connected domain.

the cut, the integrals along γ_1 and γ_2 being equal but opposite in sign will cancel, leaving

$$\int_C \frac{dz}{z - z_0} = \int_{C_1} \frac{dz}{z - z_0} + \int_{C_2} \frac{dz}{z - z_0}.$$

The integral of $1/(z - z_0)$ around the circular contour C_1 centred on z_0 was shown in Example 2.1.3 to be $2\pi i$, and to be independent of the radius of circle C_1. The same situation applies to the integral of $1/(z - z_0)$ around the circular contour C_2, though in this case the direction of integration is reversed so the result will be $-2\pi i$, so we find that

$$\int_C \frac{dz}{z - z_0} = 2\pi i - 2\pi i = 0. \qquad \diamond$$

So far the functions integrated along arcs and around contours have been functions that were analytic on the arc, analytic everywhere inside a domain D bounded by a closed contour C, or as in Example 2.1.3(i), a function analytic everywhere inside C apart from a pole. However, it is sometimes necessary to integrate a function $f(z)$ with more than one branch around a closed contour C. In this case, the value of the integral will depend on the branch that is chosen and, possibly, also on the point P on contour C from which the integration commences. The last two examples illustrate the approach that is necessary in such cases, the first of which involves a function with four branches, while the second involves a function with infinitely many branches. When the contour is closed, it is usual that the integration will be assumed to start from the point where the value of the integrand is specified.

Example 2.1.5 Integrating a Function with Four Branches
Evaluate

$$\int_C \frac{dz}{z^{1/4}},$$

when

(i) C is the circle $|z| = 1$, where the branch to be used is the one for which $1^{1/4} = -i$, with the integration around C starting from $z = -i$;
(ii) The same contour and branch as in (i), but with the integration around C starting from $z = 1$;
(iii) C is the semicircular arc $|z| = 1$ in the half plane $\text{Im}\{z\} \leqslant 0$, with integration starting from $z = -1$, where the branch to be used is the one for which $1^{1/4} = -1$.

SOLUTION
To make $z^{1/4}$ single valued a branch cut must be made along the negative real axis up to and including the origin. Then from Equation (1.37) the four branches of $w(z) = z^{1/4}$ are found to be

$$w_k(z) = r^{1/4}\exp(k\pi i/2), \qquad \text{with } r = |z| \text{ and } k = 0, 1, 2, 3.$$

Thus the value of $w(1)$ corresponding to each of the four branches is

$$w_0(1) = 1,\ w_1(1) = i,\ w_2(1) = -1, \text{ and } w_3(1) = -i.$$

(i) The branch to be selected is the one for which $1^{1/4} = -i$, so from this discussion the branch to be used is seen to be $w_3(z)$. Thus, in the integral we need to set

$$\begin{aligned} z^{1/4} &= w_3(z) = r^{1/4}\exp\left[i\left(\frac{\theta + 6\pi}{4}\right)\right] \\ &= r^{1/4}\exp(3\pi i/2)\cdot\exp(i\theta/4) \\ &= -ir^{1/4}\exp(i\theta/4). \end{aligned}$$

The integration is around $|z| = 1$ on which $r = 1$, so in the integral we must set

$$z^{1/4} = -i\exp(i\theta/4).$$

The integration is to start from $z = -i$, corresponding to $\theta = -\pi/2$, so after increasing by 2π, it will terminate when $\theta = 3\pi/2$. Thus as $dz = ie^{i\theta}\,d\theta$ on C, we have

$$\begin{aligned} \int_C \frac{dz}{z^{1/4}} &= \int_{-\pi/2}^{3\pi/2} \frac{ie^{i\theta}\,d\theta}{-ie^{i\theta/4}} \\ &= -\int_{-\pi/2}^{3\pi/2} e^{3\theta i/4}\,d\theta = \tfrac{4i}{3}(e^{-7\pi i/8} - e^{-3\pi i/8}). \end{aligned}$$

(ii) The only difference between this integral and that of (i) is that here the integration starts from $z = 1$ corresponding to $\theta = 0$, and so terminates at $\theta = 2\pi$. Thus this time

$$\int_C \frac{dz}{z^{1/4}} = \int_0^{2\pi} \frac{ie^{i\theta}\,d\theta}{-ie^{i\theta/4}} = -\int_0^{2\pi} e^{3\theta i/4}\,d$$

(iii) The branch to be used here is the one for wh
sponding to $w_2(z)$. Here the integration is aroun
in the lower half-plane, starting from $z = -1$ corr
and terminating at $z = 1$ corresponding to $\theta =$

semicircle, $z = e^{i\theta}$ and $dz = ie^{i\theta}$, and the branch to be used is $w_2(z)$, so on C we must set

$$z^{1/4} = \exp\left[i\left(\frac{\theta + 4\pi}{4}\right)\right] = -\exp(i\theta/4).$$

Thus

$$\int_C \frac{dz}{z^{1/4}} = \int_{\pi}^{2\pi} \frac{ie^{i\theta}}{-e^{i\theta/4}}\,d\theta = \tfrac{4}{3}\left[-\tfrac{1}{\sqrt{2}} + \left(1 + \tfrac{1}{\sqrt{2}}\right)i\right].$$

◇

Example 2.1.6 Integrating a Function with Infinitely Many Branches
Evaluate

$$\int_C \ln z\,dz,$$

where

(i) C is the unit circle $|z| = 1$, the branch to be used is the one for which $\ln 1 = 4\pi i$, and the integration is to start from $z = 1$.
(ii) C is the unit circle $|z| = 1$, the branch to be used is the one for which $\ln 1 = 0$, and the integration is to start from $z = i$.

SOLUTION
To make ln z single valued, a branch cut must be made along the negative real axis up to and including the origin. It follows from Equation (1.88) that if $z = re^{i\theta}$, the kth branch $w_k(z)$ is given by

$$w_k(z) = \ln r + i(\theta + 2k\pi), \quad k = 0, \pm 1, \pm 2, \dots .$$

Setting $z = 1$ and $\theta = \text{Arg}\,z = 0$ this shows that

$$w_0(1) = 0,\ w_1(1) = 2\pi i,\ w_2(1) = 4\pi i, \dots .$$

On the unit circle $|z| = 1$, $z = e^{i\theta}$ and $dz = ie^{i\theta}$, when

$$w_0(z) = i\theta,\ w_1(z) = i(\theta + 2\pi),\ w_2(z) = i(\theta + 4\pi), \dots .$$

(i) The branch required here is the one for which $\ln 1 = 4\pi i$, and so it is seen to be $w_2(z)$. Integration around the unit circle starts from $z = 1$, corresponding to $\theta = 0$, and terminates when $\theta = 2\pi$, while on the unit circle $dz = ie^{i\theta}\,d\theta$, so as $\ln z = i(\theta + 4\pi)$, the integral becomes

$$\int_C \ln z\,dz = \int_0^{2\pi} i(\theta + 4\pi)ie^{i\theta}\,d\theta = -\int_0^{2\pi} (\theta + 4\pi)e^{i\theta}\,d\theta$$
$$= -\int_0^{2\pi} \theta e^{i\theta}\,d\theta - 4\pi\int_0^{2\pi} e^{i\theta}\,d\theta = 2\pi i - 0 = 2\pi i,$$

showing that

$$\int_C \ln z\, dz = 2\pi i.$$

(ii) The branch used here is the one for which $\ln 1 = 0$, which is seen to correspond to $w_0(z)$, so for this branch $\ln z = i\theta$. This is the principal branch that is usually denoted by Ln z. As integration around the unit circle starts from $z = i$, corresponding to $\theta = \pi/2$, it will terminate when $\theta = 5\pi/2$, so using $dz = ie^{i\theta}\, d\theta$,

$$\int_C \ln z\, dz = \int_{\pi/2}^{5\pi/2} i\theta\, ie^{i\theta}\, d\theta = -\int_{\pi/2}^{5\pi/2} \theta e^{i\theta}\, d\theta = -2\pi.$$

So in this case where the principal branch has been used and integration starts from $\theta = \pi/2$,

$$\int_C \ln z\, dz = -2\pi.$$ ◇

Exercises 2.1

In Exercises 1 through 4, the integral of the given function $f(z)$ is to be evaluated along the open path obtained by using straight lines to join the given points in alphabetical order.

1. $f(z) = 3z + 1$: A ($z = 1 + i$), B ($z = 3 + 2i$), C ($z = 5 + i$).
2. $f(z) = z^2 + 1$: A ($z = 3i$), B ($z = 2 + 3i$), C ($z = 2 + i$).
3. $f(z) = \cos z$: A ($z = 0$), B ($z = \pi + i\pi$), C ($z = 2\pi + i\pi$).
4. $f(z) = \sinh z$: A ($z = 2\pi i$), B ($z = \pi + \pi i$), C ($z = \frac{3\pi}{2}(1 + i)$), D ($z = 2\pi + i\pi$).

In Exercises 5 through 8, evaluate the integral of the given function $f(z)$ around the indicated closed path in the counterclockwise (positive) direction.

5. $f(z) = z^2$ around the square with corners at A ($z = 0$), B ($z = 1$), C ($z = 1 + i$), D ($z = i$).
6. $f(z) = 1/(z - 2i)$ around the circle C given by $|z - 2| = 2$.
7. $f(z) = (z - 3)/(z - 1)$ around the circle C given by $|z - 1| = 1$.
8. $f(z) = (2z + 1)/(z + 2)$ around the circle C given by $|z + 2| = 1$.
9. Verify the integral inequality in Equation (2.16) for the integral $\int_C e^z\, dz$ where C is the straight line joining $z = 0$ and $z = 1 + i$.
10. Show that $\left|\int_C \frac{\cosh z}{z}\, dz\right| \leqslant 2\pi \cosh R$, where C is the circle $|z| = R$.

11. Evaluate $\int_C z^{1/2}\,dz$, where C is the unit circle $|z| = 1$, the branch to be used is the one for which $(-1)^{1/2} = i$, and the integration is to start from $z = i$.
12. Evaluate $\int_C \ln z\,dz$, where C is the circle $|z| = R$, the branch to be used is the one for which when $z = R$, $\ln z = \ln R + 2\pi i$, and integration starts from $z = R$.
13. Evaluate $\int_C z^\alpha\,dz$, where $\alpha \neq 0$ is an arbitrary complex number, C is the unit circle $|z| = 1$, the branch to be used is the one for which $1^\alpha = 1$, and integration is to start from $z = 1$.
14. Evaluate the integral $\int_C z^{-1/2}\,dz$, where C is the semicircular arc:
    (i) $|z| = 1$ with $y \geqslant 0$, using the branch for which $\sqrt{1} = 1$;
    (ii) $|z| = 1$ with $y \geqslant 0$, using the branch for which $\sqrt{1} = -1$;
    (iii) $|z| = 1$ with $y \leqslant 0$, using the branch for which $\sqrt{1} = 1$.

Exercises of Greater Difficulty

15. Show that $\displaystyle\int_C \frac{|\sinh z||dz|}{|z-1||z+a|} < \frac{2\pi \cosh 1}{|1-|a|^2|}$, where C is the unit circle $|z| = 1$ and $|a| \neq 1$.
16. Show that $\displaystyle\int_C \frac{|e^z||dz|}{|z^2-a^2||z^2+a^2|} < \frac{2\pi e}{|1-|a|^4|}$, where C is the unit circle $|z| = 1$ and $|a| \neq 1$.

Complex integrals can be evaluated using a numerical integration routine. When the function $f(z)$ is to be integrated is along a path C described by $z(s) = x(s) + iy(s)$, with $\alpha \leqslant s \leqslant \beta$, it follows from Equation (2.11) that

$$\int_C f(z)dz = \int_\alpha^\beta f[z(s)]\left(\frac{dz(s)}{ds}\right)ds = \int_\alpha^\beta A(s)ds + i\int_\alpha^\beta B(s)ds,$$

where

$$A(s) = \operatorname{Re}\left\{f[z(s)]\left(\frac{dz(s)}{ds}\right)\right\} \quad \text{and} \quad B(s) = \operatorname{Im}\left\{f[z(s)]\left(\frac{dz(s)}{ds}\right)\right\}.$$

So $\int_C f(z)dz$ is determined by the two real integrals $\int_\alpha^\beta A(s)ds$ and $\int_\alpha^\beta B(s)ds$, each of which can be found numerically.

In Exercises 17 through 20, use this method with Simpson's rule with six equal width intervals to find the value of the integral of the given function $f(z)$ along the specified arc C described by $z(s) = x(s) + iy(s)$.

17. $f(z) = z\cos z$, with C the arc $x(s) = s$, $y(s) = \sin s$, for $0 \leqslant s \leqslant \pi/2$.
18. $f(z) = z\exp z$, with C the arc $x(s) = s$, $y(s) = s^2$, for $0 \leqslant s \leqslant 1$.
19. $f(z) = z\sinh z$, with C the arc $x(s) = s$, $y(s) = 1 + s^2$, for $0 \leqslant s \leqslant 1$.
20. $f(z) = \sin z^2$, with C the arc $x(s) = s$, $y(s) = s$, for $0 \leqslant s \leqslant 1$.

2.2 The Cauchy Integral Theorem

Starting from first principles, the previous section developed the notion of integrating a complex function along an arc, and around a simple closed contour. Of the functions considered, some were analytic in the entire z-plane, others were analytic everywhere except at a few isolated points, while others needed branch cuts to make them analytic in the z-plane, with the exception of points on the branch cut.

To develop complex integration in a systematic way, and to make it as useful tool, it is necessary to find methods that do not need to proceed from first principles. The fundamental result needed to accomplish this is the Cauchy Integral theorem, already mentioned in the Introduction to Section 2.1, along with its extension that forms the Cauchy–Goursat theorem.

Most of this section will be concerned with the proof of the Cauchy theorem under the assumption that the function $f(z)$ to be integrated around a simple closed contour C is analytic both inside and on the contour C. This result, that is easily proved, will then be extended to integrals around multiply-connected domains, and only at the end of the section will the more general Cauchy–Goursat theorem be proved. However, apart from understanding that the Cauchy–Goursat theorem justifies applying Cauchy's theorem to integrals around nonsmooth contours, the details of its proof can be postponed.

As the theory of integration of complex functions is confined to functions that are analytic, or only cease to be analytic at distinct points, it is necessary to offer some comments about such functions.

A function $f(z)$ is said to have a *singular point* at z_0 if $f(z)$ is not analytic at z_0, but it is analytic in at least part of every neighborhood of z_0. Of special importance is the case when the singularities of $f(z)$ are separated one from the other. In this case $f(z)$ will be said to have an *isolated singularity* at z_0 if it is analytic in a neighborhood of z_0 comprising a punctured disc centered on z_0.

Example 2.2.1 Functions with Isolated Singularities

(i) The function

$$f(z) = \frac{z(z+3)}{(z^2+4)(z^2-1)}$$

has a finite number of singularities (poles) located at the zeros $z = \pm 2i$ and $z = \pm 1$ of the denominator.

(ii) The function $f(z) = 1/\cos z$ has an infinite number of isolated singularities at the zeros of the denominator $z = (n + \frac{1}{2})\pi$, for $n = 0, \pm 1, \pm 2, \ldots$. ◇

Example 2.2.2 Functions with Nonisolated Singularities

(i) The principal branch of Ln z has a singular point at $z = 0$, but this is not isolated because every neighborhood of the origin contains points on the negative real axis where Ln z is not analytic.

(ii) The function $1/\sin(1/z)$ has an infinite number of singular points corresponding to the zeros of the denominator at $z = 0$ and $z = 1/(n\pi)$ for $n = \pm 1, \pm 2, \ldots$. However, the singular point at the origin is not isolated because every neighborhood of $z = 0$, however small, will contain an infinite number of singular points at $z = 1/(n\pi)$ for $n > N$, where N is some suitably large integer. ◇

When $f(z)$ is analytic inside and on the contour C, the proof of the Cauchy integral theorem is remarkably simple. It makes use of Green's theorem, which relates a particular form of double integral over a planar area with a line integral over an oriented line integral around the boundary C of the area, that is to a line integral in a particular direction around C. The proof of the theorem will make use of the concepts of x-simple and y-simple regions. A region Ω_x will be said to be an *x-simple region* (a domain complete with its boundary) if any line in Ω_x drawn parallel to the y-axis intersects the boundary of Ω_x only twice. Similarly, Ω_y will be said to be an *y-simple region* (a domain complete with its boundary) if any line in Ω_y drawn parallel to the x-axis intersects the boundary of Ω_y only twice. A *simple region* is a one that is both x and y-simple. Examples of these three types of region are shown in Figure 2.8.

THEOREM 2.2.1 Green's Theorem for a Simple Region

Let Ω be a simple region in the (x, y)-plane with boundary curve Γ oriented in such a way that Ω lies to the left as Γ is traversed by z in the positive sense (counterclockwise). Then if $P(x, y)$ and $Q(x, y)$ and their partial derivatives are continuous in Ω and on Γ,

$$\iint_{\Omega}\left(\frac{\partial Q}{\partial x} - \frac{\partial P}{\partial y}\right)dx\,dy = \int_{\Gamma} P\,dx + Q\,dy.$$

PROOF

Consider the x-simple region Ω_x in Figure 2.8(a) that is bounded above and below by the smooth arcs

$$y = \psi(x) \quad \text{and} \quad y = \varphi(x), \qquad \text{for } a \leqslant x \leqslant b,$$

and to the left by the vertical line segments AB and CD. Then because Ω_x is an x-simple region, the double integral of $\partial P/\partial y$ over Ω_x reduces to an iterated integral and we have

$$\begin{aligned}\iint_{\Omega_x} \frac{\partial P}{\partial y}\,dx\,dy &= \int_a^b \int_{\varphi(x)}^{\psi(x)} \frac{\partial P}{\partial y}\,dy \\ &= \int_a^b \big\{P[x, \psi(x)] - P[x, \varphi(x)]\big\}dx \\ &= \int_a^b P[x, \psi(x)]dx - \int_a^b P[x, \varphi(x)]dx.\end{aligned}$$

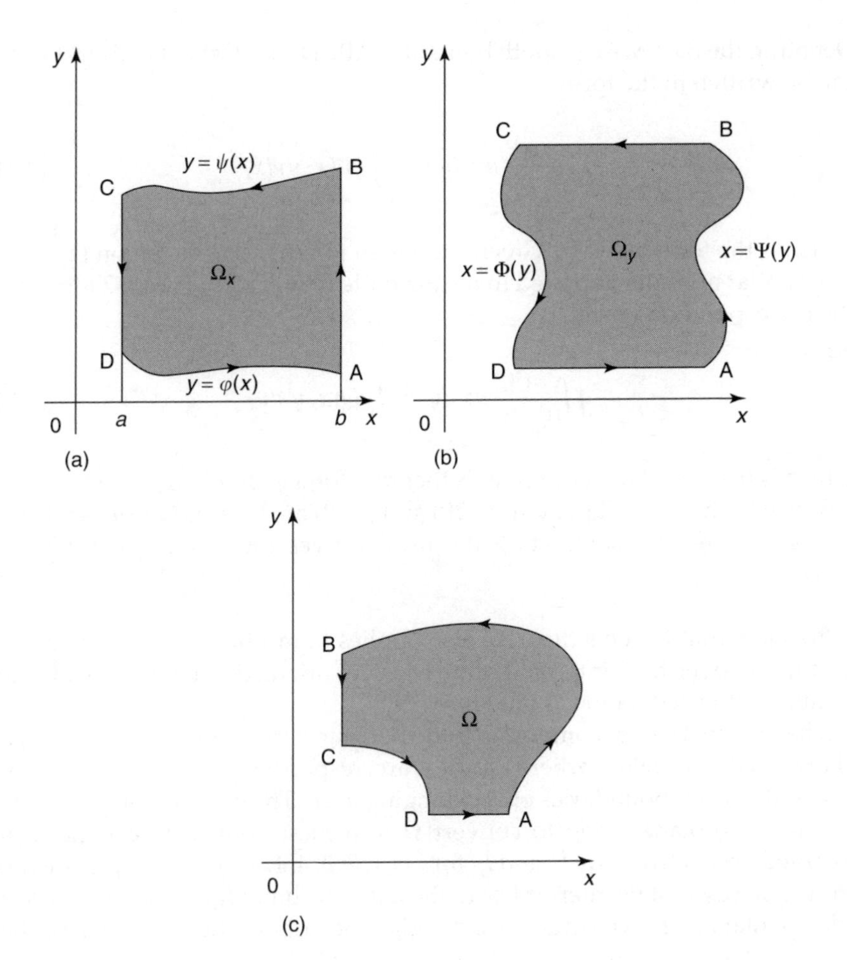

FIGURE 2.8
(a) An x-simple region Ω_x. (b) A y-simple region Ω_y. (c) A simple region (both x- and y-simple).

Expressing this result in terms of integrals along arcs, and taking account of the effect of reversing the limits, show that

$$\iint_{\Omega_x} \frac{\partial P}{\partial y}\,dx\,dy = -\int_{\text{BC}} P(x,y)dx - \int_{\text{DA}} P(x,y)dx.$$

The integral of P with respect to x along AB and CD will be zero because $x = \text{constant}$ on these lines, so we may subtract these integrals from the expression on the right without altering its validity, and as a result we find that

$$\iint_{\Omega_x} \frac{\partial P}{\partial y}\,dx\,dy = -\int_{\text{AB}} P(x,y)dx - \int_{\text{BC}} P(x,y)dx - \int_{\text{CD}} P(x,y)dx - \int_{\text{DA}} P(x,y)dx.$$

Denoting the piecewise smooth boundary ABCDA by Γ shows this last result can be written in the form

$$\iint_{\Omega_x} \frac{\partial P}{\partial y}\, dx\, dy = -\int_{\Gamma} P(x, y)dx, \tag{2.21}$$

which is the form taken by Green's theorem for an x-simple region Ω_x.

A similar argument applied to the y-simple region Ω_y in Figure 2.8(b) gives the corresponding result

$$\iint_{\Omega_y} \frac{\partial Q}{\partial x}\, dx\, dy = \int_{\Gamma} Q(x, y)dy, \tag{2.22}$$

which is the form taken by Green's theorem for a y-simple region Ω_y.

When both results Equations (2.20) and (2.22) are true, as happens when Ω is a simple region, subtracting the results gives the statement of Green's theorem. ◆

To show that Green's theorem also applies to multiply connected regions and to discover how internal boundaries are oriented, we will consider the doubly connected region Ω in Figure 2.9.

The region Ω is decomposed into the four simple regions Ω_1 to Ω_4 as shown in Figure 2.9(b), where Γ_i and γ_i are, respectively, the parts of the exterior and interior boundaries of Ω belonging to Ω_i. The cuts parallel to the axes in the (x, y)-plane made to convert Ω into these four simple regions are denoted, respectively, by l_{ix} and l_{iy} for $i = 1, 2, 3, 4$. Each boundary is oriented in such a way that its interior lies to the left as its boundary is traversed, with the orientation shown by arrows. As adjacent sides of the cuts are described

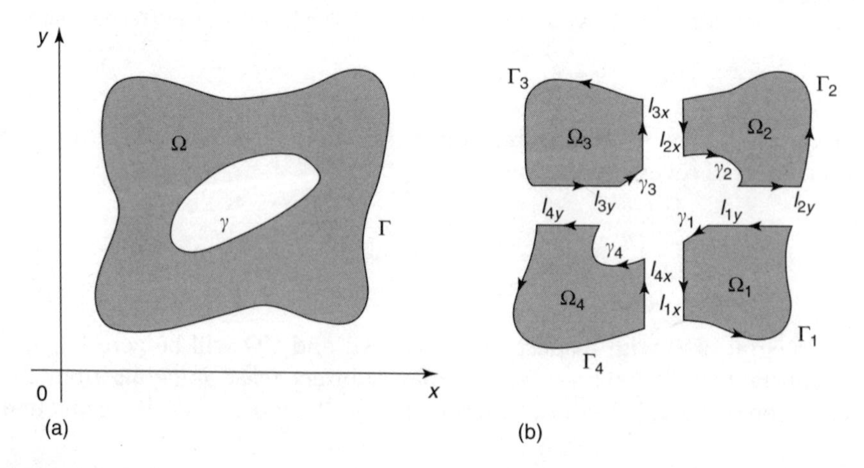

FIGURE 2.9
(a) A doubly connected region. (b) Decomposition into simple regions.

in opposite directions, the line integrals along each side of a cut will vanish because of the continuity of P and Q. Thus adding these results shows that the integral around the *total boundary* (both the exterior and interior boundaries) is

$$\iint_{\Omega}\left(\frac{\partial Q}{\partial x}-\frac{\partial P}{\partial y}\right)dx\,dy=\int_{\Gamma}P\,dx+Q\,dy. \tag{2.23}$$

Here Ω is the doubly connected region shown in Figure 2.9(a), and Γ is the total boundary of Ω, comprising the *exterior boundary* described in the positive sense and the *interior boundary* described in the negative sense.

This approach generalizes immediately to multiply connected regions like the one shown in Figure 2.10, where there are n internal boundaries $\gamma_1, \gamma_2, \ldots, \gamma_n$, with the orientation of the integration around each boundary shown by arrows.

We are now in a position to prove Cauchy's Integral theorem based on the assumption that Green's theorem applies: $f(z)$ is analytic both inside a region and on its boundary.

THEOREM 2.2.2 The Cauchy Theorem
If $f(z)$ is analytic in a simply connected domain D and on its smooth boundary C, then

$$\int_C f(z)dz = 0.$$

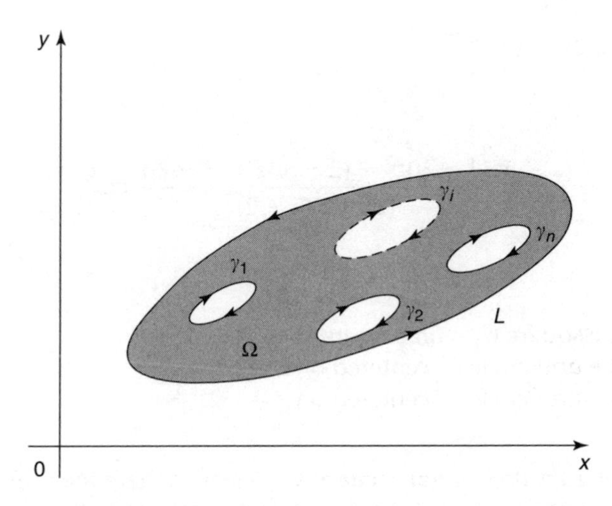

FIGURE 2.10
Integration around a multiply connected region.

PROOF

It was shown in Equation (2.8) that if $f(z) = u + iv$ is analytic in a domain D with boundary C, then

$$\int_C f(z)dx = \int_C u\,dx - v\,dy + i\int_C u\,dy + v\,dx.$$

If we assume the boundary C to be smooth, we may apply Green's theorem to each of these line integrals when we find that

$$\int_C f(z)dz = -\iint_D \left(\frac{\partial v}{\partial x} + \frac{\partial u}{\partial y}\right) dx\,dy + i\iint_D \left(\frac{\partial u}{\partial x} - \frac{\partial v}{\partial y}\right) dx\,dy.$$

However, as $f(z)$ is analytic in D, u and v satisfy the Cauchy–Riemann equations, with the result that each integrand on the right vanishes showing that

$$\int_C f(z)dz = 0,$$

and the simple form of Cauchy's Integral theorem is proved. ◆

Notice that this theorem makes no statement about the value of an integral if a singularity of $f(z)$ lies on C. More will be said about this when the theory of residues is discussed. The next example makes applications of the Cauchy Integral theorem to a function with several singularities (poles) using different contours.

Example 2.2.3
Evaluate

$$\int_C \frac{(4-2i)z^2 + (2-5i)z + (3-2i)}{(z^2+1)(z+2)}\,dz,$$

when

(i) C is the square C_1 with corners at $\frac{1}{2} + \frac{1}{2}i, -\frac{1}{2} + \frac{1}{2}i, -\frac{1}{2} - \frac{1}{2}i, \frac{1}{2} - \frac{1}{2}i$;
(ii) C is the unit circle C_2 centered on $z = i$;
(iii) C is the unit circle C_3 centered on $z = -2$.

SOLUTION

(i) The singularities of $f(z)$ occur at values of z where the numerator is finite and nonzero but the denominator vanishes. These are easily seen to occur when $z = \pm i$ and $z = -2$.

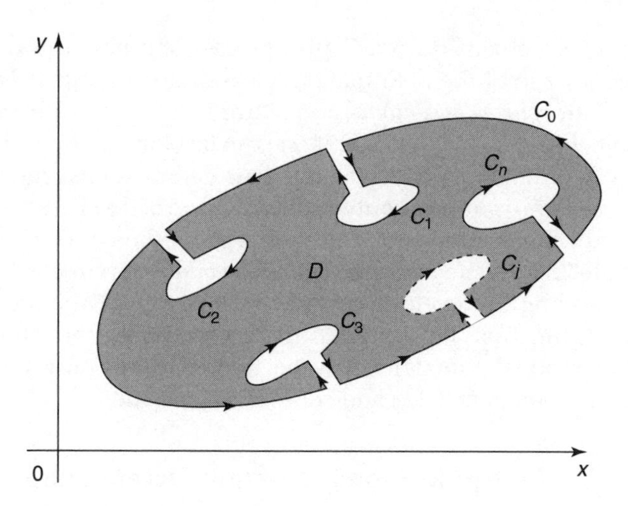

FIGURE 2.12
Cutting a multiply connected domain D to make it simply connected.

Because $f(z)$ is analytic in the cut region, by Cauchy's theorem the integral on the left vanishes, so rearranging terms and using Equation (2.24) we find that

$$\int_{C_0} f(z)dz = \int_{C_1} f(z)dz + \int_{C_2} f(z)dz + \cdots + \int_{C_n} f(z)dz, \qquad (2.26)$$

which is the required generalization. We now anticipate the result of the Cauchy–Goursat theorem to be proved later, and formulate the last result as a generalization of the Cauchy–Goursat theorem.

THEOREM 2.2.3 The Generalized Cauchy–Goursat Theorem
Let domain D be bounded externally by a simple piecewise smooth contour C_0 and internally by the n nonintersecting simple piecewise smooth closed contours $C_1, C_2, \ldots, C_n$, and let both the external and internal contours be positively oriented. Then if f(z) is analytic in D its derivative is piecewise smooth on its boundaries,

$$\int_{C_0} f(z)dz = \sum_{r=1}^{n} \int_{C_r} f(z)dz. \qquad \blacklozenge$$

An important application of this last theorem is the *deformation of a contour*. The proof of Theorem 2.2.2 made no reference to the shape of the internal boundaries, so it follows that a simple closed-contour C_r in D can be replaced by any other simple closed-contour C_r^* without altering the statement of the theorem, provided C_r^* also lies within D and does not intersect any other internal contour. The relationship between C_r and C_r^* is to be thought of as though C_r^* has been obtained from C_r by a process of continuous deformation such that every stage of the deformation always remains within D. The contours C_r and C_r^* are *equivalent contours*, and two arcs in D, each with the same initial and terminal points, are *homotopic* if one can be obtained from the other by a continuous deformation, with all intermediate deformations remaining in D.

The practical significance of contour deformation is that it allows contours with awkward shapes replaced by ones that are more convenient, as illustrated in Figure 2.13, where contour C_1 is replaced by the circular contour C_1^*.

Before proceeding to develop powerful techniques for the integration of general analytic functions $f(z)$ around contours, let us first show how contour integration of rational functions may be treated using only the result of Example 2.1.3(i), Theorem 2.2.3, and contour deformation.

Example 2.2.4 An Application of Contour Deformation

Evaluate

$$\int_C \frac{z^2 + (10 - i)z - (4 + 2i)}{(z^2 + 1)(z + 2)} dz,$$

where C is the circle $|z| = 3$.

SOLUTION

Inspection of $f(z)$ shows it is analytic everywhere except for the points $z = i, -i$ and -2, shown as P, Q, and R in Figure 2.14. Using partial fractions we find that

$$f(z) = \frac{2}{z - i} + \frac{3}{z + i} - \frac{4}{z + 2}.$$

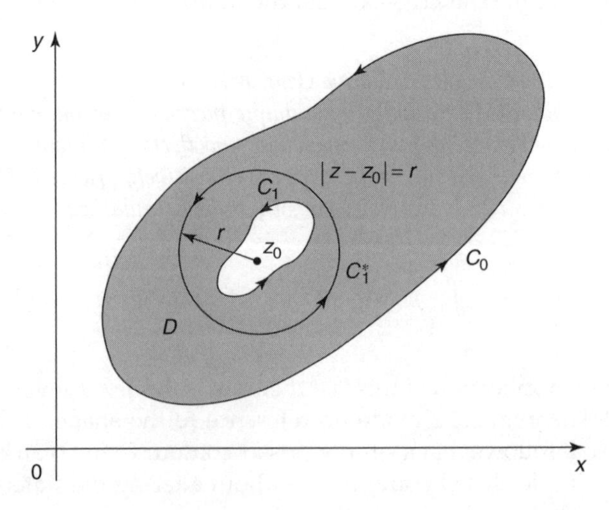

FIGURE 2.13
The deformation of contour C_1 to the circular contour C_1^*.

13. Evaluate

$$\int_C \frac{3z+1}{2z^2+3z+1}\,dz,$$

when C is (i) the circle $|z+1| = \frac{1}{4}$, and (ii) the circle $|z+\frac{1}{2}| = \frac{1}{4}$.

14. Evaluate

$$\int_C \frac{5z-1}{z^2-z-2}\,dz,$$

when C is (i) the circle $|z-2| = 1$, and (ii) the circle $|z+1| = 1$.

15. Evaluate

$$\int_C \frac{3}{z^2-4}\,dz,$$

when C is (i) the circle $|z-2| = 2$, and (ii) the circle $|z+2| = 1$.

16. Evaluate

$$\int_C \frac{(2+3i)z-4+9i}{z^2+z-6}\,dz,$$

when C is (i) the circle $|z-2| = 1$, and (ii) the circle $|z+3| = 2$.

In Exercises 17 through 20 use contour deformation to evaluate the given integrals.

17. Evaluate

$$\int_C \frac{3iz+1}{z^2-4z+3}\,dz,$$

when (i) C is the circle $|z-\frac{5}{4}| = \frac{1}{2}$, and (ii) the circle $|z-\frac{13}{4}| = \frac{1}{2}$.

18. Evaluate

$$\int_C \frac{2z-3}{z^2+1}\,dz,$$

when (i) C is the unit square centered on $z = i$ with its sides parallel to the axes, and (ii) the unit square centered on $z = -i + \frac{1}{2}$ with its sides parallel to the axes.

19. Evaluate

$$\int_C \frac{2z^2+(3-2i)z-(1+6i)}{(z^2-1)(z-2i)}\,dz,$$

when C is the cardioid $r = 2\cos^2(\theta/2)$ with $0 \leqslant \theta \leqslant 2\pi$.

20. Evaluate

$$\int_C \frac{4z^2 + 9iz + 27}{z(z^2 + 9)} dz,$$

where C is the perimeter common to the circles $|z - 3| = 4$ and $|z + 2| = 3$.

2.3 Antiderivatives and Definite Integrals

The remainder of this chapter examines the most important consequences of the Cauchy–Goursat theorem. This section begins by examining how the theorem leads to the notions of an antiderivative (indefinite integral) and a definite integral. Let $f(z)$ be analytic in a simply connected domain D to which a fixed point z_0 and an arbitrary point z belong. Let γ_1 and γ_2 be any two nonintersecting arcs in D that join z_0 and z, as shown in Figure 2.15, where the direction of integration is shown by arrows. Then it follows from Theorem 2.2.3 that as $f(z)$ is analytic in D, and so also on arcs γ_1 and γ_2, that

$$0 = \int_{\gamma_1} f(z)dz + \int_{\gamma_2} f(z)dz.$$

Reversing the direction of integration on arc γ_2 the previous result becomes

$$\int_{\gamma_1} f(z)dz = \int_{\gamma_{2-}} f(z)dz. \tag{2.27}$$

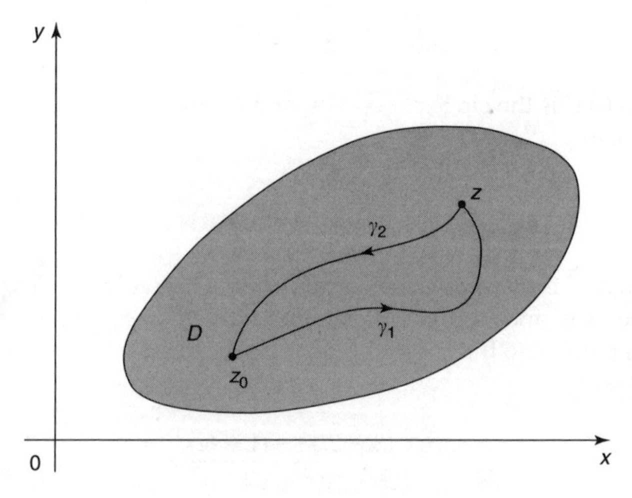

FIGURE 2.15
Two arcs γ_1 and γ_2 joining z_0 and z.

The only points common to γ_1 and γ_2 are z_0 and z, so the integrals in Equation (2.27) must be *independent* of the path joining z_0 and z. This shows that the integral of a function $f(z)$ that is analytic in D depends only on $f(z)$, the initial point z_0 and the terminal point z, and not on the path joining them. As a result we may, without ambiguity, omit any reference to the path of integration in D, identifying it only by the two points z_0 and z, and write

$$F(z) = \int_{z_0}^{z} f(\zeta)d\zeta. \tag{2.28}$$

Here, as the path of integration is assumed to be in the domain D where $f(z)$ is analytic, it follows that $F(z)$ is a function of the arbitrary point z. The function $F(z)$ is called the *antiderivative,* or *indefinite integral*, of $f(z)$.

We now relate Equation (2.28) to the equivalent concept for a function of a real variable by showing that, as in the real variable case, $F'(z) = f(z)$. This will be accomplished if we can show that $F(z)$ is such that

$$\lim_{h\to 0}\left(\frac{F(z+h)-F(z)}{h}\right) = f(z), \qquad \text{for } z \in D,$$

as the complex variable h tends to zero along *any* path γ in D. This is equivalent to requiring that

$$\lim_{h\to 0}\left|\frac{f(z+h)-F(z)}{h} - f(z)\right| = 0, \qquad \text{for } z \in D.$$

As Equation (2.28) depends only on the end points of the path of integration we may write the difference quotient as

$$\frac{F(z+h)-F(z)}{h} = \frac{1}{h}\int_{z_0}^{z+h} f(\zeta)d\zeta - \frac{1}{h}\int_{z_0}^{z} f(\zeta)d\zeta = \frac{1}{h}\int_{z}^{z+h} f(\zeta)d\zeta. \tag{2.29}$$

Now from Equation (2.28), z may be joined to $z + h$ by any arc in D, and we also have the result

$$\int_{z}^{z+h} d\zeta = h.$$

Thus we may represent $f(z)$ as an integral involving $f(\zeta)$ by writing

$$f(z) = \frac{f(z)}{h}\cdot h = \frac{f(z)}{h}\int_{z}^{z+h} d\zeta = \frac{1}{h}\int_{z}^{z+h} f(z)d\zeta, \tag{2.30}$$

where the last result follows because the integration is with respect to ζ, so $f(z)$ can be regarded as a constant, and so taken under the integral sign.

Using Equations (2.29) and (2.30) we have

$$\begin{aligned}\left|\frac{F(z+h)-F(z)}{h}-f(z)\right| &= \left|\frac{1}{h}\int_z^{z+h} f(\zeta)d\zeta - \frac{1}{h}\int_z^{z+h} f(z)d\zeta\right| \\ &= \frac{1}{|h|}\left|\int_z^{z+h}[f(\zeta)-f(z)]d\zeta\right| \\ &\leqslant \frac{1}{|h|}\max_{\zeta\in\gamma}|f(\zeta)-f(z)|\left|\int_z^{z+h} d\zeta\right| \\ &= \frac{1}{|h|}\max_{\zeta\in\gamma}|f(\zeta)-f(z)||h| \\ &= \max_{\zeta\in\gamma}|f(\zeta)-f(z)|.\end{aligned}$$

However, $f(z)$ is continuous in D, so $\lim_{h\to 0}\max_{\zeta\in\gamma}|f(\zeta)-f(z)| = 0$, and thus

$$\lim_{h\to 0}\left|\frac{F(z+h)-F(z)}{h}-f(z)\right| = 0,$$

and so proceeding to the limit this shows as required that

$$F'(z) = f(z), \quad \text{for } z \in D.$$

Clearly an *antiderivative*, or an indefinite integral, of an analytic function is unique apart from an additive complex constant c, because $dc/dz = 0$, so

$$\frac{d}{dz}[F(z)+c] = \frac{dF(z)}{dz} = f(z).$$

This important result we have proved will now be stated in the form of a theorem.

THEOREM 2.3.1 Antiderivatives or Indefinite Integrals

Let the complex function $f(z)$ be analytic in a simply connected domain D. Then if γ is any arc lying entirely within D with initial point z_0 and terminal point z, the antiderivative

$$F(z) = \int_\gamma f(\zeta)d\zeta = \int_{z_0}^{z} f(\zeta)d\zeta$$

is a single-valued analytic function of z, independent of the choice of are γ, and such that

$$F'(z) = f(z). \qquad \blacklozenge$$

Indefinite integrals (antiderivatives) of the elementary functions considered so far follow directly from knowledge of their derivatives computed in Chapter 1. So for example, from entry 26 from Section 1.5.8, integration shows that $\int dz/\sqrt{1-z^2} = \arcsin z + \text{const.}$

However, when general applications are made of the Schwarz–Christoffel transformation described in Section 4.5, some of the more complicated indefinite integrals in Table 2.1 may arise. Entries 1 through 4 in Table 2.1 follow directly after considering the appropriate derivatives listed in Table 2.1, but to derive entries 5 through 7 in Table 2.1 some extra manipulation is necessary. Entry 5 follows after using the substitution $u^2 = z^2 - 1$, while entries 6 and 7 require integration by parts before making use of the appropriate derivatives of elementary functions given in Table 2.1.

It must always be remembered that the indefinite integrals in Table 2.1 all involve functions with branches, so when these indefinite integrals are used in specific applications the branch chosen must first be specified, and thereafter used consistently.

An immediate consequence of Theorem 2.3.1 is that, as in the real variable case, we may determine a definite integral of the complex function $f(z)$ in

TABLE 2.1

Some Useful Indefinite Integrals

1. $\displaystyle\int \frac{dz}{\sqrt{1-z^2}} = \arcsin z = -i\ln\left(iz + \sqrt{1-z^2}\right)$

2. $\displaystyle\int \frac{dz}{1+z^2} = \arctan z = \tfrac{1}{2}i\ln\left(\frac{i+z}{i-z}\right)$

3. $\displaystyle\int \frac{dz}{\sqrt{z^2+1}} = \operatorname{arcsinh} z = \ln\left(z + \sqrt{z^2+1}\right)$

4. $\displaystyle\int \frac{dz}{1-z^2} = \operatorname{arctanh} z = \tfrac{1}{2}\ln\left(\frac{1+z}{1-z}\right)$

5. $\displaystyle\int \frac{dz}{\sqrt{z^2-1}} = \operatorname{arcsinh}\sqrt{z^2-1} = \ln\left(z + \sqrt{z^2-1}\right)$

6. $\displaystyle\int \frac{dz}{z\sqrt{z^2-1}} = -\arcsin\frac{1}{z} = i\ln\left[\frac{1}{z} + \left(\frac{1}{z^2}-1\right)^{1/2}\right]$

7. $\displaystyle\int \frac{dz}{z\sqrt{1+z}} = -2\operatorname{arctanh}\sqrt{1+z} = \ln\left(\frac{1-\sqrt{1+z}}{1+\sqrt{1+z}}\right)$

terms of an antiderivative $F(z)$. To see why this is, notice that we may always write

$$F(z_1) = \int_{z_0}^{z_1} f(\zeta)d\zeta + c,$$

where c is an arbitrary complex constant, so that

$$F(z_0) = \int_{z_0}^{z_1} f(\zeta)d\zeta + c = c,$$

and thus

$$\int_{z_0}^{z_1} f(z)dz = F(z_1) - F(z_0),$$

where the dummy variable has been changed from ζ to z, giving the following theorem.

THEOREM 2.3.2 Definite Integral
Let $f(z)$ be analytic in a simply connected domain D and let $F(z)$ be an antiderivative of $f(z)$. Then for any two points z_0 and z_1 in D,

$$\int_{z_0}^{z_1} f(z)dz = F(z_1) - F(z_0).$$

◆

Example 2.3.1 A Simple Definite Integral
Evaluate

$$\int_{1}^{2+3i} \sinh 3z \, dz.$$

SOLUTION
The function $\sinh 3z$ is analytic in the finite z-plane so Theorem 2.3.1 can be applied, using the fact that an antiderivative is $\frac{1}{3}\cosh 3z$. Thus we have

$$= \frac{1}{3}\cosh 3z \Big|_1^{2+3i} = \frac{1}{3}\cosh(6+9i) - \frac{1}{3}\cosh 3.$$

Using the result

$$\cosh(6+9i) = \cosh 6 \cos 9 + i \sinh 6 \sin 9,$$

the integral becomes

$$\int_1^{2+3i} \sinh 3z\,dz = \tfrac{1}{3}(\cosh 6 \cos 9 - \cosh 3) + \tfrac{1}{3} i \sinh 6 \sin 9$$
$$= -64.619 + 27.710i. \qquad \diamond$$

Example 2.3.2 A Definite Integral of a Function with Branches
Evaluate

$$\int_{-4\sqrt{2}(1+i)}^{27(1+i\sqrt{3})/2} \frac{dz}{z^{1/3}},$$

using the principal branch of $z^{1/3}$ and integrating along any path that does not pass through the negative real axis or the origin.

SOLUTION
The function $z^{1/3}$ has three branches, each analytic in the cut z-plane, with the cut extending along the negative real axis up to and including the origin, so that z must be such that $-\pi < \mathrm{Arg}(z) < \pi$. A suitable path C avoiding the cut is shown in Figure 2.16 with the path starting at $z = -4\sqrt{2}(1 + i)$ and terminating at $z = 27(1 + i\sqrt{3})/2$.

An antiderivative of $z^{-1/3}$ is $(\frac{3}{2})z^{2/3}$, so integrating along C gives

$$\int_{z_0}^{z_1} z^{-1/3}\,dz = \tfrac{3}{2}\left(z_1^{2/3} - z_0^{2/3}\right),$$

where the principal branch of $z^{-1/3}$ is to be used.

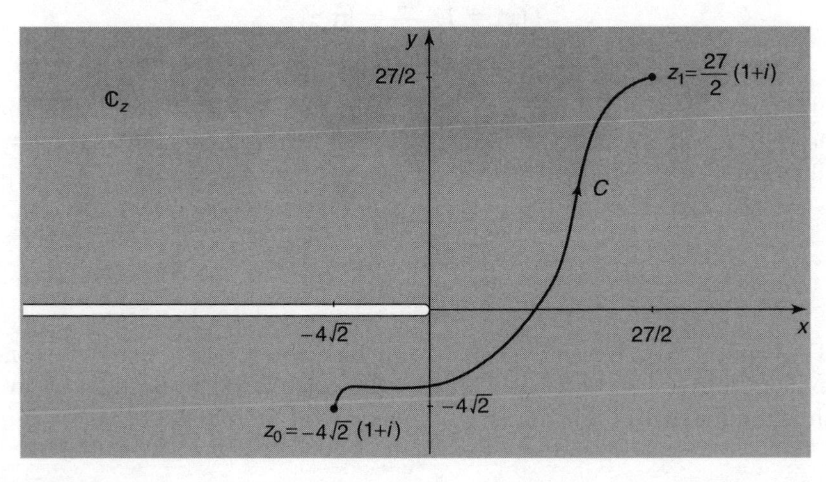

FIGURE 2.16
The cut z-plane and a possible integration path.

From Section 1.1, the principal branch of $z^{1/3}$ is $r^{1/3}e^{i\theta/3}$, so setting $z_0 = -4\sqrt{2}(1+i)$ we have $|z_0| = 8$ and $\theta_0 = \text{Arg}(z_0) = -3\pi/4$, while setting $z_1 = 27(1+i\sqrt{3})/2$ we have $|z_1| = 27$ and $\theta_1 = \text{Arg}(z_1) = \pi/3$. Thus $z_0^{1/3} = 2e^{-i\pi/4}$, $z_0^{2/3} = -4i$, while $z_1^{1/3} = 3e^{i\pi/9}$, so that $z_1^{2/3} = 9e^{2\pi i/9}$. Thus the required definite integral is

$$\int_{-4\sqrt{2}(1+i)}^{27(1+i\sqrt{3})/2} \frac{dz}{z^{1/3}} = \tfrac{3}{2}(9e^{2\pi i/9} + 4i) = 10.342 + 14.678i. \qquad \diamond$$

2.3.1 The Logarithmic Function

An important example of a function defined in terms of a definite integral with a variable upper limit z can be derived from Equation (2.28) by setting $f(\zeta) = 1/\zeta$ and $z_0 = 1$, to obtain

$$F(z) = \int_1^z \frac{d\zeta}{\zeta}. \tag{2.31}$$

The integrand is analytic everywhere except for the origin, so $F(z)$ will be a single-valued analytic function in any domain D that excludes the origin. Such a domain can be produced by cutting the z-plane along the negative real axis up to and including the origin, so ζ is restricted by the requirement that $-\pi < \text{Arg}(\zeta) < \pi$. Then the function $F(z)$ in Equation (2.31) is defined for any path C of integration that avoids the cut. If we set $z = x$ and take C to be the path from $\zeta = 1$ to $\zeta = x$ we obtain the real integral

$$F(x) = \int_1^x \frac{d\zeta}{\zeta} = \ln_e x,$$

that defines the real natural logarithmic function. It is because of this that when $z \in D$ and the principal branch of the logarithmic function is used, we will write

$$\text{Ln}\, z = \int_1^z \frac{d\zeta}{\zeta}, \qquad \text{for } z \in D.$$

The connection between the real natural logarithmic function and the complex logarithmic function has already been established in Section 1.5 in a different manner, so it is a direct consequence of Theorem 2.3.1 that

$$\frac{d}{dz}[\text{Ln}\, z] = \frac{1}{z}, \qquad \text{for } z \in D, \tag{2.32}$$

from which it follows that

$$\int_{z_0}^{z_1} \frac{dz}{z} = \text{Ln}\, z_1 - \text{Ln}\, z_0,$$

for any path in D that joins z_0 and z_1.

Exercises 2.3

Use Theorem 2.3.2 to evaluate the following definite integrals, after first verifying that their integrands are analytic. When a many-valued function is involved, use the principal branch of the function, state how the z-plane is to be cut, and explain the nature of the integration path that joins the limits of the integral.

1. $\int_0^{2+i} z^3\, dz$
2. $\int_i^{2-i} (1 + 4z - z^2) dz$
3. $\int_{1+2i}^{3-2i} (z - 2)^2\, dz$
4. $\int_{1-i}^{1+i} (1 - 2iz)^2\, dz$
5. $\int_{1-i}^{2-i} \sin 2z\, dz$
6. $\int_{1+i}^{2+i} e^{2z}\, dz$
7. $\int_{2+i}^{3-2i} \sinh z\, dz$
8. $\int_{2+i}^{4-3i} \frac{dz}{z}$
9. $\int_{-1-i}^{1+i} \frac{2z - 6}{z(z - 2)} dz$
10. $\int_1^{1+i} z^{1/2}\, dz$
11. $\int_1^{8i} z^{1/3}\, dz$
12. $\int_{1-i}^{-1-i} \frac{dz}{(z^2 - 1)}$
13. The following exercise shows that when in Theorem 2.2.2, the condition that $f(z)$ is analytic is replaced by the weaker condition that it

is continuous and bounded, the integral between two points z_0 and z_1 can become path dependent. Verify that the function of a complex variable z defined as

$$f(z) = \begin{cases} 0, & \mathrm{Re}\{z\} < 0 \\ x, & 0 \leqslant \mathrm{Re}\{z\} \leqslant 1 \\ 1, & \mathrm{Re}\{z\} > 1, \end{cases}$$

is continuous and bounded for all z. Find

$$I_1 = \int_{\Gamma_1} f(z)dz \quad \text{and} \quad I_2 = \int_{\Gamma_2} f(z)dz,$$

where the path Γ_1 from A to C is formed by the straight line segments joining the points A(0, 0), B(2, 0) and C(2, 2) in alphabetical order, and the path Γ_2 from A to C is formed by the straight line segments joining the points A(0, 0), B(0, 2) and C(2, 2) in alphabetical order.

14. Explain why, if functions $f(z)$ and $g(z)$ are analytic in a domain D in which there lie the points z_0 and z_1 and the simple arc Γ joining them, that integration by parts may be used with Theorem 2.3.2 to give the result

$$\int_{z_0}^{z_1} f'(z)g(z)dz = f(z_1)g(z_1) - f(z_0)g(z_0) - \int_{z_0}^{z_1} f(z)g'(z)dz.$$

15. Use the result of Exercise 14 to find $I = \int_0^{1+i} z\cos z\,dz$.

2.4 The Cauchy Integral Formula

In this section we derive an important and useful result called the *Cauchy integral formula*. This provides a representation of an analytic function in the form of a contour integral. The formula finds many applications, one of which will occur later when evaluating contour integrals. The theorem takes the following form.

THEOREM 2.4.1 The Cauchy Integral Formula
Let $f(z)$ be analytic in a simply connected domain D in which there lies a simple closed contour C, and let z_0 be any point inside C. Then

$$f(z_0) = \frac{1}{2\pi i}\int_C \frac{f(z)}{z - z_0}\,dz.$$

PROOF

Consider the function

$$F(z) = \frac{f(z)}{z - z_0},$$

which is analytic everywhere inside C except for the point z_0. Construct a circle C_ρ of radius ρ centered on z_0 that lies entirely inside C, as in Figure 2.17. Then as $F(z)$ is analytic in the annular region between C_ρ and C, it follows from Theorem 2.2.2 (and from its generalization to the Cauchy–Goursat theorem proved in Section 2.8), that

$$\int_C \frac{f(z)}{z - z_0}\, dz = \int_{C_\rho} \frac{f(z)}{z - z_0}\, dz.$$

Because $f(z_0)$ is a constant with respect to integration with respect to z, this last result can be rewritten as

$$\begin{aligned}\int_C \frac{f(z)}{z - z_0}\, dz &= \int_{C_\rho} \frac{[f(z) - f(z_0)] + f(z_0)}{z - z_0}\, dz \\ &= \int_{C_\rho} \left(\frac{f(z) - f(z_0)}{z - z_0}\right) dz + f(z_0)\int_{C_\rho} \frac{dz}{z - z_0}\, dz \qquad (2.33) \\ &= \int_{C_\rho} \left(\frac{f(z) - f(z_0)}{z - z_0}\right) dz + 2\pi i f(z_0),\end{aligned}$$

where use has been made of the result of Example 2.1.3(i).

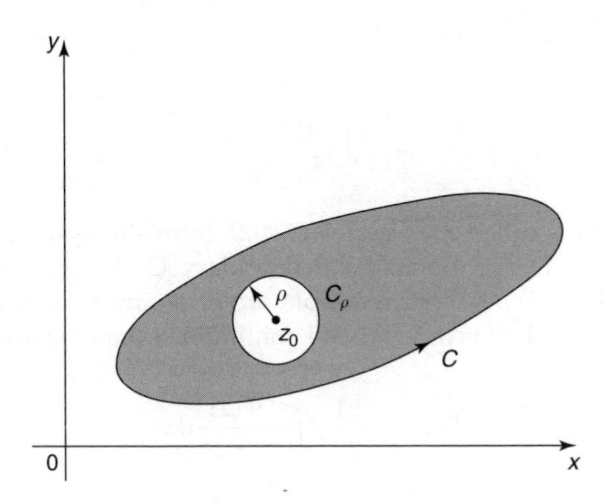

FIGURE 2.17
The function $f(z_0)$ is analytic in the shaded annular region.

The modulus of the integral on the right depends on the radius ρ of the circle C_ρ, and it can be overestimated by using the simple inequality

$$\left|\int_{C_\rho}\left(\frac{f(z)-f(z_0)}{z-z_0}\right)dz\right| \leqslant \max_{z\in C_\rho}|f(z)-f(z_0)|\left|\int_{C_\rho}\frac{dz}{z-z_0}\right|$$
$$= \max_{z\in C_\rho}|f(z)-f(z_0)|\cdot 2\pi.$$

However $f(z)$ is a continuous function of z, so letting $\rho \to 0$ in Equation (2.33) when $|f(z) - f(z_0)| \to 0$, we find that

$$f(z_0) = \frac{1}{2\pi i}\int_C \frac{f(z)}{z-z_0}\,dz,$$

and the theorem is proved. ◆

When Theorems 2.2.2 and 2.4.1 are combined (or the generalization of Theorem 2.2.2 to the Cauchy–Goursat theorem is used), this shows that if $f(z)$ is analytic in some domain D containing a simple closed-contour C, and both z and z_0 belong to D, then

$$\int_C \frac{f(z)}{z-z_0}\,dz = \begin{cases} 2\pi i f(z_0), & \text{if } z_0 \text{ lies in } D \\ 0, & \text{if } z_0 \text{ lies outside } D. \end{cases} \tag{2.34}$$

The Cauchy integral formula remains valid in multiply connected domains provided the contour C is the *total* boundary; that is the external plus the internal boundaries, where in each case the direction of integration is in the positive sense.

The integral

$$\frac{1}{2\pi i}\int_C \frac{f(z)}{z-z_0}\,dz \tag{2.35}$$

in Theorem 2.4.1 is called the *Cauchy integral*. In this integral $f(z)$ is assumed to be analytic in D and on its total boundary C. As a generalization of Equation (2.35) let Γ be either a simple arc or a simple closed contour on which the function $\Phi(z)$ is defined and continuous, then the integral

$$F(z_0) = \frac{1}{2\pi i}\int_\Gamma \frac{\Phi(z)}{z-z_0}\,dz \tag{2.36}$$

is said to be an integral of the *Cauchy type*, and it defines the function $F(z_0)$ for $z_0 \notin \Gamma$. Integrals of this type occur frequently in applications.

Example 2.4.1 A Simple Application of the Cauchy Integral Formula

Evaluate

$$\int_0^{2\pi} \frac{ze^{-z}}{z - \frac{1}{2}\pi i} dz,$$

where C is the triangular contour with vertices at $-1 - i, 1 - i$, and 2.

SOLUTION

We apply Theorem 2.4.1 with $f(z) = ze^{-z}$ and $z_0 = \pi i/2$. The point z_0 lies inside the contour C where $f(z)$ is analytic, so from result [Equation (2.34)], as $f(z_0) = \frac{1}{2} i\pi e^{-i\pi/2} = \frac{1}{2}\pi$ we find that

$$\int_0^{2\pi} \frac{ze^{-z}}{z - \frac{1}{2}\pi i} dz = (2\pi i)\left(\frac{1}{2}\pi\right) = i\pi^2. \qquad \diamond$$

Example 2.4.2 A More Complicated Application of the Cauchy Integral Formula

Evaluate

$$\int_C \frac{e^{-z}}{z^3 + 2z^2 - 3z - 10} dz,$$

where C is the triangular contour with vertices at i, $-i$ and 3.

SOLUTION

Inspection shows that a root of the denominator is 2, so a factor of the denominator is $(z - 2)$, and division of the denominator by $(z - 2)$ gives $z^2 + 4z + 5$, so the denominator can be written as

$$z^3 + 2z^2 - 3z - 10 = (z - 2)(z^2 + 4z + 5).$$

So the zeros of the denominator are $z = 2, z = 2 - i$ and $z = 2 + i$, where only the zero $z = 2$ lies inside the triangular contour. Thus the function

$$f(z) = \frac{e^{-z}}{z^2 + 4z + 5}$$

is analytic inside the triangular contour C. So applying result [Equation (2.34)] with this choice for $f(z)$ gives

$$\int_C \frac{f(z)}{z - 2} dz = 2\pi i f(2) = \frac{2\pi i}{17e^2}. \qquad \diamond$$

2.4.1 The Cauchy Integral Formula Applied to Trigonometric Functions

The Cauchy integral formula is useful when evaluating real integrals of the form

$$I = \int_0^{2\pi} F(\sin x, \cos x)dx, \tag{2.37}$$

where the integrand $F(\sin x, \cos x)$ is the quotient of two polynomials in $\sin x$ and $\cos x$. Integrals of this type are called *trigonometric integrals*, and they will be evaluated again later using a different method when the theory of residues is discussed.

The idea here is to use the familiar results that

$$\sin z = \tfrac{1}{2}(e^{iz} - e^{-iz})/i \quad \text{and} \quad \cos z = \tfrac{1}{2}(e^{iz} + e^{-iz}),$$

which suggests making the change of variable $w = e^{iz}$. If this change of variable is made, when z is purely real, so that $w = e^{ix}$, the real functions $\cos x$ and $\sin x$ can be expressed in terms of w and $1/w$, when $dw/dx = ie^{ix} = iw$, so $dx = dw/(iw)$.

In Equation (2.37) the variable x is restricted to $0 \leqslant x \leqslant 2\pi$, so as x traverses this interval w moves around the unit circle $|w| = 1$. This substitution replaces the integrand $F(\sin x, \cos x)$ by a quotient of polynomials $\Phi(w)$ in z of the form

$$\Phi(w) = \frac{p(w)}{q(w)},$$

where the polynomials $p(w)$ and $q(w)$ do not necessarily have real coefficients. So as $dx = dw/(iw)$, this change of variable allows the integral I in Equation (2.37) to be written as

$$I = \int_C \frac{\Phi(w)}{iw} dw,$$

where C is the unit circle $|w| = 1$. When this integral is sufficiently simple, it can be evaluated by using the Cauchy integral formula. Notice the integrand $F(\sin x, \cos x)$ is periodic with period 2π, so the interval of integration in Equation (2.37) can be replaced by any other interval of length 2π.

The steps involved in this approach are set out as follows:

1. Set $w = e^{ix}$, when $\cos x = \frac{1}{2}(w + 1/w)$ and $\sin x = \frac{1}{2}(w - 1/w)/i$.
2. Substitute for $\sin x$ and $\cos x$ in the integrand $F(\sin x, \cos x)$ to obtain a function $\Phi(w)$ that is a quotient of polynomials in w.
3. Make the substitution $dx = dw/(iw)$.

Then

$$I = \int_0^{2\pi} F(\sin x, \cos x)dx = \int_C \frac{\Phi(w)}{iw}\,dw,$$

where C is the unit circle $|w| = 1$.

Use the Cauchy integral formula [or equivalently Equation (2.34)] to evaluate the integral in Step 4.

If the denominator of $\Phi(w)/(iw)$ has more than one zero inside C the Cauchy integral formula must be applied to each zero inside C, after which the results must be added.

Example 2.4.3 Evaluating a Trigonometric Integral

Evaluate

$$I = \int_0^{2\pi} \frac{dx}{3 + 2\sin x}.$$

SOLUTION

Applying Steps 1 to 3 gives

$$I = \int_C \frac{dw}{iw[3 + 2(w - 1/w)/(2i)]} = \int_C \frac{dw}{w^2 + 3iw - 1}.$$

After factoring the denominator this becomes

$$I = \int_C \frac{dw}{\left[w + \frac{1}{2}(3+\sqrt{5})i\right]\left[w + (3-\sqrt{5})i\right]},$$

and only the zero of the denominator at $w = -\frac{1}{2}(3 - \sqrt{5})\,i$ lies inside C. Thus the function

$$f(w) = \frac{1}{\left[w + \frac{1}{2}(3+\sqrt{5})i\right]}$$

is analytic inside C. So from the Cauchy integral formula

$$I = \frac{1}{2\pi i}\int_C \frac{f(w)}{w - w_0}\,dw,$$

where $w_0 = -\frac{1}{2}(3 - \sqrt{5})i$. Thus $I = 2\pi i f(w_0)$, but $f(w_0) = 1/(i\sqrt{5})$, so

$$I = \int_0^{2\pi} \frac{dx}{3 + 2\sin x} = 2\pi/\sqrt{5}.$$

Because the integrand in I is periodic with period 2π, it follows at once that

$$I = \int_{0=a}^{2\pi+a} \frac{dx}{3 + 2\sin x} = 2\pi/\sqrt{5},$$

for any real a. ◇

Exercises 2.4

Evaluate the following integrals.

1. $\int_C \frac{ze^z}{z+3} dz$, with C the circle $|z| = 4$.
2. $\int_C \frac{\sinh 2z}{z - \frac{1}{4}\pi i} dz$, with C the square centered on the origin with its sides parallel to the axes.
3. $\int_C \frac{z \sin z}{z + \frac{1}{2}\pi} dz$, with C the circle $|z + 2| = 2$.
4. $\int_C \frac{e^{iz}}{(z^2 + 9)(z - 1)} dz$, with C the circle $|z - \frac{1}{2}| = \frac{3}{2}$.
5. $\int_C \frac{e^{-z}}{z^3 + 3z^2 + 8z + 6} dz$, where C is a square of side 4 centered on the origin with its sides parallel to the axes.
6. $\int_C \frac{e^{-z}}{z^3 + z^2 + 6z - 8} dz$, where C is the circle $|z - 1| = 1$.
7. Show $u(x, y) = x \cos x \cosh y + y \sin x \sinh y$ is harmonic, find its harmonic conjugate $v(x, y)$ such that $v(0, 0) = 0$, and construct the analytic function $f(z) = u + iv$ as a function of the complex variable z. Hence evaluate the integral $\int_C [f(z)/(z - 1)]dz$, where C is the circle $|z| = 2$.
8. Show $v(x, y) = y \cos 2y \cosh 2x + x \sin 2y \sinh 2x$ is harmonic, find its harmonic conjugate $u(x, y)$ such that $u(0, 0) = 0$, and construct the analytic function $f(z) = u + iv$ as a function of the complex variable z. Hence evaluate the integral

$$\int_C \frac{f(z)}{z-1} dz, \quad \text{where C is the circle } |z| = 3.$$

9. If C is a simple closed contour that does not pass through a singularity of the integrand, determine all possible values of $I = \frac{1}{2\pi i} \int_C \frac{e^z}{z(z-1)} dz$ for different choices of C.

10. Show that $v(x, y) = x^2 + 2y - y^2 + 1$ is harmonic and find its harmonic conjugate $u(x, y)$ such that $u(0, 0) = 0$. Hence find all possible values of

$$I = \frac{1}{2\pi i} \int_C \frac{f(z)}{(z+1)(z+i)} dz \quad \text{for different choices of } C.$$

Evaluate the following trigonometric integrals by using the substitution $w = e^{ix}$.

11. $\displaystyle\int_0^{2\pi} \frac{dx}{2 + \cos x}$

12. $\displaystyle\int_{-\pi}^{\pi} \frac{dx}{4 + 3\sin x}$

13. $\displaystyle\int_0^{2\pi} \frac{\cos x}{2 + \cos x} dx$

14. $\displaystyle\int_0^{2\pi} \frac{\sin x}{4 - \sin x} dx$

15. $\displaystyle\int_{-\pi}^{\pi} \frac{\cos x}{2 - \sin x} dx$

16. $\displaystyle\int_0^{2\pi} \frac{\sin x}{3 - \cos x} dx$

2.5 The Cauchy Integral Formula for Derivatives

In this section we prove an important extension of the Cauchy integral formula, called the *Cauchy integral formula for derivatives*. This formula finds many applications, including both the evaluation of integrals and the derivation of fundamental results in complex analysis. The theorem, in which the nth derivative of $f(z)$ at $z = z_0$ is denoted by $f^{(n)}(z_0)$, may be stated as follows.

THEOREM 2.5.1 The Cauchy Integral Formula for Derivatives
Let $f(z)$ be an analytic function in a simply connected domain D, and let C be a simple closed contour lying entirely within D. Then, for any point z_0 inside C,

$$f^{(n)}(z_0) = \frac{n!}{2\pi i} \int_C \frac{f(z)}{(z - z_0)^{n+1}} dz.$$

PROOF

We first establish the theorem for $n = 1$. Starting from the Cauchy integral formula

$$f(z_0) = \frac{1}{2\pi i}\int_C \frac{f(z)}{(z - z_0)}\,dz,$$

and substituting the difference quotient $[f(z_0 + h) - f(z_0)]/h$ in place of $f(z_0)$, we have

$$\begin{aligned}
\frac{f(z_0 + h) - f(z_0)}{h} &= \frac{1}{h}\frac{1}{2\pi i}\left(\int_C \frac{f(z)}{(z - z_0 - h)}\,dz - \int_C \frac{f(z)}{(z - z_0)}\,dz\right)\\
&= \frac{1}{2\pi i}\int_C \frac{f(z)}{h}\left(\frac{1}{z - z_0 - h} - \frac{1}{z - z_0}\right)dz\\
&= \frac{1}{2\pi i}\int_C \frac{f(z)}{(z - z_0 - h)(z - z_0)}\,dz.
\end{aligned}$$

Because $f(z)$ is analytic its derivative exists, but the limit of the difference quotient on the left as $h \to 0$ is $f^{(1)}(z_0)$, so proceeding to the limit we obtain

$$\lim_{h\to 0}\left(\frac{f(z_0 + h) - f(z_0)}{h}\right) = \lim_{h\to 0}\left(\frac{1}{2\pi i}\int_C \frac{f(z)}{(z - z_0 - h)(z - z_0)}\,dz\right),$$

and so

$$f^{(1)}(z_0) = \frac{1}{2\pi i}\int_C \frac{f(z)}{(z - z_0)^2}\,dz,$$

showing that the statement of the theorem is true when $n = 1$.

The general result will be established by mathematical induction. We assume the formula to be true for an arbitrary integer n, and write the difference quotient $[f^{(n)}(z_0 + h) - f^{(n)}(z_0)]/h$ as

$$\begin{aligned}
\frac{f^{(n)}(z_0 + h) - f^{(n)}(z_0)}{h} &= \frac{1}{h}\frac{n!}{2\pi i}\left(\int_C \frac{f(z)}{(z - z_0 - h)^{n+1}}\,dz - \int_C \frac{f(z)}{(z - z_0)^{n+1}}\,dz\right)\\
&= \frac{n!}{2\pi i}\int_C \frac{f(z)}{h}\left(\frac{1}{(z - z_0 - h)^{n+1}} - \frac{1}{(z - z_0)^{n+1}}\right)dz.
\end{aligned}$$

For small $|h|$, an application of the binomial theorem shows that

$$\frac{1}{(z-z_0-h)^{n+1}} - \frac{1}{(z-z_0)^{n+1}} = \frac{(z-z_0)^{n+1} - [(z-z_0)^{n+1} - h(n+1)(z-z_0)^n + \cdots]}{(z-z_0-h)^{n+1}(z-z_0)^{n+1}}$$
$$= \frac{(n+1)h + O(h^2)}{(z-z_0-h)^{n+1}(z-z_0)},$$

where $O(h^2)$, to be read "is of the order of h^2" and it represents all the remaining terms in the expansion, each of which is multiplied by at least a factor h^2. Thus a function $|f(z)| = O(h^n)$ if $|f(z)| < K|h|^n$ for some constant $K > 0$.

Using this result notation in the expression for the difference quotient and proceeding to the limit at $h \to 0$ gives

$$\lim_{h\to 0}\left(\frac{f^{(n)}(z_0+h) - f^{(n)}(z_0)}{h}\right) = \lim_{h\to 0}\left[\frac{n!}{2\pi i}\int_C \frac{f(z)}{h}\left(\frac{(n+1)h + O(h^2)}{(z-z_0-h)^{n+1}(z-z_0)}\right)dz\right]$$

that is equivalent to

$$f^{(n+1)}(z_0) = \lim_{h\to 0}\left[\frac{(n+1)!}{2\pi i}\int_C \frac{f(z)}{(z-z_0-h)^{n+1}(z-z_0)}dz\right]$$
$$+ \lim_{h\to 0}\left[\frac{n!}{2\pi i}\int_C \frac{f(z)O(h)}{(z-z_0-h)^{n+1}(z-z_0)}dz\right],$$

where we have used the fact that $O(h^2)/h = O(h)$. The second integral vanishes in the limit as $h \to 0$, because $\lim_{h\to 0} O(h) = 0$, so

$$f^{(n+1)}(z_0) = \frac{(n+1)!}{2\pi i}\int_C \frac{f(z)}{(z-z_0)^{n+2}}dz,$$

showing that the formula for $n+1$ follows from the formula for n. As the result is true for $n = 1$ it must be true for all positive integral values of n, so the theorem is proved. ◆

Two remarks are appropriate here: The first being that Theorem 2.5.1 contains the Cauchy Integral Formula 2.4.2 as a special case if we adopt the standard convention that $0! = 1$, and interpret $f^{(0)}(z_0)$ as the undifferentiated function $f(z)$ at $z = z_0$; secondly, Theorem 2.5.1 can be deduced by repeated differentiation of Theorem 2.4.2 under the integral sign with respect to z_0. However this method also makes use of mathematical induction, but this will be left as an exercise.

An almost immediate and remarkable consequence of Theorem 2.5.1 is that it shows that if a function $f(z)$ is analytic at a point P, then it possesses derivatives of all orders point P. Sometimes this result is expressed by saying that a function that is analytic at a point P is *infinitely differentiable* at P. This emphasizes the powerful restriction imposed on a function that is required to be analytic, and the result has no parallel in real variable calculus. We now state this fundamental result as a theorem.

THEOREM 2.5.2 An Analytic Function Has Derivatives of All Orders
If a function $f(z)$ is analytic at a point P in a domain D, then it has derivatives of all orders at P. ◆

Example 2.5.1 Two Important Elementary Integrals Re-Examined
Let us use Theorem 2.5.1 to re-examine the two simple but fundamental integrals considered first in Example 2.1.3(i) and (ii), namely the integrals

$$\text{(i)} \int_C \frac{dz}{z - z_0}, \quad \text{(ii)} \int_C (z - z_0)^n dz, \qquad \text{for } n = 0, \pm 1, \pm 2, \ldots,$$

where now, unlike in Example 2.1.3, C is any simple closed contour enclosing z_0.

SOLUTION

(i) Identifying the integral in (i) with the integral in Theorem 2.5.1 shows that $f(z) = 1$ and $n = 1$, so it follows directly that

$$1 = \frac{1}{2\pi i} \int_C \frac{dz}{z - z_0}, \text{ and once again we have } \int_C \frac{dz}{z - z_0} = 2\pi i,$$

but this time the result is true for *any* simple closed contour C containing z_0.

(ii) When $n = 0$ or a positive integer the function $(z - z_0)^n$ is analytic, so for any closed contour C enclosing z_0 it follows from the Cauchy–Goursat theorem that

$$\int_C (z - z_0)^n dz = 0, \qquad \text{for } n = 0, 1, 2, \ldots .$$

The case $n = -1$ has already been considered in (i), so it remains to for us to consider

$$\int_C \frac{dz}{(z - z_0)^{n+1}}, \qquad \text{for } n = 1, 2, \ldots .$$

Setting $f(z) = 1$ in Theorem 2.5.1 we find that

$$f^{(n)}(z_0)dz = \left.\frac{df}{dz}\right|_{z=z_0} = 0 = \int_C \frac{dz}{(z-z_0)^{n+1}}, \quad \text{for } n = 1, 2, \ldots,$$

and we have recovered result (ii) in Example 2.1.3, but this time for an arbitrary simple closed contour C enclosing z_0. ◇

Example 2.5.2 An Application to Two Typical Integrals
Evaluate

(i) $\int_C \frac{z^2 \sin z}{(z - \frac{1}{2}\pi)^4} dz$, for C any simple closed contour containing z_0;

(ii) $\int_C \frac{\sin 2z}{(z - \frac{1}{4}\pi)^3 (z^2+9)} dz$, for C the square with corners at $2 + 2i$, $-2 + 2i$, $-2 - 2i$ and $2 - 2i$.

SOLUTION

(i) Making the identification $f(z) = z^2 \sin z$ and $n + 1 = 4$, so $n = 3$ in Theorem 2.5.1, gives

$$\int_C \frac{z^2 \sin z}{(z - \frac{1}{2}\pi)} dz = \frac{2\pi i}{3!} f^{(3)}\left(\tfrac{1}{2}\pi\right), \quad \text{but } f^{(3)}(z) = 6\cos z - 6z\sin z - z^2 \cos z,$$

but $f^{(3)}(\frac{1}{2}\pi) = -3\pi$, so

$$\int_C \frac{z^2 \sin z}{\left(z - \frac{1}{2}\pi\right)^4} dz = -\pi^2 i.$$

(ii) Only the zero of the denominator at $z = \pi/4$ lies inside the contour C, so making the identifications $f(z) = \sin 2z/(z^2 + 9)$, and $n + 1 = 3$, when $n = 2$, we find that

$$\int_C \frac{\sin 2z}{\left(z - \frac{1}{4}\pi\right)^3 (z^2+9)} dz = \frac{2\pi i}{2!} f^{(2)}\left(\tfrac{1}{4}\pi\right),$$

but a routine calculation shows that

$$f^{(2)}\left(\tfrac{1}{4}\pi\right) = -64 \frac{(\pi^4 + 264\pi^2 + 21888)}{(\pi^2 + 144)^3} = -0.432,$$

so

$$\int_C \frac{\sin 2z}{\left(z - \frac{1}{4}\pi\right)^3 (z^2 + 9)} dz = -1.357i.$$

◇

2.5.1 The Poisson Integral Formulas for the Half-Plane and the Circular Disk

Two results that will be of importance when we consider the solution of what are called *boundary value problems* for the Laplace equation are as follows:

(i) Poisson's formula for the harmonic function $u(x, y)$ in the half-plane $y > 0$ takes the form

$$u(x_0, y_0) = \frac{y_0}{\pi} \int_{-\infty}^{\infty} \frac{u(x, 0)}{(x - x_0)^2 + y_0^2} dx,$$

where (x, y) is any point in the half-plane $y > 0$; and

(ii) Poisson's formula for the harmonic function $u(r, \theta)$ a disk of radius R centered on the origin takes the form

$$u(r, \theta) = \frac{1}{2\pi} \int_0^{2\pi} u(R, \psi) \frac{R^2 - r^2}{R^2 - 2rR\cos(\psi - \theta) + r^2} d\psi, \quad (r < R),$$

where (r, θ) are plane polar coordinates and $u(r, \theta)$ of any point in the disk of radius R centered on the origin.

To understand the significance of these results, notice first the important fact that in (i) $u(x, y)$ is harmonic in the half plane $y > 0$, and so is a solution of the Laplace equation

$$\frac{\partial^2 u}{\partial x^2} + \frac{\partial^2 u}{\partial y^2} = 0.$$

Examination of the form of the Poisson formula in (i) shows the value of $u(x, y)$ at any point in the half-plane $y > 0$ is determined by the integral in (i) that depends on the functional variation $u(x, 0)$ *assigned* to $u(x, y)$ on the x-axis. A problem like this is called a boundary value problem for the Laplace equation in the half-plane, because the assignment of the variation of $u(x, y)$ on the boundary (in this case the line $y = 0$ bounding the half-plane $y > 0$) determines the value of $u(x, y)$ throughout the entire half-plane $y > 0$.

The result in (ii) is the corresponding result that determines the value of an harmonic function $u(r, \theta)$ at any point inside the circular disk of radius R centered on the origin, in terms of the functional variation $u(R, \psi)$ assigned to $u(r, \psi)$ on the boundary $r = R$ of the disk. Thus this formula also provides the solution of a boundary value problem for the Laplace equation, but this time in terms of plane polar coordinates.

To derive (i) we need to consider a function $f(z) = u + iv$ that is analytic and bounded in the half-plane $\mathrm{Im}\{z\} > 0$, and in Theorem 2.5.1 to take for the contour C the semicircle $0 < r < R$ in the upper half-plane for which with $0 \leqslant \theta \leqslant \pi$, where $\theta = \mathrm{Arg}(z)$, diameter of the circle that coincides with the real axis (the x-axis). Then from Theorem 2.5.1 with z_0 inside C, we have

$$f(z_0) = \frac{1}{2\pi i} \int_C \frac{f(z)}{z - z_0} dz.$$

Now $\bar{z}_0$ is in the lower half-plane, and so is outside the contour C, so that

$$0 = \frac{1}{2\pi i} \int_C \frac{f(z)}{z - \bar{z}_0} dz.$$

Subtracting these results gives

$$f(z_0) = \frac{1}{2\pi i} \int_C \frac{(z_0 - \bar{z}_0) f(z)}{(z - z_0)(z - \bar{z}_0)} dz.$$

Setting $z_0 = x_0 + iy_0$, $f(z_0) = u(x_0, y_0) + iv(x_0, y_0)$ and $f(z) = u(x, y) + iv(x, y)$, and representing the integral around C as the sum of the integral along the real axis from $x = -R$ to $x = R$, and the integral around the semicircular are of radius R by C_R, we can write

$$u(x_0, y_0) + iv(x_0, y_0) = \frac{1}{2\pi i} \int_{-R}^{R} \frac{2iy_0[u(x,0) + iv(x,0)]}{[(x - x_0) - iy_0][(x - x_0) + iy_0]} dx$$
$$+ \frac{1}{2\pi i} \int_{C_R} \frac{(z - \bar{z}_0) f(z)}{(z - z_0)(z - \bar{z}_0)} dz.$$

Thus

$$u(x_0, y_0) + iv(x_0, y_0) = \frac{1}{\pi} \int_{-R}^{R} \frac{y_0[u(x,0) + iv(x,0)]}{(x - x_0)^2 + y_0^2} dx$$
$$+ \frac{1}{2\pi i} \int_{C_R} \frac{(z - \bar{z}_0) f(z)}{(z - z_0)(z - \bar{z}_0)} dz.$$

As $R \to \infty$, the fact that $f(z)$ is bounded by some constant $M > 0$ in the upper half-plane means that the modulus of the integrand is of the order

$O(M/|z - z_0|)$, so it tends to zero as $|z - z_0| \to \infty$ causing the integral around C_R to vanish, and we are left with the result

$$u(x_0, y_0) + iv(x_0, y_0) = \frac{1}{\pi}\int_{-\infty}^{\infty} \frac{y_0[u(x,0) + iv(x,0)]}{(x - x_0)^2 + y_0^2}\,dx.$$

Result (i) follows by equating the real parts on each side of the equality sign. A corresponding result for $v(x, y)$ follows by equating the imaginary parts on each side of this result, the only difference being that in this case $u(x, y)$ is replaced by $v(x, y)$. This result is not surprising because both u and v are harmonic functions, so each must satisfy an exactly similar expression.

Result (ii) can be derived in a similar fashion, so we only outline the steps involved and leave the details as an exercise. Consider the integral

$$\frac{1}{2\pi i}\int_C \frac{f(z)}{z - \zeta}\,dz,$$

where C is the circle $|z| = R$, and let $z_0 = re^{i\theta}$, where $0 < r < R$ and $\zeta = z\bar{z}/\bar{z}_0$. Then $|\zeta| = |z\bar{z}|/|\bar{z}_0| = R^2/r$ showing that the point ζ lies outside circle C, so that in fact

$$\frac{1}{2\pi i}\int_C \frac{f(z)}{z - \zeta}\,dz = 0.$$

Differencing the expression for $f(z_0)$ found from the Cauchy integral formula, this last result gives

$$f(re^{i\theta}) = \frac{1}{2\pi i}\int_C \frac{1}{z}\frac{(z\bar{z} - z_0\bar{z}_0)}{(z - z_0)(\bar{z} - \bar{z}_0)}\,dz.$$

Result (ii) then follows by setting $f(re^{i\theta}) = u(r, \theta) + iv(r, \theta)$, $z = Re^{i\theta}$ and $z_0 = re^{i\theta}$ in this result, and equating the real parts of the expressions on each side of the equality sign to obtain

$$u(r, \theta) = \frac{1}{2\pi}\int_0^{2\pi} u(R, \psi)\frac{R^2 - r^2}{R^2 - 2rR\cos(\psi - \theta) + r^2}\,d\psi, \quad (r < R).$$

Equating the imaginary parts of the expressions on each side of the equality sign gives a similar result, but with $v(r, \theta)$ in place of $u(r, \theta)$.

Exercises 2.5

1. Find

$$I = \int_C \frac{\sin^2 z}{(z^2 - \frac{1}{36}\pi^2)(z - \frac{1}{6}\pi)}\, dz,$$

where C is the circle $|z| = 4$.

2. Find

$$I = \int_C \frac{\cos z}{z^3}\, dz,$$

where C is the circle $|z| = 2$.

3. Find

$$I = \frac{1}{2\pi i}\int_C \frac{e^z}{z(1-z)^3}\, dz,$$

for all possible choices of simple closed-contour C with respect to the location of the singularities in the integrand.

4. Find

$$I = \int_{C_-} \frac{ze^{az}}{(z-1)^3}\, dz,$$

where C is the circle $|z| = 2$.

5. Find

$$I = \int_{C_-} \frac{e^{iz}}{(z^2 + a^2)(z - i)^2}\, dz,$$

where $a \neq 1$ is a positive number and C is the rectangle with corners at $z = \pm 1$ and $z = \pm 1 + (1 + a)i$.

6. Find

$$I = \int_C \frac{\sinh z}{(z^2 + 1)^2}\, dz,$$

where C is the triangle with vertices at $z = -1$, $z = 1$, and $z = 2i$.

7. Find

$$I = \int_{C_-} \frac{e^{iz}}{(z^2 + a^2)^2}\, dz,$$

where C is the circle $|z - ia| = a$, with $a > 0$ real.

8. Find

$$I = \int_C \frac{z^2}{(z+3)^n(z-2)}\,dz, \text{ for the cases when}$$

(i) $n = 2$ and C is the circle $|z + 3| = 2$;
(ii) $n = 1$ and C is the circle $|z| = 4$;
(iii) $n = 3$ and C is the circle $|z - 1| = \sqrt{2}$.

9. Show by using the Cauchy integral theorem for derivatives of a function $f(z)$ that is analytic inside a simple closed-contour C inside of which is a point z_0, that

$$m!\int_C \frac{f^{(n-m)}(z)}{(z-z_0)^{m+1}}\,dz = n!\int_C \frac{f(z)}{(z-z_0)^{n+1}}\,dz,$$

for $0 \leqslant m \leqslant n$. Verify the result by direct calculation for the case $f(z) = \cos 2z, z_0 = i, m = 2$ and $n = 4$.

Exercises of Greater Difficulty

10. The real variable polynomial $P_n(z)$ defined for $-1 \leqslant x \leqslant 1$ that is used when solving potential problems, and elsewhere, is defined by means of *Rodrigue's formula*

$$P_n(x) = \frac{1}{2^n n!}\frac{d^n}{dx^n}[(x^2-1)^n].$$

Show that in the complex z-plane

$$P_n(z) = \frac{1}{2\pi i}\int_C \frac{(z^2-1)^n}{2^n(z-z_0)^{n+1}}\,dz,$$

where C is any simple closed contour containing the point z_0. This result is called the *Schläfli integral formula* for Legendre polynomials. Use this formula to find the Legendre polynomial $P_3(x)$.

11. By using the Schläfli integral formula in Exercise 10, taking C to be the circle centered on z_0 with radius $r = |z_0^2 - 1|$, setting $(z_0^2 - 1) = re^{i\phi}$ with $0 \leqslant \phi \leqslant 2\pi$ and $z = z_0 + re^{i\theta}$, derive the *Laplace formula* for Legendre polynomials.
12. Find

$$P_n(x) = \frac{1}{\pi}\int_0^{\pi}\left(x + \sqrt{x^2-1}\cos\theta\right)^n d\theta.$$

13. Use this result to find $P_2(x)$ and confirm it is in agreement with the result obtained from Rodrigue's formula in Exercise 10.

2.6 Useful Results Deducible from the Cauchy Integral Formulas

A number of useful theorems can be deduced quite simply from the two Cauchy integral formulas in Theorems 2.4.1 and 2.5.1, the most important of which are given in this section. We start by proving the *mean value theorem* for analytic functions due to Gauss. This theorem asserts that if $f(z)$ is analytic in a domain D, then its value at any point z_0 in D is equal to the average of its values around any circle centered on z_0 that lies entirely within D. A formal statement of the theorem, including an important consequence for real valued harmonic functions now follows.

THEOREM 2.6.1 The Mean Value Theorem
Let $f(z)$ be analytic in a domain D containing a circular disk of radius R with boundary C and its center at the point z_0. Then

(a) $$f(z_0) = \frac{1}{2\pi}\int_0^{2\pi} f(z_0 + Re^{i\theta})d\theta = \frac{1}{2\pi R}\int_C f(z_0 + Re^{i\theta})ds,$$

where $ds = R\,d\theta$ is the element of arc length around C.

(b) If $u(x, y)$ is a real-valued harmonic function with continuous second-order partial derivatives, then

$$u(x_0, y_0) = \frac{1}{2\pi R}\int_C u(x_0 + R\cos\theta, y_0 + R\sin\theta)ds.$$

PROOF

(a) Starting from the Cauchy integral formula in Theorem 2.4.1 we have

$$f(z_0) = \frac{1}{2\pi i}\int_C \frac{f(z)}{z - z_0}dz.$$

Now on C, $z = z_0 + Re^{i\theta}$, for $0 \leqslant \theta \leqslant 2\pi$ and $dz = iRe^{i\theta}d\theta$. Thus the integral formula for $f(z_0)$ becomes

$$\begin{aligned} f(z_0) &= \frac{1}{2\pi i}\int_0^{2\pi} \frac{f(z_0 + Re^{i\theta})iRe^{i\theta}}{Re^{i\theta}}d\theta \\ &= \frac{1}{2\pi}\int_0^{2\pi} f(z_0 + Re^{i\theta})d\theta, \end{aligned}$$

which is the required result. The second result in (a) follows from the fact that the element ds of arc length on C is $ds = R\,d\theta$.

(b) If $u(x, y)$ is harmonic it must have a harmonic conjugate, and so may be regarded as the real part of an analytic function $f(z)$. The result in (b) follows by taking the real part of the result in (a). ◆

The next result we prove is the important and useful *Maximum Modulus theorem* for analytic functions. As well as applications throughout complex analysis, this theorem has a real variable counterpart that is used in the study of Laplace's equation.

THEOREM 2.6.2 The Maximum Modulus Theorem
Let $f(z)$ be analytic in a bounded domain D and continuous on its boundary C, which is a simple closed curve. Then,

(a) $f(z)$ is constant in D if $|f(z)|$ attains its maximum value at an interior point of D.
(b) If $f(z)$ is not constant in D, then the maximum value of $|f(z)|$ must occur on the boundary C of D.

PROOF

Let $\bar{D}$ denote the domain D together with its simple closed-boundary curve C. Then, because $f(z)$ is analytic in $\bar{D}$, it is also continuous and so also is the real-valued function $|f(z)|$ that will thus attain a maximum value M at some point $z_0 \in \bar{D}$, so we have

$$|f(z)| \leqslant |f(z_0)| = M.$$

Suppose z_0 lies inside D. Then as z_0 lies at a finite distance from C, it is possible to construct a circle C_0, of radius $\rho > 0$ and centered on z_0, that lies entirely in D. Applying the mean value theorem to $f(z)$ at z_0, and taking C_0 for the circle in the theorem, we find that

$$f(z_0) = \frac{1}{2\pi}\int_0^{2\pi} f(z_0 + \rho e^{i\theta})d\theta.$$

Taking the modulus of this result gives

$$|f(z_0)| = \frac{1}{2\pi}\left|\int_0^{2\pi} f(z_0 + \rho e^{i\theta})d\theta\right| \leqslant \frac{1}{2\pi}\int_0^{2\pi} |f(z_0 + \rho e^{i\theta})|\,d\theta,$$

but from the condition $|f(z)| \leqslant |f(z_0)| = M$ this is seen to be equivalent to

$$2\pi M \leqslant \int_0^{2\pi} |f(z_0 + \rho e^{i\theta})| d\theta \leqslant 2\pi M,$$

and so

$$\int_0^{2\pi} |f(z_0 + \rho e^{i\theta})| d\theta = 2\pi M.$$

We now show that this implies $|f(z)| = M$ for all z on the circle C_0, which is equivalent to the condition that $|f(z_0 + \rho e^{i\theta})| = M$ for $0 \leqslant \theta \leqslant 2\pi$. Now suppose, if possible, that this is not so, and for θ in an interval $\alpha \leqslant \theta \leqslant \beta$,

$$|f(z_0 + \rho e^{i\theta})| \leqslant M - \varepsilon, \qquad \text{for some } \varepsilon > 0.$$

Then,

$$\begin{aligned}\int_0^{2\pi} |f(z_0 + \rho e^{i\theta})| d\theta &= \int_0^{\alpha} |f(z_0 + \rho e^{i\theta})| d\theta + \int_\alpha^{\beta} |f(z_0 + \rho e^{i\theta})| d\theta \\ &\quad + \int_\beta^{2\pi} |f(z_0 + \rho e^{i\theta})| d\theta \\ &\leqslant M\alpha + (M - \varepsilon)(\beta - \alpha) + M(2\pi - \beta) \\ &= M[2\pi - (\beta - \alpha)] + (M - \varepsilon)(\beta - \alpha) < 2\pi M,\end{aligned}$$

but this contradicts the condition $|f(z)| \leqslant |f(z_0)| = M$. Consequently, $|f(z)| = M$ for all points z on the circle C_0, where we recall that M is the maximum value of $|f(z)|$ for $z \in \bar{D}$. This situation remains true for any other circle centered on z_0 and lying inside C_0, from which we conclude that $|f(z)| = M$ for all z inside and on C_0.

Before establishing that this result is true throughout $\bar{D}$, we first show it to be true in a domain D_1 belonging to $\bar{D}$ that is bounded by a simple closed curve containing z_0 and a point ζ on the boundary C of $\bar{D}$. Join the points z_0 and ζ by a simple arc γ, and construct overlapping circles $C_0, C_1, \ldots, C_n$ all strictly inside D_1, as shown in Figure 2.18. The circles are drawn progressively along γ with

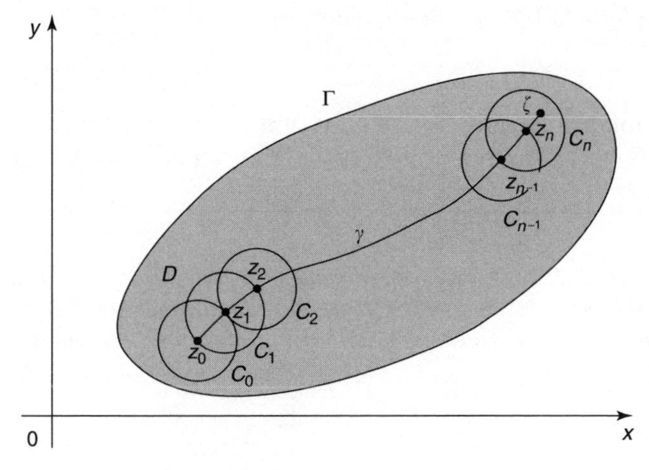

FIGURE 2.18
The arc γ and the overlapping circles $C_0, C_1, \ldots, C_n$.

C_0 having its center at z_0, C_1 its center at z_1 where C_0 intersects γ, and so on, until the process is terminated by drawing a circle with its center at z_n that contains the point ζ on C on its circumference.

We have already proved that $|f(z)| = M$ inside C_0, but disks C_0 and C_1 have a common area of intersection, so the argument that proved $|f(z)| = M$ inside C_0 also shows it is true inside C_1. A repetition of this argument shows that $|f(z)| = M$ in the domain D_1 formed by the interior and boundary points of the circles $C_0, C_1, \ldots, C_n$. As ζ was an arbitrary point on the boundary C of $\bar{D}$, we conclude that if $|f(z)|$ has its maximum point at an interior point of $\bar{D}$, then $|f(z)| = M$ for all $z \in \bar{D}$.

It remains to show that this implies $f(z) \equiv$ constant in $\bar{D}$. Setting $f = u + iv$, we have $|f(z)|^2 = u^2 + v^2$, and it has already been proved that $u^2 + v^2 = M^2$ throughout $\bar{D}$. Partial differentiation of this result, first with respect to x and then with respect to y, gives

$$u\frac{\partial u}{\partial x} + v\frac{\partial v}{\partial x} = 0 \quad \text{and} \quad u\frac{\partial u}{\partial y} + v\frac{\partial v}{\partial y} = 0.$$

Using the Cauchy–Riemann equations, these equations are easily seen to have the unique solution $u_x = u_y = v_x = v_y = 0$ throughout $\bar{D}$. This implies $u \equiv$ constant and $v \equiv$ constant, and so $f(z) \equiv$ constant for all $z \in \bar{D}$.

The remainder of the proof is straight forward because if $|f(z)|$ is not constant inside $\bar{D}$, it cannot attain a maximum value at an interior point, so the continuity of $|f(z)|$ implies that $|f(z)|$ must attain a maximum on the boundary C. ◆

COROLLARY 2.6.1(1)

If $f(z)$ is analytic and nonvanishing in a bounded domain D and continuous on its boundary C, then $|f(z)|$ attains its minimum value on C.

PROOF

The result follows directly from the maximum principle by considering the function $F(z) = 1/f(z)$, that will be analytic if $f(z)$ is analytic and such that $f(z) \neq 0$ in D. ◆

COROLLARY 2.6.2(2) The Maximum/Minimum Principle for Harmonic Functions

If $u(x, y)$ is harmonic in a bounded domain D and continuous on the boundary C of D, then either $u(x, y) \equiv$ constant in D, or it attains its maximum and minimum values on C.

PROOF

As $u(x, y)$ is harmonic, it follows that an harmonic conjugate function $v(x, y)$ exists with the same continuity and differentiability properties as $u(x, y)$. It is easily seen that the function $f(z) = e^u e^{iv}$ satisfies the conditions of Theorem 2.6.1, so either $|f(z)| = e^u \equiv$ constant in D and on its boundary C, or e^u attains its maximum value on the boundary C. Thus if $|f(z)| = e^u \equiv$ constant in D and on its boundary C, it follows immediately that $e^u \equiv$ constant, and hence that $u \equiv$ constant in D and on its boundary C.

If, however, $|f(z)| = e^u \equiv$ constant in D, then the maximum value of e^u must occur on C, but e^u is a monotonic increasing function of u, so that u itself must also attain its maximum value on C. As $f(z)$ never vanishes, Corollary 2.2.6(1) implies that when $u(x, y)$ is non constant in D, its minimum must also occur on C. The proof is complete, but it should be noticed that, unlike Corollary 2.2.6(1), when considering a real harmonic function $u(x, y)$ it is not necessary to require that $u(x, y) \neq 0$ in D. ◆

Example 2.6.1 An Application to cos *z* in the Unit Circle

Confirm that $\cos z$ in $|z| \leq 1$ obeys the maximum/minimum property of Theorem 2.2.6 and Corollary 2.2.6(1) for analytic functions, so that $\max|f(z)|$ and $\min|f(z)|$ occur on the boundary of the unit circle. Use elementary real variable methods to confirm this.

SOLUTION

The function $f(z) = \cos z$ is analytic and bounded in the unit disk, and such that $\cos z \neq 0$, so Theorem 2.2.6 and Corollary 2.2.6(1) apply showing that the extrema of $|\cos z|$ must occur on the boundary of the unit circle. To establish this by elementary real variable methods set $z = x + iy$, when

$$\begin{aligned} |\cos z| &= |\cos x \cosh y - i \sin x \sinh y| \\ &= (\cos^2 x + \sinh^2 y)^{1/2}. \end{aligned}$$

To simplify the analysis, we square this result and consider the real variable function $w = \cos^2 x + \sinh^2 y$. The extrema of $|\cos z|$ and of w will correspond, because $|\cos z|$ is non-negative.

Any stationary points of w inside $|z| = 1$ can only occur when $w_x = w_y = 0$. We have

$$\frac{\partial w}{\partial x} = -2\cos x \sin x = -\sin 2x,$$

and

$$\frac{\partial w}{\partial y} = 2\sinh y \cosh y = \sinh 2y.$$

Thus the only stationary point inside $|z| = 1$ occurs when $\sin 2x = 0$ and $\sinh 2y = 0$, corresponding to the origin. From the study of extrema in elementary calculus, in order to identify the nature of this stationary point, it is necessary to consider the discriminant $\Delta(x, y) = (\partial^2 w/\partial x^2)(\partial^2 w/\partial y^2) - \partial^2 w/\partial x\partial y$. This turns out to be $\Delta(x, y) = -4\cos 2x \cosh 2y$, so at the origin $\Delta(0, 0) = -4 < 0$, showing that the origin is a saddle point and so neither a maximum or a minimum of $|\cos z|$.

The unit circle has the equation $x^2 + y^2 = 1$, so on this boundary

$$w = \cos^2 x + \sinh^2(1 - x^2)^{1/2}, \quad \text{for } -1 \leqslant x \leqslant 1.$$

Both $\cos^2 x$ and $\sinh^2(1 - x^2)^{1/2}$ are monotonic decreasing functions of $|x|$ for $0 \leqslant |x| \leqslant 1$, so the maxima of both w and $|\cos z|$ occur when $x = 0$ and $y = \pm 1$, while the minima occur when $x = \pm 1$ and $y = 0$.

On the boundary $|f(z)| = w^{1/2}$, so at the maxima $|f(z)| = 1.5431$, while at the minima $|f(z)| = 0.5403$ and at the saddle point at the origin $|f(z)| = 1$, confirming the statements in Theorem 2.2.6 and Corollary 2.2.6(1). ◇

Example 2.6.2 An Application to cos *z* in a Circle of Radius π

Confirm that the function $f(z) = \cos z$ satisfies the maximum principle in the disk $|z| \leqslant \pi$, but not the minimum principle.

SOLUTION

The function $\cos z$ is analytic bounded and continuous in the disk $|z| \leqslant \pi$, but it has zeros at $z = \pm\frac{1}{2}\pi$. Thus although the maximum principle in Theorem 2.2.6 applies, the vanishing of $\cos z$ inside the disk means that Corollary 2.2.6(1) is not applicable, so only the maximum of $|\cos z|$ will occur on the boundary of the disk.

To examine the situation further, we proceed as in Example 2.6.1, but now the stationary points occurs inside the disk if $\sin 2x = 0$ and $\sinh 2y = 0$ for $|z| \leqslant \pi$. This shows that the stationary points occur at the origin and at the points $(\pm\frac{1}{2}\pi, 0)$. As before, the origin is a saddle point at which $|\cos z| = 1$, but at the points $(\pm\frac{1}{2}\pi, 0)$ the discriminant $\Delta\,(\pm\frac{1}{2}\pi, 0) = 4 > 0$, so as $\partial^2 w/\partial x^2 = 2 > 0$ at these points, it follows that they must be minima of w, and so also minima of $|\cos z|$. We have already seen that $|\cos z| = 0$ at these minima, so it remains for us to examine what happens on the boundary of the disk. Reasoning as in Example 2.6.1, we find that maxima of $|\cos z|$ occur at the points $(0, \pm\pi)$ where $|\cos z| = 1.592$, while minima of $|\cos z|$ occur at the points $(\pm\pi, 0)$ where $|\cos z| = 1$. So, although $|\cos z|$ still attains its maximum value on the boundary, its minimum value on the boundary is greater than its minimum values inside the disk, confirming the fact that Corollary 2.2.6(1) is not applicable. ◇

Example 2.6.3 The Maximum/Minimum Principle

Verify that $u(x, y) = e^x \cos y$ for $0 \leqslant x \leqslant 1$, $-\frac{1}{2}\pi \leqslant y \leqslant \frac{1}{2}\pi$ obeys Corollary 2.2.6(2).

SOLUTION

It is easily confirmed that $u(x, y)$ is harmonic with continuous second-order partial derivatives in the given rectangle, so it satisfies the conditions of Corollary 2.2.6(2). Clearly the function has no stationary points inside this rectangle because as $u_x = e^x \cos y$ and $u_y = -e^x \sin y$, the conditions for a stationary point $u_x = u_y = 0$ can never be satisfied.

Inspection shows that the minimum value of u is attained all along the boundaries $0 \leqslant x \leqslant 1, y = -\frac{1}{2}\pi$ and on $0 \leqslant x \leqslant 1, y = \frac{1}{2}\pi$ where $u = 0$, while the maximum value occurs at the point $(1, 0)$ where $u = e$, in agreement with Corollary 2.2.6(2). ◇

The theorem that follows, called *Morera's theorem*, is the converse of the Cauchy–Goursat theorem, and it provides sufficient conditions for a function to be analytic in a domain.

THEOREM 2.6.3 Morera's Theorem

Let $f(z)$ be a continuous function in a simply connected domain D, and such that

$$\int_C f(z)dz = 0$$

for every simple closed curve C in D. Then $f(z)$ is analytic in D.

PROOF

It has already been established that if points z_0 and $z \in D$ and the path joining them lies in D, then for a function $f(z)$ satisfying the conditions of the theorem the function

$$F(z) = \int_{z_0}^{z} f(\zeta)d\zeta$$

is analytic in D and such that $F'(z) = f(z)$. Thus $f(z)$ is seen to be the derivative of an analytic function and so must itself be analytic, so the proof is complete. ◆

The next theorem, due to *Liouville*, has a number of applications one of which we will use to establish the *fundamental theorem of algebra.*

THEOREM 2.6.4 Liouville's Theorem

If a function $f(z)$ is analytic and bounded throughout the complex plane, then $f(z)$ is constant.

PROOF

Let C be a circle of arbitrary radius R centred on an arbitrary point z_0 in the complex plane. Then it follows from the Cauchy integral theorem for derivatives that

$$f'(z_0) = \frac{1}{2\pi i}\int_C \frac{f(z)}{(z-z_0)^2}\,dz.$$

By hypothesis, if $f(z)$ is bounded for all z, then so also is $|f(z)|$, and thus a positive constant M *exists* such that $|f(z)| < M$ for all z. Setting $z = z_0 + Re^{i\theta}$, for $0 \leqslant \theta \leqslant 2\pi$ so that as $dz = iRe^{i\theta}\,d\theta$, we have

$$\begin{aligned} f'(z_0) &= \frac{1}{2\pi i}\int_0^{2\pi} \frac{f(z)iRe^{i\theta}}{R^2e^{2i\theta}}\,d\theta, \\ &= \frac{1}{2\pi i}\int_0^{2\pi} \frac{f(z)}{e^{i\theta}}\,d\theta. \end{aligned}$$

Taking the modulus of each side of this equation we find that

$$\begin{aligned} |f'(z_0)| &< \frac{1}{2\pi R}\int_0^{2\pi} |f(z)|\,dz \\ &< \frac{M}{2\pi R}\int_0^{2\pi} d\theta = \frac{M}{R}. \end{aligned}$$

This result must be true for any R and any z_0, so by taking R arbitrarily large we conclude that $|f'(z)| = 0$ for all z_0 in the complex plane. However, this is only possible if $f(z) \equiv$ constant, so the theorem is proved. ◆

An immediate implication of Liouville's theorem is that if a nonconstant function is analytic in the entire z-plane, then it cannot be bounded. Simple examples of this type are provided by the complex trigonometric functions $\sin z$ and $\cos z$. When z is purely real, these functions are bounded for all real x, but in the complex plane we have

$$|\sin z| = (\cosh^2 y - \cos^2 x)^{1/2}, \quad \text{and} \quad |\cos z| = (\cos^2 x + \sinh^2 y)^{1/2},$$

which become unbounded when $|z|$ becomes large. These functions are, of course, examples of the *entire functions* encountered previously, another example of which is a complex polynomial.

The next theorem, called the *fundamental theorem of algebra*, is a key algebraic result that says every polynomial equation of degree n has n roots. This result will already have been used many times without question in more elementary work, and indeed the explicit forms of the two roots of complex quadratic equations have already been found, but a formal proof that the result is true for an arbitrary complex polynomial equation of degree n is lacking, and so needs to be given. Liouville's theorem provides one of the ways of proving this important general result.

THEOREM 2.6.5 The Fundamental Theorem of Algebra
Any polynomial

$$P(z) = z^n + a_{n-1}z^{n-1} + a_{n-2}z^{n-2} + \cdots + a_0,$$

where the coefficients $a_0, a_1, \ldots, a_n$ may be either real or complex, has precisely n zeros when repeated zeros are counted according to their multiplicity. The zeros of $P(z)$ may be either real or complex.

PROOF

This proof, based on Liouville's theorem, is a proof by contradiction. Suppose, if possible, that $P(z)$ has no zeros in the finite complex plane, and consider the function

$$f(z) = \frac{1}{P(z)} = \frac{1}{z^n}\left(\frac{1}{1 + \dfrac{a_{n-1}}{z} + \dfrac{a_{n-2}}{z^2} + \cdots + \dfrac{a_0}{z^n}}\right).$$

Then if $P(z)$ has no zeros in the finite z-plane it follows that $f(z)$ must be an entire function. Taking the modulus of $f(z)$ gives

$$|f(z)| = \frac{1}{|P(z)|} = \frac{1}{|z|^n}\left(\frac{1}{\left|1 + \dfrac{a_{n-1}}{z} + \dfrac{a_{n-2}}{z^2} + \cdots + \dfrac{a_0}{z^n}\right|}\right).$$

Now for z on the circle $|z| = R$,

$$\lim_{R\to\infty} \frac{1}{|z|^n} = 0 \quad \text{and} \quad \lim_{R\to\infty}\left(\frac{1}{\left|1 + \dfrac{a_{n-1}}{z} + \dfrac{a_{n-2}}{z^2} + \cdots + \dfrac{a_0}{z^n}\right|}\right) = 1,$$

so that

$$\lim_{|z|\to\infty} |f(z)| = 0 \cdot 1 = 0.$$

This limit asserts that for any $\varepsilon > 0$ there is a number $R > 0$ such that

$$|f(z)| < \varepsilon \quad \text{for } |z| > R,$$

and so $f(z)$ must be bounded outside $|z| = R$. However, $f(z)$ is continuous so it is also bounded inside $|z| = R$, and hence it is bounded everywhere in the finite complex z-plane. Liouville's theorem asserts that such a function $f(z)$

must be an absolute constant, but clearly this is not so, and thus the assumption that $P(z)$ has no zeros in the finite complex z-plane is false.

If z_1 is a zero of $P(z)$, then $(z - z_1)$ is a factor and

$$P(z) = (z - z_1)Q_{n-1}(z),$$

where $Q_{n-1}(z)$ is a polynomial in z of degree $n - 1$. The same form of argument shows that $Q_{n-1}(z)$ must also have a zero, say z_2, and a corresponding factor $(z - z_2)$, so

$$P(z) = (z - z_1)(z - z_2)Q_{n-2}(z),$$

where now $Q_{n-2}(z)$ is a polynomial in z of degree $n - 2$. Proceeding in this way, it follows that

$$P(z) = (z - z_1)(z - z_2)\cdots(z - z_n),$$

showing that there are n zero $z_1, z_2, \ldots, z_n$ of $P(z)$, and these may be either real or complex. When these zeros are counted according to their multiplicity, it follows that every polynomial of degree n has precisely n zeros and, equivalently, that every polynomial equation of degree n has precisely n roots. The proof is complete. ◆

On occasion it is necessary to estimate the modulus of a derivative of an analytic function, and this can be done with the help of the Cauchy formula for derivatives, as established in the next theorem.

THEOREM 2.6.6 The Cauchy Inequality

Let $f(z)$ be analytic in a simply connected domain D containing a circular contour C of radius R centered on z_0. If at each point z on $C|f(z)| \leq M$, then

$$|f^{(n)}(z_0)| \leq \frac{n!M}{R^n}, \quad \text{for } n = 1, 2, \ldots .$$

PROOF

The starting point is the Cauchy integral formula for derivatives in Theorem 2.5.1. On the contour C, we set $z = z_0 + Re^{i\theta}$ for $0 \leq \theta \leq 2\pi$, in which case $dz = iRe^{i\theta}\,d\theta$. As a result we find that

$$f^{(n)}(z_0) = \frac{n!}{2\pi i}\int_C \frac{f(z)}{(z - z_0)^{n+1}}\,dz = \frac{n!}{2\pi i}\int_0^{2\pi} \frac{f(z_0 + Re^{i\theta})iRe^{i\theta}}{R^{n+1}e^{i(n+1)\theta}}\,d\theta.$$

Taking the modulus of this result gives

$$|f^{(n)}(z_0)| \leqslant \frac{n!}{2\pi R^n}\int_0^{2\pi} |f(z_0 + Re^{i\theta})|\, d\theta \leqslant \frac{n!M}{2\pi R^n}\int_0^{2\pi} d\theta = \frac{n!M}{R^n},$$

and the theorem is proved. ◆

Example 2.6.4 An Application of the Cauchy Inequality
Given $f(z) = 1/(z+1)$, verify the Cauchy inequality if z lies on the circle $|z-4| = 3$.

SOLUTION
The function $f(z)$ is analytic for $z \neq -1$, so because it is analytic both inside and on the circle C given by $|z-4| = 3$ and shown in Figure 2.19, the Cauchy inequality may be applied. First, let us find the true value of $|f^{(n)}(z)|$ at the center of the circle. Differentiation of $f(z)$ gives

$$f^{(1)}(z) = \frac{-1}{(z+1)^2},\quad f^{(2)}(z) = \frac{2!}{(z+1)^3},\quad f^{(3)}(z) = \frac{-3!}{(z+1)^4},\ldots,$$

so in general

$$f^{(n)}(z) = \frac{(-1)^n n!}{(z+1)^{n+1}}.$$

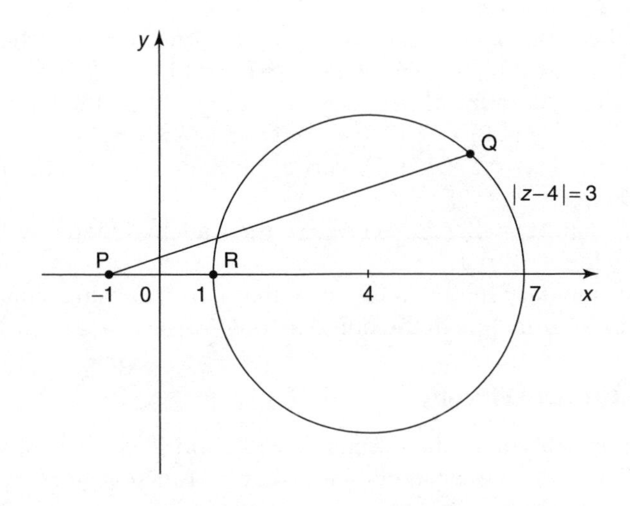

FIGURE 2.19
The circle C.

Thus

$$|f^{(n)}(z)| = \frac{n!}{|z+1|^n}, \quad \text{and hence when } z = 4, \quad |f^{(n)}(4)| = \frac{n!}{5^{n+1}}.$$

We now apply Theorem 2.6.6 to find the over estimate it provides of the true value of $|f^{(n)}(4)|$. Identifying the parameters in the theorem with the problem shows that $z_0 = 4$, $R = 3$ and M is an upper bound of $|f(z)| = 1/|z+1|$ when z lies on the circle C. Inspection of Figure 2.17 shows that $|z+1| = PQ$, so an over estimate of $|f(z)|$ when z is on C will be given by using the smallest possible value of $|z+1|$, that is seen to be $\min_{z \in C} |z+1| = PR = 2$. Thus $|f(z)| \leqslant 1/2$ when z lies on C, and so from the Cauchy inequality we have $|f^{(n)}(4)| \leqslant n!/(2 \cdot 3^n)$. Forming the quotient of this estimate provided by the theorem and the true value gives

$$\frac{n!}{2 \cdot 3^n} \Big/ \frac{n!}{5^{n+1}} = \frac{5^{n+1}}{2 \cdot 3^n} = \frac{3}{2}\left(\frac{5}{3}\right)^{n+1} > 1.$$

As this quotient is greater than 1, it confirms that the theorem provides an over estimate of the modulus of the derivative for $n = 1, 2, \ldots$. ◇

Exercises 2.6

In Exercises 1 and 2, verify the mean value theorem for the function $f(z)$, using the given value of z_0.

1. $f(z) = 3 + 5z$ and $z_0 = 0$.
2. $f(z) = 2 + 3z$ and $z_0 = 1$.
3. Find the maxima and minima of $u(x, y) = \sin x \cosh y$ on the square with corners at $(0, 0)$, $(\frac{1}{2}\pi, 0)$, $(\frac{1}{2}\pi, \frac{1}{2}\pi)$ and $(0, \frac{1}{2}\pi)$.
4. Find the maximum of $u(x, y) = \cos 2x \sinh 2y$ on the square with corners at $(-\frac{1}{4}\pi, 0)$, $(\frac{1}{4}\pi, 0)$, $(\frac{1}{4}\pi, \frac{1}{2}\pi)$ and $(-\frac{1}{4}\pi, \frac{1}{2}\pi)$.
5. Given that $f(z) = 1/(2z + 1)$, verify the Cauchy inequality for z on the circle $|z - 3| = 2$.
6. Given that $f(z) = 2/(3z - 1)$, verify the Cauchy inequality for z on the circle $|z + 1| = 1$.
7. Use elementary methods to verify that $f(z) = 1 + \sin z$ satisfies the maximum principle in the unit disk $|z + \frac{1}{2}\pi| \leqslant 1$.

Exercises of Greater Difficulty

8. If $g(z)$ is analytic in the domain $|z| < R$, and if $|a| < r < R$, where r and R are positive constants, show that the function $g(z)/(r^2 - z\bar{a})$ is analytic for $|z| \leqslant r$.

Hence prove that

$$g(a) = \frac{(r^2 - a\bar{a})}{2\pi i} \int_C \frac{g(z)}{(z-a)(r^2 - \bar{a}z)} dz,$$

where C is the circle $|z| = r$.

9. By considering the function $g(z) = f(z)/z$, and using the maximum modulus principle, prove that if $f(0) = 0$, $f(z)$ is analytic for $|z| < 1$ and $|f(z)| < 1$, then $|f(z)| \leqslant |z|$ and $|f'(0)| \leqslant 1$. This result is known as the *Schwarz lemma*. Prove that if $|f(z)| = |z|$ at an interior point of $|z| = 1$, then $f(z) = ze^{i\alpha}$ for some α.
10. Let $f(z)$ be analytic for $|z| < R$ and such that $|z| \leqslant M$ on the circle $|z| = R$. Prove that

$$|f^{(m)}(re^{i\theta})| \leqslant \frac{m!MR}{(R-r)^{m+1}} \quad \text{for } m = 0, 1, 2, \ldots, \quad \text{and} \quad 0 \leqslant r < R.$$

11. Let $f(z)$ be analytic everywhere in the complex plane, and take a and b to be any two distinct points inside the circle C with equation $|z| = R$, so by the Cauchy integral theorem

$$f(a) = \frac{1}{2\pi i} \int_C \frac{f(z)}{z-a} dz \quad \text{and} \quad f(b) = \frac{1}{2\pi i} \int_C \frac{f(z)}{z-b} dz.$$

By considering the difference $f(a) - f(b)$, prove that if $f(z)$ is everywhere bounded, then $f(z) \equiv$ constant.

12. Let $f(z)$ be analytic for all z and such that $\lim_{z\to\infty} [f(z)/z] = 0$. By taking z_0 arbitrarily and expressing $f'(z_0)$ in terms of the Cauchy integral theorem for derivatives, prove that $f(z) \equiv$ constant.
13. Use the Cauchy integral theorem for derivatives to prove that if $f(z)$ is analytic for all z, and such that $|f(z)| \leqslant M|z|^n$ for large $|z|$, where M is a constant and n is an integer, then $f(z)$ is a polynomial in z whose degree does not exceed n.
14. Prove that if $f(z)$ is continuous in a finite simply connected domain D, and is analytic in the domain D^* obtained from D by deleting a single point z_0, then $\int_C f(z)dz = 0$ for all simple contours C in D. Use this result with Morera's theorem to prove the following *extension of Morera's theorem*. If $f(z)$ is continuous in D and analytic in D^*, then $f(z)$ is analytic everywhere in D. (*Hint*: Deform any contour C containing z_0 into a circle of radius δ centered on z_0, and use the fact that the continuity of $f(z)$ in D implies the boundedness of $|f(z)|$.)

2.7 Evaluation of Improper Definite Integrals by Contour Integration

Contour integration finds many applications; one of the simplest being the evaluation of convergent improper integrals of the form

$$\int_{-\infty}^{\infty} g(x)dx \quad \text{and} \quad \int_{0}^{\infty} g(x)dx, \tag{2.38}$$

provided the real function $g(x)$ can be related to the real part of a suitable complex function $f(z)$. In its simplest form, the approach involves integrating a complex function $f(z)$ around a simple closed-contour Γ_R bounding a domain D of the type shown in either Figure 2.20(a) or (b), and then using either the Cauchy integral formula or the Cauchy integral formula for derivatives to determine the value of the complex integral. The result follows by letting $R \to \infty$ and, if appropriate, $r \to 0$, in order to arrive at a result for a real integral such as Equation (2.38) after equating corresponding real or imaginary parts of the result.

In Figure 2.20(a) the contour Γ_R is traversed counterclockwise (positively) first along the line OA on the real axis, then around the arc C_R from A to B and, finally, along the radial line B to O. In Figure 2.20(b) the contour Γ_R is along the real axis from A to B, then around the circular arc C_R, back along the radial line CD and, finally, around the circular arc γ_R from D to A.

The contour Γ_R in Figure 2.20(a) is used when $f(z)$ has no singularities on any part of the contour, while the contour Γ_R in Figure 2.20(b) is used when a singularity of $f(z)$ occurs at the origin. The modification of the contour at the origin in Figure 2.20(b) is in order to exclude the singularity at the origin and is called *indenting* the contour.

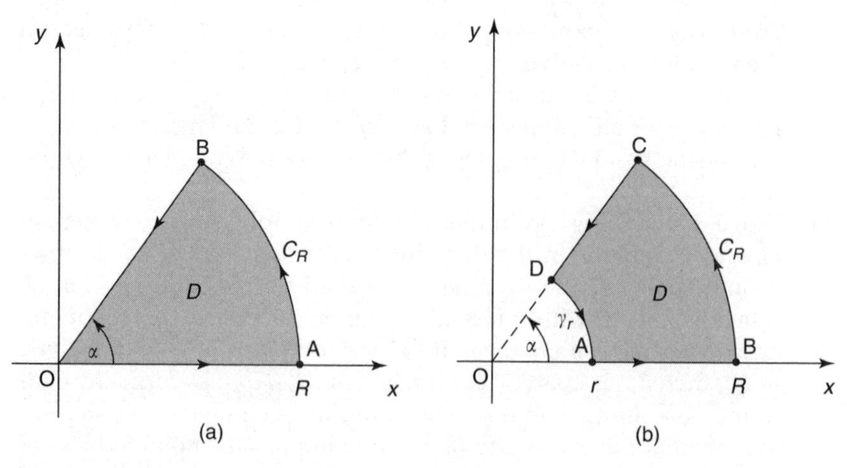

FIGURE 2.20
(a) D is a sector. (b) D is part of an annulus.

The value of the integral of $f(z)$ around Γ_R determined by the Cauchy–Goursat theorem is equal to the sum of the integrals around each part of the contour Γ_R, and for suitable integrals the integral around C_R tends to zero as $R \to \infty$, and if the contour is indented at the origin, the integral around the arc γ_R can be found in a straight forward manner. Combining these results and allowing $R \to \infty$ and, when necessary, $r \to 0$, leads to the determination of an improper integral involving a real function $g(x)$ of the type shown in Equation (2.38) as the real part of the complex result.

Not every convergent real integral can be evaluated by means of contour integration, and a result of this type that is often needed when working with contour integration is related to *error function*. For the sake of completeness we now derive this result before proceeding to the derivation of *Jordan's lemma* which, if satisfied, by a complex integrand $f(z)$, is sufficient to guarantee that the integral of $f(z)$ around a circular arc C_R like those in Figure 2.20 will vanish as $R \to \infty$. In the process we will also derive the *Jordan inequality*.

2.7.1 A Fundamental Definite Integral

$$\int_{-\infty}^{\infty} \exp[-\tfrac{1}{2}(x/\sigma)^2]dx = \sigma\sqrt{2\pi}, \tag{2.39}$$

from which it follows that

$$\int_0^{\infty} \exp(-x^2)dx = \tfrac{1}{2}\sqrt{\pi}. \tag{2.40}$$

PROOF

Consider the integral

$$I_R = \int_{-R}^{R} \exp[-\tfrac{1}{2}(x/\sigma)^2]dx, \quad (R > 0).$$

Notice that by changing the dummy variable of integration from x to y, I_R can also be written as

$$I_R = \int_{-R}^{R} \exp[-\tfrac{1}{2}(y/\sigma)^2]dy, \quad (R > 0),$$

so I_R^2 can be written as the product of the two ordinary integrals

$$I_R^2 = \left(\int_{-R}^{R} \exp[-\tfrac{1}{2}(x/\sigma)^2]dx\right)\left(\int_{-R}^{R} \exp[-\tfrac{1}{2}(y/\sigma)^2]dy\right),$$

or, equivalently, as the double integral

$$I_R^2 = \int_{-R}^{R}\int_{-R}^{R} \exp[-(1/2\sigma^2)(x^2 + y^2)]dx\,dy.$$

Examination of Figure 2.21 shows that I_R^2 may be interpreted as the volume above the square $|x| \leq R$, $|y| \leq R$ in the (x, y)-plane and below the surface $\exp[-(1/2\sigma^2)(x^2 + y^2)]$. The integrand of I_R^2 is non-negative, so it follows that the integral representing I_R^2 will be bounded below by the integral of the same integrand over the inscribed circle C_1 of radius R, and bounded above by the integral of the same integrand over the circumscribed circle C_2 of radius $R\sqrt{2}$.

In terms of plane polar coordinates (r, θ) the integral over C_1 is

$$I_1 = \int_0^R \int_0^{2\pi} \exp[-\tfrac{1}{2}(r/\sigma)^2 r\, d\theta\, dr = \int_0^{2\pi} d\theta \int_0^R r \exp[-\tfrac{1}{2}(r/\sigma)^2]dr,$$

while the integral over C_2 is

$$I_2 = \int_0^{R\sqrt{2}} \int_0^{2\pi} \exp[-\tfrac{1}{2}(r/\sigma)^2 r\, d\theta\, dr = \int_0^{2\pi} d\theta \int_0^{R\sqrt{2}} r \exp[-\tfrac{1}{2}(r/\sigma)^2]dr.$$

Now

$$I_1 = 2\pi \int_0^R r \exp[-\tfrac{1}{2}(r/\sigma)^2]dr = 2\pi\sigma^2\left\{1 - \exp\left[-\tfrac{1}{2}(R/\sigma)^2\right]\right\},$$

and

$$I_2 = 2\pi \int_0^{R\sqrt{2}} r \exp[-\tfrac{1}{2}(r/\sigma)^2]dr = 2\pi\sigma^2\left\{1 - \exp\left[-\tfrac{1}{2}(R\sqrt{2}/\sigma)^2\right]\right\},$$

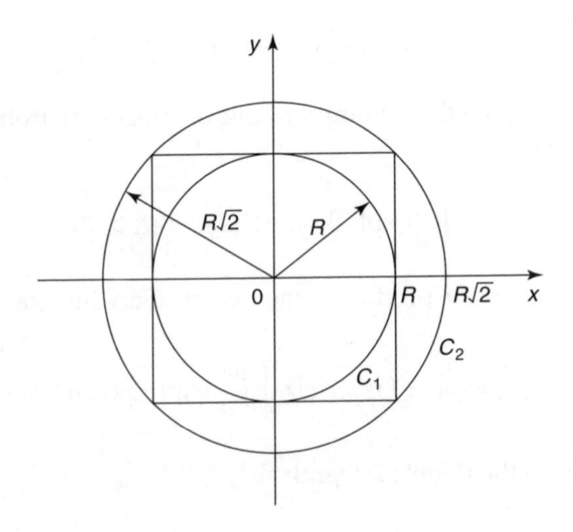

FIGURE 2.21
The area of integration.

and so

$$2\pi\sigma^2\left\{1 - \exp\left[-\tfrac{1}{2}(R/\sigma)^2\right]\right\} < \left\{\int_{-\infty}^{\infty} \exp\left[-\tfrac{1}{2}(x/\sigma)^2\right] dx\right\}$$
$$< 2\pi\sigma^2\left\{1 - \exp\left[-\tfrac{1}{2}(R\sqrt{2}/\sigma)^2\right]\right\}.$$

Allowing $R \to \infty$ causes the left and right expressions in this inequality to tend to $2\pi\sigma^2$, so taking the square root we find that

$$\int_{-\infty}^{\infty} \exp\left[-\tfrac{1}{2}(x/\sigma)^2\right] dx = \sigma\sqrt{2\pi}.$$

The integrand is an even function, so this can be written as

$$\int_{-\infty}^{\infty} \exp\left[-\tfrac{1}{2}(x/\sigma)^2\right] dx = 2\int_{0}^{\infty} \exp\left[-\tfrac{1}{2}(x/\sigma)^2\right] dx = \sigma\sqrt{2\pi},$$

and so

$$\int_{0}^{\infty} \exp\left[-\tfrac{1}{2}(x/\sigma)^2\right] dx = \sigma\sqrt{\tfrac{1}{2}\pi}.$$

Setting $\sigma = \frac{1}{2}\sqrt{\pi}$ this becomes

$$\int_{0}^{\infty} \exp(-x^2) dx = \tfrac{1}{2}\sqrt{\pi},$$

and our derivation of Equations (2.39) and (2.40) is complete. ◆

In preparation for what is to follow, we now prove the simple but important *Jordan inequality*

$$\frac{\sin\theta}{\theta} > \frac{2}{\pi}, \quad \text{for } 0 < \theta < \tfrac{1}{2}\pi. \tag{2.41}$$

PROOF

This inequality is obvious graphically if the graphs of $\sin\theta$ and $2\theta/\pi$ are superimposed for $0 < \theta < \frac{1}{2}\pi$, because the graph of $2\theta/\pi$ represents the chord joining the points (0, 0) and $(\frac{1}{2}\pi, 1)$ on the graph of $\sin\theta$. However, an analytical proof takes the following form. Consider $f(\theta) = (\sin\theta)/\theta$, then

$$f'(\theta) = \frac{\theta\cos\theta - \sin\theta}{\theta^2} = \left(\frac{\cos\theta}{\theta^2}\right)(\theta - \tan\theta).$$

So for $0 < \theta < \frac{1}{2}\pi$, the sign of $f'(\theta)$ is determined by the sign of $g(\theta) = \theta - \tan\theta$, which is negative, as can be seen from the fact that $g'(\theta) = 1 - \sec^2\theta = -\tan^2\theta < 0$ and $g(0) = 0$, so $g(\theta) < 0$ for $0 < \theta < \frac{1}{2}\pi$.

The fact that $f'(\theta) < 0$ for $0 < \theta < \frac{1}{2}\pi$ shows $f(\theta)$ to be a strictly decreasing function of θ in this interval, but we have the well known limits

$$\lim_{\theta \to 0}\left(\frac{\sin\theta}{\theta}\right) = 1 \quad \text{and} \quad \lim_{\theta \to \pi/2}\left(\frac{\sin\theta}{\theta}\right) = \frac{2}{\pi},$$

so we have proved that

$$1 \geqslant \frac{\sin\theta}{\theta} > \frac{2}{\pi} \quad \text{for } 0 < \theta < \tfrac{1}{2}\pi,$$

from which the Jordan inequality in Equation (2.41) then follows. ◆

To show that the integral of $f(z)$ around contours like C_R in Figure 2.20 vanishes in the limit as $R \to \infty$, necessitates over estimating the value of the modulus of the integral on C_R and then showing that it tends to zero as $R \to \infty$. Sometimes elementary arguments suffice for this, but usually a more delicate estimate is needed which is provided by Jordan's lemma.

THEOREM 2.7.1 Jordan's Lemma

Let C_R be a semicircle of radius R centered on the origin and lying in the upper half of the complex plane, and let the function $f(z)$ be such that

(i) It is analytic in the upper half of the complex plane with the exception of a finite number of singularities

(ii) $|f(z)| \to 0$ uniformly as $|z| \to \infty$ for $0 \leqslant \mathrm{Arg}(z) \leqslant \pi$.

Then, if $m > 0$,

$$\lim_{R \to \infty} \int_{C_R} e^{imz} f(z)dz = 0.$$

PROOF

Condition (i) implies that for suitably large $R_0 > 0$, all singularities of $f(z)$ will lie inside the domain $|z| < R_0$, $\mathrm{Im}\{z\} > 0$, and that $f(z)$ will be continuous on C_R for $R > R_0$. The uniformity condition in (ii) means that for any arbitrary small number $\varepsilon_R > 0$, it is always possible to find an $R > R_0$ such that on C_R $|f(z)| \leqslant \varepsilon_R$, where $\varepsilon_R \to 0$ as $R \to \infty$.

On C_R, $z = Re^{i\theta}$ and $dz = iRe^{i\theta}\,d\theta$, so as $|e^{i\theta}| = 1$ and $|e^{imz}| = e^{-mR\sin\theta}$, we have

$$\left|\int_{C_R} e^{imz}\, f(z)dz\right| = \left|\int_{C_R} e^{imz}\, f(z) iRe^{i\theta}\, d\theta\right| \leqslant R\int_0^{\pi} |f(z)|\, e^{-mR\sin\theta}\, d\theta$$
$$\leqslant \varepsilon_R R\int_0^{\pi} e^{-mR\sin\theta}\, d\theta.$$

However,

$$\int_0^{\pi} e^{-mR\sin\theta}\, d\theta = \int_0^{\pi/2} e^{-mR\sin\theta}\, d\theta + \int_{\pi/2}^{\pi} e^{-mR\sin\theta}\, d\theta,$$

but by making the change of variable $u = \pi - \theta$ in the second integral is a simple matter to show it equals the first integral, so that

$$\int_0^{\pi} e^{-mR\sin\theta}\, d\theta = 2\int_0^{\pi/2} e^{-mR\sin\theta}\, d\theta.$$

Combining results we have

$$\left|\int_{C_R} e^{imz} f(z)dz\right| \leqslant 2\varepsilon_R R\int_0^{\pi/2} e^{-mR\sin\theta}\, d\theta,$$

and after applying the Jordan inequality to the integral on the right the result becomes

$$\left|\int_{C_R} e^{imz} f(z)dz\right| \leqslant 2\varepsilon_R R\int_0^{\pi/2} \exp[-(2mR/\pi)\theta]d\theta = \pi\varepsilon_R(1 - e^{-mR}) < \frac{\pi\varepsilon_R}{m}.$$

However, $\varepsilon_R \to 0$ as $R \to \infty$, so the result is proved. ◆

Notice that Jordan's lemma is only a *sufficient* condition for $\int_{C_R} e^{imz} f(z)dz \to 0$ as $R \to \infty$, so it is to be expected that the result may remain true under somewhat weaker conditions than those stated in the lemma, but such conditions will not be considered here.

In the theorems considered so far the function $f(z)$ has been required to be continuous on the contour C around which integration is to be performed. This condition can be relaxed if a singularity of $f(z)$ occurs on C at a point z_0, provided the point is excluded from the domain D inside C by indenting the contour around z_0 by means of a circular arc γ_r of radius r centered on z_0. The contribution to the total contour integral around C made by integrating around the arc γ_r and then letting $r \to 0$ must then be found separately. This approach can lead to attributing a finite value to some real improper integrals that are divergent in the ordinary sense.

To clarify the situation, suppose the real function $f(x)$ becomes infinite at $z = c$, where $a < c < b$. Then what is called the *Cauchy principal value* of the integral

$$\int_a^b f(x)dx,$$

denoted by writing

$$\text{P.V.} \int_a^b f(x)dx,$$

is defined as

$$\text{P.V.} \int_a^b f(x)dx = \lim_{\varepsilon \to 0} \left(\int_a^{c-\varepsilon} f(x)dx + \int_{c+\varepsilon}^b f(x)dx \right), \tag{2.42}$$

provided the limit exists.

This definition should be compared with the corresponding real variable definition of an improper integral with a singularity at $x = c$

$$\int_a^b f(x)dx = \lim_{\varepsilon \to 0} \int_a^{c-\varepsilon} f(x)dx + \lim_{\delta \to 0} \int_{c+\delta}^b f(x)dx.$$

The essential difference is that in this definition the limits to the left and right of c are evaluated *independently*, whereas in Equation (2.42) they are evaluated *symetrically* about c.

To see how the Cauchy principal value can sometimes attribute a finite value to an otherwise divergent integral, consider the integral $\int_{-2}^{3} 1/(t-1)^2 dt$ in which the integrand becomes infinite in the interval of integration when $x = 1$. That this is divergent in the ordinary sense follows from the fact that using the real variable definition for the value of an improper integral with a singularity at $x = 1$, we have

$$\begin{aligned}
\int_{-2}^{3} \frac{1}{(t-1)^3} dt &= \lim_{\varepsilon \to 0} \int_{-2}^{1-\varepsilon} \frac{1}{(t-1)^3} dt + \lim_{\delta \to 0} \int_{1+\delta}^{3} \frac{1}{(t-1)^3} dt \\
&= \lim_{\varepsilon \to 0} \left(-\frac{1}{2\varepsilon^2} + \frac{1}{18} \right) + \lim_{\delta \to 0} \left(-\frac{1}{8} + \frac{1}{2\delta^2} \right) \\
&= -\frac{5}{72} - \lim_{\varepsilon \to 0} \left(\frac{1}{2\varepsilon^2} \right) + \lim_{\delta \to 0} \left(\frac{1}{2\delta^2} \right).
\end{aligned}$$

Because ε and δ are allowed to tend to zero independently, this result has no limit because although the last two terms have opposite signs, each is infinite so their difference is indeterminate. Consequently, this integral is divergent in the ordinary sense as its value is undefined. However, if the Cauchy principal value is determined, we find that

$$\text{P.V.} \int_{-2}^{3} \frac{1}{(t-1)^3}\,dt = -\frac{5}{72} - \lim_{\varepsilon \to 0}\left(\frac{1}{2\varepsilon^2}\right) + \lim_{\varepsilon \to 0}\left(\frac{1}{2\varepsilon^2}\right) = -\frac{5}{72},$$

showing that in this case a finite value has been assigned.

That the Cauchy principal value does not always attribute a finite value to an improper integral can be seen by replacing the integrand in the example just considered by $1/(t-1)^2$ because then

$$\text{P.V.} \int_{-2}^{3} \frac{1}{(t-1)^2}\,dt = \lim_{\varepsilon \to 0}\left(\frac{1}{\varepsilon} - \frac{1}{3}\right) + \lim_{\varepsilon \to 0}\left(-\frac{1}{2} + \frac{1}{\varepsilon}\right) = \infty.$$

Clearly, if an improper integral is convergent, its value must coincide with the Cauchy principal value, so in that case the P.V. symbol can be omitted. The Cauchy principal value extends to situations where an improper integral involves integration over an infinite interval, when the Cauchy principal value of the integral $\int_{-\infty}^{\infty} f(x)dx$ is defined as

$$\text{P.V.} \int_{-\infty}^{\infty} f(x)dx = \lim_{R \to \infty} \int_{-R}^{R} f(x)dx.$$

In the remainder of this section we present a number of examples chosen to illustrate how the Cauchy theorems may be used to evaluate various types of improper integral. Not until complex series have been discussed in the next chapter can we develop the powerful *theory of residues* that greatly simplifies the evaluation of contour integrals.

The first example shows how by choosing a suitable contour, the Cauchy–Goursat theorem may be used to evaluate two nontrivial integrals that find application in diffraction theory.

Example 2.7.1 The Fresnel Integrals

By evaluating

$$\int_C \exp(iz^2)dz,$$

where C is the limit of the contour Γ_R in Figure 2.22 as $R \to \infty$, show that the *Fresnel integrals*

$$\int_0^{\infty} \cos x^2\,dx = \int_0^{\infty} \sin x^2\,dx = \tfrac{1}{4}\sqrt{2\pi}.$$

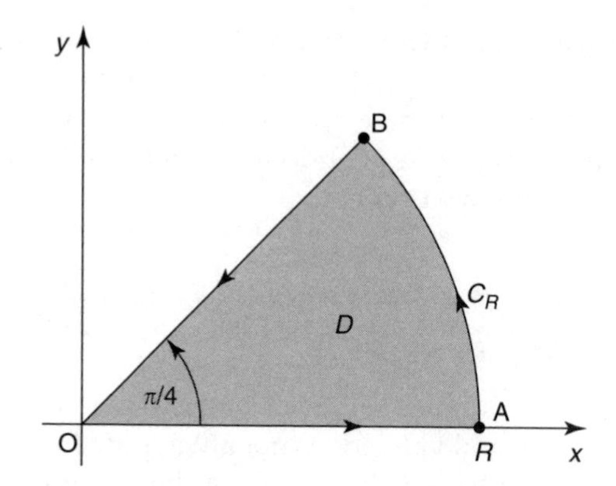

FIGURE 2.22
The contour Γ_R for the Fresnel integrals.

SOLUTION
The contour C is the limit of the contour Γ_R shown in Figure 2.22 which comprises the interval OA of the real axis of length R, the circular arc C_R of radius R from A to B, centered on the origin, and the radial line BO inclined to the real axis at an angle $\pi/4$.

The function $f(z) = \exp(iz^2)$ is analytic everywhere within the domain D bounded by Γ_R and continuous on Γ_R. Thus the conditions of the Cauchy–Goursat theorem are satisfied, so applying the theorem gives

$$\int_{\Gamma_R} \exp(iz^2)dz = \int_{\text{OA}} \exp(iz^2)dz + \int_{C_R} \exp(iz^2)dz + \int_{\text{BO}} \exp(iz^2)dz = 0,$$

so as $z = x$ on OA,

$$\int_0^R \exp(ix^2)dx + \int_{C_R} \exp(iz^2)dz + \int_{\text{BO}} \exp(iz^2)dz = 0.$$

Our next step will be to estimate the magnitude of the integral around C_R, and to find the form of the integral along BO. On arc C_R, $z = Re^{i\theta}$ and $dz = iRe^{i\theta}\,d\theta$ for $0 \leqslant \theta \leqslant \pi/4$, so

$$|\exp(iz^2)| = \exp(-R^2 \sin 2\theta).$$

Now $0 \leqslant \theta \leqslant \pi/4$, so $0 \leqslant 2\theta \leqslant \pi/2$, and so Jordan's inequality in Equation (2.41) in the form $\sin 2\theta \geqslant 4\theta/\pi$ may be applied to the previous result to yield the inequality

$$|\exp(iz^2)| = \exp[-(4R^2/\pi)\theta].$$

As a result we have

$$\left|\int_{C_R} \exp(iz^2)dz\right| \leqslant \int_0^{\pi/4} |\exp(iz^2)||iRe^{i\theta}|d\theta \leqslant R\int_0^{\pi/4} \exp[-(4R^2/\pi)\theta]d\theta$$
$$= \frac{\pi}{4R}[1 - \exp(-R^2)],$$

and so

$$\lim_{R\to\infty}\left|\int_{C_R} \exp(iz^2)dz\right| = 0.$$

On the radial line BO, $z = re^{i\pi/4}$, $iz^2 = -r^2$, and $dz = e^{i\pi/4}dr$, but the integration is from B to O, so we have

$$\int_{\mathrm{BO}} \exp(iz^2)dz = \int_R^0 \exp(-r^2)e^{i\pi/4}dr = -e^{i\pi/4}\int_0^R \exp(-r^2)dr.$$

Combining results and proceeding to the limit as $R \to \infty$, this becomes

$$\int_0^\infty \exp(ix^2)dx = e^{i\pi/4}\int_0^\infty \exp(-r^2)dr.$$

The integral on the right was shown in Equation (2.40) to equal $\frac{1}{2}\sqrt{\pi}$, so

$$\int_0^\infty \exp(ix^2)dx = \tfrac{1}{2}e^{i\pi/4}\sqrt{\pi},$$

and equating corresponding real and imaginary parts on each side of this equation gives the Fresnel integrals

$$\int_0^\infty \cos x^2\,dx = \int_0^\infty \sin x^2\,dx = \tfrac{1}{4}\sqrt{2\pi}.$$

Notice that, as is typical of improper integrals evaluated in this way, two results are always obtained, one by equating real parts on either side of the complex result, and the other by equating the imaginary parts. ◇

The next example involves a contour integral in which the integrand has a single singularity inside the contour C around which the integral is to be evaluated.

Example 2.7.2 An Integral with a Single Singularity inside *C*
By evaluating the integral

$$\int_C \frac{e^{ikz}}{z^2 + a^2}dz,$$

where $k > 0$, a is real, and C is the limit of the contour Γ_R in Figure 2.23 as $R \to \infty$, show that

$$\int_0^\infty \frac{\cos kx}{x^2 + a^2}\,dx = \frac{\pi e^{-ka}}{2a}.$$

SOLUTION

The contour Γ_R comprises the segment of the real axis $-R \leqslant x \leqslant R$, together with the semicircle C_R in the upper half-plane of radius R centered on the origin. Thus

$$\int_{\Gamma_R} \frac{e^{ikz}}{z^2 + a^2}\,dz = \int_{-R}^{R} \frac{e^{ikx}}{x^2 + a^2}\,dx + \int_{C_R} \frac{e^{ikz}}{z^2 + a^2}\,dz. \tag{2.43}$$

To evaluate the integral on the left of Equation (2.43) notice that the function $g(z) = e^{ikz}/(z^2 + a^2)$ in the integrand is analytic in domain D within Γ_R, except for the single point ia, and it is continuous on Γ_R. By writing $g(z)$ as

$$g(z) = \frac{e^{ikz}}{(z - ia)(z + ia)},$$

so by setting

$$f(z) = \frac{e^{ikz}}{(z + ia)},$$

the integral on the left of Equation (2.43) can be written as

$$\int_{\Gamma_R} \frac{e^{ikz}}{z^2 + a^2}\,dz = \int_{\Gamma_R} \frac{f(z)}{z - ia}\,dz,$$

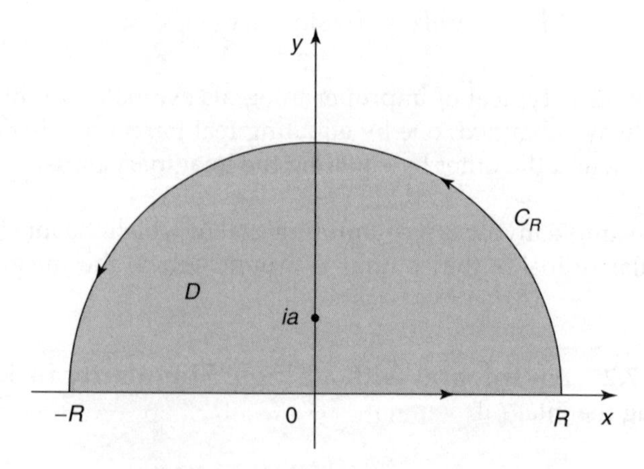

FIGURE 2.23
The contour Γ_R containing a singularity $z = ia$.

where now $f(z)$ is analytic throughout D. When expressed in this form the Cauchy integral formula can be applied to the expression on the right to give

$$\int_{\Gamma_R} \frac{e^{ikz}}{z^2 + a^2} dz = 2\pi i f(ia) = \frac{\pi e^{-ka}}{a}.$$

Combining this with Equation (2.43) gives

$$\int_{-R}^{R} \frac{e^{ikx}}{x^2 + a^2} dx + \int_{C_R} \frac{e^{ikz}}{z^2 + a^2} dz = \frac{\pi e^{-ka}}{a}. \tag{2.44}$$

Let us now examine the integral around the semicircular arc C_R. Rewriting the integrand as

$$e^{ikz}\left(\frac{1}{z^2 + a^2}\right),$$

and identifying $1/(z^2 + a^2)$ with the function $f(z)$ in Jordan's lemma, we see that $|1/(z^2 + a^2)| \to 0$ uniformly as $|z| \to \infty$, as required by the lemma. As the integrand is analytic everywhere in D apart from the single point $z = ia$, the conditions of the lemma are satisfied, showing that this integral will vanish as $R \to \infty$. Consequently, taking the limit of Equation (2.44) as $R \to \infty$ gives

$$\text{P.V.}\int_{-\infty}^{\infty} \frac{e^{ikx}}{x^2 + a^2} dx = \frac{\pi e^{-ka}}{a}.$$

Equating the respective real and imaginary parts of this result gives

$$\text{P.V.}\int_{-\infty}^{\infty} \frac{\cos kx}{x^2 + a^2} dx = \frac{\pi e^{-ka}}{a} \quad \text{and} \quad \text{P.V.}\int_{-\infty}^{\infty} \frac{\sin kx}{x^2 + a^2} dx = 0.$$

Comparison of these two improper integrals with the known convergent improper integral $\int_{-\infty}^{\infty} 1/(x^2 + a^2)dx$ shows them to be convergent in the ordinary sense, so the symbols P.V. may be omitted when we obtain

$$\int_{-\infty}^{\infty} \frac{\cos kx}{x^2 + a^2} dx = \frac{\pi e^{-ka}}{a} \quad \text{and} \quad \int_{-\infty}^{\infty} \frac{\sin kx}{x^2 + a^2} dx = 0.$$

As the cosine is an even function the first result can be rewritten as

$$\int_{0}^{\infty} \frac{\cos kx}{x^2 + a^2} dx = \frac{\pi e^{-ka}}{2a} \quad \text{for } k > 0.$$

Notice that, as usual, two results have been obtained from the single complex result. ◇

The next Example makes use of the Cauchy integral theorem for derivatives when evaluating a contour integral.

Example 2.7.3 An Integral with a Repeated Singularity inside C
By evaluating the integral

$$\int_C \frac{dz}{(z^2+1)^2}$$

around the semicircular contour C in Example 2.7.2, shows that

$$\int_{-\infty}^{\infty} \frac{dx}{(x^2+1)^2} = \tfrac{1}{2}\pi.$$

SOLUTION
The contour C is the limit of the finite contour Γ_R as $R \to \infty$, so separating the integral around Γ_R to the integral along the real axis over $-R \leqslant x \leqslant R$, and the integral around the semicircle C_R, gives

$$\int_{\Gamma_R} \frac{dz}{(z^2+1)^2} = \int_{-R}^{R} \frac{dx}{(x^2+1)^2} + \int_{C_R} \frac{dz}{(z^2+1)^2}. \tag{2.45}$$

Let us first evaluate the integral on the left with the help of the Cauchy integral theorem for derivatives. Consider the integrand $g(z) = 1/(z^2+1)^2$ that can be written as

$$g(z) = \frac{1}{(z+i)^2(z-i)^2}$$

showing that only the repeated singularity at $z = i$ lies inside the contour Γ_R. Then by defining $f(z) = 1/(z+i)^2$, we can write

$$\int_{\Gamma_R} \frac{dz}{(z^2+1)^2} = \int_{\Gamma_R} \frac{f(z)}{(z-i)^2}, \tag{2.46}$$

where now $f(z)$ is analytic throughout the interior of contour Γ_R.
Setting $n = 1$ in the Cauchy integral formula for derivatives shows this last result can be written as

$$\int_{\Gamma_R} \frac{dz}{(z^2+1)^2} = 2\pi i f'(i),$$

Equating corresponding real parts on each side of this equation gives

$$\int_0^{\infty} \exp(-mx^2)dx \cos(2\text{max})dx = \exp(-ma^2)\int_0^{\infty} \exp(-mx^2)dx,$$

but $\int_0^{\infty} \exp(-mx^2)dx = \frac{1}{2}\sqrt{\pi/m}$, so that

$$\int_0^{\infty} \exp(-mx^2)dx \cos(2\text{max})dx = \exp(-ma^2)\sqrt{\pi/m}.$$

Equating imaginary parts gives

$$\int_0^{\infty} \exp(-mx^2)\sin(2\text{max})dx = \exp(-ma^2)\int_0^{a} \exp(my^2)dy,$$

where the definite integral on the right cannot be further simplified. ◇

In the examples considered so far, neither a singularity nor a branch point has occurred on the path of integration. When this occurs at a point $z = a$ on the contour, the singularity or branch point must be excluded from the domain bounded by the contour by means of a circular arc indentation C_ρ of radius ρ centered on $z = a$, and the result of integrating around C_ρ found in the limit as $\rho \to 0$. The following theorem determines the contribution made by integrating around C_ρ when the singularity in the integrand $f(z)$ occurs at $z = a$ and is such that $\lim_{z\to a}[(z - a)f(z)] = k$, with k a constant. Later we will see that when $f(z)$ has this type of singularity it will be said to have a *simple pole* at $z = a$.

THEOREM 2.7.2 Integration around an Indentation to Exclude a Simple Pole
Let $f(z)$ have a singularity at $z = a$ such that

$$\lim_{z\to a}[(z - a)f(z)] = k \quad (k = \text{constant}),$$

and take C_ρ to be the circular arc of radius ρ centered on $z = a$ such that $\alpha \leqslant \text{Arg}(z - a) \leqslant \beta$, as shown in Figure 2.25.
Then

$$\lim_{\rho\to 0}\int_{C_\rho} f(z)dz = ik(\beta - \alpha),$$

where the integral around C_ρ is taken counterclockwise (in the positive sense).

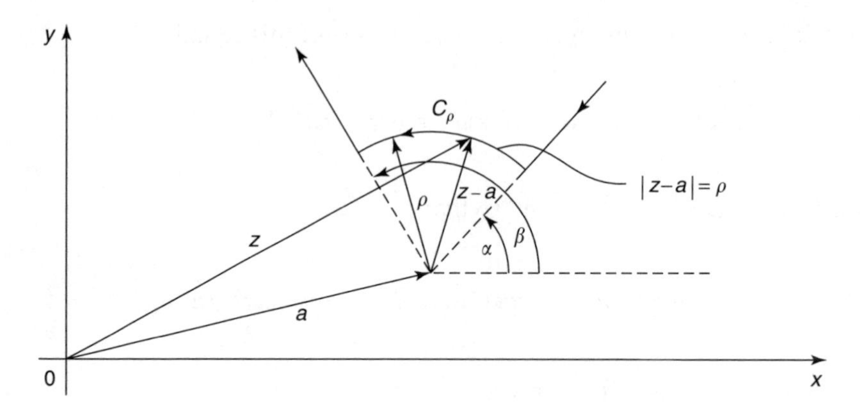

FIGURE 2.25
The contour indentation C_ρ: $|z-a| = \rho$, $\alpha \leqslant \text{Arg}(z-a) \leqslant \beta$.

PROOF

The existence of the limit

$$\lim_{z \to a} [(z-a)f(z)] = k$$

implies that for any arbitrary $\varepsilon > 0$, however small, it is always possible to find a suitably small $\delta > 0$ such that if $|z-a| < \delta$, then

$$(z-a)f(z) - k = \mu(z),$$

where $|\mu(z)| < \varepsilon$.

Expressed differently, this means that by taking z close enough to a, the magnitude of the difference $(z-a)f(z) - k$ will not exceed ε. Thus, for these ε and δ,

$$f(z) - \frac{k}{z-a} = \frac{\mu(z)}{z-a}.$$

Thus

$$\int_{C_\rho} f(z)dz - k\int_{C_\rho} \frac{dz}{z-a} = \int_{C_\rho} \frac{\mu(z)}{z-a} dz,$$

and so

$$\left| \int_{C_\rho} f(x)dx - k\int_{C_\rho} \frac{dz}{z-a} dz \right| = \left| \int_{C_\rho} \frac{\mu(z)}{z-a} dz \right|.$$

However, on C_ρ, $z = a + \rho e^{i\theta}$ for $\alpha \leqslant \theta \leqslant \beta$, and $dz = i\rho e^{i\theta}\, d\theta$, so

$$\int_{C_\rho} \frac{dz}{z-a} = \int_\alpha^\beta \frac{i\rho e^{i\theta}}{\rho e^{i\theta}}\, d\theta = i\int_\alpha^\beta d\theta = i(\beta - \alpha).$$

We also know that for z on C_ρ, $|\mu(z)| < \varepsilon$, and so

$$\left|\int_{C_\rho} \frac{\mu(z)}{z-a}\, dz\right| < \varepsilon \left|\int_{C_\rho} \frac{dz}{z-a}\right| = \varepsilon(\beta - \alpha).$$

If we now take $\rho < \delta$, then since $\delta \to 0$ as $\varepsilon \to 0$, it follows that

$$\lim_{\rho \to 0} \int_{C_\rho} f(z) dz = ik(\beta - \alpha),$$

and the proof is complete. ◆

The examples that follow illustrate the application of this theorem when either a pole or a branch point occurs on the contour of integration.

Example 2.7.5 The Dirichlet Integral
By evaluating

$$\int_C \frac{e^{iz}}{z}\, dz,$$

where C is the limit as $\rho \to 0$ and $R \to \infty$ of the contour Γ_R in Figure 2.26 show that

$$\text{P.V.} \int_{-\infty}^{\infty} \frac{\cos x}{x}\, dx = 0,$$

and that the Dirichlet integral

$$\int_{-\infty}^{\infty} \frac{\sin x}{x}\, dx = \tfrac{1}{2}\pi.$$

SOLUTION
The only singularity of $f(z) = e^{iz}/z$ occurs at the origin, where from Figure 2.26, it is seen to be excluded from the contour of integration Γ_R by an indentation C_ρ about the origin. As $f(z)$ is analytic in the semi-annular domain D in Figure 2.26, it follows from the Cauchy–Goursat theorem that

$$\int_{\Gamma_R} \frac{e^{iz}}{z}\, dz = 0 = \int_\rho^R \frac{e^{ix}}{x}\, dx + \int_{C_R} \frac{e^{iz}}{z}\, dz + \int_{-R}^{\rho} \frac{e^{ix}}{x}\, dx + \int_{C_\rho} \frac{e^{iz} z}{dz}. \quad (2.48)$$

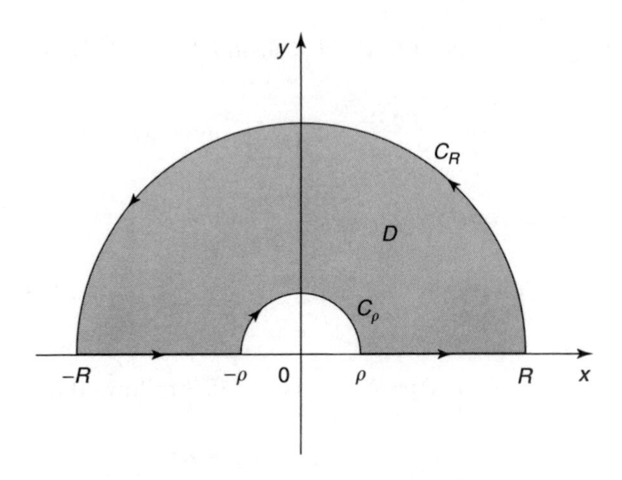

FIGURE 2.26
The contour Γ_R indented at the origin by C_ρ.

The only singularity of $f(z)$ occurs at the origin $z = 0$, where $\lim_{z\to 0}[zf(z)] = \lim_{z\to 0}[e^{iz}] = 1$, so the conditions of Theorem 2.7.8 are satisfied. At the point $z = \rho$ we see that $\alpha = \text{Arg}(z) = 0$ while at the point $z = -\rho$, $\beta = \text{Arg}(z) = \pi$, so $\beta - \alpha = \pi$. Thus from Theorem 2.7.8 the contribution made by the integral around C_ρ is $-i(\beta - \alpha) = -i\pi$, where the negative sign has been introduced because integration around C_ρ is in the negative sense (the clockwise direction), so

$$\lim_{\rho\to 0}\int_{C_\rho}\frac{e^{iz}z}{z}\,dz = -i\pi.$$

To find the contribution made by the integral around C_R as $R \to \infty$, notice that the integrand satisfies the conditions of Jordan's lemma, so that

$$\lim_{R\to\infty}\int_{C_R}\frac{e^{ix}}{x}\,dx = 0.$$

Thus combining these results gives

$$\int_{-\infty}^{\infty}\frac{e^{ix}}{x}\,dx = i\pi.$$

Equating the respective real and imaginary parts on each side of this equation gives

$$\text{P.V.}\int_{-\infty}^{\infty}\frac{\cos x}{x}\,dx = 0, \tag{2.49}$$

and

$$\int_{-\infty}^{\infty} \frac{\sin x}{x}\, dx = \pi. \tag{2.50}$$

where the symbol P.V. has been omitted from the second result because the singularity of $(\sin x)/x$ at the origin is removable because $\lim_{x\to 0}[(\sin x)/x] = 1$. As the integrand in the second integral is even, the result can be written as

$$\int_{-\infty}^{\infty} \frac{\sin x}{x}\, dx = 2\int_{0}^{\infty} \frac{\sin x}{x}\, dx = \pi \quad \text{and so} \quad \int_{0}^{\infty} \frac{\sin x}{x}\, dx = \tfrac{1}{2}\pi. \tag{2.51}$$

Result from Equation (2.49) could have been anticipated by using the fact that $(\cos x)/x$ is an odd function together with the symmetrical manner in which a Cauchy principal value is defined. ◇

Example 2.7.6 Integration around an Indented Rectangle

By evaluating the integral

$$\int_C \frac{e^{az}}{\sinh \pi z}\, dz \quad (-\pi < a < \pi),$$

show that

$$\int_0^a \frac{\sinh ax}{\sinh \pi x}\, dx = \tfrac{1}{2}\tan\left(\tfrac{1}{2}a\right),$$

where C is the limit as $R \to \infty$ of the contour Γ_R shown in Figure 2.27.

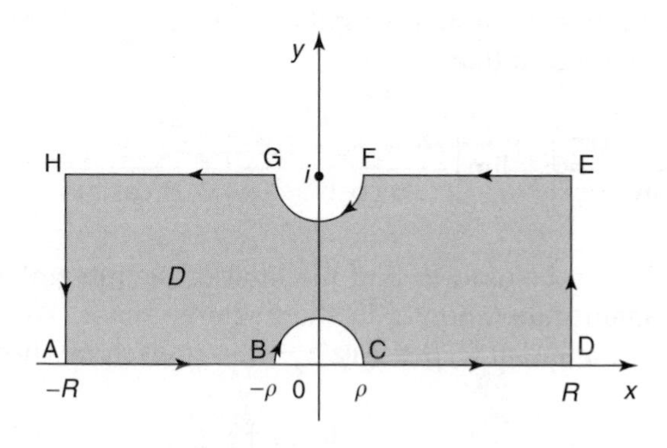

FIGURE 2.27
The contour Γ_R and the doubly indented rectangle.

SOLUTION

The zeros of $\sinh \pi z$ that concern us here are those that occur at $z = 0$ and $z = i$, so the function $f(z) = e^{az}/\sinh \pi z$ is analytic inside the indented contour Γ_R shown in Figure 2.27. So from the Cauchy–Goursat theorem

$$\int_{\Gamma_R} \frac{e^{az}}{\sinh \pi z} dz = 0 = \int_{-R}^{-\rho} \frac{e^{ax}}{\sinh \pi x} dx + \int_{\text{BC}} \frac{e^{az}}{\sinh \pi z} dz + \int_{\rho}^{R} \frac{e^{ax}}{\sinh \pi x} dx$$
$$+ \int_{\text{DE}} \frac{e^{az}}{\sinh \pi z} dz + \int_{R}^{\rho} \frac{e^{a(x+i)}}{\sinh \pi(x+i)} dx + \int_{\text{FG}} \frac{e^{az}}{\sinh \pi z} dz$$
$$+ \int_{-\rho}^{-R} \frac{e^{a(x+i)}}{\sinh \pi(x+i)} dx + \int_{\text{HA}} \frac{e^{az}}{\sinh \pi z} dz.$$

Let us first determine the limiting values of the integrals around the semicircular arcs BC and FG as $\rho \to 0$ by means of Theorem 2.7.2. The integrand $f(z) = e^{az}/\sinh \pi z$ has a singularity at the origin, so if Theorem 2.7.2 is applied, we see from the conditions of the theorem that $\lim_{z\to 0} [zf(z)]$ must have a finite value. This limit is an indeterminate form, so applying L'Hospital's rule we find that

$$\lim_{z\to 0}\left(\frac{ze^{az}}{\sinh \pi z}\right) = \lim_{z\to 0}\left(\frac{e^{az} + aze^{az}}{\pi \cosh \pi z}\right) = \frac{1}{\pi}.$$

As this limit is finite, Theorem 2.7.2 may be applied with $k = 1/\pi$ in order to determine the limit of the integral around BC as $\rho \to 0$. The integral around BC is in the negative sense, so at B, $\alpha = \operatorname{Arg}(z) = \pi$, and at C, $\beta = \operatorname{Arg}(z) = 0$, so that

$$\lim_{\rho\to 0} \int_{\text{BC}} \frac{e^{az}}{\sinh \pi z} dz = i\frac{1}{\pi}(0 - \pi) = -i.$$

The integrand has a similar singularity at $z = i$, where again using L'Hospital's rule, we find that

$$\lim_{z\to i}\left(\frac{(z-i)e^{az}}{\sinh \pi z}\right) = \lim_{z\to i}\left(\frac{e^{az} + a(z-i)e^{az}}{\pi \cosh \pi z}\right) = \frac{e^{ia}}{\pi \cosh i\pi} = -\frac{e^{ia}}{\pi},$$

so Theorem 2.7.2 can be used to find the limit of the integral around FG as $\rho \to 0$, where again integration is in the negative sense. We see that at F, $\alpha = \operatorname{Arg}(z - i) = 2\pi$ and at G, $\beta = \operatorname{Arg}(z - i) = \pi$, so from Theorem 2.7.2,

$$\lim_{\rho\to 0} \int_{\text{FG}} \frac{e^{az}}{\sinh \pi z} dz = i\left(-\frac{e^{ia}}{\pi}\right)(\pi - 2\pi) = ie^{ia}.$$

Next we consider the behavior of the integrals along DE and HA as $R \to \infty$. Along DE we have

$$\int_{\mathrm{DE}} \frac{e^{az}}{\sinh \pi z} dz = \int_0^1 \frac{e^{a(R+iy)}}{\sinh \pi(R+iy)} i dy$$
$$= i \int_0^1 \frac{e^{aR} e^{iay}}{\sinh \pi R \cos \pi y + i \cosh \pi R \sin \pi y} dy.$$

The functions e^{iy}, $\cos \pi y$ and $\sin \pi y$ are bounded, but $\sinh \pi R$ and $\cosh \pi R$ are asymptotic to $\frac{1}{2} e^{\pi R}$, so for large R the integrand is asymptotic to

$$\frac{2e^{aR} e^{iay}}{e^{\pi R} \cos \pi y + i e^{\pi R} \sin \pi y} = 2e^{(a-\pi)R} e^{i(a-\pi)y}.$$

However, as $-\pi < a < \pi$ we see that $a - \pi < 0$, so the integrand tends to zero exponentially as $R \to \infty$, from which we have

$$\lim_{R \to \infty} \int_{\mathrm{DE}} \frac{e^{az}}{\sinh \pi z} dz = 0.$$

A similar form of argument applied to the integral along HA shows that

$$\lim_{R \to \infty} \int_{\mathrm{HA}} \frac{e^{az}}{\sinh \pi z} dz = 0.$$

Proceeding to the limits as $\rho \to 0$ and $R \to \infty$ in the sum of integrals around the respective segments of Γ_R gives

$$\int_{-\infty}^0 \frac{e^{ax}}{\sinh \pi x} dx - i \int_0^\infty \frac{e^{ax}}{\sinh \pi x} dx - e^{ia} \int_\infty^0 \frac{e^{ax}}{\sinh \pi x} dx - e^{ia} \int_0^{-\infty} \frac{e^{ax}}{\sinh \pi x} dx = 0.$$

Replacing x by $-x$ in the first and last integrals, interchanging the order of the limits where necessary after making corresponding changes of sign, and using the definition $\sinh ax = \frac{1}{2}(e^{ax} - e^{-ax})$, we find that

$$2 \int_0^\infty \frac{\sinh ax}{\sinh \pi x} dx + i(e^{ax} - 1) + 2e^{ia} \int_0^\infty \frac{\sinh ax}{\sinh \pi x} dx = 0.$$

Rearranging terms, this becomes

$$\int_0^\infty \frac{\sinh ax}{\sinh \pi x} dx = \frac{i}{2} \left(\frac{1 - e^{ia}}{1 + e^{ia}} \right) = \frac{1}{2} \frac{\left(\dfrac{e^{ia/2} - e^{-ia/2}}{2i} \right)}{\left(\dfrac{e^{ia/2} + e^{ia/2}}{2} \right)}$$
$$= \frac{1}{2} \frac{\sin \frac{1}{2} a}{\cos \frac{1}{2} a} = \frac{1}{2} \tan \frac{1}{2} a.$$

◇

2.7.1 Contour Integrals Involving Functions with Branch Points

We have already seen that by means of a suitable *branch cut*, a *many-valued* complex function can be replaced by an equivalent set of different single-valued functions called ***branches*** of the original function. Every branch has the same intrinsic mathematical properties, but different branches have different arguments. When evaluating definite integrals by contour integration and a many-valued function is involved, the choice of branch is unimportant. But once the choice has been made, the branch must not be changed. The branches can be finite in number, as with the square root function that has two branches, or infinite in number, as with the logarithmic function. Crossing a branch cut is equivalent to changing from one branch of the function to another, and so to changing from a function with one argument to a function with a different argument. As a result, when contour integration is involved, not only is it necessary to specify the branch to be used, but also the contour around which integration is to be performed, that must always be chosen in such a way that it *avoids* crossing the branch cut and does not pass through the branch point.

The only many-valued functional behavior to be considered here will be due to a function of the form $z^{\alpha-1}$ in the integrand, with $\alpha < 1$ a real constant, or to the function $\ln z$ in the integrand.

Let us consider the real integral

$$\int_0^{\infty} f(x)dx$$

associated with a contour integral involving a many-valued function $f(z)$ with a branch point at the origin and a branch cut along the positive real axis up to and including the origin. Then this integral will exist provided:

(i) $f(z)$ has no singularities on the positive real axis.

(ii) The integral around the circle or circular arc C_R of radius R centered on the origin used to close the contour in the usual way vanishes in the limit as $R \to \infty$.

(iii) The integral around the circle or circular arc indentation C_ρ of radius ρ centered on the origin (around the branch point) vanishes as $\rho \to 0$, because then the many-valued behavior of $f(z)$ at the origin will make no contribution to the integral.

In order to arrive at conditions that ensure the vanishing of the integrals around C_ρ and C_R in the limit as $\rho \to 0$ and $R \to \infty$ we proceed in essentially the same way as when deriving Theorem 2.7.2.

In order to consider both a circular arc and a circle, let C_r be the arc AB in Figure 2.28 with radius r and its center at the origin, where $\alpha \leqslant \text{Arg}(z) \leqslant \beta$, with $0 \leqslant \alpha < \beta \leqslant 2\pi$. If $M(r) = \max|f(z)|$ for z on C_r, on which $z = re^{i\theta}$. Then we have

$$\left|\int_{C_r} f(z)dz\right| < \int_{\alpha}^{\beta} |f(z)|r\,d\theta < rM(r)(\beta - \alpha).$$

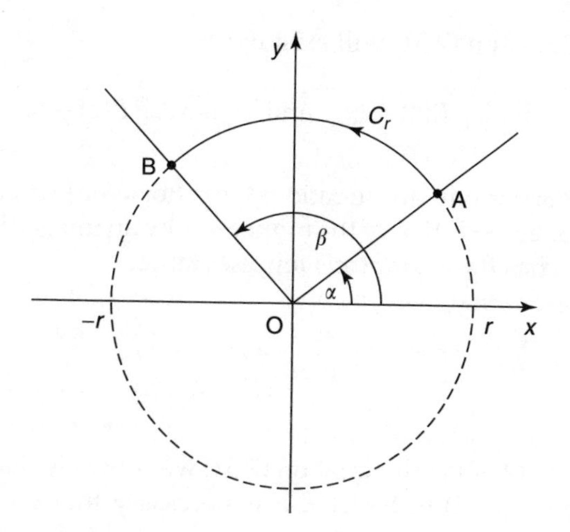

FIGURE 2.28
The circular arc C_r: $|z| = r, \alpha \leqslant \text{Arg}(z) \leqslant \beta$.

If now, we identify C_r with either C_ρ or C_R, we conclude that

(i) $\lim_{\rho \to 0} \int_{C_\rho} f(z)dz = 0 \quad \text{if } \lim_{\rho \to 0} \rho M(\rho) = 0,$

(ii) $\lim_{R \to \infty} \int_{C_R} f(z)dz = 0 \quad \text{if } \lim_{R \to \infty} RM(R) = 0,$

and we have proved the following theorem.

THEOREM 2.7.3 Conditions for the Vanishing of an Integral around an Arc
Let $M(r) = \max|f(z)|$ on C_r, which may be any arc of the circle $|z| = r$. Then if

$$\lim_{\rho \to 0} \rho M(\rho) = 0 \quad \text{and} \quad \lim_{R \to \infty} RM(R) = 0,$$

it follows that

$$\lim_{\rho \to 0} \int_{C_\rho} f(z)dz = 0 \quad \text{and} \quad \lim_{R \to \infty} \int_{C_R} f(z)dz = 0.$$

◆

We now need to know when integrals of the form

$$\int_0^\infty x^{\alpha - 1} R(x) dx \tag{2.52}$$

exist, where α is a real (nonintegral) number and $R(x)$ is a rational function of x whose denominator has no zeros on the positive x-axis. From Theorem 2.7.3

we know that Equation (2.51) will exist if

$$\lim_{|z|\to 0} |z|^{\alpha} R(z) = 0 \quad \text{and} \quad \lim_{R\to\infty} |z|^{\alpha} R(z) = 0. \tag{2.53}$$

If we now consider only those rational functions $R(z)$ where the degree of the denominator exceeds that of the numerator by an integer $K > 0$, it follows at once that $R(z)$ has the asymptotic representation

$$R(z) \sim \frac{A}{z^k} \quad \text{as } |z| \to \infty, \tag{2.54}$$

where $A \neq 0$.

Using Equation (2.54) with Equation (2.53) we conclude that for the second result in Equation (2.52) to be true, it is necessary that $k - \alpha > 0$, while if $R(0) \neq 0$ for the first result to be true it is necessary that $0 \leqslant \alpha \leqslant k$. These two conditions are *sufficient* in order that integrals of the form in Equation (2.53) exist.

The next example illustrates the evaluation of an integral with a branch cut in which the contour is closed by a large circle C_R, and also how the same integral may be evaluated using only a semicircular arc C_R to close the contour, in conjunction with an indented real axis and a cut along the positive real axis up to and including the origin.

Example 2.7.7 Two Different Contours Used to Evaluate an Integral Involving a Many-Valued Function

By integrating the function $z^{\alpha-1}/(z+1)$ around the limiting forms of contours $\Gamma_R^{(1)}$ and $\Gamma_R^{(2)}$ illustrated in Figure 2.29(a) and (b) as $\rho \to 0$ and $R \to \infty$,

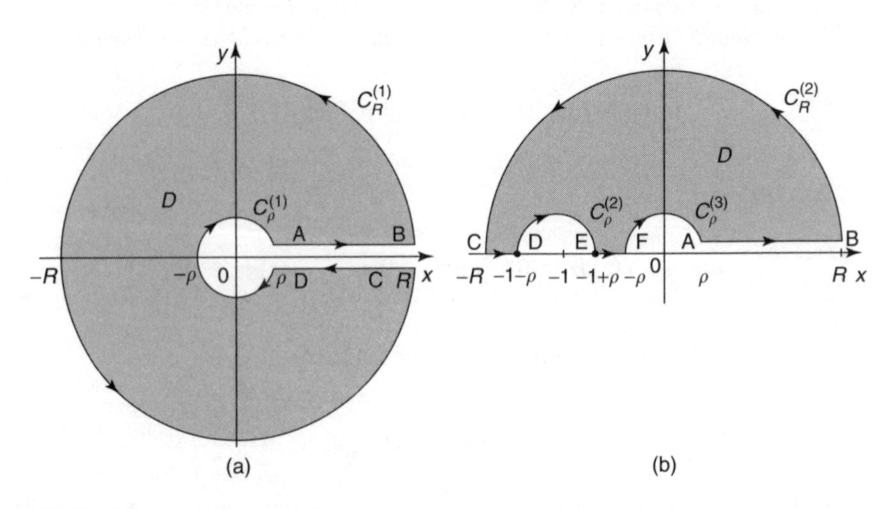

FIGURE 2.29
(a) The contour $\Gamma_R^{(1)}$. (b) The contour $\Gamma_R^{(2)}$.

using the branch of $z^{\alpha-1}$ that is purely real when z lies on the upper edge of the cut along the positive real axis and taking $0 < \alpha < 1$, find

$$I_1 = \int_0^\infty \frac{x^{\alpha-1}}{x+1}dx \quad \text{and} \quad I_2 = \text{P.V.}\int_0^\infty \frac{e^{\alpha-1}}{x-1}dx$$

SOLUTION USING CONTOUR $\Gamma_R^{(1)}$

Integral I_1 exists because $R(z) = 1/(z+1)$ has no singularity on the positive real axis, and Equation (2.54) applies with $k = 1$, so as $0 < \alpha < 1$ we have $k - \alpha < 0$, while $\lim_{|z|\to 0}|z|^\alpha R(z) = 0$ and $\lim_{|z|\to\infty}|z|^\alpha R(z) = 0$.

If the function $g(z) = z^{\alpha-1}/(z+1)$ is integrated around the contour $\Gamma_R^{(1)}$ it is analytic inside the contour with the exception of the point $z = -1$ on the negative real axis. The cut along the positive real axis makes $g(z)$ single-valued, with the path of integration taken to lie along the upper edge AB of the cut, around C_R, back along the lower edge CD of the cut, and then around the circle C_ρ used to exclude the branch point from the domain D inside $\Gamma_R^{(1)}$.

The function $g(z)$ is many valued due to the presence of the term $z^{\alpha-1}$, but we are instructed to use the branch of $z^{\alpha-1}$ that is purely real on the upper edge of the cut, so we must use the principal branch of the function. Working in the cut plane and in modulus-argument form, the point -1 on the negative real axis, where $g(z)$ has a singularity, has the unique representation $e^{i\pi}$. Thus setting $g(z) = f(z)/(z - z_0)$, with $f(z) = z^{\alpha-1}$ and $z_0 = e^{i\pi}$, it follows from Cauchy's integral theorem for derivatives that

$$\int_{\Gamma_R^{(1)}} \frac{z^{\alpha-1}}{z+1}dz = \int_{\Gamma_R^{(1)}} \frac{f(z)}{z-z_0}dz = 2\pi i f(z_0)$$
$$= 2\pi i e^{(\alpha-1)\pi i} = -2\pi i e^{\alpha\pi i}.$$

Integration around the various segments of $\Gamma_R^{(1)}$ gives

$$\int_{AB} g(z)dz = \int_{C_R^{(1)}} g(z)dx + \int_{CD} g(z)dz + \int_{C_\rho^{(1)}} g(z)dz = -2\pi i e^{\alpha\pi i}.$$

We observed at the outset that $g(z)$ satisfies the conditions of Theorem 2.7.3, so in the limit as $\rho \to 0$ and $R \to \infty$, the second and fourth integrals vanish. The path AB lies along the upper edge of the cut, so on AB we must set $z = re^{i0} = r$, whereas as the path CD lies on the lower edge of the cut we must set $z = re^{2\pi i}$. Making these substitutions and proceeding to the limits we find that

$$\int_0^\infty \frac{r^{\alpha-1}}{r+1}dr + \int_\infty^0 \frac{r^{\alpha-1}e^{2(\alpha-1)\pi i}}{r+1}dr = -2\pi i e^{\alpha\pi i}.$$

Because $e^{2\pi i} = 1$, this simplifies to

$$(1 - e^{2\alpha\pi i})\int_0^\infty \frac{r^{\alpha-1}}{r+1}\,dr = -2\pi i e^{\alpha\pi i}.$$

Replacing the dummy variable of integration r by x, this becomes

$$\int_0^\infty \frac{x^{\alpha-1}}{x+1}\,dx = -\frac{2\pi i e^{\alpha\pi i}}{1 - e^{2\alpha\pi i}}$$

$$= \pi\left(\frac{2i}{e^{\alpha\pi i} - e^{-\alpha\pi i}}\right) = \frac{\pi}{\sin\alpha\pi}.$$

Thus, when using the principal branch of $z^{\alpha-1}$, we have shown that

$$I_1 = \int_0^\infty \frac{x^{\alpha-1}}{x+1}\,dx = \frac{\pi}{\sin\alpha\pi}, \qquad 0 < \alpha < 1.$$

So use of this contour has determined the value of I_1, but not the value of I_2. ◇

SOLUTION USING CONTOUR $\Gamma_R^{(2)}$

Here again integral I_1 exists for the reasons already given. The contour $\Gamma_R^{(2)}$ contains no singularities of the integrand, so by the Cauchy–Goursat theorem

$$\int_{\Gamma_R^{(2)}} \frac{z^{\alpha-1}}{z+1}\,dz = 0.$$

Using the same notation as before, and integrating around the segments of $\Gamma_R^{(2)}$, we have

$$\int_{AB} g(z)dz + \int_{C_R^{(2)}} g(z)dz + \int_{CD} g(z)dx + \int_{C_\rho^{(2)}} g(z)dx + \int_{EF} g(z)dz + \int_{C_\rho^{(3)}} g(z)dz = 0.$$

As before, the integrals around $C_\rho^{(3)}$ and $C_R^{(2)}$ vanish as $\rho \to 0$ and $R \to \infty$. We have already seen that the integration in the positive sense around the singularity at $z = -1$ yields the value $-2\pi i e^{\alpha\pi i}$, but now the integration is in the negative sense around a semicircle, so in this case as $\rho \to 0$ the integration will yield $\pi i e^{\alpha\pi i}$.

The path AB is along the top if the cut, so on AB we must set $z = re^{i0} = r$. The paths CD and EF lie along the negative real axis, so on these we must set $z = re^{i\pi}$, where in each case r is positive because it is the modulus of z.

Proceeding to the limits in the usual way we have

$$\int_0^{\infty} \frac{r^{\alpha-1}}{r+1} dr + \int_{\infty}^{1} \frac{r^{\alpha-1} e^{(\alpha-1)\pi i}}{re^{\pi i}+1} dr + i\pi e^{\alpha\pi i} + \int_0^{1} \frac{r^{\alpha-1} e^{(\alpha-1)\pi i}}{re^{\pi i}+1} dr = 0,$$

where on CD and EF we have used the result that $dz = e^{i\pi} dr$ and made appropriate changes of sign. When combine the last two integrals become

$$\int_0^{\infty} \frac{r^{\alpha-1}}{r+1} dr + i\pi e^{\alpha\pi i} = e^{\alpha\pi i} \int_0^{\infty} \frac{r^{\alpha-1}}{1-r} dr.$$

Replacing the dummy variable r by x and equating corresponding real and imaginary parts shows first that

$$I_1 = \int_0^{\infty} \frac{x^{\alpha-1}}{x+1} dx = \frac{\pi}{\sin \alpha\pi}, \quad 0 < \alpha < 1,$$

and then that

$$I_2 = \text{P.V.} \int_0^{\infty} \frac{x^{\alpha-1}}{1-x} dx = \pi \cot \alpha\pi.$$

The symbol P.V. is necessary because the integrand has a singularity at $x = 1$ on the real axis where it changes sign, causing the ordinary improper integral to be undefined.

◇

In the final example, we integrate the product of the principal branch of the logarithmic function and a simple rational function around an indented semicircle with a branch cut. The method used may be extended to other integrals of similar type.

Example 2.7.8 An Integral Involving Ln z

By integrating $g(z) = \ln z/(z^2 + a^2)$ around the contour Γ_R in Figure 2.30, show that if $a > 0$ and the principal branch $\text{Ln}\, z$ of the logarithmic function is uses

$$\int_0^{\infty} \frac{\ln_e x}{x^2 + a^2} dx = \frac{\pi \ln_e a}{2a}.$$

SOLUTION

The function $\ln z$ has infinitely many branches and its branch point is located at the origin. To make $g(z)$ single-valued we must make a branch cut along the negative real axis up to and including the origin, that must itself be excluded from the path of integration by an indentation, the contour being closed by the semicircular arc C_R shown in Figure 2.30.

We are to use the principal branch Ln z of ln that is defined by the condition that Ln z is purely real when z lies on the positive real axis, that is when $\operatorname{Arg}(z) = 0$. Thus, for example, this branch will convert the point $z = c$ with $c > 0$ to the real number $\ln_e c$.

Because the denominator $z^2 + a^2$ of $g(z)$ has zeros at $z = \pm ia$, it follows that $g(z)$ is analytic everywhere inside Γ_R with the exception of the point $z = ia$. Using the modulus-argument representation for the zeros, in which $ia = e^{i\pi/2}$ and $-ia = e^{-i\pi/2}$, allows us to write

$$z^2 + a^2 = (z - ae^{-i\pi/2})(z - ae^{i\pi/2}).$$

Thus, if we set

$$f(z) = \frac{\operatorname{Ln} z}{(z - ae^{-i\pi/2})},$$

we can write

$$\int_{\Gamma_R} \frac{\operatorname{Ln} z}{z^2 + a^2}\,dz = \int_{\Gamma_R} \frac{f(z)}{z - z_0}\,dz,$$

where $z_0 = ae^{i\pi/2}$.

It then follows from the Cauchy integral theorem that

$$\begin{aligned}\int_{\Gamma_R} \frac{\operatorname{Ln} z}{z^2 + a^2}\,dz &= 2\pi i f(z_0) = 2\pi i \frac{\operatorname{Ln}(ae^{i\pi/2})}{ae^{i\pi/2} - ae^{-i\pi/2}} \\ &= \frac{\pi}{a}(\ln_e a + \tfrac{1}{2} i\pi).\end{aligned}$$

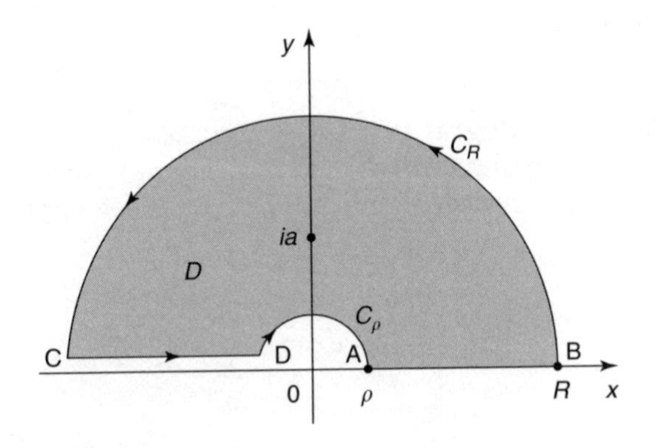

FIGURE 2.30
The contour Γ_R with a cut along the negative real axis and an indentation at the origin.

Expressing the integral around Γ_R as the sum of integrals over its respective segments gives

$$\int_{AB} g(z)dz + \int_{C_R} g(z)dz + \int_{CD} g(z)dz + \int_{C_\rho} g(z)dz = \frac{\pi}{a}(\ln_e a + \tfrac{1}{2}\pi i).$$

Because $g(z)$ satisfies the conditions of Theorem 2.7.3, the integrals around C_R and C_ρ will vanish in the limit as $\rho \to 0$ and $R \to \infty$, leaving only the integrals along AB and CD on the left side of the equation. On AB, $z = re^{0i} = r$ so $dz = dr$, while on CD, $z = re^{i\pi}$, so $dz = re^{i\pi}\, dr$. Proceeding to the limits and making these substitutions we find that

$$\int_0^\infty \frac{\ln_e r}{r^2 + a^2} dr - \int_\infty^0 \frac{\text{Ln}(re^{i\pi})}{r^2 + a^2} dr = \frac{\pi}{a}\left(\ln_e a + \tfrac{1}{2}\pi i\right).$$

After simplification this becomes

$$\int_0^\infty \frac{\ln_e r}{r^2 + a^2} dr + \int_0^\infty \frac{(\ln_e r + \pi i)}{r^2 + a^2} dr = \frac{\pi}{a}\left(\ln_e a + \tfrac{1}{2}\pi i\right),$$

or

$$2\int_0^\infty \frac{\ln_e r}{r^2 + a^2} dr + \pi i \int_0^\infty \frac{dr}{r^2 + a^2} = \frac{\pi}{a}\left(\ln_e a + \tfrac{1}{2}\pi i\right).$$

The second integral is elementary and has the value $\pi/(2a)$, causing the result to reduce to

$$2\int_0^\infty \frac{\ln_e r}{r^2 + a^2} dr + \frac{i\pi^2}{2a} = \frac{\pi}{a}\left(\ln_e a + \tfrac{1}{2}\pi i\right),$$

and so to

$$\int_0^\infty \frac{\ln_e r}{r^2 + a^2} dr = \frac{\pi \ln_e a}{2a},$$

which becomes the required result when the dummy variable r is replaced by x. ◇

Exercises 2.7

1. Use real variable integration to show that

$$\text{P.V.}\int_2^8 \frac{dx}{x - 4} = \ln_e 2, \text{ but that } \int_2^8 \frac{dx}{x - 4} \text{ is divergent in the usual sense.}$$

2. Evaluate

$$\int_{\Gamma_R} \exp(iz^2)dz,$$

where Γ_R is the perimeter OABO of the sector shown in Figure 2.30 and by letting $R \to \infty$ and using the Fresnel integrals in Example 2.7.1, show that for $0 \le \alpha < \pi/4$ $\int_0^\infty \exp(-x^2 \sin 2\alpha)\cos(x^2 \cos 2\alpha)dx = \frac{1}{4}(\sin\alpha + \cos\alpha)\sqrt{2\pi}$.

3. By considering the integral

$$\int_{\Gamma_R} z\exp(-z^2)dz,$$

show that if Γ_R is the contour in Figure 2.31 and $0 < \alpha < \pi/4$, that in the limit $R \to \infty$,

$$\int_0^\infty x\exp(-x^2 \cos 2\alpha)\cos[x^2 \sin 2\alpha - 2\alpha]dx = \tfrac{1}{2}.$$

4. Evaluate

$$\int_{\Gamma_R} \exp(-az^2)dz \qquad (a > 0)$$

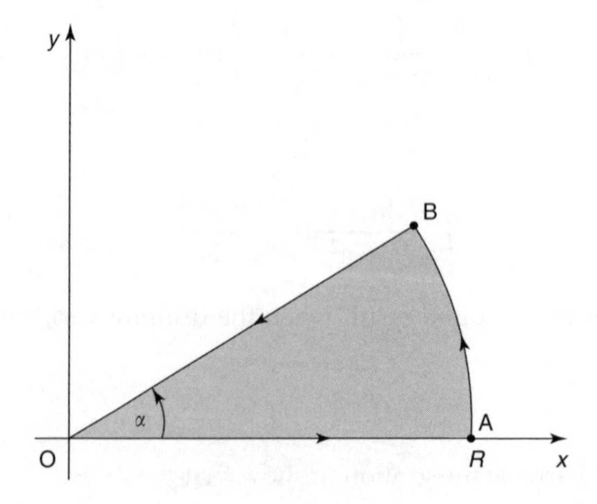

FIGURE 2.31
The contour Γ_R.

where Γ_R is the contour ABCDA shown in Figure 2.32 that in the limit as $R \to \infty$,

$$\int_0^{\infty} \exp(-ax^2)\cos(2bx)dx = \frac{1}{2}\sqrt{\frac{\pi}{a}}\exp(-b^2/a).$$

5. Evaluate

$$\int_{\Gamma_R} \frac{(z+1)e^{iz}}{z^2 - \pi z + 1 + \frac{1}{4}\pi^2}dz,$$

where Γ_R is the contour ABCA shown in Figure 2.33 and by letting $R \to \infty$ find

$$I_1 = \int_{-\infty}^{\infty} \frac{(x+1)\cos x}{x^2 - \pi x + 1 + \frac{1}{4}\pi^2}dx \quad \text{and} \quad I_2 = \int_{-\infty}^{\infty} \frac{(x+1)\sin x}{x^2 - \pi x + 1 + \frac{1}{4}\pi^2}dx.$$

6. Show by integrating $f(z) = 1/(1 + z^{2n})$ around the contour Γ_R forming the perimeter OABO of the sector in Figure 2.34 that in the limit as $R \to \infty$

$$\int_0^{\infty} \frac{dx}{1 + x^{2n}}dx = \frac{\pi}{2n\sin(\pi/2n)}.$$

7. Show by integrating

$$f(z) = \frac{z^{2m}}{1 + z^{2n}} \quad (m, n \text{ integers with } 0 \leqslant m < n)$$

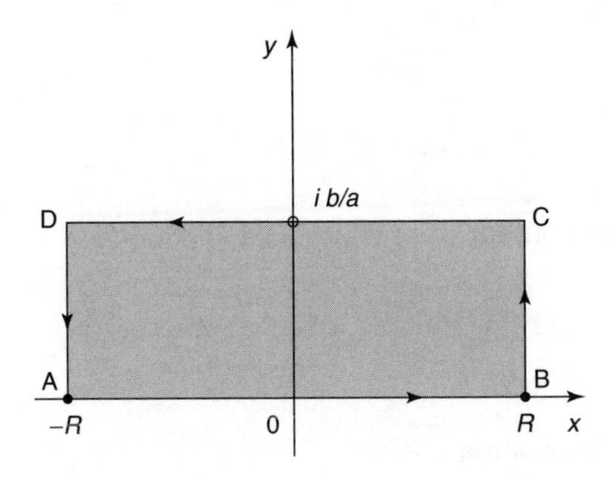

FIGURE 2.32
The rectangular contour Γ_R.

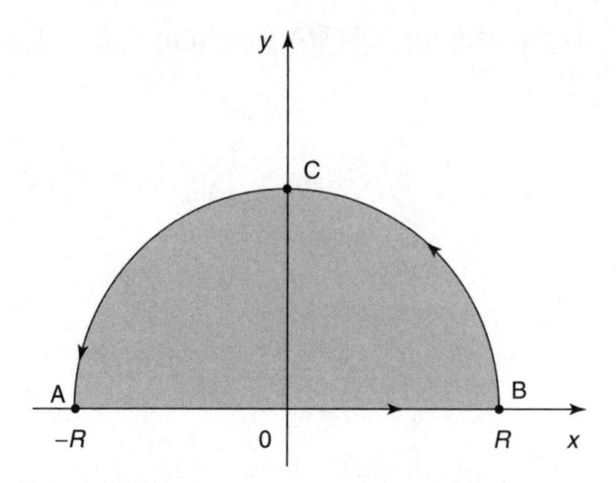

FIGURE 2.33
The semicircular contour Γ_R.

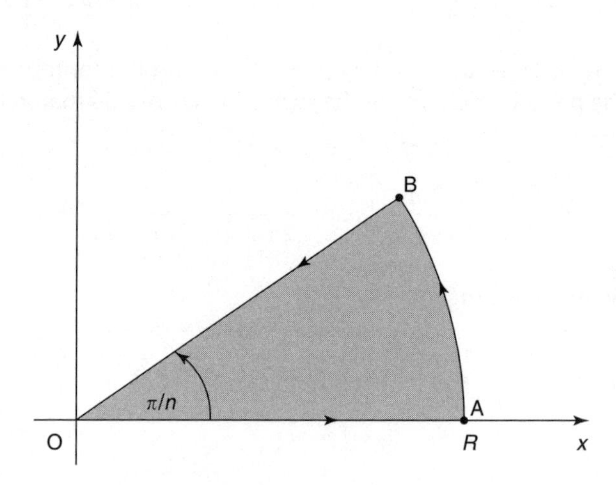

FIGURE 2.34
The sector Γ_R.

around the contour Γ_R in Figure 2.34 and letting $R \to \infty$, that

$$\int_0^\infty \frac{x^{2m}}{1+x^{2n}}\,dx = \frac{\pi}{2n\sin[\pi(2m+1)/2n]}.$$

8. Show by integrating

$$f(z) = \frac{z^2}{1+z^4}$$

around the contour Γ_R in Figure 2.35 and letting $R \to \infty$, that

$$\int_0^\infty \frac{x^2}{1+x^4}\,dx = \frac{\pi}{2\sqrt{2}}.$$

9. By integrating

$$f(z) = \frac{e^{imz}}{(z^2+a^2)(z^2+b^2)} \qquad (a > 0,\ b > 0,\ a \neq b)$$

around the contour Γ_R in Figure 2.35 and letting $R \to \infty$, find

$$I = \int_{-\infty}^\infty \frac{\cos mx}{(x^2+a^2)(x^2+b^2)}\,dx.$$

10. By integrating the function

$$f(z) = \frac{z^2+1}{z^4+1}$$

around the contour Γ_R in Figure 2.35 and letting $R \to \infty$, show that

$$\int_0^\infty \frac{x^2+1}{x^4+1}\,dx = \pi/\sqrt{2}.$$

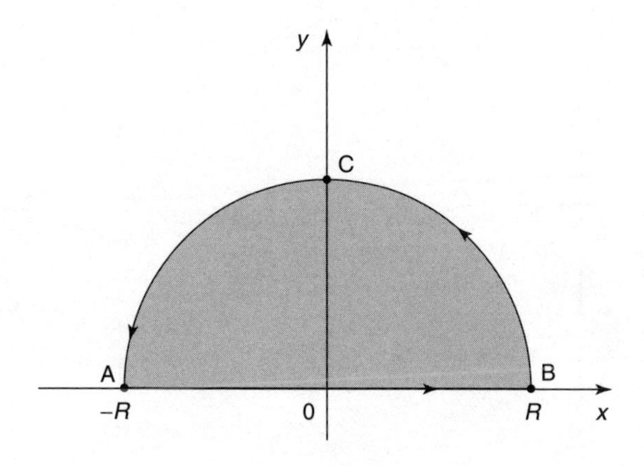

FIGURE 2.35
The semicircle Γ_R.

11. Making the change of variable $x = mt$ ($m \neq 0$ a constant) in the *Dirichlet integral* of Example 2.7.5 shows that

$$\int_0^\infty \frac{\sin mx}{x}\,dx = \begin{cases} \frac{1}{2}\pi, & \text{for } m > 0, \\ 0, & \text{for } m = 0, \\ -\frac{1}{2}\pi, & \text{for } m < 0. \end{cases}$$

Prove this by integrating $f(z) = e^{imz}/z$ around the contour Γ_R in Figure 2.36(a) and then letting $\rho \to 0$ and $R \to \infty$. Why is it not possible to use this contour when $m < 0$, so how may the result be obtained?

12. Evaluate

$$\int_{\Gamma_R} \left(\frac{e^{3iz} - 3e^{iz} + 2}{z^3} \right) dz$$

where Γ_R is the indented contour shown in Figure 2.36(a) and by letting $\rho \to 0$ and $R \to \infty$ show that

$$\int_0^\infty \left(\frac{\sin x}{x} \right)^3 dx = \tfrac{3}{8}\pi.$$

13. Show by integrating

$$f(z) = \frac{e^{imz}}{z(z^2 + a^2)} \qquad (m > 0, a > 0)$$

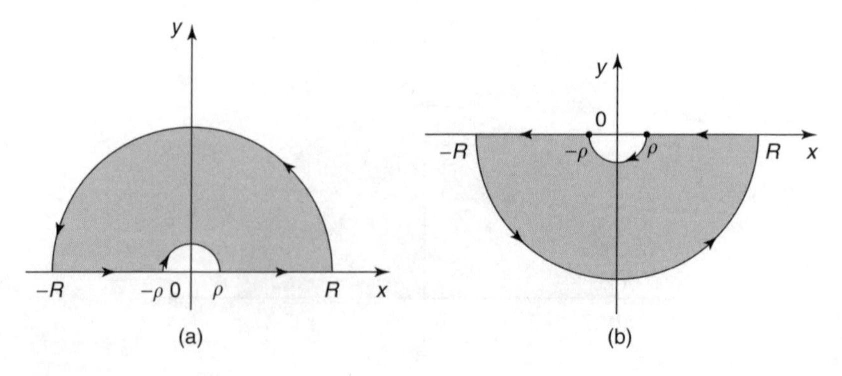

FIGURE 2.36
(a) The contour Γ_R. (b) The alternative contour Γ_R.

around the contour Γ_R shown in Figure 2.36(a) and then letting $\rho \to 0$ and $R \to \infty$, that

$$\int_0^\infty \frac{\sin mx}{x(x^2 + a^2)}\, dx = \frac{\pi}{2a}(1 - e^{-ma}).$$

14. By considering

$$\int_{\Gamma_R} \frac{e^{iz}}{(z^2 + 4)(z - 1)}\, dz$$

where Γ_R is the contour shown in Figure 2.37. Show by letting $\rho \to 0$ and $R \to \infty$, that

$$\text{P.V.} \int_{-\infty}^{\infty} \frac{\cos x}{(x^2 + 4)(x - 1)}\, dx = -\frac{\pi}{5}\left(\sin 1 + \frac{1}{2e^2}\right),$$

and

$$\text{P.V.} \int_{-\infty}^{\infty} \frac{\sin x}{(x^2 + 4)(x - 1)}\, dx = \frac{\pi}{5}\left(\cos 1 - \frac{1}{2e^2}\right).$$

15. Let

$$g(z) = \frac{e^{iz}}{(z^2 + 9)(z - 1)},$$

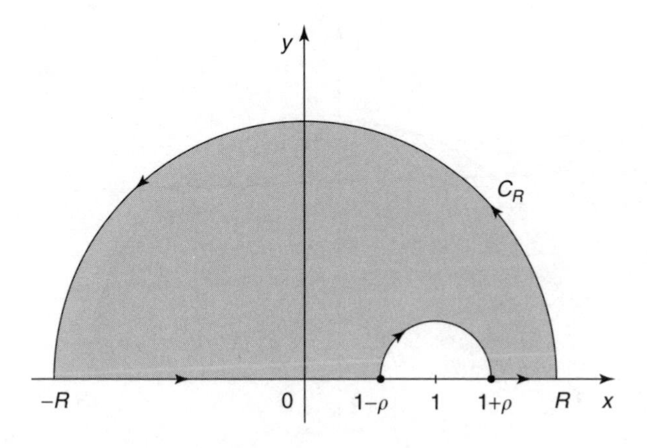

FIGURE 2.37
The contour Γ_R indented on the real axis and the semicircle C_R.

and Γ_R is the contour shown in Figure 2.37. By letting $\rho \to 0$ and $R \to \infty$, find

$$I_1 = \text{P.V.}\int_{-\infty}^{\infty} \frac{\cos x}{(x^2+9)(x-1)}\,dx \quad \text{and} \quad I_2 = \text{P.V.}\int_{-\infty}^{\infty} \frac{\sin x}{(x^2+9)(x-1)}\,dx.$$

16. Show by integrating

$$f(z) = \frac{e^{i\alpha z} - e^{i\beta z}}{z^2}$$

around the contour Γ_R is the contour shown in Figure 2.38 and letting $\rho \to 0$ and $R \to \infty$, that

$$\int_0^{\infty} \frac{\cos \alpha x - \cos \beta x}{x^2}\,dx = \tfrac{1}{2}\pi(\beta - \alpha) \quad (\alpha \geqslant 0,\ \beta \geqslant 0).$$

17. Show by integrating

$$f(z) = \frac{1}{(z^2+1)^4}$$

around the semicircle ABCA shown in Figure 2.39 and letting $R \to \infty$ that

$$\int_0^{\infty} \frac{dx}{(x^2+1)^4} = \tfrac{5}{32}\pi.$$

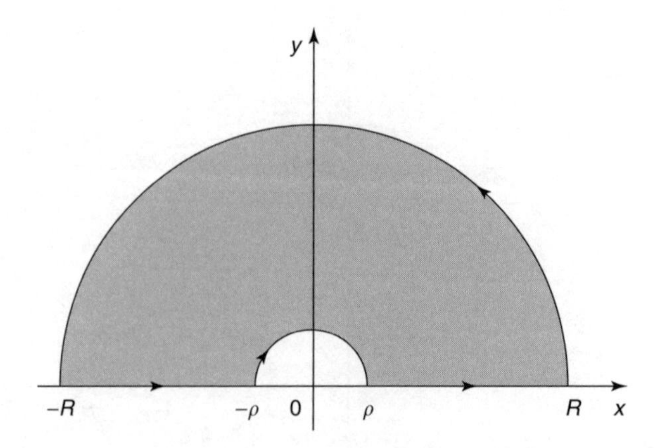

FIGURE 2.38
The indented contour Γ_R.

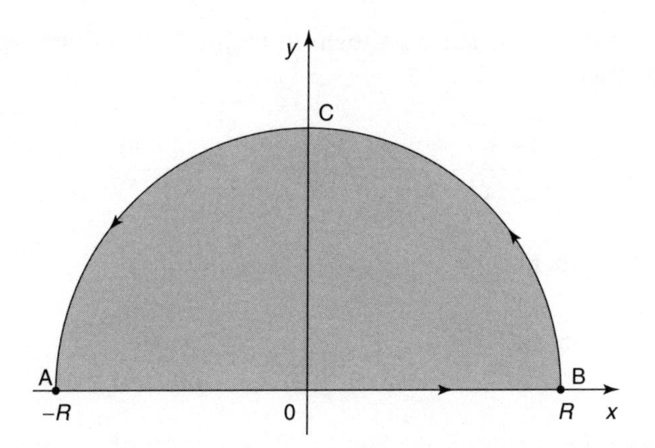

FIGURE 2.39
The semicircle Γ_R.

18. By integrating

$$f(z) = \frac{z^2}{(z^2 + a^2)^2}$$

around the semicircle ABCA shown in Figure 2.39 and letting $R \to \infty$, find

$$I = \int_0^\infty \frac{x^2}{(x^2 + a^2)^2}\, dx.$$

19. By integrating

$$f(z) = \frac{z^2}{(z^2 + a^2)}$$

around the semicircle ABCA shown in Figure 2.39 and letting $R \to \infty$, show that

$$\int_0^\infty \frac{x^2}{(x^2 + 1)^2}\, dx = \tfrac{1}{4}\pi.$$

20. By integrating

$$f(z) = \frac{1}{(z^2 + a^2)^2(z^2 + b^2)} \qquad (a > 0, b > 0)$$

around the semicircle Γ_R shown in Figure 2.39 and letting $R \to \infty$, show that

$$\int_{-\infty}^{\infty} \frac{dx}{(x^2 + a^2)^2(x^2 + b^2)} = \frac{\pi(2a + b)}{2a^3 b(a + b)^2}.$$

21. Show by integrating

$$f(z) = \frac{z^\alpha}{1 + z^2} \qquad (0 < \alpha < 1)$$

around the contour Γ_R in Figure 2.38 using the principal; branch of z^α, and letting $\rho \to 0$ and $R \to \infty$, that

$$\int_0^{\infty} \frac{x^\alpha}{1 + x^2}\, dx = \frac{\pi}{2\cos(\pi a/2)}.$$

22. Let

$$g(z) = \frac{z^{\alpha-1}}{z + c}, \qquad (c > 0, 0 < \alpha < 1).$$

Take the circular contour C to be the limiting contour Γ_R in (a) of Example 2.7.7, and consider $\int_C g(z)dz$. By using the principal branch of $z^{\alpha-1}$ and taking the appropriate limits to show that

$$\int_0^{\infty} \frac{x^{\alpha-1}}{x + c}\, dx = \frac{\pi c^{\alpha-1}}{\sin \alpha\pi}.$$

23. Let $g(z) = z^{\alpha-1} e^{-z}$ with $\alpha > 0$. By considering $\int_{\Gamma_R} g(z)dz$, using the principal branch of $z^{\alpha-1}$ and taking Γ_R to be the sector of the annulus shown in Figure 2.40 with $-\frac{1}{2}\pi < \beta < \frac{1}{2}\pi$, show by letting $\rho \to 0$ and $R \to \infty$ that

$$\int_0^{\infty} x^{\alpha-1} e^{-x\cos\beta} \cos(x \sin\beta) dx = \Gamma(\alpha)\cos(\alpha\beta)$$

and

$$\int_0^{\infty} x^{\alpha-1} e^{-x\cos\beta} \sin(x \sin\beta) dx = \Gamma(\alpha)\sin(\alpha\beta),$$

where $\Gamma(\alpha)$ is the gamma function.

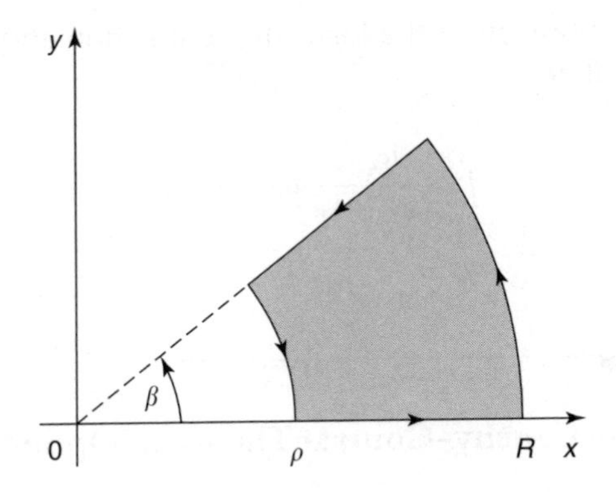

FIGURE 2.40
The sector Γ_R of an annulus.

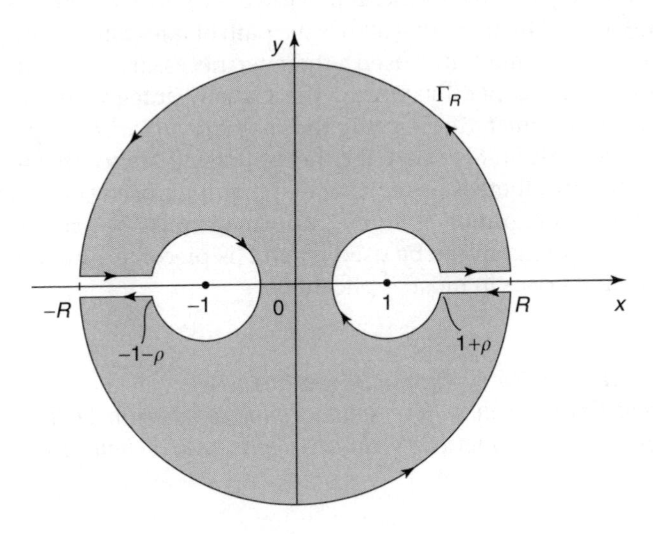

FIGURE 2.41
The cut and indented contour Γ_R.

24. Repeat Example 2.7.8, but this time using the branch $w_1(z)$ of $\ln z$ for which $\ln 1 = 2\pi i$, and hence show that using this branch of the logarithmic function also gives

$$\int_0^{\infty} \frac{\ln_e x}{x^2 + a^2}\, dx = \frac{\pi \ln_e a}{2a}.$$

25. Show by integrating

$$f(z) = \frac{\ln z}{(z^2 + 1)^2}$$

around the contour Γ_R in Figure 2.41, cut along the negative real axis and indented by a small semicircle to exclude the origin, using the principal branch of the logarithmic function and letting $\rho \to 0$ and $R \to \infty$, that

$$\int_0^\infty \frac{\ln_e x}{(x^2+1)^2}dx = -\tfrac{1}{4}\pi$$

2.8 Proof of the Cauchy–Goursat Theorem (Optional)

In this section, we prove the Cauchy–Goursat theorem mentioned in Section 2.2, where only the weaker form of the theorem due to Cauchy was established with the help of Green's theorem. However, subsequently, the result of the Cauchy–Goursat theorem that allows the path of integration to be piecewise continuous was assumed and used wherever necessary in what followed. Here we prove the generalization of the Cauchy Integral theorem to the *Cauchy–Goursat Integral theorem*, using the *proof by contradiction* approach due to Goursat, although first we state the theorem itself. Notice that in the statement of the theorem there is *no* requirement that the closed continuous curve Γ around which integration is to be performed must be *smooth*. Thus the Cauchy–Goursat theorem can be used when Γ is piecewise smooth, which is the situation that arises in most applications.

THEOREM 2.8.1 The Cauchy–Goursat Integral Theorem
Let a function $f(z)$ be analytic in a simply connected domain D. Then the value of the integral of $f(z)$ around every closed simple curve Γ lying entirely in D is such that

$$\int_\Gamma f(z)dz = 0.$$

PROOF

In effect, the proof involves approximating a region Δ in D by a set of n contiguous triangles with outer boundary γ_n, and after considering the effect of the partition of a single triangle, the limit is taken as $n \to \infty$. The result of the theorem follows because any closed-curve Γ, forming the boundary of the region Δ around which integration is to be performed, can be regarded as the limit as $n \to \infty$ of the polygonal path γ_n formed by the outer boundaries of the contiguous triangles that approximate Δ. So, although the boundary Γ must be continuous, it need only be piecewise continuous in order for it to be approximated by γ_n as $n \to \infty$.

Step (i) Consider first the case when Γ is the boundary of the triangle with interior Δ shown in Figure 2.42. As the area of the triangle is finite and $f(z)$ is

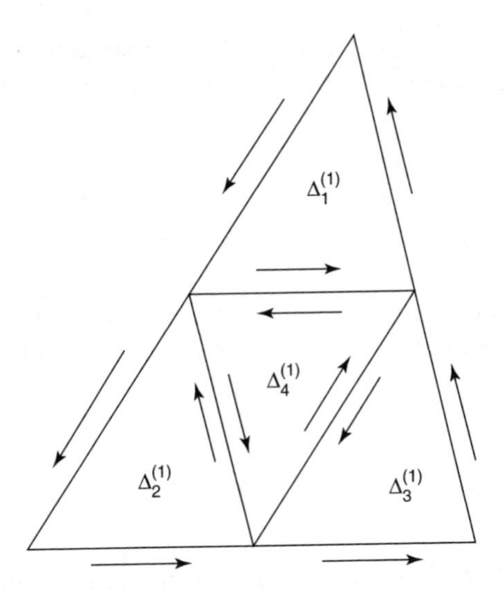

FIGURE 2.42
Triangle Δ and its subdivision into four subtriangles.

analytic in D, and so bounded in Δ, it follows that for some $m > 0$

$$\left|\int_{\Delta} f(z)dz\right| = m. \tag{2.55}$$

The triangle Δ is now partitioned into four subtriangles $\Delta_1^{(1)}$ to $\Delta_4^{(1)}$ by joining the midpoint of each side by straight lines, as shown in Figure 2.42, where the superscript (1) indicates the first partitioning of Δ, so that

$$\int_{\Delta} f(z)dz = \sum_{k=1}^{4} \int_{\Delta_k^{(1)}} f(z)dz. \tag{2.56}$$

The integrals on the right involve integration in opposite directions along each boundary of the inner triangle, so as $f(z)$ is continuous these integrals cancel. Results in Equations (2.55) and (2.56) imply that at least for one of these integrals, say the one over $\Delta_i^{(1)}$, we have

$$\left|\int_{\Delta_i^{(1)}} f(z)dz\right| \geqslant \tfrac{1}{4}m,$$

because if this is not so then

$$m = \left|\int_{\Delta} f(z)dz\right| \leqslant \sum_{k=1}^{4}\left|\int_{\Delta_k} f(z)dz\right| < 4 \cdot \tfrac{1}{4}m = m,$$

which is a contradiction because it implies that $m < m$.

If, now the triangle $\Delta_i^{(1)}$ is partitioned into four subtriangles in a similar way, the previous argument shows that for at least one of these subtriangles, say $\Delta_i^{(2)}$ we have

$$\left|\int_{\Delta_i^{(2)}} f(z)dz\right| \geqslant \frac{m}{4^2},$$

where now the superscript (2) indicates the second partition. Proceeding in this way, after n partitions we arrive at the following lower estimate for I_n, namely

$$I_n = \left|\int_{\Delta_i^{(n)}} f(z)dz\right| \geqslant \frac{m}{4^n}. \tag{2.57}$$

To proceed further, we must now find an upper estimate for I_n, so that I_n can be squeezed between the two estimates as $n \to \infty$. To do this, suppose the length of the boundary of the first triangle is L, then the length of the boundary of triangle $\Delta_i^{(n)}$ is $L_n = L/2^n$, so $L_n \to 0$ as $n \to \infty$. Thus the sequence of triangles is nested, with each triangle containing all subsequent triangles. Hence only one point z_0 can be, either inside Δ or on its boundary Γ, that belongs to all of these triangles.

As $z_0 \in D$, and $f(z)$ is differentiable, we can write

$$f(z) = f(z_0) + (z - z_0)f'(z_0) + R(z - z_0),$$

where $R(z - z_0)$ is the remainder term in this form of Taylor's theorem. Thus

$$\int_{\Delta_i^{(n)}} f(z)dz = f(z_0)\int_{\Delta_i^{(n)}} dz + f'(z_0)\int_{\Delta_I^{(N)}} z\,dz - z_0 f'(z_0)\int_{\Delta_i^{(n)}} dz + \int_{\Delta_I^{(N)}} R(z - z_0)dz. \tag{2.58}$$

It is known from elementary calculus that the remainder term $R(z - z_0)$ in Taylor's theorem that follows the term in $(z - z_0)$ is such that for every $\varepsilon > 0$, $\delta = \delta(\varepsilon) > 0$ exists such that

$$|R(z - z_0)| < \varepsilon|z - z_0|.$$

So using this result in Equation (2.58), and taking n sufficiently large that $\Delta_i^{(n)}$ lies in the circle $|z - z_0| < \delta$, it follows that

$$I_n = \left|\int_{\Delta_i^{(n)}} f(z)dz\right| \leqslant \int_{\Delta_i^{(n)}} |z - z_0||dz| < \varepsilon(L_n)^2 = \varepsilon L^2/4^n.$$

Thus I_n has the upper bound

$$I_n < \varepsilon \frac{L^2}{4^n}. \tag{2.59}$$

A comparison of Equations (2.57) and (2.59) gives $m/4^n < \varepsilon L^2/4^n$, or $m < \varepsilon L^2$. However this is impossible because $\varepsilon > 0$ is arbitrarily small and $m > 0$ is a constant, so the result can only be true if $m = 0$. Thus we have shown that $\int_\Gamma f(z)dz = 0$ for *all* triangles Δ inside D.

(ii) Now let γ_n be an arbitrary n sided polygon in D, and partition it as in (i). Then if γ_n is the outside perimeter of this set of triangles, $I_n = \int_{\gamma_n} f(z)dz$ is the sum of integrals around the boundary of each partitioned triangle, and it follows that $I_n = 0$. So as any polygon can always be partitioned into a finite number of convex polygons, it is always true that

$$\int_{\gamma_n} f(z)dz = 0. \tag{2.60}$$

The final stage in the proof involves appealing to integral inequality Equation (2.16) in Section 2.1. This can be used to show if γ_n is an n-sided polygonal approximation to Γ, then by taking n sufficiently large in such a way that the length of each polygonal side tends to zero, the integral $\int_\Gamma f(z)dz$ can be approximated as closely as desired by the integral $\int_{\gamma_n} f(z)dz$. This means that when n is sufficiently large, for some arbitrary $\varepsilon(n) > 0$ we can write

$$\left| \int_\Gamma f(z)dz - \int_{\gamma_n} f(z)dz \right| < \varepsilon.$$

Because $\varepsilon > 0$ is arbitrary and because of Equation (2.60), we have $\int_\Gamma f(z)dz = 0$; and the proof is complete. ◆

3

Taylor and Laurent Series: Residue Theorem and Applications

3.1 Sequences, Series, and Convergence

This chapter re-examines complex integration from the viewpoint of the representation of complex functions in terms of series. It shows how, by extending the concept of a Taylor series expansion of a function about a point z_0, a Laurent series expansion arises involving a sum of both positive and negative powers of $(z - z_0)$, where a term of the form $(z - z_0)^{-n}$ is called a *pole* of order n. Laurent series expansions describe functions that are analytic at all points of a domain with the exception of certain points, called *singularities*, where they cease to be differentiable. Laurent series, in turn, lead to the residue theorem that both simplifies the task of evaluating definite integrals and, in particular, makes it possible to evaluate definite integral that are too difficult to determine directly by using only the Cauchy integral theorem, as shown in Chapter 2.

As in the case of real variables, the starting point for this approach to complex analysis involves considering the relationship between sequences and series. An infinite set of complex numbers $z_1, z_2, \ldots, z_n, z_{n+1}, \ldots$, enumerated in the specific order indicated by the suffixes, is called an *infinite complex sequence*. Sometimes, an infinite sequence is indicated by listing the first few terms in order and writing $\{z_1, z_2, z_3, \ldots\}$, with the remaining terms are represented by the three dots ..., called an *ellipsis*. However this notation is often abbreviated to $\{z_n\}$, it being understood that the suffix n runs successively through all of the positive integers, in which case the term z_n is called the *nth term*, or *general term* of the sequence. The order in which terms occur in a sequence is important, and changing the order produces a different sequence with different properties.

A sequence $\{z_n\}$ with the property that $|z_n| < K$ for some fixed positive number K and all n is said to be *bounded*. Geometrically, this means that when the terms of a bounded sequence are plotted in the complex plane they will all lie inside a circle of radius K centered on the origin. When no such number K exists the sequence is said to be *unbounded*.

It can happen that the general term z_n of a sequence is determined explicitly in terms of n by some rule, and when this is so, the precise form of z_n can be written down for any given n. However, z_n may be generated in a more complicated fashion, perhaps by the repeated application of a mathematical operation, or algorithm, in which case is usually impossible to express z_n explicitly in terms of n. These ideas are now illustrated by means of some example.

Example 3.1.1 Some Complex Sequences Exhibiting Different Types of Behavior

(a) Write down and graph the first few terms of the sequence $\{z_n\}$ with the general term

$$z_n = 1 + 2i + r^n e^{\pi n i/10}, n = 1, 2, 3, \ldots,$$

first with $r = 0.9$ and then with $r = 1.1$. Comment on the behavior of the terms in each case as n increases.

(b) Write down and graph the first few terms of the sequence $\{z_n\}$ with the general term

$$z_n = (i)^n, n = 1, 2, 3, \ldots,$$

and comment on the behavior of the sequence.

(c) Examine the behavior of the sequence $\{z_n\}$ with the general term $z_n = z_{n-1}/(3z_{n-1} - |z_{n-1}|)$, for $n = 2, 3, 4, \ldots$, with $z_1 = \alpha$ an arbitrary complex number.

SOLUTION

(a) When $r = 0.9$, $z_n = 1 + 2i + (0.9)^n e^{n\pi i/10}$ for $n = 1, 2, 3, \ldots$, and it is seen from Figure 3.1(a) that the successive points spiral *inward* in a counterclockwise sense around the point $1 + 2i$. The arrow on the graph indicates the direction in which z_n moves as n increases. The factor 0.9^n decreases monotonically as n increases, so any circle centered on $1 + 2i$ must always contain an infinite number of points of the sequence, while a finite number will be outside it. Because the points of the sequence can always be contained within a circle of sufficiently large radius centered on the origin this sequence is *bounded*.

The situation is different when $r = 1.1$, because 1.1^n increases without bound as n increases, so that when points of this sequence are plotted in Figure 3.1(b) the points spiral *outward* in such a way that their distance from the origin increases without bound, so that this sequence is *unbounded*.

(b) A routine calculation shows that $z_1 = i$, $z_2 = -1$, $z_3 = -i$, and $z_4 = 1$, after which this same set of values is repeated periodically as n increases. This sequence, illustrated in Figure 3.1(c) experiences a cyclic periodicity as n increases and because $|z_n| = 1$ it follows that the sequence is *bounded*.

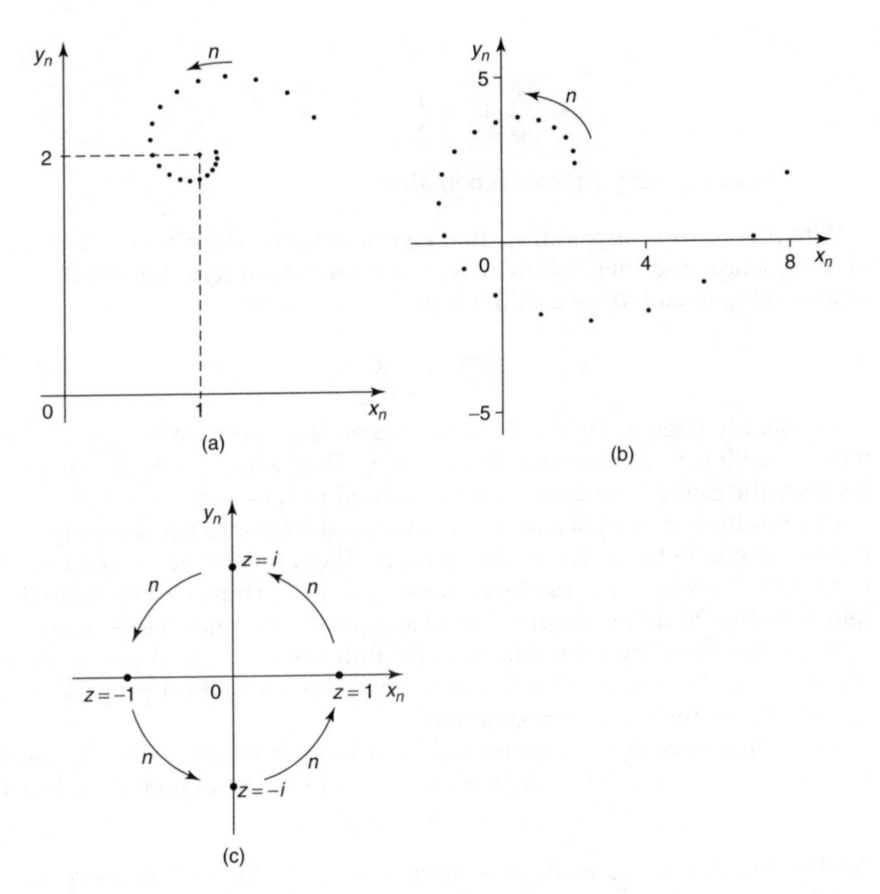

FIGURE 3.1
Graphical representations of sequences (a) $r = 0.9$ and (b) $r = 1.1$.

(c) It is not possible to determine z_n explicitly in terms of n and α, but something can still be deduced about the general behavior of the terms of the sequence. Taking the modulus of the algorithm defining z_n gives

$$|z_n| = \left|\frac{z_{n-1}}{3z_{n-1} - |z_{n-1}|}\right|$$
$$= \frac{|z_{n-1}|}{\left|3z_{n-1} - |z_{n-1}|\right|}, \quad n = 2, 3, \ldots, \quad \text{with } z_0 = \alpha \text{ arbitrary.}$$

However,

$$\left|3z_{n-1} - |z_{n-1}|\right| \geqslant \left|3|z_{n-1}| - |z_{n-1}|\right| = 2|z_{n-1}|,$$

so

$$|z_n| \leqslant \frac{|z_{n-1}|}{2|z_{n-1}|} = \frac{1}{2}, \quad n = 2, 3, \ldots,$$

for any α, so the sequence is bounded. ◇

With this example in mind, we first give an intuitive definition of the limit of a sequence, and then follow it with a formal definition. Informally, the sequence $\{z_n\}$ is said to have a limit α as $n \to \infty$, written

$$\lim_{n\to\infty} z_n = \alpha, \tag{3.1}$$

if for suitably large n, say $n > N$ with N some large positive integer, all the terms z_n with $n > N$ are arbitrarily close to α. Thus when $n > N$, the larger n becomes the closer the points z_n cluster around the point α.

This intuitive concept of a limit conveys the general idea satisfactorily, but it is too vague to be useful mathematically. The difficulty lies in the use of terms like "the closer points cluster about point α." Thus, we are led to the following precise definition of a limit of a sequence, in which the closeness of z_n to α is measured by the modulus of the difference $|z_n - \alpha|$. A precise definition of the concept of a limit is necessary if the analytical properties of sequences and series are to be examined.

We say that the complex sequence $\{z_n\}$ has a *limit* α if, for any arbitrarily small real number $\varepsilon > 0$, a positive integer N exists (possibly depending on ε), such that

$$|z_n - \alpha| < \varepsilon \quad \text{for all } n > N. \tag{3.2}$$

See Figure 3.2. This means that the number α will be the limit of a sequence $\{z_n\}$ if, no matter how small a circle is drawn about α, an infinite number of

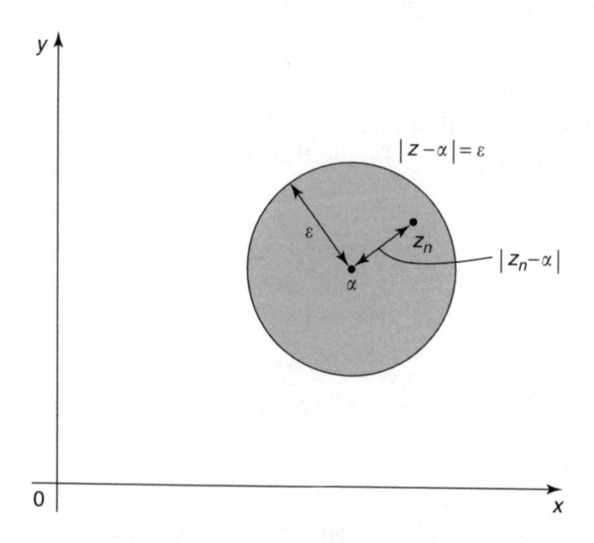

FIGURE 3.2
A point z_0 close to α, in the sense that $|z_0 - \alpha| < \varepsilon$ with $\varepsilon > 0$ arbitrary.

points of the sequence will lie inside the circle, but only a *finite* number will lie outside it. If the sequence $\{z_n\}$ has the limit α we say the sequence is *convergent* and that z_n converges to α as $n \to \infty$. A sequence that is not convergent is said to be *divergent*. Notice that the limit of a sequence does not necessarily belong to the sequence itself because it may happen, and indeed usually does, that for no integer n does z_n equal the value α.

With this definition of convergence to a limit in mind, it is easy to see that Example 3.1.1(a) is *convergent* when $r = 0.9$, but *divergent* when $r = 1.1$. The sequence in Example 3.1.1(b) only cycles through the four distinct values $z_1 = i$, $z_2 = -1$, $z_3 = -i$ and $z_4 = 1$ so clearly the sequence has no limit, and so it is *divergent*. In the case of Example 3.1.1(b) each of the values $z_1 = i$, $z_2 = -1$, $z_3 = -i$ and $z_4 = 1$ occurs infinitely often, so each of these points is called a *cluster point* of the sequence because any small circle about any one of these points contains infinitely many points of the sequence, but also excludes infinitely many points of the sequence, none of these cluster points can be a limit point. In general, a point β of a sequence is called a *cluster point* of the sequence if every deleted neighborhood of β contains an infinite number of points belonging to the sequence. A deleted neighborhood is used here because in general, unlike in Example 3.1.1(b), a cluster point is not a member of the sequence. Thus a convergent sequence has only one cluster point, namely its limit, while a divergent sequence may either have no cluster point, or more than one.

Notice that a convergent sequence is, of necessity, bounded, although the converse is not necessarily true, as can be seen from Example 3.1.1(b) where the sequence was bounded, but divergent.

Two important results will now be proved. The first being that when a limit of a sequence exists it is unique, while the second shows how the limit of a sequence can be deduced from the limits of the real and imaginary parts of the sequence.

THEOREM 3.1.1 Uniqueness of the Limit of a Sequence
If a sequence $\{z_n\}$ converges to the limit α, then the limit is unique.

PROOF

The proof will be by contradiction. Suppose, if possible, that a convergent sequence has two distinct limits α and α^*, so that

$$\lim_{n\to\infty} z_n = \alpha \quad \text{and} \quad \lim_{n\to\infty} z_n = \alpha^*,$$

with $\alpha \neq \alpha^*$. Then setting

$$\varepsilon = \tfrac{1}{2}|\alpha - \alpha^*|,$$

it follows from the definition of a limit that corresponding to this there is some positive integer $N = N(\varepsilon)$ (it *may* depend on ε) such that

$$|z_n - \alpha| < \varepsilon \quad \text{and} \quad |z_n - \alpha^*| < \varepsilon$$

for $n > N$. Using the triangle inequality we have

$$\begin{aligned}|\alpha - \alpha^*| &= \left|(\alpha - z_n) + (z_n - \alpha^*)\right| \\ &\leqslant |z_n - \alpha| + |z_n - \alpha^*| \\ &< \varepsilon + \varepsilon = 2 \cdot \tfrac{1}{2}|\alpha - \alpha^*|.\end{aligned}$$

Thus if two distinct limits α and α^* exist we have shown that $|\alpha - \alpha^*| < |\alpha - \alpha^*|$, which is impossible, so the limit α must be unique. ◆

The next theorem reduces the problem of finding the limit of a complex sequence $\{z_n\}$ with $z_n = x_n + iy_n$, to finding the limits of the two real sequences $\{x_n\}$ and $\{y_n\}$.

THEOREM 3.1.2 A Complex Sequence Has a Limit if Its Real and Imaginary Parts Have Limits

Let the general term of a complex sequence $\{z_n\}$ be of the form $z_n = x_n + y_n$, and set $\alpha = a + ib$ with a, b real numbers. Then

$$\lim_{n\to\infty} z_n = \alpha$$

if and only if,

$$\lim_{n\to\infty} x_n = a \quad \text{and} \quad \lim_{n\to\infty} y_n = b.$$

PROOF

The proof is in two parts. Let us suppose first that $\{z_n\}$ converges to the limit α. Then for any $\varepsilon > 0$ a positive integer $N = N(\varepsilon)$ exists such that $|z_n - \alpha| < \varepsilon$ if $n > N$. However, $z_n - \alpha = (x_n - a) + i(y_n - b)$, so $|z_n - \alpha|^2 = (x_n - a)^2 + (y_n - b)^2$ and thus $(x_n - a)^2 + (y_n - b)^2 < \varepsilon^2$, showing that

$$|x_n - a| < \varepsilon \quad \text{and} \quad |y_n - b| < \varepsilon \text{ if } n > N.$$

These results are simply the definitions of the convergence of $\{x_n\}$ to the limit a and $\{y_n\}$ to the limit b, which together form the last statements in the theorem. To complete the proof, let us suppose that

$$\lim_{n\to\infty} x_n = a \quad \text{and} \quad \lim_{n\to\infty} y_n = b,$$

and use these assumptions to prove that then $\lim_{n\to\infty} z_n = \alpha$. Restricting the definition of the limit of a sequence to the real variable case we see that these two limits imply that for any $\varepsilon > 0$, positive integers $M = M(\varepsilon)$ and $M^* = M^*(\varepsilon)$ exist such that

$$|x_n - a| < \varepsilon/\sqrt{2} \text{ if } n > M \quad \text{and} \quad |y_n - b| < \varepsilon/\sqrt{2} \text{ if } n > M^*.$$

Next we prove the sufficiency of the condition. Let $|z_n - z_m| < \varepsilon$ for all $n > m > N$. This implies that

$$[(x_n - x_m)^2 + (y_n - y_m)^2]^{1/2} < \varepsilon,$$

for all $n > m > N$. Thus $|x_n - x_m| < \varepsilon$ and $|y_n - y_m| < \varepsilon$ for $n > m > N$, so the sequences $\{x_n\}$ and $\{y_n\}$ have limits, so as Theorem 3.1.2 implies $\lim_{n\to\infty} z_n$ exists the proof is complete. ◆

The Cauchy convergence principle can be rephrased to give the following *Cauchy test for convergence* of a sequence.

THEOREM 3.1.5 The Cauchy Convergence Test
The sequence $\{z_n\}$ converges if and only if

$$\lim_{m,n\to\infty} |z_n - z_m| = 0.$$

PROOF

The test follows as the limit of Theorem 3.1.4 as $\varepsilon \to 0$, for then $N \to \infty$, and thus m and n must tend to infinity independently because m and n in Theorem 3.1.4 are independent integers. ◆

A sequence that satisfies the Cauchy convergence test is called a *Cauchy sequence*.

Example 3.1.3 An Application of the Cauchy Convergence Test

Use the Cauchy convergence test to prove the convergence of the sequence $\{z_n\}$ with the general term

$$z_n = \frac{(n+1) + i(2n-1)}{n}.$$

SOLUTION
We have

$$\begin{aligned} |z_n - z_m| &= \left| \frac{(n+1) + i(2n-1)}{n} - \frac{(m+1) + i(2m-1)}{m} \right| \\ &= \left| \frac{(m-n) + i(n-m)}{mn} \right| = \left| \frac{1}{n} - \frac{1}{m} + i\left(\frac{1}{m} - \frac{1}{n}\right) \right| \leqslant 2\left(\frac{1}{n} + \frac{1}{m}\right), \end{aligned}$$

showing that

$$\lim_{m,n\to\infty} |z_n - z_m| = \lim_{m,n\to\infty} 2\left(\frac{1}{n} + \frac{1}{n}\right) = 0.$$

Thus $\{z_n\}$ is a Cauchy sequence, and so converges to a limit. In this case the sequence is simple enough for both the existence and value of the limit to be obvious once z_n is written as

$$z_n = \left(1 + \frac{1}{n}\right) + i\left(2 - \frac{1}{n}\right),$$

from which it follows that $\lim_{n\to\infty} z_n = 1 + 2i$. ◇

Having finished discussing complex sequences, we are now in a position to develop the concept of infinite series and to construct tests for their convergence. If $z_1, z_2, z_3, \ldots,$ are terms of an infinite sequence of complex numbers $\{z_n\}$, the sum of N terms $z_1 + z_2 + \cdots + z_N$, and the sum of an infinite number of terms $z_1 + z_2 + z_3, \ldots,$ denoted, respectively by

$$\sum_{n=1}^{N} z_n \quad \text{and} \quad \sum_{n=1}^{\infty} z_n,$$

are called a *finite* and an *infinite series*, where the term *infinite* refers to the number of terms summed, and not to the value of the sum, where the numbers z_n are called the *terms* of the series. The summation index n is a *dummy index* because its occurrence in the limits of the summation symbol Σ merely serve to indicate with which term the summation starts and with which term it finishes, so n may be replaced by any other symbol without changing the meaning of the summation, so that

$$\sum_{n=1}^{N} z_n = \sum_{r=1}^{N} z_r = \cdots = \sum_{k=1}^{N} z_k \quad \text{and} \quad \sum_{n=1}^{\infty} z_n = \sum_{p=1}^{\infty} z_p = \cdots = \sum_{q=1}^{\infty} z_q.$$

The finite sum

$$S_N = z_1 + z_2 + \cdots + z_N \tag{3.3}$$

is called the *Nth partial sum of the sequence* $\{z_n\}$ because it is simply the sum of the first N terms of the sequence.

The connection between sequences and series follows from the fact that we say an infinite series converges (has a finite sum) if the sequence of partial sums has a limit L, so that

$$\lim_{N\to\infty} S_N = \lim_{N\to\infty}\left(\sum_{n=1}^{N} z_n\right) = L.$$

When the number L exists it is called the sum of the series, and a series that does not converge is said to diverge. The divergence of a series may be due to the sum being infinite, or to the fact that the limit L does not exist.

Associated with every infinite series of complex numbers $\Sigma_{n=1}^{\infty} z_n$ is an infinite series of real numbers $\Sigma_{n=1}^{\infty}|z_n|$, and we say the series $\Sigma_{n=1}^{\infty} z_n$ is absolutely convergent if the associated series $\Sigma_{n=1}^{\infty}|z_n|$ is convergent.

THEOREM 3.1.6 Absolute Convergence Implies Convergence
If the series $\Sigma_{n=1}^{\infty}|z_n|$ is convergent, then $\Sigma_{n=1}^{\infty} z_n$ is convergent or, when expressed in words, the absolute convergence of a series implies its convergence.

PROOF

Let $S_N = \Sigma_{n=1}^{N} z_n$ and $\Sigma_{n=1}^{N}|z_n|$, so that S_N is the nth partial sum of Σz_n and s_N is the Nth partial sum of $\Sigma|z_n|$ (the summation limits are often omitted when discussing general properties).

Then, by hypothesis, the sequence $\{S_N\}$ is convergent, so for any $\varepsilon > 0$ an integer $N = N(\varepsilon)$ exists such that

$$|S_n - S_m| < \varepsilon \quad \text{for all } n > m > N,$$

which is equivalent to

$$|z_{m+1}| + |z_{m+2}| + \cdots + |z_n| < \varepsilon.$$

However,

$$s_n - s_m = z_{m+1} + z_{m+2} + \cdots + z_n,$$

but

$$|z_{m+1} + z_{m+2} + \cdots + z_n| \leqslant |z_{m+1}| + |z_{m+2}| + \cdots + |z_n|,$$

and so

$$|s_n - s_m| < \varepsilon \quad \text{for all } n > m > N.$$

This is simply the statement that the sequence $\{s_n\}$ converges, so the theorem is proved. ◆

It is important to understand that the converse of this theorem is not necessarily true and that convergence does not necessarily imply absolute convergence. This can be seen by considering the series $\Sigma_{n=1}^{\infty}(-1)^{n+1}((1/n) + i(3/n))$, because the real series $\Sigma_{n=1}^{\infty}((-1)^{n+1}/n)$ occurring in both the real and imaginary parts is convergent by the alternating series test[1], but the series of absolute values $\Sigma_{n=1}^{\infty} 1/n$ is the divergent harmonic series. So the complex series is convergent, but not absolutely convergent.

Because the sum of the first M terms of an infinite series is finite in number, and so has a finite sum, the omission of these terms from an infinite series will not affect its convergence, though it will of course change its sum. This observation is important, because it happens frequently that the summation of an infinite series does not start with the summation index equal to unity. Thus the series $\Sigma_{n=1}^{\infty} z_n$ and the series $\Sigma_{n=M}^{\infty} z_n$ will converge or diverge together.

It is an immediate consequence of Theorem 3.1.2 that when a complex series Σz_n converges to the sum L, the sum of the real parts of z_n converge to the real part of L, while the sum of the imaginary parts of z_n converge to the imaginary part of L. This result forms the statement of the next theorem.

THEOREM 3.1.7 The Sum of a Complex Series
Let $z_n = x_n + iy_n$ and $L = a + ib$. Then

$$L = \sum_{n=1}^{\infty} z_n \quad \text{if and only if, } a = \sum_{n=1}^{\infty} x_n \quad \text{and} \quad b = \sum_{n=1}^{\infty} y_n. \qquad \blacklozenge$$

Example 3.1.4 An Application of Theorem 5.7
Sum the series $\Sigma_{n=1}^{\infty} z_n$ where the general term is $z_n = (\frac{1}{2})^n + i(\frac{1}{3})^n$.

SOLUTION
In this case $x_n = (\frac{1}{2})^n$ and $y_n = (\frac{1}{3})^n$, so $S = \Sigma_{n=1}^{\infty}(\frac{1}{2})^n + i\Sigma_{n=1}^{\infty}(\frac{1}{3})^n$. These are geometric series with the respective sums $\Sigma_{n=1}^{\infty}(\frac{1}{2})^n = 1$ and $\Sigma_{n=1}^{\infty}(\frac{1}{3})^n = (\frac{1}{2})$, so that $\Sigma_{n=1}^{\infty} z_n = 1 + (\frac{1}{2})i$. $\lozenge$

A necessary, but not sufficient, condition for the convergence of a series follows from Theorem 3.1.5 by setting $m = n - 1$ and applying the result to the sequence of partial sums $\{S_n\}$. Because $S_n - S_{n-1} = z_n$, it follows that a necessary condition for convergence is that $\lim_{n\to\infty} z_n = 0$. That this condition is not a sufficient one for convergence follows from the fact that we have related m to n, whereas in Theorem 3.1.5 the necessary and sufficient conditions followed by allowing m and n to tend to infinity independently. We have established the next result.

THEOREM 3.1.8 A Necessary Condition for Convergence
If the complex series $\Sigma_{n=1}^{\infty} z_n$ is convergent, then $\lim_{n\to\infty} z_n = 0$.

An immediate consequence of this theorem is that a series $\Sigma_{n=1}^{\infty} z_n$ for which $\lim_{n\to\infty} z_n \neq 0$ is divergent. $\blacklozenge$

[1] A real *alternating series* has the form $\Sigma_{n=1}^{\infty}(-1)^n a_n$ with $a_n > 0$, so the signs of the terms alternate. The *alternating series test* for real series says that this series converges if $a_{n+1} < a_n$ for all n, and $\lim_{n\to\infty} a_n = 0$.

Example 3.1.5 A Divergent Series for which $\lim_{n\to\infty} z_n \neq 0$

If $z_n = (1+i)/n$, show that, although $\lim_{n\to\infty} z_n = 0$, the series $\Sigma_{n=1}^{\infty} z_n$ is divergent.

SOLUTION

The result is almost immediate because clearly $\lim_{n\to\infty} z_n = 0$, but the series $\Sigma_{n=1}^{\infty} 1/n$ is the divergent harmonic series so the series $\Sigma_{n=1}^{\infty} z_n$ is divergent. ◇

When working with series, it can happen that two convergent series need to be scaled by complex multipliers and then added. The justification for these operations is provided by applying Theorem 3.1.3(i) to the respective partial sums, when it yields the following result.

THEOREM 3.1.9 Scaling and Addition of Complex Series

Let $\Sigma_{n=1}^{\infty} z_n$ and $\Sigma_{n=1}^{\infty} z_n$ be two convergent series with the respective sums L_1 and L_2, and let λ, μ be any two complex numbers. Then the series $\Sigma_{n=1}^{\infty} (\lambda z_n + \mu Z_n)$ converges to the sum $L_1 + L_2$. ◆

It was proved in Theorem 3.1.6 that the absolute convergence of complex series implies its convergence, so as the absolute convergence of a complex series involves investigating the convergence of a real series of positive terms, we may test for convergence by using any real variable convergence test that applies to series with positive terms. Thus we may use the *comparison test*, the *ratio test* (also called the *D'Alembert ratio test*), and the *nth root test*. The proofs of these theorems are found in any text offering a first course in calculus, so we merely state the tests in a form appropriate to a complex series. It should be remembered that, in general, convergence tests only give information about the convergence of a series, and not about its sum if it is convergent.

THEOREM 3.1.10 Comparison Test

Let $\Sigma_{n=1}^{\infty} a_n$ be a convergent series of real numbers a_n with $a_n \geqslant 0$ for all n, and let $\Sigma_{n=1}^{\infty} z_n$ be a complex series. Then if $|z_n| \leqslant a_n$, the series $\Sigma_{n=1}^{\infty} z_n$ is both convergent and absolutely convergent. ◆

THEOREM 3.1.11 The Ratio Test

If the complex series $\Sigma_{n=1}^{\infty} z_n$ is such that the sequence of real numbers $\{|z_{n+1}/z_n|\}$ is defined for all n and has a limit

$$\lim_{n\to\infty} \left| \frac{z_{n+1}}{z_n} \right| = L,$$

then the series $\Sigma_{n=1}^{\infty} z_n$

(i) *Is both convergent and absolutely convergent if* $L < 1$
(ii) *Is divergent if* $L > 1$
(iii) *The test offers no information about convergence if* $L = 1$. ◆

THEOREM 3.1.12 The n*th Root Test*
If the complex series $\Sigma_{n=1}^{\infty} z_n$ *is such that*

$$\lim_{n\to\infty} |z_n|^{1/n} = L$$

exists, then the series $\Sigma_{n=1}^{\infty} z_n$

(i) *Is both convergent and absolutely convergent if* $L < 1$
(ii) *Is divergent if* $L > 1$
(iii) *The test offers no information about convergence if* $L = 1$. ◆

When testing series for convergence, it is often helpful to make use of the following properties of series and limits.

1. A series Σa_n is divergent if $\lim_{n\to\infty} a_n \neq 0$
2. $\Sigma_{n=1}^{\infty} r^{n-1}$ is convergent for $|r| < 1$ and divergent for $|r| \geqslant 1$ **(geometric series)**
3. $\Sigma_{n=1}^{\infty} 1/n^p$ is convergent for $p > 1$ and divergent for $0 < p \leqslant 1$
4. $\Sigma_{n=1}^{\infty} 1/n!$ is convergent
5. $\Sigma_{n=1}^{\infty} 1/(2n-1)!$ is convergent
6. $\Sigma_{n=1}^{\infty} 1/(2n)!$ is convergent
7. $\lim_{n\to\infty} ((\ln_e n/n)) = 0$ or, equivalently, $\lim_{n\to\infty} n^{1/n} = 1$
8. $\lim_{n\to\infty} (a^n/n!) = 0$ for all real a
9. $\lim_{n\to\infty} (n!/n^n) = 0$
10. $\lim_{n\to\infty} (\ln_e n)^{1/n} = 1$
11. $\lim_{n\to\infty} (1 + k/n)^n = e^k$ (the definition of e^k in terms of a limit).

Example 3.1.6 Testing Series for Convergence
Test the following series for convergence

(i) $\displaystyle\sum_{n=1}^{\infty} \frac{\cos(2n-3i)}{n^p}$ (ii) $\displaystyle\sum_{n=1}^{\infty} \frac{(-1)^n(3+i)}{n!}$ (iii) $\displaystyle\sum_{n=1}^{\infty} \left(\frac{3n-2}{np+1}\right)^n (3-4i)^n$ with $p > 0$.

SOLUTION

(i) We will use the comparison test. The general term is

$$z_n = \frac{\cos(2n-3i)}{n^p} = \frac{\cos(2n)\cosh 3 + i\sin(2n)\sinh 3}{n^p},$$

so

$$|z_n| = \frac{\left[\cos^2(2n)\cosh^2 3 + \sin^2(2n)\sinh^2 3\right]^{1/2}}{n^p} \leqslant \frac{[\cosh^2 3 + \sinh^2 3]^{1/2}}{n^p}$$

Thus in the comparison test we may set $a_n = [\cosh^2 3 + \sinh^2 3]^{1/2}/n^p$, when the convergence properties of the complex series with p as a parameter are determined by the convergence properties of $\Sigma_{n=1}^{\infty} a_n$. However,

$$\sum_{n=1}^{\infty} a_n = \left[\cosh^2 3 + \sinh^2 3\right]^{1/2} \sum_{n=1}^{\infty} 1/n^p.$$

Entry 3 in the reference set of series shows that this converges for $p > 1$ and it diverges for $0 < p \leqslant 1$, so the series $\Sigma_{n=1}^{\infty} (\cos(2n - 3i)/n^p)$ is both convergent and absolutely convergent if $p > 1$, but divergent if $0 < p \leqslant 1$.

(ii) The convergence of this series can be determined by the ratio test. Because

$$z_n = (-1)^n \frac{(3+i)^n}{n!}$$

we have

$$\left|\frac{z_{n+1}}{z_n}\right| = \left|\frac{(-1)^{n+1}(3+i)^n}{(n+1)!} \cdot \frac{n!}{(-1)^n(3+i)^n}\right| = \frac{|3+i|}{(n+1)}.$$

Thus

$$L = \lim_{n\to\infty}\left|\frac{z_{n+1}}{z_n}\right| = \lim_{n\to\infty}\left[\frac{|3+i|}{(n+1)}\right] = 0,$$

so as $L < 1$ the complex series is both convergent and absolutely convergent.

(iii) In this case it is necessary to use the nth root test. Setting

$$z_n = \left(\frac{3n-2}{np+1}\right)^n (3-4i)^n,$$

we have

$$|z_n| = \left(\frac{3n-2}{np+1}\right)^n \cdot 5^n = \left(\frac{15n-10}{np+1}\right)^n,$$

and so

$$L = \lim_{n\to\infty} |z_n|^{1/n} = \lim_{n\to\infty}\left(\frac{15n-10}{np+1}\right) = \frac{15}{p}.$$

So if $p > 15$, then $L < 1$, and the complex series is both convergent and absolutely convergent, but if $0 < p < 15$, then $L > 1$, and the series is divergent. The test fails to give information when $p = 15$, because then $L = 1$. ◇

Exercises 3.1

For Exercises 1 through 6, write down the terms z_1 to z_6 of the sequence $\{z_n\}$ for the given general term z_n, and comment on the behavior of the sequence.

1. $z_n = \left(\frac{1+i}{2}\right)^n$
2. $z_n = (-1)^n + e^{n\pi i/2}$
3. $z_n = \left(\frac{3-3i}{2}\right)^n$
4. $z_n = \frac{n^2}{2n^2+1} + \frac{3ni}{2n^2-1}$
5. $z_n = \sin(\frac{1}{2}\pi + ni)$
6. $z_n = \ln_e\left(\frac{n}{2n+1}\right) + \left(\frac{n}{n+1}\right)i$

In Exercises 7 through 11, determine if the sequence $\{z_n\}$ with the given general term converges or diverges, and locate any cluster or limit points that may exist.

7. $z_n = [1 + (-1)^n](1 + 1/n) + i[1 - (-1)^n](3 - 1/n)$
8. $z_n = \frac{2n^2 + (5n^2+1)i}{6n^2+3}$
9. $z_n = \sinh[n(1+i)]$
10. $z_n = e^{n\pi i/3} + \left(-\frac{1}{2} + \frac{\sqrt{3}}{2}i\right)^n$
11. $z_n = \mathrm{Ln}(ke^{n\pi i/2})$, with $k \neq 0$ an arbitrary complex constant.
12. Given that the sequences $\{z_n\}$ and $\{Z_n\}$ have the general terms

    $$z_n = 1 + i(3 + 1/n) \quad \text{and} \quad Z_n = (n + 2n^2 i)/n^2,$$

    show they converge and find
    (i) $\lim_{n\to\infty}(3z_n - 2Z_n)$ and $\lim_{n\to\infty}(z_n^2/Z_n^2)$

These are not abstract mathematical questions, but very practical and fundamental ones. The theory of analytic functions requires functions to be infinitely differentiable, so if an analytic function is to be represented by a series like the one in Equation (3.5), it is necessary to know under what conditions the series will possess the same properties as the function itself. The answer to each of these basic questions is to be found in the study of the uniform convergence of complex functional series.

Each theorem established in this section is of fundamental importance for what is to follow and its meaning is easily understood. Readers should study carefully the statement of each of these theorems. The proofs of the theorems have been included for the sake of completeness, but the proofs are not always simple, so as an understanding of the proof is not essential when applying a theorem, a reader may omit study of the proofs until later if this seems desirable.

3.2.1 Uniform Convergence of a Complex Functional Series

Let a function $f(z)$ be represented by the series $\Sigma_{n=0}^{\infty} f_n(z)$ for all z in some domain D, so that

$$f(z) = \sum_{n=0}^{\infty} f_n(z). \tag{3.6}$$

Then the series $\Sigma_{n=0}^{\infty} f_n(z)$ is said to **converge uniformly** to $f(z)$ in D if, for any $\varepsilon > 0$, there exists a positive integer $N = N(\varepsilon)$, independent of z, such that

$$\left| f(z) - \sum_{r=0}^{n} f_r(z) \right| < \varepsilon \quad \text{for all } n > N. \tag{3.7}$$

The condition in Equation (3.6) can be re-expressed in a way that emphasizes its importance in what follows. If we introduce the *remainder* $R_n(z)$ of series Equation (3.6) after n terms, then

$$R_n(z) = \sum_{r=n+1}^{\infty} f_r(z), \tag{3.8}$$

inequality [Equation (3.7)] becomes

$$|R_n(z)| < \varepsilon. \tag{3.9}$$

Thus the uniform convergence of Equation (3.6) is seen to ensure that when $n > N$, the modulus of the error made when $f(z)$ is approximated by the partial sum

$$s_n(z) = \sum_{r=0}^{n} f_r(z) \tag{3.10}$$

never exceeds ε for any z in D. It is this placing of a uniform bound ε on the magnitude of the error when $f(z)$ is approximated by the partial sum $s_n(z)$ that leads to the name *uniform convergence*. In general, when a series is only convergent, and not uniformly convergent, the number of terms to be summed in Equation (3.10) in order that the magnitude of the error will be less than ε will depend on both ε and z, and so will *not* be independent of z.

Example 3.2.1 The Geometric Series Is Not Uniformly Convergent
It is not difficult to see that when $|z| < 1$ although the geometric series $f(z) = \Sigma_{n=0}^{\infty} z^n$ is convergent, it is *not* uniformly convergent.

SOLUTION
It is a familiar result from elementary algebra that the sum of the first $n + 1$ terms of the geometric series is given by

$$\sum_{r=0}^{n} z^r = \frac{1 - z^{n+1}}{1 - z},$$

so provided $|z| < 1$, the sum of an infinite number of terms is

$$\sum_{r=0}^{\infty} z^r = \frac{1}{1 - z},$$

showing that the geometric series is convergent. If the function $1/(1 - z)$ is approximated by $\Sigma_{r=0}^{n} z^r$, the modulus of the remainder is $|(1/(1 - z)) - \Sigma_{r=0}^{n} z^r| = |(z^{n+1})/(1 - z)|$. This result should be compared with Equation (3.4), when we see that in this case the magnitude of the remainder depends not only on n, but also on z, so the series is *not* uniformly convergent. So in this case, if the modulus of the remainder is to remain less than some arbitrary $\varepsilon > 0$, the closer $|z|$ approaches the value 1 from below, the more terms must be summed for this condition to be satisfied. This illustrates the difference between convergence and uniform convergence. ◇

The analog of the Cauchy convergence principle in the case of uniform convergence takes the following form.

THEOREM 3.2.1 Cauchy Convergence Principle for Uniform Convergence
For the complex functional series $\Sigma_{n=0}^{\infty} f_n(z)$ to be uniformly convergent for all z in a domain D, it is necessary and sufficient that for any $\varepsilon > 0$ a positive integer $N = N(\varepsilon)$ exists such that

$$|f_m(z) + f_{m+1}(z) + \cdots + f_n(z)| < \varepsilon$$

for all z in D and all $n > m > N$.

PROOF

First we prove the necessity of the condition. If $\Sigma_{n=0}^{\infty} f_n(z)$ converges uniformly to $f(z)$ for all z in D, then for any $\varepsilon > 0$ a positive integer $N = N(\varepsilon)$ exists, independent of z, such that when $n > m > N$, $|R_n(z)| < \varepsilon$. Now for all z in D and $n > m > N$ we have

$$\begin{aligned}|f_m(z) + f_{m+1}(z) + \cdots + f_n(z)| &= |s_n(z) - s_m(z)| \\ &= |s_n(z) - f(z) + f(z) - s_n(z)| \\ &\leqslant |s_n(z) - f(z)| + |f(z) - s_m(z)| \\ &= |R_n(z)| + |R_m(z)| < 2\varepsilon.\end{aligned}$$

Thus, as ε was arbitrary, the necessity of the condition has been established.

Next we prove the sufficiency of the condition. If the Cauchy convergence principle for uniform convergence is satisfied

$$\lim_{n\to\infty} s_n(z) = f(z) \quad \text{for all } z \text{ in } D.$$

However,

$$|s_n(z) - s_m(z)| < |f(z) - s_m(z)| = |R_m(z)| < \varepsilon$$

for all z in D and all $m > N$, so we have established the sufficiency of the condition and the proof is complete. ◆

It is often difficult to use the definition of uniform convergence as an actual test for uniform convergence, so for practical purposes the definition needs to be replaced by a test that is simpler. The consequence of seeking such a test turns out to be that although when its conditions are satisfied the test will certainly establish the uniform convergence of a series, there will be some uniformly convergent series that fail the test. Thus the conditions of such a simpler test will be *sufficient* to establish uniform convergence, but not *necessary*. The most frequently used test of this nature forms the basis of the next theorem.

THEOREM 3.2.2 The Weierstrass M-Test

Let $\{f_n(z)\}$ be a sequence of complex functions defined in a domain D, and let $\{M_n\}$ be a sequence of positive numbers such that $\Sigma_{n=0}^{\infty} M_n$ is convergent. Then if $|f_n(z)| < M_n$ for all positive integers n and all z in D, the complex functional series $\Sigma_{n=0}^{\infty} f_n(z)$ is uniformly convergent for all z in D.

PROOF

As the real numerical series $\Sigma_{n=0}^{\infty} M_n$ is convergent, it follows from the Cauchy convergence principle that given an $\varepsilon > 0$, a positive integer

$N = N(\varepsilon)$ exists such that

$$\sum_{r=m+1}^{n} M_r < \varepsilon \quad \text{for } n > m > N.$$

Now, for all z in D

$$\begin{aligned} |f_{m+1}(z) + f_{m+2}(z) + \cdots + f_n(z)| &\leqslant \sum_{r=m+1}^{n} |f_r(z)| \\ &\leqslant \sum_{r=m+1}^{n} M_r < \varepsilon \qquad \text{for } n > m > N, \end{aligned}$$

and the result is proved. ◆

Example 3.2.2 Testing for Uniform Convergence by the *M*-Test
Use the Weierstrass M-test to show that for any $k > 0$ the series

$$\sum_{n=0}^{\infty} \frac{e^{-inz}}{n^2 + 1}$$

is uniformly convergent for all z in the half plane $\text{Im}\{z\} < -k < 0$.

SOLUTION
We have

$$\left|\frac{e^{-inz}}{n^2+1}\right| = \left|\frac{e^{-inx}e^{ny}}{n^2+1}\right| = \frac{e^{ny}}{n^2+1}.$$

Then if $k > 0$ and $\text{Im}\{z\} < -k < 0$, it follows that $e^{ny} < e^{-nk}$, so that

$$\left|\frac{e^{-inz}}{n^2+1}\right| \leqslant \frac{e^{-nk}}{n^2+1}.$$

Setting $M_n = e^{-nk}/(n^2 + 1)$, and testing the series $\Sigma_{n=0}^{\infty} M_n$ for convergence using the ratio test, gives

$$\begin{aligned} \lim_{n\to\infty}\left(\frac{M_{n+1}}{M_n}\right) &= \lim_{n\to\infty}\left[\left(\frac{e^{-(n+1)k}}{n^2+2n+2}\right)\left(\frac{n^2+1}{e^{-nk}}\right)\right] \\ &= \lim_{n\to\infty}\left[\left(\frac{n^2+1}{n^2+2n+2}\right)e^{-k}\right] = e^{-k} < 1. \end{aligned}$$

Thus the series $\Sigma_{n=0}^{\infty} M_n$ is convergent, and so by the Weierstrass M-test the series $\Sigma_{n=0}^{\infty}(e^{-ink}/(n^2 + 1))$ is uniformly convergent for all z such that $\text{Im}\{z\} < -k < 0$. ◇

The three theorems that complete this section describe the most important properties of uniformly convergent complex functional series.

THEOREM 3.2.3 *An Infinite Series of Uniformly Convergent Functions Is a Continuous Function*

Let the sequence of complex functions $\{f_n(z)\}$ be continuous for all z in domain D. Then if the complex functional series $\Sigma_{n=1}^{\infty} f_n(z)$ is uniformly convergent to the function f(z) for all z in D, the function f(z) is continuous in D.

PROOF

If the points z_0 and ζ_0 lie in the domain D, then the uniform convergence of $\Sigma_{n=1}^{\infty} f_n(z)$ implies that for any $\varepsilon > 0$, a positive integer $N = N(\varepsilon)$ exists, independent of z, such that for all $n > N$,

$$|R_n(z_0)| = |f(z_0) - s_n(z_0)| < \varepsilon \quad \text{and} \quad |R_n(\zeta_0)| < |f(\zeta_0) - s_n(\zeta_0)| < \varepsilon.$$

Because the functions $f_n(z)$ are continuous in D, the finite sums $s_n(z)$ and $s_n(\zeta)$ will themselves be continuous functions of their arguments. So for any $\varepsilon > 0$ and a fixed $n^* > N$, we choose a number $\delta = \delta(\varepsilon) > 0$ such that

$$|s_{n^*}(\zeta_0) - s_{n^*}(z_0)| < \varepsilon \quad \text{for } |z_0 - \zeta_0| < \delta.$$

Writing

$$|f(\zeta_0) - f(z_0)| = |f(\zeta_0) - s_{n^*}(\zeta_0) + s_{n^*}(\zeta_0) - s_{n^*}(z_0) + s_{n^*}(z_0) - f(z_0)|,$$

and using the triangle inequality we find that

$$|f(\zeta_0) - f(z_0)| \leq |f(\zeta_0) - s_{n^*}(\zeta_0)| + |s_{n^*}(\zeta_0) - s_{n^*}(z_0)| + |s_{n^*}(z_0) - f(z_0)| < 3\varepsilon,$$

for $|z_0 - \zeta_0| < \delta$. Consequently, as ε was arbitrary, the continuity of $f(z)$ has been established for all z in D, so the proof is complete. ◆

THEOREM 3.2.4 *A Uniformly Convergent Series of Complex Functions May Be Integrated Term by Term*

Let the sequence of complex functions $\{f_n(z)\}$ be continuous for all z in domain D, and let each function $f_n(z)$ be integrable along a simple are C in D. Then if the complex functional series $\Sigma_{n=0}^{\infty} f_n(z)$ is integrated along C, the operations of integration and summation may be interchanges to yield the result

$$\int_C \left[\sum_{n=0}^{\infty} f_n(z)\right] dz = \sum_{n=0}^{\infty} \left(\int_C f_n(z) dz\right).$$

PROOF

Because the series $\Sigma_{n=0}^{\infty} f_n(z)$ is uniformly convergent for all z in D, it follows that for any $\varepsilon > 0$ a positive integer $N = N(\varepsilon)$ exists such that for all z in D

$$|R_n(z)| = \left| \sum_{r=n+1}^{\infty} f_r(z) \right| < \varepsilon \quad \text{for } n > N.$$

Then if C is a simple arc of length L, by denoting the element of arc length along C by ds we have the estimate

$$\left| \int_C R_n(z) dz \right| \leqslant \int_C |R_n(z)| \, dz < \varepsilon L \quad \text{for all } n > N.$$

The statement of the theorem now follows by proceeding to the limit as $n \to \infty$, because then

$$\lim_{n \to \infty} \left| \int_C R_n(z) dz \right| = 0,$$

from which we conclude that

$$\int_C \left[\sum_{n=0}^{\infty} f_n(z) \right] dz = \sum_{n=0}^{\infty} \left(\int_C f_n(z) dz \right). \quad \blacklozenge$$

The final theorem in this section concerns the differentiation of complex functional series and it is the most difficult to prove.

THEOREM 3.2.5 *A Uniformly Convergent Series of Complex Functions May Be Differentiated Term by Term*

Let $\{f_n(z)\}$ be a sequence of single-valued functions analytic in a simply connected domain D bounded by a simple closed-curve C, and let the series $\Sigma_{n=0}^{\infty} f_n(z)$ converge uniformly to a function $f(z)$ inside and on the boundary C^* of a simply connected domain D^* contained entirely within D. Then $f(z)$ is an analytic function, its derivative may be found by differentiating the series term by term as often as required, and the series

$$f^{(r)}(z) = \sum_{n=0}^{\infty} f_n^{(r)}(z),$$

is uniformly convergent inside and on the boundary of D^*.

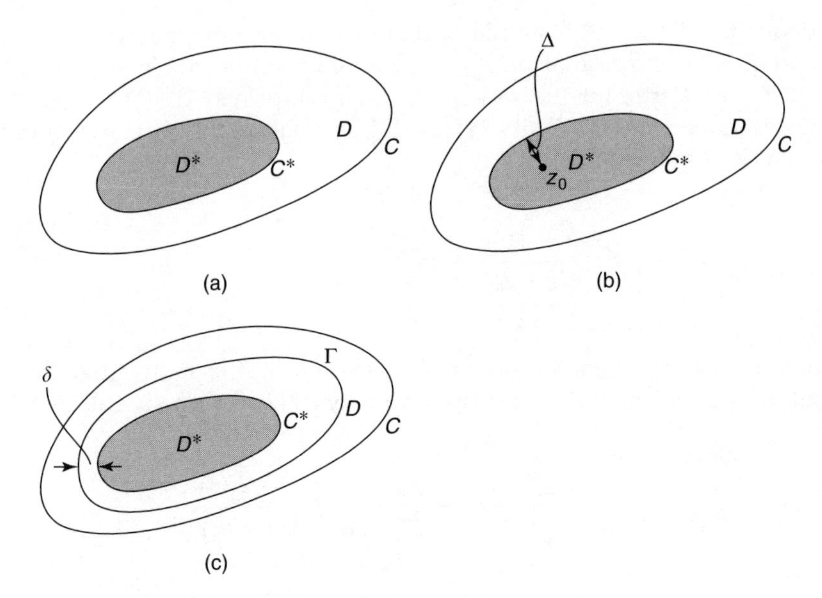

FIGURE 3.3
The simple closed-curves C, C^*, and Γ in D.

PROOF

Take any simple contour C^* inside D, as shown in Figure 3.3(a) and denote its interior by D^*. Then as the series is uniformly convergent in D^* it follows from Theorem 5.2.4 that

$$\int_{C^*}\left[\sum_{n=0}^{\infty} f_n(z)\right]dz = \sum_{n=0}^{\infty}\left(\int_{C^*} f_n(z)dz\right).$$

The functions $f_n(z)$ are all analytic in D, and so also in D^*, so by the Cauchy–Goursat theorem $\int_{C^*} f_n(z)dz = 0$, and hence $\int_{C^*} f(z)dz = 0$.

Theorem 3.2.3 shows that $f(z)$ is continuous in D, so because C^* lies inside D, an application of Morera's theorem shows that $f(z)$ is analytic in D, so the first part of the theorem is proved.

We must now show that the series can be differentiated term-wise r times to obtain $f^{(r)}(z)$, where r is any positive integer. Let z be any point on C^* and take z_0 to be an interior point of D^*, as shown in Figure 3.3(b). Then for z on C^*, define the function $h(z)$ to be

$$h(z) = \frac{1}{(z - z_0)^{r+1}},$$

where r is a positive integer, and let Δ be the least distance from z_0 to C^*, as shown in Figure 3.3(b). Then we see that $|h(z)| \leq 1/\Delta^{r+1}$, showing that $h(z)$ is bounded for all z on C^*. A simple extension of the argument used to establish Theorem 3.2.1 shows that if $\Sigma_{n=0}^{\infty} f_n(z)$ is uniformly convergent in D^*, the result remains true for the series $\Sigma_{n=0}^{\infty} h(z) f_n(z)$, provided $h(z)$ is bounded, which is indeed the case. Thus if $f(z) = \Sigma_{n=0}^{\infty} f_n(z)$ is uniformly convergent for all z on C^*, it follows that

$$\frac{f(z)}{(z - z_0)^{r+1}} = \sum_{n=0}^{\infty} \frac{f_n(z)}{(z - z_0)^{r+1}}$$

is uniformly convergent for all z on C^*. Theorem 3.2.4 us to integrate the last result term by term, so after multiplication by $r!/(2\pi i)$ we arrive at the result

$$\frac{r!}{2\pi i} \int_C \frac{f(z)}{(z - z_0)^{r+1}} dz = \sum_{n=0}^{\infty} \left(\frac{r!}{2\pi i} \int_{C^*} \frac{f_n(z)}{(z - z_0)^{r+1}} dz \right).$$

An application of the Cauchy integral theorem for derivatives reduces this to

$$f^{(r)}(z_0) = \sum_{n=0}^{\infty} f_n^{(r)}(z_0).$$

because z_0 was any point in D^*, we have established that term by term differentiation of the series r times is permissible for any z in D^*, and that it yields $f^{(r)}(z_0)$.

Finally we must prove the uniform convergence of this last result. Enclose C^* by any simple closed-curve Γ lying entirely inside D, and denote by δ the least distance from C^* to Γ, as shown in Figure 3.3(c). Then because the remainder term $R_n(z)$ is analytic in D, it must be analytic on Γ and an application of the Cauchy integral theorem for derivatives gives

$$R_n(z) = \frac{r!}{2\pi i} \int_{\Gamma} \frac{R_n(\zeta)}{(\zeta - z)^{r+1}} d\zeta.$$

The uniform convergence of $\Sigma_{n=0}^{\infty} f_n(z)$ on Γ implies that for any $\varepsilon > 0$ a positive integer $N = N(\varepsilon)$ exists, independent of z, such that $|R_n(\zeta)| < \varepsilon$ for $n > N$.

Denoting the length of the perimeter of Γ by L and letting the element of arc length be ds, we arrive at the inequality

$$|R_n^{(r)}(z)| \leqslant \frac{r!}{2\pi i} \int_{\Gamma} \frac{|R_n(\zeta)|}{|z - \zeta|^{r+1}} ds \leqslant \frac{r! L\varepsilon}{2\pi \delta^{r+1}} \quad \text{for all } n > N \text{ and all } z$$

inside or on the boundary of the domain inside Γ. The uniform convergence of the differentiated series is thus established for z inside or on Γ, so the theorem is proved. ◆

3.3 Power Series

Section 3.1 has already defined a power series about the point z_0 to be a series of the form

$$\sum_{n=0}^{\infty} a_n(z - z_0)^n \tag{3.11}$$

where z is a complex variable, z_0 is a complex number about which the series is said to be *expanded*, and the coefficients a_n are complex numbers. A power series expanded about the origin ($z_0 = 0$) has the form

$$\sum_{n=0}^{\infty} a_n z^n \tag{3.12}$$

and Equation (3.11) may always be reduced to this form by setting $\zeta = z - z_0$.

When series Equation (3.11) converges for all z in some domain, D it defines a function $f(z)$ in D. It is clear that Equation (3.11) always converges at the single point $z = z_0$, for then the series reduces to the constant term a_0. The question we now consider is how to determine the largest domain in which Equation (3.11) is convergent. The next theorem provides useful information of a qualitative nature about the convergence and divergence of power series. We build on the results of this theorem and use the tests for the convergence of numerical series introduced earlier to find the precise domain in which the power series Equation (3.11) is convergent.

THEOREM 3.3.1 *Qualitative Behavior of Power Series*

(i) *Let the power series*

$$\sum_{n=0}^{\infty} a_n(z - z_0)^n$$

be convergent at the point $z_1 \neq z_0$. Then it converges at any point z inside the disk $|z - z_0| < \rho_1$, where $\rho_1 = |z_1 - z_0|$, and it converges uniformly at all points of the disk $|z - z_0| \leqslant d$, where $d < \rho_1$.

(ii) *If the power series diverges at the point z_2, it diverges everywhere outside the disk $|z - z_0| > \rho_2$, where $\rho_2 = |z_2 - z_0|$.*

PROOF

(i) The presupposed convergence of the series at z_1 implies that the nth term of the numerical series $\Sigma_{n=0}^{\infty} a_n(z_1 - z_0)^n$ must vanish, and so

$$\lim_{n\to\infty} a_n(z_1 - z_0) = 0.$$

Thus because $\{a_n(z_1 - z_0)^n\}$ is a convergent sequence its terms are bounded by some real number K, so that

$$|a_n(z_1 - z_0)^n| \leqslant K, n = 0, 1, 2, \ldots$$

Rewriting the series we have

$$\sum_{n=0}^{\infty} a_n(z - z_0)^n = \sum_{n=0}^{\infty} a_n(z_1 - z_0)^n \left(\frac{z - z_0}{z_1 - z_0}\right)^n,$$

so taking the modulus of this result and using the bound K we find that

$$\left|\sum_{n=0}^{\infty} a_n(z - z_0)^n\right| \leqslant \sum_{n=0}^{\infty} |a_n(z - z_0)^n| = \sum_{n=0}^{\infty}\left[|a_n(z_1 - z_0)^n| \left|\frac{z - z_0}{z_1 - z_0}\right|^n\right]$$
$$\leqslant M\sum_{n=0}^{\infty}\left|\frac{z - z_0}{z_1 - z_0}\right|^n.$$

As the last series is a geometric series with common ratio $r = |(z - z_0)/(z_1 - z_0)| < 1$, it is convergent, thereby establishing the absolute convergence of the series $\Sigma_{n=0}^{\infty} a_n(z_1 - z_0)^n$. It is now necessary to show that when z is restricted to the closed disk $|z - z_0| \leqslant d$, which is contained entirely within the open disk $|z - z_0| < \rho_1$, the series is uniformly convergent. Within this disk $|(z - z_0)/(z_1 - z_0)| \leqslant r/|z_1 - z_0|$, so $|a_n(z - z_0)^n| \leqslant r/|z_1 - z_0|$ for all $n = 0, 1, 2, \ldots$, and independently of z. The Weierstrass M-test then shows the series to be uniformly convergent for $|z - z_0| \leqslant d$, so the proof of (i) is complete.

(ii) The result in (ii) will be proved by contradiction. By hypothesis, the series diverges at z_2. So, if possible, suppose that the series converges at a point z_3 such that $|z_3 - z_0| > \rho_2$, where $\rho_2 = |z_2 - z_0|$. By part (i) of this theorem the series must be convergent at z_2, but this is a contradiction, so the series diverges for z such that $|z - z_0| > \rho_2$, and the proof of (ii) is complete. ◆

The implications of Theorem 3.3.1 can be understood by inspection of Figure 3.4. This shows that the power series Equation (3.11) is convergent inside circle C_1 and divergent outside circle C_2, though nothing can be said about the behavior of the series in the annular domain between C_1 and C_2. Let z^* be the point closest to z_0 at which the series is divergent, then the real number

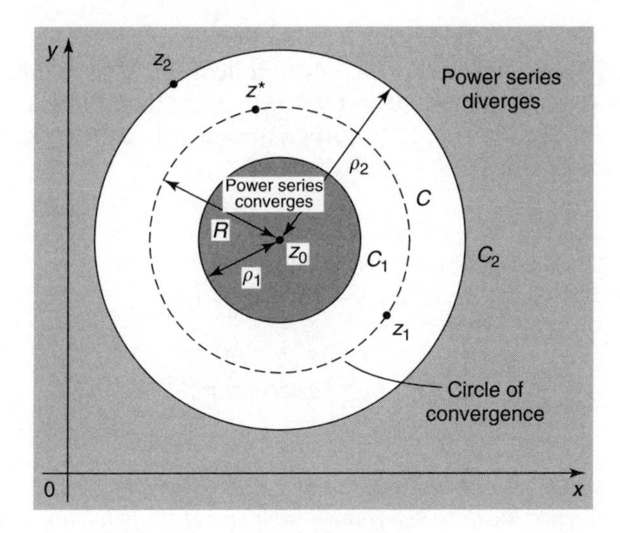

FIGURE 3.4
Domains of convergence, divergence, and the circle of convergence.

$R = |z^* - z_0|$ is called the *radius of convergence* of power series Equation (3.11), and the circle C centered on z_0 with radius R is called the *circle of convergence* of the series inside of which the series converges.

The circle of convergence C is shown in Figure 3.4, and it has the property that series Equation (3.11) converges at all points inside C, and it diverges at all points outside C. The behavior of the series on C must be determined separately, because it may converge on C, diverge on C, or converge on parts of C while diverging on other parts of C.

When series Equation (3.11) is convergent in the entire complex plane, the radius of convergence is said to be infinite and we write $R = \infty$. If the series only converges at a single point z_0 the radius of convergence $R = 0$.

To improve on this qualitative understanding of power series, it is necessary to find how to determine the radius of convergence R. This forms the content of the next theorem, though before stating it we must define the notion of the *limit superior* of a sequence of real numbers $\{c_n\}$, written lim sup c_n.

Given a sequence of real numbers $\{c_n\}$, we define $L = \limsup c_n$ to be the smallest real number L such that for any $\varepsilon > 0$ only a finite number of terms of the sequence are greater than $L + \varepsilon$. If no such number exists, we set $\limsup c_n = \infty$. The meaning of $\limsup c_n$ is illustrated by the following examples.

(i) If $c_n = 2 - (1/3)^n$ for $n = 0, 1, 2, \ldots$, then $\limsup c_n = 2$ because however small $\varepsilon > 0$ may be, there are no terms $c_n > 2$.

(ii) If $c_n = \begin{cases} 3 + \left(\frac{1}{4}\right)^n & \text{when } n \text{ is even} \\ 2 + \left(-\frac{1}{2}\right)^n & \text{when } n \text{ is odd} \end{cases}$

with $n = 0, 1, 2, \ldots$, then $\limsup c_n = 3$ because however small $\varepsilon > 0$ may be, there will always be only a finite number of terms $c_n > 3$.

(iii) If $c_n = (1 + a)^n$ with $a > 0$ and $n = 0, 1, 2, \ldots$, then $\limsup c_n = \infty$.

(iv) If $c_n = 2 - n$ for $n = 0, 1, 2, \ldots$, then $\limsup c_n = 2$, because there are no terms $c_n > 2$.

THEOREM 3.3.2 Finding the Radius of Convergence R
If $R \geqslant 0$ is the radius of convergence of the power series

$$f(z) = \sum_{n=0}^{\infty} a_n (z - z_0)^n,$$

the series converges absolutely for $|z - z_0| < R$ and diverges for $|z - z_0| > R$. The radius of convergence R can be determined from any of the following formulas:

1. $R = \lim_{n\to\infty} \left| \dfrac{a_n}{a_{n+1}} \right|$, when the limit exists (from the *ratio test*) (3.13)

2. $R = \dfrac{1}{\lim_{n\to\infty} |a_n|^{1/n}}$, when the limit exists (from the *nth root test*) (3.14)

3. $R = \dfrac{1}{\limsup |a_n|^{1/n}}$, which is always defined (the *Cauchy–Hadamard formula*). (3.15)

PROOF

1. The result follows by considering a fixed value of z and applying the ratio test to the resulting numerical series $\Sigma_{n=0}^{\infty} a_n (z - z_0)^n$. If the limit L exists, where $L = \lim_{n\to\infty} |a_{n+1}/a_n|$, an application of the ratio test to the power series shows the series converges absolutely if

$$\lim_{n\to\infty} \left| \frac{a_{n+1}(z - z_0)^{n+1}}{a_n (z - z_0)^n} \right| = |z - z_0| \lim_{n\to\infty} \left| \frac{a_{n+1}}{a_n} \right| = |z - z_0| L < 1,$$

and diverges if $|z - z_0| L > 1$. Consequently, the power series converges absolutely for all z such that

$$|z - z_0| < \frac{1}{L} = \lim_{n\to\infty} \left| \frac{a_n}{a_{n+1}} \right|,$$

and diverge for all z such that

$$|z - z_0| > \lim_{n\to\infty}\left|\frac{a_n}{a_{n+1}}\right|.$$

Thus the radius of convergence R is given by

$$R = \lim_{n\to\infty}\left|\frac{a_n}{a_{n+1}}\right|,$$

and the series is absolutely convergent for all $|z - z_0| < R$.

2. The result follows in similar fashion by applying the nth root test to the power series, when it is found that provided the limit L exists, where $L = \lim_{n\to\infty}|a_n|^{1/n}$ the power series converges absolutely if

$$\lim_{n\to\infty}|a_n(z - z_0)^n|^{1/n} = |z - z_0| \lim_{n\to\infty}|a_n|^{1/n} = |z - z_0|L < 1,$$

and diverges if $|z - z_0|L > 1$. From which it then follows that the radius of convergence R is given by

$$R = \frac{1}{\lim_{n\to\infty}|a_n|^{1/n}},$$

so the series is absolutely convergent for all z in $|z - z_0| < R$ and divergent for $|z - z_0| > R$.

3. We have $\limsup|a_n(z - z_0)^n|^{1/n} = |z - z_0| \limsup|a_n|^{1/n}$, because $(z - z_0)$ does not depend on n. An application of the nth root test then shows that

$$\limsup|a_n(z - z_0)^{1/n}| = |z - z_0|\limsup|a_n|^{1/n} = |z - z_0|L.$$

So the series is absolutely convergent for $|z - z_0|L < 1$, and divergent for $|z - z_0|L > 1$, where now the radius of convergence R is given by

$$R = \frac{1}{\limsup|a_n|^{1/n}}. \qquad \blacklozenge$$

The theoretical importance of the *Cauchy–Hadamard formula* in Equation (3.15) of Theorem 3.3.2 arises from the fact that $\limsup|a_n|^{1/n}$ is defined for *every* sequence $\{a_n\}$. This means that the formula can be applied to *every* power series, even when some powers of $(z - z_0)$ are missing, which will cause case the determination of R from the ratio test in Equation (3.13), and from the nth root test in Equation (3.14), to fail. However, the disadvantage of the Cauchy–Hadamard formula is that it is often difficult to apply.

Example 3.3.1 A Case where the Cauchy–Hadamard Formula Can Be Used

Consider the general power series $\Sigma_{n=0}^{\infty} a_n z^n$, where some of the coefficients a_n may be missing and find its radius of convergence. Differentiate the series term by term and show the differentiated series has the same radius of convergence as the original series.

SOLUTION

As some of the coefficients a_n may be missing it is necessary to find the radius of convergence R from the Cauchy–Hadamard formula, when we find that $R = 1/(\limsup|a_n|^{1/n})$. Differentiating the series term by term produces the series $\Sigma_{n=1}^{\infty} n a_n z^{n-1}$. Applying the Cauchy–Hadamard formula to the differentiated series shows its radius of convergence is $R_1 = 1/(\lim \sup|na_n|^{1/n}) = 1/(\lim \sup[n^{1/n}|a_n|^{1/n}])$. However, $\lim_{n\to\infty}(n^{1/n}) = 1$, so $R_1 = R$, showing that both the original series and its derivative are absolutely convergent for $|z| < R$. ◇

The next example illustrates the application of Theorem 3.3.2 to several different types of power series and, in particular, it shows how when the radius of convergence is difficult to find from the Cauchy–Hadamard formula the result can sometimes be found from first principles.

Example 3.3.2 Finding the Radius of Convergence

Find the radius of convergence for the following power series:

(i) $$\sum_{n=1}^{\infty} \frac{(-1)^{n+1} n(z - z_0)^n}{3n^2 + 1}$$

(ii) $$\sum_{n=0}^{\infty} \frac{z^{2n}}{(1 + 2i)^n}$$

(iii) $$\sum_{n=0}^{\infty} \left(\frac{2n + 3}{5n + 6}\right)^n (z - z_0)^n$$

(iv) $$\sum_{n=0}^{\infty} a_n (z - z_0)^n, \quad \text{with } a_n = \begin{cases} 3^n & \text{for even } n \\ -4^n & \text{for odd } n \end{cases}$$

(v) $$\sum_{n=0}^{\infty} \frac{z^n}{n^n}$$

(vi) $$\sum_{n=0}^{\infty} n^n z^n$$

(vii) $$\sum_{n=1}^{\infty} \frac{n z^{n^2}}{(n - 1)!}.$$

SOLUTION

(i) No coefficients of the power series vanish, so the obvious choice for the determination of R is Formula (3.13) in Theorem 3.3.2. Setting $a_n = ((-1)^{n+1}n)/(3n^2+1)$, it follows from Formula (3.13) that

$$R = \lim_{n\to\infty}\left|\frac{a_n}{a_{n+1}}\right| = \lim_{n\to\infty}\left|\frac{(-1)^{n+1}n}{3n^2+1}\cdot\frac{3(n+1)^2+1}{(-1)^{n+2}(n+1)}\right|$$
$$= \lim_{n\to\infty}\left(\frac{n}{n+1}\cdot\frac{3n^2+6n+4}{3n^2+1}\right) = 1.$$

So the series is absolutely convergent for $|z-z_0|<1$ and divergent for $|z-z_0|>1$.

(ii) This power series is expanded about the origin, but we must proceed with caution because all odd powers of z are missing. So although Formula (3.13) of Theorem 3.3.2 would seem to be the natural choice with which to determine R, our approach must be modified to allow for the missing odd powers of z. Setting $\zeta = z^2$ reduces the series to $\Sigma_{n=0}^{\infty}\zeta^n/(1+2i)^n$, where now every power of ζ is present. Setting $a_n = 1/(1+2i)^n$ it follows that

$$R = \lim_{n\to\infty}\left|\frac{a_n}{a_{n+1}}\right| = \lim_{n\to\infty}\left|\frac{(1+2i)^{n+1}}{(1+2i)^n}\right| = |1+2i| = \sqrt{5}.$$

Thus the series in ζ converges for $|z|<\sqrt{5}$ and diverges for $|z|>\sqrt{5}$. As $\zeta = z^2$ the original series converges for $|z|<5^{1/4}$ and diverges for $|z|>5^{1/4}$.

(iii) Formula (3.14) in Theorem 3.2.2 is the natural choice with which to determine R, so as all powers of $(z-z_0)$ are present we set $a_n = ((2n+3)/(5n+6))^n$, when $R = 1/K$ with $K = \lim_{n\to\infty}|((2n+3)/(5n+6))^n|^{1/n} = \lim_{n\to\infty}|((2n+3)/(5n+6))| = (2/5)$, so $R = 5/2$. Thus the series converges for $|z-z_0|<5/2$ and diverges for $|z-z_0|>5/2$.

(iv) In this case $|a_n|^{1/n}$ oscillates finitely, so we must use Formula (3.15) of Theorem 3.3.2, when we find that $\lim\sup|a_n|^{1/n} = 4$, so $R = 1/4$, and the series converges for $|z-z_0|<1/4$ and diverges for $|z-z_0|>1/4$.

(v) Applying Formula (3.14) of Theorem 3.3.2 with $a_n = 1/n^n$ gives

$$R = \frac{1}{\lim_{n\to\infty}|1/n^n|^{1/n}} = \lim_{n\to\infty} n = \infty.$$

So this series converges for all z.

(vi) Formula (3.14) of Theorem 3.3.2 can be applied with $a_n = n^n$, when we find that

$$R = \frac{1}{\lim_{n\to\infty}|n^n|^{1/n}} = \lim_{n\to\infty}\left(\frac{1}{n}\right) = 0.$$

So this series only converges at the single point $z = 0$.

(vii) As powers of z are missing neither of Formulas (3.13) nor (3.14) of Theorem 3.3.2 can be applies, but a direct application of Formula (3.15) is difficult, so we will use a different approach by making a direct application of the ratio test. Fixing z we will consider the numerical series $\Sigma_{n=0}^{\infty} c_n z^n$, with $c_n = nz^{n^2}/(n+1)!$. We know from the ratio test that this numerical series will converge for this fixed value of z if

$$\lim_{n\to\infty}\left|\frac{c_{n+1}}{c_n}\right| = \lim_{n\to\infty}\left|\frac{(n+1)z^{n^2+2n+1}}{n!}\cdot\frac{(n-1)!}{nz^{n^2}}\right|$$
$$= \lim_{n\to\infty}\left[\left(\frac{n+1}{n^2}\right)|z|^{2n+1}\right] < 1.$$

This is only possible if $|z| < 1$, so the radius of convergence of this series $R = 1$. ◇

3.3.1 Properties of Power Series

The properties of power series that will be used late follow almost immediately from the theorems of Section 3.2. To relate a power series of the form

$$f(z) = \sum_{n=0}^{\infty} a_n(z - z_0)^n \tag{3.16}$$

to the functional series considered previously it is only necessary to set $f_n(z) = a_n(z - z_0)^n$. It follows from Theorem 3.3.1 that the poser series in Equation (3.16) is uniformly convergent in any disk $|z - z_0| \leqslant d$ where $d < R$ [the radius of convergence of series Equation (3.16)]. A direct application of Theorem 3.2.5 gives rise to the following theorem, the results of which are used throughout complex analysis and its applications.

THEOREM 3.3.3 Continuity and Repeated Differentiability of Power Series
Let the power series

$$f(z) = \sum_{n=0}^{\infty} a_n(z - z_0)^n$$

have a radius of convergence $R > 0$. Then it follows that:

1. *$f(z)$ has for its domain of convergence the open disk $|z - z_0| < R$.*
2. *For any number d such that $0 < d < R$, the function $f(z)$ is analytic in the closed disk $|z - z_0| \leqslant d$.*

3. *The power series may be differentiated term by term as often as required within its circle of convergence* $|z - z_0| < R$ *to give*

$$f^{(r)}(z) = \sum_{n=0}^{\infty} \frac{d^r}{dz^r}[a_n(z - z_0)^n],$$

and the differentiated series will have the same circle of convergence as the series from which it was derived, namely, the open disk $|z - z_0| < R$. ◆

Example 3.3.3 Using Repeated Differentiation to Develop a Required Power Series

Use the geometric series

$$\frac{1}{1-z} = \sum_{n=0}^{\infty} z^n \quad \text{for } |z| < 1$$

to deduce the power series for $1/(1 - z)^3$.

SOLUTION

Item 3 of Theorem 3.3.3 allows this series to be differentiated term by term within its circle of convergence $|z| < 1$. Inspection shows that to obtain the function $1/(1 - z)^3$ it is necessary to differentiate the function $1/(1 - z)$ twice. We have

$$\frac{d}{dz}\left(\frac{1}{1-z}\right) = \sum_{n=0}^{\infty} \frac{d}{dz}(z^n)$$

and so

$$\frac{1}{(1-z)^2} = \sum_{n=1}^{\infty} nz^{n-1}.$$

Notice that summation in the new series must now start from $n = 1$ because the first power of z to occur in the series must be z^0. A second differentiation gives

$$\frac{d}{dz}\left(\frac{1}{(1-z)^2}\right) = \sum_{n=1}^{\infty} \frac{d}{dz}(nz^{n-1}) = \sum_{n=2}^{\infty} n(n-1)z^{n-2},$$

and so

$$\frac{2}{(1-z)^3} = \sum_{n=2}^{\infty} n(n-1)z^{n-2},$$

where now the summation must start from $n = 2$. Thus one form of the required series is

$$\frac{1}{(1-z)^3} = \frac{1}{2}\sum_{n=2}^{\infty} n(n-1)z^{n-2}.$$

We can, if we wish, alter the lower summation limit so that it starts from zero. To achieve this set $m = n - 2$ when the result becomes

$$\frac{1}{(1-z)^3} = \frac{1}{2}\sum_{m=0}^{\infty} (m+1)(m+2)z^m,$$

but the summation index is a dummy variable, so we may replace m by n to obtain

$$\frac{1}{(1-z)^3} = \frac{1}{2}\sum_{n=0}^{\infty} (n+1)(n+2)z^n.$$

◇

Example 3.3.4 Using Theorem 3.3.3 to Sum a Series
Sum the series $\Sigma_{n=1}^{\infty} (n+2)z^n$.

SOLUTION
Starting from the known geometric series

$$\frac{1}{1-z} = \sum_{n=0}^{\infty} z^n \quad \text{for } |z| < 1,$$

Multiplying by z^2 gives

$$\frac{z^2}{1-z} = \sum_{n=0}^{\infty} z^{n+2} \quad \text{for } |z| < 1.$$

Differentiation of this result with respect to z, which is permitted by Theorem 3.3.3, gives

$$\frac{d}{dz}\left(\frac{z^2}{1-z}\right) = \sum_{n=0}^{\infty} \frac{d}{dz}(z^{n+2}), \quad \text{and so} \quad \frac{2z - z^2}{(1-z)^2} = \sum_{n=0}^{\infty} (n+2)z^{n+1} \quad \text{for } |z| < 1.$$

A factor z may be canceled throughout, so after separating out the constant term corresponding to $n = 0$ this becomes

$$\frac{2-z}{(1-z)^2} = 2 + \sum_{n=1}^{\infty} (n+2)z^{n+1} \quad \text{for } |z| < 1,$$

leading to the required summation

$$\sum_{n=1}^{\infty}(n+2)z^n = \frac{z(3-2z)}{(1-z)^2} \quad \text{for } |z|<1. \qquad \diamond$$

THEOREM 3.3.4 Term-by-Term Integration of Power Series
Let the power series

$$f(z) = \sum_{n=0}^{\infty} a_n(z-z_0)^n$$

have a radius of convergence $R > 0$. Then the series may be integrated term by term along any simple arc C in $|z - z_0| < R$ to give

$$\int_C f(z)dz = \sum_{n=0}^{\infty} \int_C a_n(z-z_0)^n dz,$$

and the resulting power series will have the same circle of convergence as the series from which it was derived.

PROOF

It is only necessary to prove that the original and the integrated series have the same circle of convergence. This follows directly from item 3 of Theorem 3.3.2 because

$$R = \frac{1}{\limsup |a_n/a_{n+1}|^{1/n}} = \frac{1}{\limsup |a_n|^{1/n}},$$

where we have used the fact that $\lim_{n\to\infty}(n+1)^{1/n} = \lim_{n\to\infty}(n^{1/n}) = 1$. ◆

Example 3.3.5 Integration of a Power Series
Deduce the Maclaurin series for $\text{Ln}(1+z)$ from the geometric series

$$\frac{1}{1-z} = \sum_{n=0}^{\infty} z^n \quad \text{for } |z|<1.$$

SOLUTION
It follows from Theorem 3.3.4 that the geometric series may be integrated term by term along any arc C inside its circle of convergence $|z| < 1$. Take any point z_0 inside the circle of convergence and join it to the origin by an arbitrary

simple arc C lying entirely within the circle of convergence. Then integrating along C gives

$$\int_0^{z_0} \frac{dz}{1+z} = \sum_{n=0}^{\infty} \left((-1)^n \int_0^{z_0} z^n \, dz \right),$$

so after performing the indicated integrations we obtain

$$\mathrm{Ln}(1+z)\Big|_0^{z_0} = \sum_{n=0}^{\infty} \frac{(-1)^n z^{n+1}}{(n+1)}\Bigg|_0^{z_0},$$

and thus

$$\mathrm{Ln}(1+z_0) = \sum_{n=0}^{\infty} \frac{(-1)^n z_0^{n+1}}{n+1} \quad \text{for } |z_0| < 1.$$

Because z_0 was an arbitrary point in the circle of convergence, we may drop the suffix zero to obtain the general result

$$\mathrm{Ln}(1+z) = \sum_{n=0}^{\infty} \frac{(-1)^n z^{n+1}}{n+1} \quad \text{for } |z| < 1. \qquad \diamond$$

Exercises 3.3

In Exercises 1 through 14, find the radius and circle of convergence of the given series.

1. $\displaystyle\sum_{n=0}^{\infty} \frac{z^n}{n+1}$

2. $\displaystyle\sum_{n=0}^{\infty} \frac{z^n}{n^2+3}$

3. $\displaystyle\sum_{n=0}^{\infty} \frac{z^n}{n!}$

4. $\displaystyle\sum_{n=0}^{\infty} \left(1+i\sqrt{3}\right)^n z^n$

5. $\displaystyle\sum_{n=1}^{\infty} (n^2+1)^n z^n$

6. $\sum_{n=0}^{\infty} \frac{e^{in\pi} z^n}{n+1}$

7. $\sum_{n=0}^{\infty} [3 + (-1)^n]^n (z-1)^n$

8. $\sum_{n=1}^{\infty} \left(1 + \frac{3}{n}\right)^n (z+3)^n$

9. $\sum_{n=0}^{\infty} \frac{(z-i)^n}{4^n}$

10. $\sum_{n=1}^{\infty} \frac{k^n (z+i)^n}{n} \qquad (k > 0)$

11. $\sum_{n=0}^{\infty} \frac{n(n-1)(z+2i)^n}{n^2+3}$

12. $\sum_{n=1}^{\infty} \left(\frac{2n+1}{8n+3}\right)^n (z+1-3i)^n$

13. $\sum_{n=0}^{\infty} a_n (z-2i)^n \quad \text{where } a_n = \begin{cases} 2^n & \text{for } n \text{ even} \\ -3^n & \text{for } n \text{ odd} \end{cases}$

14. $\sum_{n=0}^{\infty} a_n (z+1+6i)^n \quad \text{where } a_n = \begin{cases} 2 + (-1)^n & \text{for } n \text{ even} \\ -5^n & \text{when } n \text{ is odd} \end{cases}$

15. Use repeated differentiation and a change of variable to derive a power series for $1/(1-3z)^2$ from the geometric series

$$\frac{1}{1-z} = \sum_{n=0}^{\infty} z^n \quad \text{for } |z| < 1.$$

16. Use repeated differentiation and a change of variable to derive a power series for $1/(1+2z)^3$ from the geometric series

$$\frac{1}{1-z} = \sum_{n=0}^{\infty} z^n \quad \text{for } |z| < 1.$$

17. Given that

$$\sin z = \sum_{n=1}^{\infty} (-1)^n \frac{z^{2n-1}}{(2n-1)!},$$

show the series is uniformly convergent and so can be differentiated term by term and hence obtain the series for $\cos z$.

18. Given that

$$\cosh z = \sum_{n=1}^{\infty} \frac{z^{2n}}{(2n)!},$$

show the series is uniformly convergent and so can be differentiated term by term and hence obtain the series for $\sinh z$.

19. Given that

$$\frac{1}{1+z^2} = \sum_{n=0}^{\infty} (-1)^n z^{2n} \quad \text{for } |z^2| < 1,$$

justify term-by-term integration of this series to obtain a series for

$$\text{Arctan}\, z = \int_0^z (dt/(1+t^2)).$$

20. Justify integrating term by term the series for $\sin z$ in Exercise 17 and using the condition $\cos 0 = 1$ to obtain the series

$$\cos z = \sum_{n=0}^{\infty} (-1)^n \frac{z^{2n}}{(2n)!}.$$

3.4 Taylor Series

In the previous section, it was shown that when expanded about a point z_0, a convergent power series with radius of convergence $R > 0$ define a function of z within the circle of convergence $|z - z_0| < R$, after which the properties of such series were investigated. In this section this process will be reversed because the question we now ask is "How can a given function $f(z)$ be represented by a power series expanded about a point z_0 where $f(z)$ is analytic?" Thus, given a function $f(z)$, we wish to find the coefficients a_n in the series representation of the function

$$f(z) = \sum_{n=0}^{\infty} a_n (z - z_0)^n \tag{3.17}$$

that converges to $f(z)$ for $|z - z_0| < R$, for some $R > 0$.

Let us be quite clear about our objective which is to use a given analytic function $f(z)$ to find the coefficients a_n on the right of Equation (3.17), so that for some $R > 0$ the series on the right will converge to the given function $f(z)$ on the left for $|z - z_0| < R$. For convenience, the function to which the series on the right of Equation (3.17) converges is called the *sum function* of the series. So our question reduces to finding how to find the series on the right of Equation (3.17), and then determining when the sum function of the series on the right converges to the function $f(z)$.

We know from Theorem 3.3.3 that the series in Equation (3.17) can be repeatedly differentiated term by term within its circle of convergence, so performing the first few differentiations gives

$$f^{(1)}(z) = \sum_{n=1}^{\infty} na_n(z - z_0)^{n-1}, \tag{3.18}$$

$$f^{(2)}(z) = \sum_{n=2}^{\infty} n(n-1)a_n(z - z_0)^{n-2}, \tag{3.19}$$

$$f^{(3)}(z) = \sum_{n=3}^{\infty} n(n-1)(n-2)a_n(z - z_0)^{n-3}, \tag{3.20}$$

$$\vdots$$

$$f^{(r)}(z) = \sum_{n=r}^{\infty} n(n-1)\cdots(n-r+1)a_n(z - z_0)^{n-r}, \tag{3.21}$$

$$\vdots$$

Setting $z = z_0$ in these results gives

$$f(z_0) = a_0, f^{(1)}(z_0) = a_1, f^{(2)}(z_0) = 2!a_2,$$
$$f^{(3)}(z_0) = 3!a_3, \ldots, f^{(r)}(z_0) = r!a_r, \ldots.$$

So it follows from this that the general coefficient a_n in series Equation (3.17) is related to the function $f(z)$ through the expression

$$a_n = \frac{f^{(n)}(z_0)}{n!}, \quad \text{for } n = 0, 1, 2, \ldots, \tag{3.22}$$

it being understood that $f^{(0)}(z_0) \equiv f(z_0)$.

This suggests that the answer to the question posed earlier is that the function $f(z)$ can be represented by the power series

$$\sum_{n=0}^{\infty} \frac{f^{(n)}(z_0)}{n!}(z - z_0)^n. \tag{3.23}$$

This *formal representation* of the function $f(z)$ in terms of a powe[illegible] called the *Taylor series* of $f(z)$ expanded about the point z_0, and [illegible] $f^{(n)}(z_0)/n!$ in Equation (3.23) are called the *Taylor series coeffici*[illegible] term *formal* has been used here because we have still to[illegible] what conditions it is permissible to equate $f(z)$ to the sum function of the series in Equation (3.23). The answer to this question is provided by Taylor's theorem which we now state and prove.

THEOREM 3.4.1 Taylor's Theorem
Let $f(z)$ be analytic at the point z_0. Then the Taylor series expansion of $f(z)$ about the point z_0 is given by

$$f(z) = \sum_{n=0}^{\infty} \frac{f^{(n)}(z_0)}{n!}(z - z_0)^n,$$

where the Taylor coefficients are $f^{(n)}(z_0)/n!$ for $n = 0, 1, 2, \ldots$.

The series converges uniformly to $f(z)$ at all points of the disk $|z - z_0| < R_S$, where R_S is the distance from z_0 to the point (singularity) nearest to z_0 where $f(z)$ ceases to be analytic.

PROOF

Let $f(z)$ be analytic in the disk $|z - z_0| < R_S$ shown in Figure 3.5, where P is the singularity (the point where $f(z)$ becomes non-differentiable) closest to z_0. Then, for any point z inside the disk, draw a circle C_r of radius r centered on z_0 such that z lies in its interior and C_r is contained within the circle $|z - z_0| < R_S$. Constrain ζ to lie on C_r, so that $|z - z_0| < |\zeta - z_0| = r < R_S$.

Then by the Cauchy integral theorem we have

$$f(z) = \frac{1}{2\pi i}\int_{C_r} \frac{f(\zeta)d\zeta}{(\zeta - z)}.$$

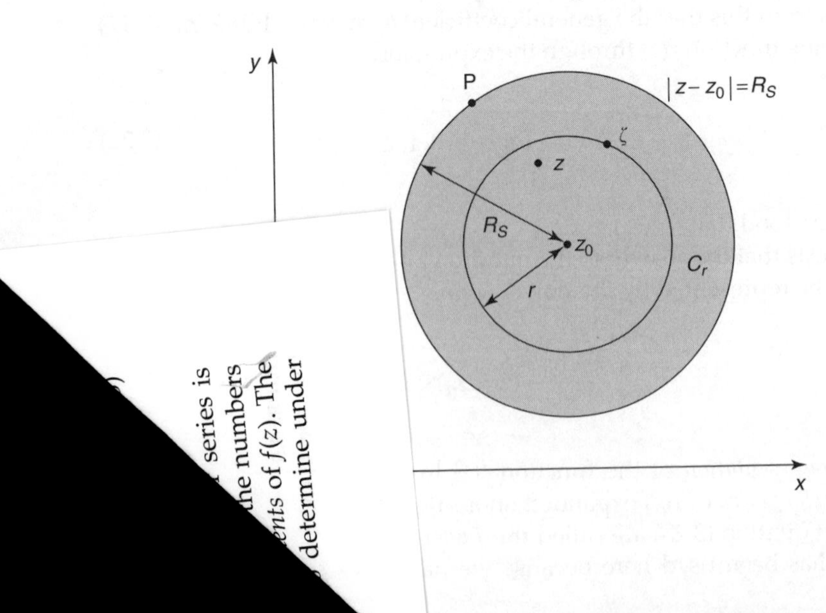

Let us now rewrite the factor $1/(\zeta - z)$ in the integrand as follows

$$\frac{1}{\zeta - z} = \frac{1}{\zeta - z_0} \frac{1}{\left[1 - \left(\dfrac{z - z_0}{\zeta - z_0}\right)\right]} = \frac{1}{\zeta - z_0}\left[1 - \left(\frac{z - z_0}{\zeta - z_0}\right)\right]^{-1},$$

which is permissible because ζ lies on C_r, so $\zeta \neq z_0$. Now as $|z - z_0| < |\zeta - z_0|$, we have $|(z - z_0)/(\zeta - z_0)| < 1$, so the last factor on the right may be expanded by the binomial theorem when the expression for $1/(\zeta - z)$ becomes

$$\frac{1}{\zeta - z} = \frac{1}{\zeta - z_0}\left[1 + \left(\frac{z - z_0}{\zeta - z_0}\right) + \left(\frac{z - z_0}{\zeta - z_0}\right)^2 + \left(\frac{z - z_0}{\zeta - z_0}\right)^3 + \cdots\right],$$

which simplifies to

$$\frac{1}{\zeta - z} = \frac{1}{\zeta - z_0} + \frac{(z - z_0)}{(\zeta - z_0)^2} + \frac{(z - z_0)^2}{(\zeta - z_0)^3} + \cdots.$$

This series is uniformly convergent when ζ lies on C_r, because the modulus $|(z - z_0)/(\zeta - z_0)| = |z - z_0|/r < 1$ is independent of ζ.

Multiplying this series by $f(\zeta)/2\pi i$, and then integrating each term around C_r which is permissible because the series is uniformly convergent, gives

$$\frac{1}{2\pi i}\int_{C_r} \frac{f(\zeta)}{\zeta - z}\, d\zeta = \frac{1}{2\pi i}\int_{C_r} \frac{f(\zeta)}{\zeta - z_0}\, d\zeta + \frac{(z - z_0)}{2\pi i}\int_{C_r} \frac{f(\zeta)}{(\zeta - z_0)^2}\, d\zeta + \frac{(z - z_0)^2}{2\pi i}\int_{C_r} \frac{f(\zeta)}{(\zeta - z_0)^3}\, d\zeta + \cdots.$$

Applying the Cauchy integral formula to the terms on both the left and right, this becomes

$$f(z) = \sum_{n=0}^{\infty} \frac{f^{(n)}(z_0)}{n!}(z - z_0)^n \quad \text{for } |z - z_0| < R_S.$$

This is the Taylor series representation of $f(z)$ expanded about the point z_0 that was to be established and it is uniformly convergent inside every circle C_r such that $0 < r < R_S$. Thus the theorem has been proved. ◆

As a Taylor series is a power series it follows immediately that all of the properties of power series that have been established so far are also properties of Taylor series. Thus Taylor series are continuous functions that may be scaled, differentiated term by term and also integrated term by term within their circle of convergence. When $z_0 = 0$, the expansion is about the origin, and in this form the series becomes

$$f(z) = \sum_{n=0}^{\infty} \frac{f^{(n)}(0)}{n!} z^n \quad \text{for } |z| < R_S, \tag{3.24}$$

when it is called the *Maclaurin series* expansion of $f(z)$. If $f(z)$ is analytic in all of complex plane (it is an entire function) then the radius of convergence of both the Taylor and Maclaurin expansions will be infinite.

It is important to recognize that the properties of Taylor and Maclaurin series require $f(z)$ to be an *analytic* function, and not merely a function that is infinitely differentiable. An example that illustrates this involves the function

$$f(z) = \begin{cases} \exp(-1/z^2) & \text{for } z \neq 0 \\ 0 & \text{for } z = 0. \end{cases}$$

Clearly $f(z)$ is analytic for $z \neq 0$, and so is infinitely differentiable for $z \neq 0$. To show that $f(z)$ is also infinitely differentiable at the origin, let's start by computing $f'(0)$. Because of the definition of $f(z)$ at the origin it is necessary to compute $f'(0)$ from first principles, when we find that

$$f'(0) = \lim_{z \to 0} \left(\frac{f(z) - f(0)}{z - 0} \right) = \lim_{z \to 0} \left(\frac{\exp(-1/z^2)}{z} \right).$$

To evaluate this limit set $z = 1/\zeta$, when it becomes

$$f'(0) = \lim_{\zeta \to \infty} [\zeta \exp(-\zeta^2)] = 0.$$

so both $f(0) = 0$ and $f'(0) = 0$, causing the first two terms in the Maclaurin series to vanish. To compute $f^{(n)}(0)$ we notice that when $z \neq 0$, repeated differentiation gives

$$f^{(n)}(z) = \exp(-1/z^2) \times (\text{Polynomial in } 1/z),$$

from which it follows by induction that $f^{(n)}(0) = 0$ for all positive integers n. Thus, although $f(z)$ is infinitely differentiable for all z, its Maclaurin series expansion is $f(z) = 0 + 0 + 0 + \cdots$, showing that the series only converges to $\exp(-1/z^2)$ when $z = 0$.

The next property of a Taylor series established here is simple, but it has far-reaching consequences. It says that a Taylor series representation of a function expanded about a point z_0 is *unique*, so however such an expansion is obtained it must be a Taylor series expansion. This means that if instead of repeated differentiation of a function to find the Taylor series coefficients, the expansion is obtained in some other simpler way, the result will always be a Taylor series. The different ways of obtaining a Taylor series include using a binomial expansion, modifying a known series, scaling and combining known series and the multiplication of series. Such approaches often simplify the derivation of a Taylor series, as will be seen from subsequent examples.

THEOREM 3.4.2 Uniqueness of a Taylor Series
If $f(z)$ is analytic at z_0, then its Taylor Series expansion about that point is unique.

PROOF

Suppose, if possible, that $f(z)$ has the two different Taylor series expansions about the point z_0

$$f(z) = \sum_{n=0}^{\infty} a_n (z - z_0)^n \quad \text{and} \quad f(z) = \sum_{n=0}^{\infty} b_n (z - z_0)^n$$

for $|z - z_0| < R$, where $a_n \neq b_n$. Then as $f(z)$ is analytic in the disk $|z - z_0| < R$, it follows from Theorem 3.4.2 that

$$a_n = \frac{f^{(n)}(z_0)}{n!} \quad \text{and} \quad b_n = \frac{f^{(n)}(z_0)}{n!}, \quad \text{for } n = 0, 1, 2, \ldots,$$

so the series is unique. ◆

THEOREM 3.4.3 Linear Combinations and Products of Taylor Series
Let the functions $f(z)$ and $g(z)$ both be analytic at z_0 and have the Taylor series expansions

$$f(z) = \sum_{n=0}^{\infty} a_n (z - z_0)^n \quad \text{for } |z - z_0| < R_1,$$

and

$$g(z) = \sum_{n=0}^{\infty} b_n (z - z_0)^n \quad \text{for } |z - z_0| < R_2.$$

Then if α *and* β *are arbitrary complex numbers and* $R = \min\{R_1, R_2\}$,

$$\text{(i)}\quad \alpha f(z) + \beta g(z) = \sum_{n=0}^{\infty} (\alpha a_n + \beta n_n)(z - z_0)^n \quad \text{for } |z - z_0| < R,$$

and

$$\text{(ii)}\quad f(z)g(z) = \sum_{n=0}^{\infty} \left(\sum_{r=0}^{n} a_r b_{n-r} \right) (z - z_0)^n \quad \text{for } |z - z_0| < R.$$

PROOF

(i) The details of the proof of this result are omitted since they are a direct extension of Theorem 3.1.3 when applied to series. The restriction on the radius of convergence of the series $\Sigma_{n=0}^{\infty} (\alpha a_n + \beta b_n)(z - z_0)^n$ arises because although scaling a series cannot alter its radius of convergence, the region in which it is analytic is the interior of a circle centered on z_0 that passes through the singularity of the function that is nearest to z_0. Thus, a linear combination of the scaled sum of series has a radius of convergence determined by whichever of the singularities of $f(z)$ and $g(z)$ is closest to z_0.

(ii) To determine the product rule for the series for $f(z)g(z)$ expanded about z_0 we will differentiate the product using Leibniz's rule to obtain

$$\frac{d^n[f(z)g(z)]}{dz^n} = \sum_{r=0}^{n} \frac{n!}{r!(n-r)!} f^{(r)}(z) g^{(n-r)}(z).$$

Then, since from Taylor's theorem

$$a_n = \frac{f^{(n)}(z_0)}{n!} \quad \text{and} \quad b_{n-r} = \frac{g^{(n-r)}(z_0)}{(n-r)!}$$

we see that

$$\frac{d^n[f(z)g(z)]}{dz^n} = n! \sum_{r=0}^{n} a_r b_{n-r},$$

showing that the Taylor series expansion about z_0 is

$$f(z)g(z) = \sum_{n=0}^{\infty} \left(\sum_{r=0}^{n} a_r b_{n-r} \right) (z - z_0)^n \quad \text{for } |z - z_0| < R,$$

where the radius of convergence is restricted for the same reason as in (i), so the proof is complete. ◆

Example 3.4.2 A Linear Combination of Series
Find the Maclaurin series for $\sinh z$ and $\cosh z$.

SOLUTION
By definition $\sinh z = \frac{1}{2}(e^z - e^{-z})$ and $\cosh z = \frac{1}{2}(e^z + e^{-z})$, so the Maclaurin series for these functions can be found by combining the Maclaurin series for e^z and e^{-z}. In the case of $\sinh z$. the even powers of z cancel, while in the case of $\cosh z$ it is the odd powers of z that cancel, giving

$$\sinh z = \sum_{n=0}^{\infty} \frac{z^{2n+1}}{(2n+1)!} \quad \text{and} \quad \cosh z = \sum_{n=0}^{\infty} \frac{z^{2n}}{(2n)!}.$$

As e^z and e^{-z} are entire functions, so also are $\sinh z$ and $\cosh z$. ◇

Example 3.4.3 Using a Binomial Expansion to Obtain a Maclaurin Series
Find the Maclaurin series expansion of $f(z) = 1/(1+z^2)^2$.

SOLUTION
We start from the binomial expansion

$$(1+a)^n = 1 + na + \frac{n(n-1)}{2!}a^2 + \cdots + \frac{n(n-1)(n-2)\cdots(n-r+1)}{r!} + \cdots.$$

When n is a positive integer this reduces to a polynomial of degree n, but when n is a negative integer it becomes an infinite series that converges for $|a| < 1$. The required Maclaurin series follows by setting $n = -2$ and $a = z^2$ to obtain

$$f(z) = \frac{1}{(1+z^2)^2} = 1 - 2z^2 + 3z^4 - 4z^6 + \cdots = \sum_{n=0}^{\infty}(-1)^n(n+1)z^{2n}.$$

The binomial expansion converges for $|a| < 1$, so this series converges for $|z| < 1$. ◇

Example 3.4.4 Finding a Maclaurin Series Expansion Using Term-by-Term Integration
Find the Maclaurin series expansion of the *error function*

$$\operatorname{erf}(z) = \frac{2}{\sqrt{\pi}} \int_0^z \exp(-t^2)dt.$$

This important special function occurs in statistics and also throughout applications of mathematics.

SOLUTION
The result will be obtained by expanding the analytic function $\exp(-t^2)$ as a Maclaurin series, joining the origin to a point z by any simple arc C, and then integrating the Maclaurin series term by term along C. Proceeding in this manner gives

$$\operatorname{erf}(z) = \frac{2}{\sqrt{\pi}} \int_0^z \left(\sum_{n=0}^{\infty} (-1)^n \frac{t^{2n}}{n!} \right) dt = \frac{2}{\sqrt{\pi}} \left(\sum_{n=0}^{\infty} \int_0^z (-1)^n \frac{t^{2n}}{n!} dt \right)$$

$$= \frac{2}{\sqrt{\pi}} \sum_{n=0}^{\infty} (-1)^n \frac{z^{2n+1}}{n!(2n+1)}.$$

The error function is an entire function because the exponential function from which it is derived is itself an entire function. ◇

Example 3.4.5 Obtaining a Maclaurin Series by Multiplication of Series

Find the Maclaurin series expansion of

$$f(z) = \frac{1}{(1-z)(1+z^2)^2}.$$

SOLUTION
We start from the binomial expansion

$$\frac{1}{1-z} = \sum_{n=0}^{\infty} z^n = 1 + z + z^2 + \cdots,$$

and use the result from Example 3.4.3 that

$$f(z) = \frac{1}{(1+z^2)^2} = 1 - 2z^2 + 3z^4 - 4z^6 + \cdots = \sum_{n=0}^{\infty} (-1)^n (n+1) z^{2n},$$

both of which series converge for $|z| < 1$.

Some care must be exercised when applying Theorem 3.4.3(i) because all powers of z occur in the series for $1/(1-z)$, but only even powers of z occur in the series for $1/(1+z^2)^2$. In the notation of the theorem $a_n = 1$ and $b_{2n} = (-1)^n(n+1)$, while $b_{2n+1} = 0$, for $n = 0, 1, 2, \ldots$. Thus

$$f(z)g(z) = \sum_{n=0}^{\infty} \left(\sum_{r=0}^{n} a_r b_{n-r} \right) z^n = \sum_{n=0}^{\infty} \left(\sum_{r=0}^{n} b_{n-r} \right) z^n$$

$$= \sum_{n=0}^{\infty} (b_0 + b_1 + \cdots + b_n) z^n$$

$$= 1 + z - z^2 - z^3 + 2z^4 + 2z^5 - 2z^6 - 2z^7 + 3z^8 + 3z^9 + \cdots, \text{ for } |z| < 1.$$

◇

Exercises 3.4

1. Show that the Maclaurin series for $\sin z$ is

$$\sin z = \sum_{n=0}^{\infty} (-1)^n \frac{z^{2n+1}}{(2n+1)!},$$

and hence show that $\sin z$ is an entire function.
2. Find the Maclaurin series for $\sin^2 z$, and determine its radius of convergence.
3. Find the Maclaurin series for $1/(2z-1)$, and determine its radius of convergence.
4. Find the Maclaurin series for $z^2/(1+z)^2$, and determine its radius of convergence.
5. Find the Maclaurin series for $\text{Ln}(1+z^2)$, and determine its radius of convergence.
6. Find the Taylor series expansion of $z/(z+2)$ about the point $z=1$, and determine its radius of convergence.
7. Find the Taylor series expansion of $z^2/(1+z)^2$ about the point $z=1$, and determine its radius of convergence.
8. Find the Taylor series expansion of $\cosh z$ about the point $z=i$, and determine its radius of convergence.
9. Find the Taylor series expansion of $\text{Ln}\, z$ about the point $z=i$, and determine its radius of convergence.
10. Find the Maclaurin series for $1/(z-2)$, and hence derive a series expansion for $1/[z^3(z-2)]$, and determine the domain in which it converges.
11. Use Theorem 3.4.3 to show that

$$\frac{1}{(1-z)^2} = \sum_{n=0}^{\infty} (n+1)z^n \quad \text{for } |z| < 1.$$

12. Use the identity $\cosh 2z = 1 + 2\sinh^2 z$ to find the Maclaurin series for $\sinh^2 z$ from the Maclaurin series for $\cosh z$.

Exercises of Greater Difficulty

When only the first few terms of the quotient of two power series are needed they can be found as follows. Let the series be

$$f(z) = a_0 + a_1 z + a_2 z^2 + \cdots \quad \text{for } |z| < R_1$$

and

$$g(z) = b_0 + b_1 z + b_2 z^2 + \cdots \quad \text{for } |z| < R_2,$$

and set

$$f(z)/g(z) = c_0 + c_1 z + c_2 z^2 + \cdots.$$

Then

$$\frac{f(z)}{g(z)} = \frac{a_0 + a_1 z + a_2 z^2 + \cdots}{b_0 + b_1 z + b_2 z^2 + \cdots} \equiv c_0 + c_1 z + c_2 z^2 + \cdots.$$

Multiplication by the series for $g(z)$ then gives

$$a_0 + a_1 z + a_2 z^2 + \cdots \equiv (b_0 + b_1 z + b_2 z^2 + \cdots)(c_0 + c_1 z + c_2 z^2 + \cdots).$$

This will be an identity if the coefficients of corresponding powers of z on each side of this expression are equal. Thus the coefficients $c_0, c_1, c_2, \ldots$, follow sequentially by solving the equations

$$\begin{aligned} a_0 &= b_0 c_0 \\ a_1 &= b_0 c_1 + b_1 c_0 \\ a_2 &= b_0 c_2 + b_1 c_1 + b_2 c_0 \\ a_3 &= b_0 c_3 + b_1 c_2 + b_2 c_1 + b_3 c_0 \\ &\vdots \end{aligned}$$

The series for $f(z)/g(z)$ is valid provided $g(z) \neq 0$ and $|z| < R$, where $R = \min\{R_1, R_2\}$.

The approach must be modified if the series for $g(z)$ begins with the power z^m, so that $g(z) = b_0 z^m + b_1 z^{m+1} + b_2 z^{m+2} + \cdots$. All that is necessary is to write $f(z)/g(z)$ in the form

$$\frac{f(z)}{g(z)} = \left(\frac{1}{z^m}\right)\left(\frac{f(z)}{b_0 + b_1 z + b_2 z^2 + \cdots}\right),$$

and then to treat the quotient $f(z)/(b_0 + b_1 z + b_2 z^2 + \cdots)$ as before.

Use this approach in Exercises 13 through 18 to find the required terms in the expansions of the given quotients.

13. $\tan z = \dfrac{\sin z}{\cos z}$ up to and including the term in z^7.

14. $\sec z = \dfrac{1}{\cos z}$ up to and including the term in z^6.

15. $\tanh z = \dfrac{\sinh z}{\cosh z}$ up to and including the term in z^7.

16. $\dfrac{\cosh z}{\cosh 2z}$ up to and including the term in z^4.

17. The function $f(z) = z \operatorname{ctn} z$ can be written in the form

$$f(z) = \cos z\left(\frac{z}{\sin z}\right) = \cos z \bigg/ \left(\frac{\sin z}{z}\right).$$

Divide the series for $\cos z$ by the series for $\sin(z/z)$ and hence find the series for $f(z)$ up to and including the term in z^6. Deduce from this the corresponding terms in the series for ctn z.

18. The function $f(z) = z \csc z$ can be written

$$f(z) = \frac{z}{\sin z} = 1\Big/\left(\frac{\sin z}{z}\right).$$

Use this result to find the series for $f(z)$ up to and including the term in z^4. Deduce from this the corresponding terms in the series for $\csc z$.

19. Let $f(z)$ have the Maclaurin series expansion $f(z) = \Sigma_{n=0}^{\infty} a_n z^n$, where $f(z)$ satisfies the equation $f(z) = 1 + zf(z) + z^2 f(z)$. By substituting the series for $f(z)$ into the equation, and then equating the coefficients of terms with corresponding powers of z on each side of the equation, show that the coefficients a_n are given by $a_0 = 1$, $a_1 = 1$, and thereafter by the recurrence relation $a_{n+2} = a_n + a_{n+1}$. The sequence of numbers $\{a_n\}$ generated, namely, $\{1, 1, 2, 3, 5, 8, 13, 21, \ldots\}$, is called the ***Fibonacci*** sequence. Verify the coefficients a_0 through a_4 by determining the first few terms of the series representing the quotient $f(z) = 1/(1 - z - z^2)$. Comment on the difficulty of finding the radius of convergence R of the series for $f(z)$, but do not attempt to find R.

3.5 Laurent Series

We now introduce an important generalization of the Taylor series expansion of a function $f(z)$ about a point where the function is analytic, to a series expansion about a point z_0 where the function may or may not be analytic. In general, a point z_0 where $f(z)$ fails to be analytic, and so $f'(z)$ fails to exist, is called a *singular point* of the function, when $f(z)$ is then said to have a *singularity* at z_0. In this section our concern is with *isolated singularities*, which are points where although $f'(z)$ fails to exist, the function $f(z)$ itself is analytic in a deleted neighborhood of the singularity. For example, the function $f(z) = \sin z/(z-1)$ has an isolated singularity at $z = 1$ because although $f'(z)$ is not defined at $z = 1$, the function is analytic for all $z \neq 1$. In the next section, the different types of singularity that can arise are defined, but for the time being we only consider functions with isolated singularities. In effect, as the previous example shows, an isolated singular point of $f(z)$ may be taken to be one with the property that in its neighborhood the function $f(z)$ becomes unbounded. A singularity of the form $1/(z - z_0)^n$ is called a *pole* of *order* n, and later we find that when working with contour integrals the most important type of pole occurs when $n = 1$, which is usually called a *simple pole*.

The generalization of a series to be considered here is called a *Laurent series* expansion of $f(z)$ about a point z_0, where the complex number z_0 is called the *center* of expansion. The Laurent series can contain both positive and negative powers of $(z - z_0)$, and its general form is

$$f(z) = \cdots + \frac{a_{-3}}{(z-z_0)^3} + \frac{a_{-2}}{(z-z_0)^2} + \frac{a_{-1}}{(z-z_0)} + a_0 + a_1(z-z_0) + a_2(z-z_0)^2 + \cdots,$$

which can be written more concisely as

$$f(z) = \sum_{n=-\infty}^{\infty} a_n(z-z_0)^n. \tag{3.25}$$

The constants a_n in series Equation (3.25) are called the *coefficients* of the Laurent series, and the coefficients $a_0, a_1, a_2, \ldots,$ of the positive powers of $(z - z_0)$ may either be finite or infinite in number, as may be the coefficients $a_{-1}, a_{-2}, a_{-3}, \ldots,$ of the negative powers of $(z - z_0)$. It is these powers, when present, that cause $f(z)$ to become unbounded as $z \to z_0$.

If we set

$$f_1(z) = \sum_{n=1}^{\infty} \frac{a_{-n}}{(z-z_0)^n}, \tag{3.26}$$

and

$$f_2(z) = \sum_{n=0}^{\infty} a_n(z-z_0)^n, \tag{3.27}$$

the Laurent series Equation (3.25) becomes

$$f(z) = f_1(z) + f_2(z). \tag{3.28}$$

In representation Equation (3.28) of $f(z)$ the series $f_1(z)$ is called the *principal part* of the Laurent series expansion of $f(z)$ about the point (center) z_0. The series $f_2(z)$ is called the *regular part* of the Laurent series expansion about z_0. Notice that when the principal part of a Laurent series expansion is present, so that not all of the coefficients $a_{-1}, a_{-2}, a_{-3}, \ldots$ are zero, the *principal part* of a Laurent series will become *unbounded* as $z \to z_0$, while when the series $f_2(z)$ is convergent the regular part will remain bounded as $z \to z_0$.

The Laurent series Equation (3.25) is said to *converge* at a point z if both series Equations (3.26) and (3.27) converge at that point. Series Equation (3.27) is an ordinary power series, so its domain of convergence is a disk $|z - z_0| < R$, for some nonnegative number R. If $R = 0$ the regular part of the

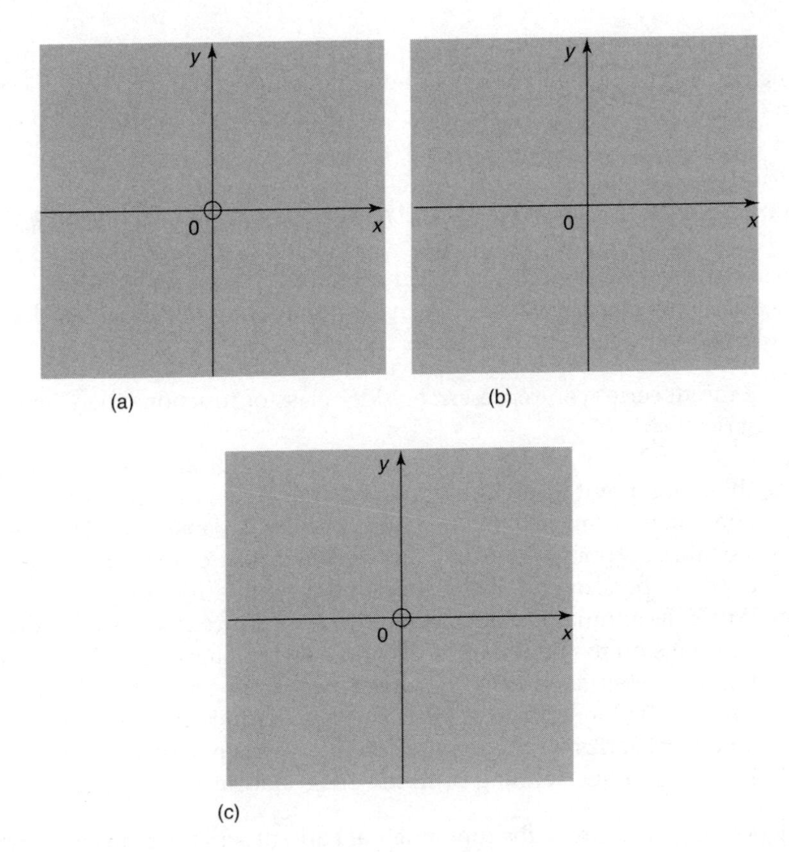

FIGURE 3.7
(a) Principal part of $f(z)$ converges for all $z \neq 0$, (b) Regular part of $f(z)$ converges for all z, (c) Laurent series for $f(z)$ converges for all $z \neq 0$.

exponential function, and then multiply by z^2 to arrive at the series

$$g(z) = z^2 \sum_{n=0}^{\infty} \frac{1}{n!z^n} = \sum_{n=0}^{\infty} \frac{1}{n!z^{n-2}}$$

$$= \underbrace{\left(z^2 + z + \frac{1}{2!}\right)}_{\text{regular part}} + \underbrace{\left(\frac{1}{3!z} + \frac{1}{4!z^2} + \frac{1}{5!z^3} + \cdots\right)}_{\text{principal part}}. \tag{3.31}$$

This process is purely *formal* because a Maclaurin series expansion requires a function to be analytic at the origin, and clearly $\exp(1/z)$ is not analytic at $z = 0$. In actual fact, Equation (3.31) is the Laurent series expansion of $g(z)$ about a center located at the origin, and the principal part of $g(z)$ is the infinite series involving negative powers of z

$$f_1(z) = \frac{1}{3!\,z} + \frac{1}{4!\,z^2} + \frac{1}{5!\,z^3} + \cdots,$$

that converges for all $z \neq 0$. Setting $z_0 = 0$ in Equation (3.26) and identifying the coefficients a_{-n} of the Laurent series shows them to be infinite in number, with $a_{-1} = 1/3!, a_{-2} = 1/4!, a_{-3} = 1/5!, \ldots$. The regular part of the expansion is

$$f_2(z) = \tfrac{1}{2} + z + z^2,$$

which converges for all z. So for the regular part of the expansion $a_0 = \frac{1}{2}$, $a_1 = 1, a_2 = 1$, and $a_n = 0$ for $n = 3, 4, \ldots$. Thus the Laurent series for $g(z)$ converges in the same degenerate annular region as $f(z)$ shown in Figure 3.7.

The following comparisons between Taylor and Laurent series will help to provide motivation for the rest of this section.

1. Laurent series can represent a wider class of functions than Taylor series.
2. Unlike a Taylor series, a Laurent series for $f(z)$ may be expanded about any point z_0, including one where z_0 is a singularity of $f(z)$.
3. There may be more than one Laurent series expansion of $f(z)$ about a center z_0 (point) each with its own annular domain of convergence, depending on the location of the singularities of $f(z)$.
4. Within its annular region of convergence, a Laurent series is unique.
5. At points on the boundary of its annular domains of convergence a Laurent series (such as Taylor series) may either converge or diverge.
6. As with Taylor series, it is often easier to deduce the coefficients of a Laurent series by the manipulation of known series than to find them from their defining complex integral formula (to be given).

We now state and prove the fundamental Laurent series expansion theorem.

THEOREM 3.5.1 Laurent's Theorem
Let $f(z)$ be analytic in the annular domain D of convergence $r < |z - z_0| < R$ with its center at z_0, and r and R such that $r \geqslant 0$ and $R > r$. Then inside the annulus $f(z)$ is represented by the convergent Laurent series

$$f(z) = \sum_{n=-\infty}^{\infty} a_n (z - z_0)^n,$$

where the Laurent coefficients a_n are given by

$$a_n = \frac{1}{2\pi i} \int_{\Gamma} \frac{f(\zeta)}{(\zeta - z_0)^{n+1}} d\zeta, \quad \text{for } n = 0, \pm 1, \pm 2, \ldots,$$

and Γ is any circle centered on z_0 with ζ moving around Γ in the positive (counterclockwise) sense and radius ρ such that $r < \rho < R$.

PROOF

Let C_1 and C_2 be two circles of radius r_1 and r_2, respectively, each centered on z_0 with a positive (counterclockwise) orientation, with $r < r_1 < r_2 < R$, as shown in Figure 3.8. Then it follows from an application of the Cauchy integral formula to the annular region that

$$f(z) = \frac{1}{2\pi i}\int_{C_1} \frac{f(\zeta)}{\zeta - z}\,d\zeta + \frac{1}{2\pi i}\int_{C_2-} \frac{f(\zeta)}{\zeta - z}\,d\zeta,$$

where z is any point between C_1 and C_2 and, it is recalled, the notation C_{2-} indicates the reversal of the orientation of C_2 to the negative (clockwise) sense. So, reversing the direction of integration in the second integral, we have

$$f(z) = \frac{1}{2\pi i}\int_{C_1} \frac{f(\zeta)}{\zeta - z}\,d\zeta - \frac{1}{2\pi i}\int_{C_2} \frac{f(\zeta)}{\zeta - z}\,d\zeta. \tag{3.32}$$

In the first integral on the left of Equation (3.32) ζ lies on circle C_2, so

$$\left|\frac{z - z_0}{\zeta - z_0}\right| = \frac{|z - z_0|}{r_2} < 1.$$

Thus if for any fixed z we write

$$\frac{1}{\zeta - z} = \frac{1}{(\zeta - z_0)\left[1 - \left(\dfrac{z - z_0}{\zeta - z_0}\right)\right]} = \frac{1}{(\zeta - z_0)}\left[1 - \left(\frac{z - z_0}{\zeta - z_0}\right)\right]^{-1}, \tag{3.33}$$

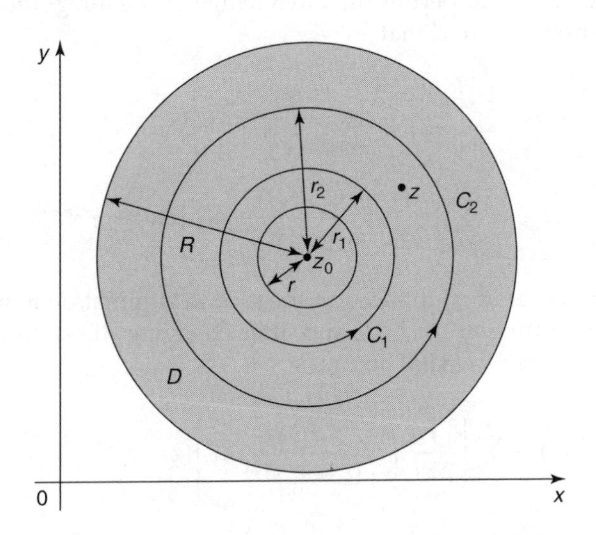

FIGURE 3.8
Circles C_1 and C_2 in the annular domain of convergence D with z any point between C_1 and C_2.

it follows that we may expand the last term by the binomial theorem as a uniformly convergent series to obtain

$$\frac{1}{\zeta - z} = \frac{1}{(\zeta - z_0)} \sum_{n=0}^{\infty} \left(\frac{z - z_0}{\zeta - z_0} \right)^n = \sum_{n=0}^{\infty} \frac{(z - z_0)^n}{(\zeta - z_0)^{n+1}}.$$

In the second integral on the right ζ lies on C_1, so that

$$\left| \frac{\zeta - z_0}{z - z_0} \right| = \frac{r_1}{|z - z_0|} < 1,$$

so if for any fixed z we write

$$\frac{1}{\zeta - z} = \frac{-1}{(z - z_0)\left[1 - \left(\dfrac{\zeta - z_0}{z - z_0}\right)\right]} = \frac{1}{(z - z_0)} \left[1 - \left(\frac{\zeta - z_0}{z - z_0}\right)\right]^{-1},$$

expanding the last term by the binomial theorem in a uniformly convergent series gives

$$\frac{1}{\zeta - z} = \frac{-1}{(z - z_0)} \sum_{n=0}^{\infty} \left(\frac{\zeta - z_0}{z - z_0} \right)^n = -\sum_{n=0}^{\infty} \frac{(\zeta - z_0)^n}{(z - z_0)^{n+1}}. \tag{3.34}$$

Substituting Equations (3.33) and (3.34) into Equation (3.32) and using the uniform convergence to permit the interchange of the integration and summation operations, we find that

$$\begin{aligned} f(z) = {} & \sum_{n=0}^{\infty} \left(\frac{1}{2\pi i} \int_{C_2} \frac{f(\zeta)}{(\zeta - z_0)^{n+1}} d\zeta \right) (z - z_0)^n \\ & + \sum_{n=0}^{\infty} \left(\frac{1}{2\pi i} \int_{C_1} \frac{f(\zeta)}{(\zeta - z_0)^{-n}} d\zeta \right) (z - z_0)^{-(n+1)}. \end{aligned} \tag{3.35}$$

We now shift the summation index in the second summation to make it start with $n = 1$, by setting $m = n + 1$, and after changing the summation index from m to n in Equation (3.35) becomes

$$\begin{aligned} f(z) = {} & \sum_{n=0}^{\infty} \left(\frac{1}{2\pi i} \int_{C_2} \frac{f(\zeta)}{(\zeta - z_0)^{n+1}} d\zeta \right) (z - z_0)^n \\ & + \sum_{n=1}^{\infty} \left(\frac{1}{2\pi i} \int_{C_1} \frac{f(\zeta)}{(\zeta - z_0)^{-(n-1)}} d\zeta \right) (z - z_0)^{-n}. \end{aligned} \tag{3.36}$$

Inside the annulus $r < |z - z_0| < R$ the function $f(\zeta)/(\zeta - z_0)^n$ is analytic for all n, so it follows by contour deformation that each of the contours C_1 and C_2 may be deformed into the contour C in the statement of the theorem.

If, now, we define the constants a_n by means of the integral

$$a_n = \frac{1}{2\pi i}\int_C \frac{f(\zeta)}{(\zeta - z_0)^{n+1}}\,d\zeta, \quad \text{for } n = 0, \pm 1, \pm 2, \ldots, \tag{3.37}$$

result in Equation (3.36) reduces to

$$f(z) = \sum_{n=-\infty}^{\infty} a_n (z - z_0)^n, \tag{3.38}$$

and the theorem is proved. ◆

We mention in passing that a slight modification of the above argument shows that the Laurent series converges *uniformly* to $f(z)$ in any closed region within the domain of convergence, though the details will be omitted. This fact allows a Laurent series to be integrated and differentiated term by term within its domain of convergence.

THEOREM 3.5.2 Uniqueness of Laurent Series
Within its annular domain of convergence a Laurent series is unique.

PROOF

Suppose, if possible, that a function $f(z)$ analytic in an annular region $r < |z - z_0| < R$ has the two different Laurent expansions

$$f(z) = \sum_{n=-\infty}^{\infty} a_n (z - z_0)^n \quad \text{and} \quad f(z) = \sum_{n=-\infty}^{\infty} A_n (z - z_0)^n.$$

Multiply each series by $(z - z_0)^{-(N+1)}$, where N is a fixed integer, and equate the results to obtain

$$\sum_{n=-\infty}^{\infty} a_n (z - z_0)^{n-N-1} = \sum_{n=-\infty}^{\infty} A_n (z - z_0)^{n-N-1}.$$

Now integrate this result around a circle C with radius ρ centered on z_0, where $r < \rho < R$. Then, from Example 2.1.3, $\int_C (z - z_0)^m dz = \begin{cases} 0, & m \neq -1 \\ 2\pi i, & m = -1, \end{cases}$ so $a_N = A_N$ for $N = 0, \pm 1, \pm 2, \ldots$, and the uniqueness is established. ◆

This simple but important result is of great practical significance because it means that the form of a Laurent series expanded about a given center z_0 in a specific domain of convergence is *independent* of the method of its derivation. Thus, simple techniques such as the manipulation of a known series, and the use of the binomial theorem, may be used to find the coefficients a_n of a Laurent series in place of the integral in Equation (3.37), which is often difficult to compute.

Example 3.5.1 Finding a Laurent Series Expansion Using Formula (3.37)

Find all possible Laurent series expansions of $f(z) = 1/(1 - 2z)$ about the center $z_0 = 0$.

SOLUTION

As the Laurent series expansions of $f(z)$ are made about the origin, they are expansions in powers of z, and their annular domains of convergence are centered on the origin. The only singularity of $f(z)$ occurs when $z = \frac{1}{2}$. So the circle C with the equation $|z| = \frac{1}{2}$ passing through the singularity represents the boundary separating the annular domain D_1 inside $|z| = \frac{1}{2}$ and the annular domain D_2 outside $|z| = \frac{1}{2}$, in each of which $f(z)$ is analytic. So one expansion is made inside C, where for the time being, the origin is excluded, and another is made outside C, where the domains D_1 and D_2 are shown in Figure 3.9. Later, however, we find that the point at the origin can be included in D_1, making it a complete disk.

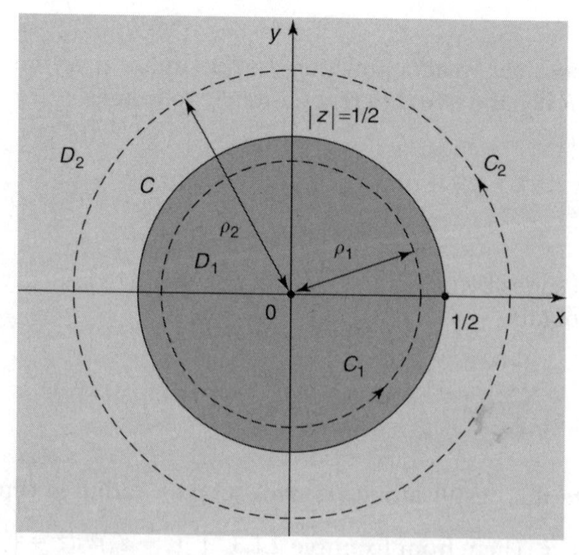

FIGURE 3.9
The annular domains of convergence D_1 and D_2.

Let us denote the coefficients of the Laurent expansions in D_1 and D_2 by $a_n^{(1)}$ and $a_n^{(2)}$, respectively. Starting with the expansion in D_1 we set $f(\zeta) = 1/(1-2\zeta)$ in Formula (3.37), giving

$$a_n^{(1)} = \frac{1}{2\pi i}\int_\Gamma \frac{d\zeta}{(1-2\zeta)\zeta^{n+1}}, \qquad \text{for } n = 0, \pm 1, \pm 2, \ldots,$$

where Γ is identified with any circle C_1 in D_1 with the equation $|z| = \rho_1$ such that $0 < \rho_1 < \frac{1}{2}$ as shown in Figure 3.9 where the integration is in the positive sense around C_1.

If $n > -1$ (n integral), the only singularity of the integrand inside C_2 occurs at the origin, and it follows from the Cauchy integral theorem for derivatives that

$$a_n^{(1)} = \frac{1}{n!}\frac{d^n}{d\zeta^n}(1-2\zeta)^{-1}\bigg|_{\zeta=0} = 2^n, \qquad \text{for } n = 0, 1, 2, \ldots.$$

However, if $n \leqslant -1$, the integrand is analytic inside C_1 so from the Cauchy–Goursat theorem

$$a_n^{(1)} = 0, \quad \text{for } n = -1, -2, \ldots.$$

Substituting these coefficients into the general expression for the Laurent series gives

$$\frac{1}{1-2z} = 1 + 2z + 2^2z^2 + 2^3z^3 + \cdots.$$

This series converges for all points z in D_1, including the origin, so the previous series is the Laurent series for $f(z)$ at all points such that $0 \leqslant |z| < \frac{1}{2}$

Next, we consider the expansion in the annular region D_2. The Laurent series coefficients are again given by

$$a_n^{(2)} = \frac{1}{2\pi i}\int_\Gamma \frac{d\zeta}{(1-2\zeta)\zeta^{n+1}}, \qquad \text{for } n = 0, \pm 1, \pm 2, \ldots,$$

although now we may identify Γ with any circle C_2 in D_2 with the equation $|z| = \rho_2$, oriented in the positive sense, and such that $\rho_2 > \frac{1}{2}$. Because C_2 contains C, a singularity exists at $\zeta = \frac{1}{2}$ and, if $n > -1$ (n integral), one also exists at the origin. So in this case

$$\begin{aligned} a_n^{(2)} &= \underbrace{\lim_{\zeta \to 1/2}\left(\frac{-(\zeta - \frac{1}{2})}{2(\zeta - \frac{1}{2})\zeta^{n+1}}\right)}_{\substack{\text{contribution from} \\ \text{singularity at } \zeta = 1/2}} + \underbrace{\frac{1}{n!}\frac{d^n}{d\zeta^n}(1-2\zeta)^{-1}\bigg|_{\zeta=0}}_{\substack{\text{contribution from} \\ \text{singularity at } \zeta = 0}} \\ &= -2^n + 2^n = 0 \qquad \text{for } n = 0, 1, 2, \ldots. \end{aligned}$$

However, if $n \leqslant -1$, the integral only has a singularity at $\zeta = \frac{1}{2}$, so then

$$a_n^{(2)} = \lim_{\zeta \to 1/2} \left(\frac{-(\zeta - \frac{1}{2})}{2(\zeta - \frac{1}{2})\zeta^{n+1}} \right) = -2^n, \quad \text{for } n = -1, -2, \ldots .$$

Substituting these coefficients in the Laurent expansion gives

$$\frac{1}{1-2z} = -\left(\frac{1}{2z} + \frac{1}{2^2 z^2} + \frac{1}{2^3 z^3} + \cdots \right),$$

for all z in D_2.

Thus we have arrived at the two different Laurent expansions for $f(z) = 1/(1 - 2z)$ about the origin, namely

$$\frac{1}{1-2z} = 1 + 2z + 2^2 z^2 + 2^3 z^3 + \cdots, \quad \text{for } 0 \leqslant |z| < \frac{1}{2}, \tag{3.39}$$

and

$$\frac{1}{1-2z} = -\left(\frac{1}{2z} + \frac{1}{2^2 z^2} + \frac{1}{2^3 z^3} + \cdots \right), \quad \text{for } |z| > \frac{1}{2}. \tag{3.40}$$

Notice that Equation (3.39) is simply the Maclaurin series for $f(z)$ for $0 \leqslant |z| < \frac{1}{2}$, as would be expected. It is important to recognize that result from Equation (3.39) could have been obtained with less effort by simply expanding $(1 - 2z)^{-1}$ by the binomial theorem, which is permitted because the uniqueness of the Laurent series expansion allows it to be found by any method, and not only from Formula (3.37).

Result from Equation (3.40) may be derived in similar fashion, but now $|z| > \frac{1}{2}$, so for the binomial theorem to be applied (for it to converge) it is necessary to rewrite $f(z)$ as

$$f(z) = \frac{-1}{2z\left(1 - \dfrac{1}{2z}\right)} = \frac{-1}{2z}\left(1 - \frac{1}{2z}\right)^{-1},$$

where now the last term can now be expanded by the binomial theorem. ◇

Example 3.5.2 Finding Laurent Series Expansions about the Origin
Find all possible Laurent series expansions of

$$f(z) = \frac{2}{(z-1)(3-z)}$$

about the origin.

SOLUTION
The only singularity of $f(z)$ is located at $z = -1$ and the center of the expansion is to be at $z = i$. The circle C centered on $z = i$ passing through the singularity with the equation $|z - 1| = \sqrt{2}$ is shown in Figure 3.11 so one annular domain D_1 of convergence is inside this circle and the other D_2 is outside it. To expand $f(z)$ about $z = i$ we need to expand it in powers of $(z - i)$, so we write $f(z)$ as

$$f(z) = \frac{1}{(1+i)+(z-i)}.$$

Setting $\zeta = z - i$, this becomes

$$\begin{aligned} f(\zeta + i) &= \frac{1}{(1+i)+\zeta} \\ &= \frac{1}{(1+i)\left[1+\dfrac{\zeta}{1+i}\right]} = \frac{1}{(1+i)}\left[1+\frac{\zeta}{1+i}\right]^{-1}. \end{aligned}$$

The last factor in the previous expression may be expanded by the binomial theorem as a uniformly convergent series in powers of ζ provided

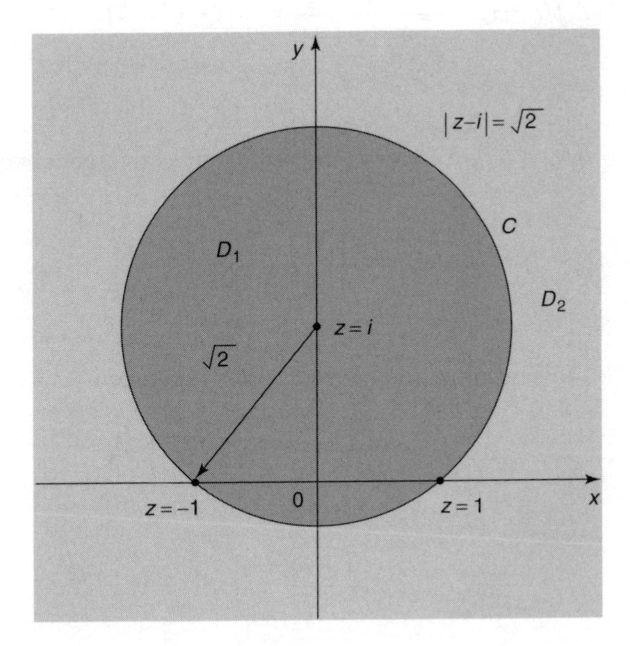

FIGURE 3.11
The annular domains of convergence D_1 and D_2.

$|\zeta/(1+i)| < 1$, which is equivalent to the condition $|z-i| < \sqrt{2}$, corresponding to the interior of circle C. This expansion is

$$f(\zeta + i) = \frac{1}{1+i}\left[1 - \left(\frac{\zeta}{1+i}\right) + \left(\frac{\zeta}{1+i}\right)^2 - \left(\frac{\zeta}{1+i}\right)^3 + \cdots\right],$$

and after reverting to the variable z, this becomes

$$f(z) = \frac{1}{1+i}\left[1 - \left(\frac{z-i}{1+i}\right) + \left(\frac{z-i}{1+i}\right)^2 - \left(\frac{z-i}{1+i}\right)^3 + \cdots\right],$$

$$= \frac{1}{1+i}\sum_{n=0}^{\infty}(-1)^n\left(\frac{z-i}{1+i}\right)^n \quad \text{for } |z-1| < \sqrt{2}.$$

This is the Laurent series expansion of $f(z)$ in domain of convergence D_1.

When the expansion is made in D_2, $|\zeta/(1+i)| > 1$ so apply the binomial theorem to the function $f(\zeta + i)$ it is necessary to rewrite it in the form

$$f(\zeta + i) = \frac{1}{\zeta\left[1 + \left(\frac{1+i}{\zeta}\right)\right]} = \frac{1}{\zeta}\left[1 + \left(\frac{1+i}{\zeta}\right)\right]^{-1}.$$

Because $|(1+i)/\zeta| < 1$, we arrive at the uniformly convergent expansion

$$f(\zeta + i) = \frac{1}{\zeta}\left[1 - \left(\frac{1+i}{\zeta}\right) + \left(\frac{1+i}{\zeta}\right)^2 - \left(\frac{1+i}{\zeta}\right)^3 + \cdots\right],$$

so in terms of z this becomes the Laurent series expansion

$$f(z) = \frac{1}{z-i}\left[1 - \left(\frac{1+i}{z-i}\right) + \left(\frac{1+i}{z-i}\right)^2 - \left(\frac{1+i}{z-i}\right)^3 + \cdots\right],$$

$$= \frac{1}{z-i}\sum_{n=0}^{\infty}(-1)^n\left(\frac{1+i}{z-i}\right)^n \quad \text{for } |z-1| > \sqrt{2}.$$

This is the Laurent series expansion of $f(z)$ in D_2. ◇

To find the expansion in D_4, we write

$$f(z) = f\left[\tfrac{1}{2}(1-\zeta)\right] = \frac{-2}{\zeta\left(1-\dfrac{1}{\zeta}\right)} + \frac{2}{\zeta},$$

so that

$$f\left[\tfrac{1}{2}(1-\zeta)\right] = -\frac{2}{\zeta}\left(1-\frac{1}{\zeta}\right)^{-1} + \frac{2}{\zeta}.$$

After expanding the first term by the binomial theorem this becomes

$$f\left[\tfrac{1}{2}(1-\zeta)\right] = -\frac{2}{\zeta}\left(1+\frac{1}{\zeta}+\frac{1}{\zeta^2}+\cdots\right) = -2\sum_{n=2}^{\infty}\left(\frac{1}{\zeta}\right)^n \quad \text{for } |\zeta| > 1.$$

In terms of the variable z, we then have

$$f(z) = -2\sum_{n=2}^{\infty}\left(\frac{1}{1-2z}\right)^n \quad \text{for } |z-\tfrac{1}{2}| > \tfrac{1}{2}.$$

Consideration of the four Laurent expansions obtained in (i) and (ii) shows that, in each case, at each point of the complex plane $f(z)$ is represented by a combination of two different expansions. This does *not* contradict the uniqueness of Laurent series expansions, because each of these expansions is about a different center and each has its own different domain of convergence. ◇

Example 3.5.5 Finding a Laurent Series Using an Elementary Hyperbolic Identity

Find the Laurent series expansion of

$$f(z) = \frac{\sinh 2z \cosh 5z}{z^2}$$

that involves powers of z.

SOLUTION

As in the previous example, this expansion must be abou[illegible] the Maclaurin series of sinh 2z and cosh 5z can be used. H[illegible] multiplying the series for $\sinh 2z$ and $\cosh 5z$, it is easier to use the hyperbolic identity

$$\sinh(A + B) + \sinh(A - B) = 2\sinh A \cosh B$$

to simplify the product $\sinh 2z \cosh 3z$. By setting $A = 2z$ and $B = 5z$ we find that

$$\sinh 2z \cosh 5z = \tfrac{1}{2}(\sinh 7z - \sinh 3z),$$

and so

$$f(z) = \frac{\sinh 7z - \sin 3z}{2z^2}.$$

Replacing $\sinh 7z$ and $\sinh 3z$ by their Maclaurin series gives

$$f(z) = \frac{\left(7z + \frac{(7z)^3}{3!} + \frac{(7z)^5}{5!} + \cdots\right) - \left(3z + \frac{(3z)^3}{3!} + \frac{(3z)^5}{5!} + \cdots\right)}{2z^2},$$

and after grouping terms and simplification this becomes

$$f(z) = \frac{2}{z} + \frac{79}{3}z + \frac{4141}{60}z^3 + \frac{205339}{2520}z^5 + \cdots.$$

Because $\sinh z$ and $\cosh z$ are entire functions, and the only singularity of $f(z)$ occurs at the origin, it follows that this expansion is valid for $|z| > 0$. ◇

Example 3.5.6 Finding a Laurent Series of an Algebraic Function in a Neighborhood of the Point at Infinity

Find the Laurent series of

$$f(z) = \frac{1}{(1 + z^2)^2}$$

(a) about the point $z = i$ and (b) in the neighborhood of the point at infinity.

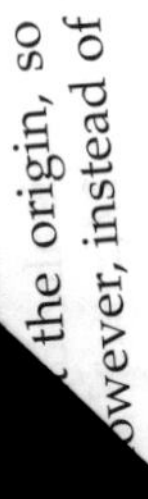

singularity of $f(z)$, so this influences the way the
nsion in terms of powers of $(z - i)$ is to be found.

$$f(z) = \frac{1}{(z - i)^2(z + i)^2},$$

to be about the point $z = i$, and the factor $(z - i)$
he denominator, it remains to expand the term
oint $z = i$. Accordingly, using a purely algebraic

approach, in order to introduce the expression $z - i$ into the term $1/(z+i)^2$, we rewrite it as

$$\frac{1}{[2i+(z-i)]^2} = \frac{-1}{4\left[1+\left(\dfrac{z-i}{2i}\right)\right]^2} = -\frac{1}{4}\left[1+\left(\frac{z-i}{2i}\right)\right]^{-2}.$$

The last term can be expanded by the binomial theorem provided $|(z-i)/2i| < 1$, corresponding to $|z-i| < 2$. After performing this expansion we find

$$\frac{1}{(z+i)^2} = -\frac{1}{4} - \frac{i}{4}(z-i) + \frac{3}{16}(z-i)^2 + \frac{i}{8}(z-i)^3 - \frac{5}{64}(z-i)^4 - \frac{3i}{64}(z-i)^5 + \cdots.$$

Using this result in the expression for $f(z)$ then gives the required Laurent series expansion about the point $z = i$,

$$f(z) = -\frac{1}{4(z-i)^2} - \frac{i}{4(z-i)} + \frac{3}{16} + \frac{i}{8}(z-i) - \frac{5}{64}(z-i)^2 - \frac{3i}{64}(z-i)^3 + \cdots,$$

for $0 < |z-i| < 2$. After consideration of the general term in the binomial expansion, this Laurent series can be written more concisely as

$$f(z) = -\frac{1}{4(z-i)^2} - \frac{i}{4(z-i)} + \sum_{n=0}^{\infty}\frac{i^n(n+3)(z-i)^n}{2^{n+4}} \quad \text{for } 0 < |z-i| < 2.$$

Of course, instead of using the binomial series to expand $(z+i)^{-2}$ about the point $z = i$ a Maclaurin series expansion of the function could have been used instead, although the resulting Laurent series expansion would have been the same.

(b) Finding a Laurent series expansion of a function in a neighborhood of the point at infinity involves finding the form of the expansion when $|z|$ is large. However, this is equivalent to setting $z = 1/\zeta$, finding the expansion when $|\zeta| << 1$, and then returning to the original variable by setting $\zeta = 1/z$. Making the substitution $z = 1/\zeta$ we have

$$f(1/\zeta) = \frac{\zeta^4}{(1+\zeta^2)^2} = \zeta^4(1+\zeta^2)^{-2}.$$

The factor $(1+\zeta^2)^{-2}$ can be expanded by the binomial theorem because $|\zeta| << 1$, so when this is done we find that

$$f(1/\zeta) = \zeta^4(1 - 2\zeta^2 + 3\zeta^4 - 4\zeta^6 + 5\zeta^8 - \cdots).$$

Setting $\zeta = 1/z$, we finally arrive at the Laurent series expansion in a neighborhood of the point at infinity

$$f(z) = \sum_{n=1}^{\infty} (-1)^{n+1} \frac{n}{z^{2n+2}} \quad \text{for } |z| > 1.$$

The domain of convergence follows from the fact that the singularities of $f(z)$ all lie on the unit circle, so $|z|$ must be large enough for it always to lie outside the unit circle. ◇

Example 3.5.7 Using an Elementary Trigonometric Identity and a Maclaurin Expansion when Finding Laurent Series Expansions

Find the Laurent series expansion of

$$f(z) = \sin\left(\frac{z}{1-z}\right)$$

(a) about the singularity at $z = 1$ and (b) in a neighborhood of the point at infinity.

SOLUTION

(a) $f(z)$ has its only singularity at $z = 1$, so a Laurent series expansion must be developed in terms of powers of $(z-1)$. Using partial fractions we have

$$\frac{z}{1-z} = -\left(\frac{1}{z-1} + 1\right)$$

The trigonometric identity $\sin(A+B) = \sin A \cos B + \cos A \sin B$, coupled with the previous result shows that

$$f(z) = -\sin\left[1 + \left(\frac{1}{z-1}\right)\right] = -\sin 1 \cos\left(\frac{1}{z-1}\right) - \cos 1 \sin\left(\frac{1}{z-1}\right).$$

Representing the trigonometric functions $\cos z$ and $\sin z$ by their Maclaurin series representations with z replaced by $1/(z-1)$ we arrive at the Laurent series

$$f(z) = -\sin 1\left(1 - \frac{1}{2!(z-1)^2} + \frac{1}{4!(z-1)^4} + \cdots\right)$$
$$- \cos 1\left(\frac{1}{(z-1)} - \frac{1}{3!(z-1)^3} + \frac{1}{5!(z-1)^5} + \cdots\right),$$

which can be written more concisely as

$$f(z) = -\sum_{n=0}^{\infty} \frac{\sin(1 + n\pi/2)}{n!(z-1)^n}.$$

Since $z = 1$ is the only singularity of $f(z)$, the domain of convergence of this Laurent series expansion is seen to be $0 < |z - 1| < \infty$.

(b) To find the Laurent series expansion in a neighborhood of the point at infinity $z = \infty$ we proceed as in Example 3.5.6 by making the substitution $z = 1/\zeta$. Setting $f(1/\zeta) = F(\zeta)$ gives

$$f(1/\zeta) = F(\zeta) = \sin\left(\frac{1}{\zeta - 1}\right).$$

This time there is no simple algebraic way of expanding $F(\zeta)$ in powers of ζ, but this does not matter because the required expansion is simply the Maclaurin series expansion of $\sin[1/(\zeta - 1)]$. Routine calculations show that

$$F(0) = -\sin 1, F'(0) = -\cos 1, F''(0) = \sin 1 - 2\cos 1, F'''(0) = 6\sin 1 - 5\cos 1,$$

so the Maclaurin series expansion of $F(\zeta)$ is

$$F(\zeta) = -\sin 1 - (\cos 1)\zeta + \frac{1}{2!}(\sin 1 - 2\cos 1)\zeta^2 + \frac{1}{3!}(6\sin 1 - 5\cos 1)\zeta^3 + \cdots.$$

Returning to the original variable z by means of the transformation $\zeta = 1/z$, the Laurent series expansion about $z = \infty$ is seen to be given by

$$f(z) = -\sin 1 - \frac{\cos 1}{z} + \frac{\sin 1 - 2\cos 1}{2!\, z^2} + \frac{6\sin 1 - 5\cos 1}{3!\, z^3} + \cdots.$$

Since by supposition $|z|$ is large, and $f(z)$ has a singularity at $z = 1$, it follows that the domain of convergence must be such that $|z| > 1$. ◇

We close this section by proving a set of inequalities satisfied by the coefficients of a Laurent series, and in the final example, show one of the many ways in which it may be used by proving a simple property of entire functions.

THEOREM 3.5.3 Cauchy Inequalities for the Coefficients of Laurent Series
Let $f(z)$ be analytic and possess the Laurent series expansion about the point z_0 of the form

$$f(z) = \sum_{n=-\infty}^{\infty} a_n (z - z_0)^n,$$

with the annular domain of convergence $R_0 < |z - z_0| < R_1$. Then the Laurent series coefficients satisfy the Cauchy inequalities

$$|a_n| \leqslant \frac{M_R}{R^n} \quad \text{for } n = 0, \pm 1, \pm 2, \ldots,$$

where M_R is the maximum value of $|f(z)|$ on the circle $|z - z_0| = R$, with R is such that $R_0 < R < R_1$.

PROOF

From Theorem 3.5.1 we have

$$a_n = \frac{1}{2\pi i} \int_\Gamma \frac{f(\zeta)}{(\zeta - z_0)^{n+1}} \, d\zeta,$$

where Γ may be taken to be the circle $|z - z_0| = R$.
Taking the modulus of this result gives

$$|a_n| \leqslant \frac{1}{2\pi} \int_\Gamma \frac{|f(\zeta)|}{|\zeta - z_0|^{n+1}} \, ds,$$

where $ds = R\,d\theta$ is the element of arc length around Γ. Thus $M_R = \max\{|f(z)|\}_{|z-z_0|=R}$, and using the fact that ζ lies on Γ, so that $|\zeta - z_0| = R$, it follows that

$$|a_n| \leqslant \frac{M_R}{2\pi R^{n+1}} \int_0^{2\pi} R\,d\theta = \frac{M_R}{R^n} \quad \text{for } n = 0, \pm 1, \pm 2, \ldots,$$

and the theorem is proved. ◆

Example 3.5.8 A Condition that Ensures an Entire Function Is a Polynomial

Let $f(z)$ be an entire function with the representation $f(z) = \Sigma_{n=0}^{\infty} a_n z^n$. Show that if for $|z| > R_0$ and a given nonnegative integer N

$$|f(z)| < M|z|^N \quad (M > 0 \text{ a constant}),$$

that $f(z)$ is a polynomial, and its degree cannot exceed N.

SOLUTION

Identify the circle Γ in Theorem 3.5.3 with $|z| = R$, where $R > R_0$. Then, applying Theorem 3.5.3 to $f(z)$ gives

$$|a_n| \leqslant \frac{M}{R^n} R^N = MR^{N-n} \quad \text{for } n = 1, 2, \ldots .$$

Thus $|a_n| \leqslant M/R^{N-n}$ for $n = 0, 1, \ldots, N$, and $|a_N| \leqslant M/R^{n-N}$ if $n > N$. As R may be made arbitrarily large, the second inequality shows that $|a_n| = 0$ for $n > N$, and hence $a_n = 0$ for $n > N$. Thus $f(z)$ is a polynomial, and its degree cannot exceed N. ◇

Exercises 3.5

1. Find the Laurent series expansion of

$$f(z) = \frac{2z - 5}{(z-2)(z-3)}$$

(a) in the disk $0 \leqslant |z| < 2$, (b) in the annulus $2 < |z| < 3$, (c) in the domain $|z| < 3$.

2. Find the Laurent series expansion of

$$f(z) = \frac{1}{z(2z-1)}$$

about the point $z = 2$ and determine its domain of convergence.

3. Find two possible Laurent series expansion of

$$f(z) = \frac{2z}{z^2 - 1}$$

about the point $z = i$.

4. Find the Laurent series expansion of

$$f(z) = \frac{1}{(z-2)^3}$$

(a) about the point $z = 0$, and (b) near the point at infinity by setting $z = 1/\zeta$ and considering the situation when $|\zeta|$ is small.

5. Find the Laurent series expansion of

$$f(z) = \frac{z^2 - 2z + 5}{(z-2)(z^2+1)}$$

in powers of z in the annulus $1 < |z| < 2$.

6. Find the Laurent series expansion of

$$f(z) = \frac{\sin z \cos 3z}{z^4}$$

in powers of z.

7. Find the Laurent series expansion of

$$f(z) = \frac{\sinh z \cosh 5z}{z^3}$$

in powers of z.

8. Find the Laurent series expansion of

$$f(z) = \sin\left(\frac{2z}{2z+1}\right)$$

(a) about the point $z = -\frac{1}{2}$ and (b) in a neighborhood of the point at infinity.

9. Find the Laurent series expansion of

$$f(z) = z^3 e^{1/z}$$

(a) near the point $z = 0$, and (b) near the point at infinity by setting $z = 1/\zeta$ and considering the situation when $|\zeta|$ is small.

10. Find the Laurent series expansion of

$$f(z) = (3 - z)\sin\left(\frac{z}{1-z}\right)$$

about the point $z = 1$.

11. Find the Laurent series expansion of

$$f(z) = \mathrm{Ln}\left(\frac{z^2}{1-4z^2}\right)$$

in powers of z for $|z| > \frac{1}{2}$.

12. Find the Laurent series expansion of

$$f(z) = \mathrm{Ln}\left(\frac{z-a}{z-b}\right)$$

in powers of z.

3.6 Classification of Singularities and Zeros

So far, a point z_0 has been called a *singularity* of a complex function $f(z)$ if it is not differentiable at z_0. The time has now come for singularities to be classified more precisely and, in particular, to pay special attention to what are called *isolated singularities*.

A point z_0 is called a *singularity*, or a *singular point*, of the complex function $f(z)$ if the function is not analytic at z_0, but is such that every neighborhood of z_0 contains at least one point where $f(z)$ is analytic. This definition is rather general because it makes no distinction between functions whose singularities occur at isolated (separated) points, which may be finite or infinite in number, and functions whose singularities occur at every point of the continuum of points forming a simple arc.

Typical examples of functions with isolated singularities are $f(z) = 1/(1 - z^2)$, with only two distinct singularities at $z = \pm 1$, and $g(z) = 1/\sin z$, with an infinite number of isolated singularities at $z = n\pi$ with $n = 0, \pm 1, \pm 2, \ldots$.

An example of a function with singularities at every point on a simple arc is the logarithmic function Ln z, for which every point on the negative real axis up to and including the origin is a singularity.

To distinguish between these two different situations we first give a formal definition of an isolated singularity. The function $f(z)$ is said to have an *isolated singularity* at z_0 if it is analytic in the punctured disk $0 < |z - z_0| < R$ for some $R > 0$, but not differentiable at z_0. Thus the singularities of $f(0) = 1/(1 - z^2)$ and $g(z) = 1/\sin z$ are isolated, but those of Ln z on the negative real axis are not isolated. Points that are *not* singularities of an analytic function $f(z)$ are called *regular points*.

This brings us to Laurent series expansions of functions because an isolated singularity of a function $f(z)$ at z_0 is classified by reference to its Laurent series expansion about z_0. Three different types of singularities arise and these are described in what follows.

3.6.1 Classification of Singularities

Let $f(z)$ have an isolated singularity at z_0, where it has the Laurent series expansion

$$f(z) = \sum_{n=-\infty}^{\infty} a_n (z - z_0)^n, \tag{3.41}$$

which converges in a punctured disk $0 < |z - z_0| < R$ for some $R > 0$. Then the singularity at z_0 is classified as being one of the following three types:

1. An isolated singularity of $f(z)$ at z_0 is called a *removable singularity* if the principal part of the Laurent series expansion about z_0 is identically zero. Thus, at a removable singularity z_0 of the function $f(z)$, the coefficients $a_n = 0$ for $n = -1, -2, \ldots$, causing the Laurent series expansion of $f(z)$ to become

$$f(z) = a_0 + a_1(z - z_0) + a_2(z - z_0)^2 + \cdots = \sum_{n=0}^{\infty} a_n (z - z_0)^n,$$

which is convergent in the punctured disk $0 < |z - z_0| < R$.

2. An isolated singularity of $f(z)$ is called a *pole* of *order* k if the principal part of the Laurent series expansion of $f(z)$ about z_0 contains only a finite number of terms forming a polynomial of degree k in $(z - z_0)^{-1}$. Thus, if $f(z)$ has a pole of order k at z_0, the Laurent series expansion takes the form

$$f(z) = \frac{a_{-k}}{(z - z_0)^k} + \frac{a_{-(k-1)}}{(z - z_0)^{k-1}} + \cdots + \frac{a_{-1}}{(z - z_0)} + \sum_{n=0}^{\infty} c_n (z - z_0)^n,$$

where $a_{-k} \neq 0$, though some or all of $a_{-(k-1)}, a_{-(k-2)}, \ldots, a_{-1}$ may be zero, and $f(z)$ is convergent in a punctured disk $0 < |z - z_0| < R$, for some $R > 0$. An important frequently occurring case arises if $f(z)$ only has a pole of order 1 at z_0, when the Laurent series expansion of $f(z)$ becomes

$$f(z) = \frac{a_{-1}}{(z - z_0)} + \sum_{n=0}^{\infty} a_n (z - z_0)^n.$$

A pole located at z_0, where the only singularity is of the form $a_{-1}/(z - z_0)$, is called a *simple pole*.

3. An isolated singularity of $f(z)$ at z_0 is called an *essential singularity* if the principal part of the Laurent series expansion of $f(z)$ about z_0 is an infinite series in powers of $(z - z_0)^{-1}$, so that then

$$f(z) = \sum_{n=1}^{\infty} \frac{a_{-n}}{(z - z_0)^n} + \sum_{n=0}^{\infty} a_n (z - z_0)^n,$$

where an infinite number of the coefficients $a_{-1}, a_{-2}, a_{-3}, \ldots,$ are nonzero. This expansion is convergent in a punctured disk $0 < |z - z_0| < R$, for some $R > 0$.

Removable Singularities

Usually when z_0 is a removable singularity of $f(z)$, the function is not defined at z_0 and the annular domain of convergence of $f(z)$ is a punctured disk $0 < |z - z_0| < R$ for some $R > 0$. However, it follows from the definition of a removable singularity that the limit of $f(z)$ as $z \to z_0$ exists and is finite, so by defining the value of the function $f(z)$ at z_0 by

$$f(z_0) = \lim_{z \to z_0} f(z),$$

the function $f(z)$ becomes continuous and analytic at z_0, and the singularity has been "removed." Examples of this, the weakest of singularities, are provided by

$$f(z) = \frac{\sin z}{z} \quad \text{and} \quad g(z) = \frac{1-\cos(z-1)}{(z-1)^2}.$$

The Laurent series expansions of these functions about the origin and about $z = 1$, respectively, where each has an isolated singularity, are

$$f(z) = \frac{\left(z - \frac{1}{3!}z^3 + \frac{1}{5!}z^5 - \cdots\right)}{z} = 1 - \frac{z^2}{3!} + \frac{z^4}{5!} - \cdots, \quad \text{for } |z| > 0,$$

and

$$\begin{aligned} g(z) &= \frac{1 - \left(1 - \frac{1}{2!}(z-1)^2 + \frac{1}{4!}(z-1)^4 - \frac{1}{6!}(z-1)^6 + \cdots\right)}{(z-1)^2} \\ &= \frac{1}{2} - \frac{(z-1)^2}{4!} + \frac{(z-1)^4}{6!} - \cdots, \quad \text{for } |z-1| > 0. \end{aligned}$$

Thus

$$\lim_{z\to 0} f(z) = \lim_{z\to 0} \frac{\sin z}{z} = 1 \quad \text{and} \quad \lim_{z\to 1} g(z) = \lim_{z\to 1} \frac{1-\cos(z-1)}{(z-1)^2} = \frac{1}{2}.$$

Hence, by defining the functions $f(z)$ and $g(z)$ as

$$f(z) = \begin{cases} \dfrac{\sin z}{z} & \text{for } |z| > 0 \\ 1 & \text{for } z = 0, \end{cases} \quad \text{and} \quad g(z) = \begin{cases} \dfrac{1-\cos(z-1)}{(z-1)^2} & \text{for } |z-1| > 0 \\ \dfrac{1}{2} & \text{for } z = 1, \end{cases}$$

the singularities have been removed, and they become analytic (entire) functions.

Poles

If $f(z)$ has a pole of order k at z_0, the definition of a pole implies that the dominant singularity of $f(z)$ at z_0 is proportional to $(z - z_0)^{-k}$, though weaker singularities may also be present if not all of the coefficients $a_{-(k-1)}, a_{-(k-2)}, \ldots,$ a_{-1} vanish.

Consider the functions $f(z) = (\cos z)/z$ and $g(z) = e^z/z$. When $\cos z$ and e^z have been replaced by their Maclaurin series expansions for $|z| > 0, f(z)$ and $g(z)$ become

$$f(z) = \frac{\cos z}{z} = \frac{1}{z} - \frac{z}{2!} + \frac{z^3}{4!} - \frac{z^5}{6!} + \cdots,$$

and

$$g(z) = \frac{e^z}{z^3} = \frac{1}{z^3} + \frac{1}{z^2} + \frac{1}{2z} + \frac{1}{3!} + \frac{z}{4!} + \frac{z^2}{5!} + \cdots$$

This shows that $f(z)$ has a simple pole at the origin, while $g(z)$ has a pole of order 3 at the origin.

The following simple test is an immediate consequence of the definition of a pole of order k.

Test for a Pole of Order k

An isolated singularity of $f(z)$ at $z = z_0$ is a pole of order k if

$$\lim_{z\to z_0} [(z - z_0)^k f(z)] = L, \quad \text{where } L \neq 0. \tag{3.42}$$

Applying this test to the functions $f(z) = (\cos z)/z$ and $g(z) = e^z/z$ just considered confirms that $f(z)$ has a simple pole at the origin, while $g(z)$ has a pole of order 3 at the origin because

$$\lim_{z\to 0}\left[z\left(\frac{\cos z}{z}\right)\right] = \lim_{z\to 0} \cos z = 1\ (\neq 0) \quad \text{and} \quad \lim_{z\to 0}\left[z^3\frac{e^z}{z^3}\right] = \lim_{z\to 0} e^z = 1\ (\neq 0).$$

A different and less trivial illustration of the use of test Equation (3.42) is provided by the function

$$f(z) = \frac{1}{(z+1)^2 \sin(z+1)},$$

which has an isolated singularity at $z = -1$ due to the vanishing of the factor $(z + 1)^2$ and the zero of $\sin(z + 1)$ in the denominator, and isolated simple poles at $z = n\pi - 1$ for $n = \pm 1, \pm 2, \ldots,$ because of the zeros of the function $\sin(z + 1)$ in the denominator. The isolated singularity at $z = -1$ is a pole of order 3, because

$$\lim_{z\to -1}\left[(z+1)^k\left(\frac{1}{(z+1)^2\sin(z+1)}\right)\right] = \lim_{z\to -1}\left[\frac{(z+1)^{k-2}}{\sin(z+1)}\right] = 1, \text{ if } k = 3, \text{ and } 0 \text{ if } k > 3.$$

The singularities at $z = n\pi - 1$ for $n = \pm 1, \pm 2, \ldots$, are all simple poles because

$$\lim_{z \to n\pi - 1}\left[(z - n\pi + 1)\left(\frac{1}{(z+1)^2 \sin(z+1)}\right)\right] = \frac{1}{n^2\pi^2} \lim_{z \to n\pi - 1}\left(\frac{z - n\pi + 1}{\sin(z+1)}\right)$$
$$= \frac{(-1)^n}{n^2\pi^2} \neq 0.$$

3.6.2 Essential Singularities

The definition of an essential singularity shows it to be stronger than that due to any pole, whatever its order. The following are typical examples of functions with essential singularities at the origin

$$f(z) = \sin(1/z) = \frac{1}{z} - \frac{1}{3!z^3} + \frac{1}{5!z^5} - \cdots,$$

and

$$g(z) = \exp(1/z) = 1 + \frac{1}{z} + \frac{1}{2!z^2} + \frac{1}{3!z^3} + \cdots,$$

each of which converges for $|z| > 0$ because each has infinitely many terms involving powers of $1/z$.

The behavior of a function close to an essential singularity is extremely irregular and a theorem due to French mathematician Charles Emile Picard (1856–1941) shows that in a neighborhood of an essential singularity a function attains any given value infinitely many times, with one possible exception. An example illustrating this erratic behavior is provided by

$$f(z) = \exp(1/z)$$

which, as we have already seen, has an essential singularity at the origin.

For any given complex number $K \neq 0$, if $\exp(1/z) = K$, then the points z_n where this result holds true are such that

$$\frac{1}{z_n} = \ln|K| + i(\operatorname{Arg} K + 2n\pi), \quad \text{for } n = 0, \pm 1, \pm 2, \ldots,$$

and so

$$z_n = \frac{1}{\ln|K| + i(\operatorname{Arg} K + 2n\pi)}, \quad \text{for } n = 0, \pm 1, \pm 2, \ldots.$$

Thus at infinitely many points z_n it follows that $\exp(1/z_n) = K$, for all K with the exception of $K = 0$, which is then called the *Picard exceptional value*.

To complete our examination of the classification of singularities, we consider the behavior of $f(z)$ at the point at infinity ($z = \infty$) in the extended complex plane. The *point at infinity* is said to be an *isolated singular point* of $f(z)$ if, after setting $z = 1/\zeta$ and writing $F(\zeta) = f(1/\zeta)$, the function $F(\zeta)$ has an isolated singular point at $\zeta = 0$.

To illustrate matters, we consider the behavior of

$$f(z) = \frac{3z^3 + 2z^2 + 1}{z + 3}$$

at the point at infinity. Setting $z = 1/\zeta$ we have

$$f(1/\zeta) = F(\zeta) = \frac{1}{\zeta^2}\left(\frac{3 + 2\zeta + \zeta^3}{1 + 3\zeta}\right),$$

showing that $\zeta = 0$ is an isolated singularity of $F(\zeta)$. An application of test Equation (3.42) gives

$$\lim_{\zeta \to 0}\left[\zeta^2\left(\frac{3 + 2\zeta + \zeta^3}{\zeta^2(1 + 3\zeta)}\right)\right] = 3 \neq 0,$$

showing that $F(\zeta)$ has a pole of order 2 at the origin, and hence that $f(z)$ has a pole of order 2 at infinity.

If the bracketed term in $F(\zeta)$ is replaced by the first few terms of its Maclaurin series expansion, $F(\zeta)$ becomes

$$F(\zeta) = \frac{3}{\zeta^2} - \frac{7}{\zeta} + 21 - 62\zeta + 186\zeta^2 + \cdots,$$

so in terms of z this becomes

$$f(z) = 3z^2 - 7z + 21 - \frac{62}{z} + \frac{186}{z^2} - \cdots.$$

This shows that in a neighborhood of the point at infinity, the behavior of $f(z)$ is dominated by the polynomial $3z^2 - 7z + 21$, which is the *principal part* of the Laurent series expansion of $f(z)$ in a neighborhood of the point at infinity.

As a final example we consider the behavior of the function $f(z) = 1/[z(1 + z)]$ in a neighborhood of the point at infinity. Setting $z = 1/\zeta$ and $f(1/\zeta) = F(\zeta)$, after expanding the result for small $|\zeta|$ we find that

$$F(\zeta) = \zeta^2 - \zeta^3 + \zeta^4 - \cdots.$$

Thus this function has no singularity at the point of infinity, which is thus a regular point of the function.

We mention in passing that the functions e^z, $\sin z$ and $\cos z$ all have essential singularities at $z = \infty$, because replacing z by $1/\zeta$ each of these functions is seen to have an essential singularity at the origin.

3.6.3 Zeros of Functions

The *zeros* of an analytic function $f(z)$ play an important role throughout complex analysis, so it is necessary to classify them.

An analytic function $f(z)$ is said to have a *zero* of *order k* at z_0, or a *zero of multiplicity* **k**, if

$$f(z) = (z - z_0)^k g(z), \tag{3.43}$$

and $g(z_0) \neq 0$. When $k = 1$ the zero at z_0 is said to be a *simple zero*.

Written out, this definition says that z_0 is a zero of order k, or of multiplicity k (repeated k times), if when k factors $(z - z_0)$ have been removed from $f(z)$ the remaining function $g(z)$ has no zero at z_0.

Thus the function

$$f(z) = (z - 1)^3 \sin(z - 1)$$

has a zero of order 4 (multiplicity 4) at $z = 1$, because when $\sin(z - 1)$ is expanded about the point $z = 1$ we find that

$$\begin{aligned} f(z) &= (z-1)^3 \left((z-1) - \frac{(z-1)^3}{3!} + \frac{(z-1)^5}{5!} - \cdots \right) \\ &= (z-1)^4 \underbrace{\left(1 - \frac{(z-1)^2}{3!} + \frac{(z-1)^4}{5!} - \cdots \right)}_{g(z)} \end{aligned}$$

where the factor $(z - 1)$ is repeated 4 times, and $g(1) \neq 0$.

This is an appropriate point at which to introduce two important classes of complex functions. The first of these occurs when $g(z)$ and $h(z)$ are two analytic functions defined in some domain D in which $h(z) \neq 0$, in which case the class of the functions $f(z)$ defined as the quotient

$$f(z) = \frac{g(z)}{h(z)}$$

is said to be *holomorphic* in D. Thus an entire function is holomorphic throughout the extended complex plane. The second important class of functions occurs when $f(z)$ is holomorphic in D except for poles, in which case the class of functions with this property is said to be *meromorphic* in D. So a function $f(z)$ that is *meromorphic* in a domain D is a function that is analytic in D apart from points where $f(z)$ has poles. An important special class of meromorphic functions that appear in many applications arises when $g(z)$ and $h(z)$ are both polynomials with no common factors. Then the *rational function*

$$f(z) = \frac{g(z)}{h(z)}$$

is meromorphic in the entire complex plane. It is clear from this that the only singularities of a rational function $f(z)$ are poles of different orders that occur at the zeros of $h(z)$, where the order of a pole at a zero z_0 of $f(z)$ is determined by the multiplicity of the zero at z_0.

Exercises 3.6

Locate and identify the nature of the zeros of the following functions.

1. $z^3(\cosh z - 1)$
2. $z^2 \operatorname{Ln}\left(\dfrac{1-z}{1+z}\right)$
3. $z \sin 3z$
4. $\cos(2z - 1)$
5. $\sinh z \sin z$

Locate and identify the singularities of the following functions.

6. $1/(4z - z^3)$
7. $(z^4 + 3)/(1 + z^4)$
8. $z \exp(-z^2)$
9. $\exp z/(3 - z)$
10. $\sin[1/(3 - z)]$
11. $3/(e^z - 1) - (1/z)$
12. $(\cos 3z)/z^2$
13. $(\cot z)/z^2$
14. $\cot z - 1/z$

3.7 Residues and the Residue Theorem

Chapter 2 showed how the Cauchy theorems may be used to integrate analytic functions around a simple closed-contour C containing isolated singularities, at each of which $f(z)$ has only a pole. We now introduce a new and more powerful method for the evaluation of such integrals that both simplifies

PROOF

From Lemma 3.7.2 we have

$$\begin{aligned}\text{Res}[f(z), z_0] &= \lim_{z\to z_0}\left((z - z_0)\frac{g(z)}{h(z)}\right)\\ &= g(z_0)\lim_{z\to z_0}\left[\frac{(z - z_0)}{h(z)}\right] = \frac{g(z_0)}{\lim\limits_{z\to z_0} h(z)/(z - z_0)},\end{aligned}$$

so proceeding to the limit we have

$$\text{Res}[f(z), z_0] = \frac{g(z_0)}{h'(z_0)}.$$

◆

Because it is possible for $z = \infty$ to be a regular point or an isolated singularity of $f(z)$, it is necessary to define a residue at the point at infinity. This is accomplished by adapting definition in Equation (3.44). When making this adaptation, it is important to recognize that in the definition in Equation (3.44) z_0 is the *only* singularity of $f(z)$ *inside* the simple closed-curve Γ. This implies that when the point at infinity is considered, the domain to which a neighborhood of $z = \infty$ belongs may always be considered to be *outside* a sufficiently large circle Γ defined by $|z| = R$. So to move around Γ in such a way that $z = \infty$ lies *inside* and to the left it is necessary to move around Γ in the *clockwise* sense. Consequently, the residue of $f(z)$ at infinity is defined as

$$\text{Res}[f(z), \infty] = \frac{-1}{2\pi i}\int_{\Gamma_+} f(\zeta)d\zeta, \tag{3.46}$$

where Γ_+ indicates that the positive direction around Γ is counterclockwise, while the negative sign has been introduced because when the point at infinity is considered the sense around Γ is taken to be *clockwise*. Thus the residue at infinity is defined as

$$\text{Res}[f(z), \infty] = -a_{-1}. \tag{3.47}$$

The simplest example with which to illustrate the residue at the point at infinity involves considering the function $f(z) = 1/z$, which is its own Laurent series expansion for $|z| > 0$. We see from this that $a_{-1} = 1$, so from Equation (3.47)

$$\text{Res}[1/z, \infty] = -1,$$

eventhough $f(z) = 1/z$ has no singularity at the point at infinity.

A less trivial example is provided by considering the function $f(z) = \exp(3/z)$. This has the Laurent series expansion about the origin

$$f(z) = 1 + \frac{3}{z} + \frac{1}{2!}\left(\frac{3}{z}\right)^2 + \frac{1}{3!}\left(\frac{3}{z}\right)^3 + \cdots \qquad \text{for } |z| > 0.$$

So as $a_{-1} = 3$, it follows from Equation (3.47) that

$$\text{Res}\,[\exp(3/z), \infty] = -3.$$

Thus, although $f(z) = \exp(3/z)$ has no singularity at the point at infinity, its residue there is non-zero, although the function has an essential singularity at the origin.

Example 3.7.1 Residues at Simple Poles
Given

$$f(z) = \frac{z^2 - 5z + 7}{z^2 + 3z + 2},$$

find the residue at each of its singularities.

SOLUTION
The zeros of the denominator are $z = -1$ and $z = -2$. So from Lemma 3.7.2,

$$\text{Res}[f(z), -1] = \lim_{z \to -1} [(z+1)f(z)] = \lim_{z \to -1} \left(\frac{z^2 - 5z + 7}{z + 2} \right) = 13,$$

and

$$\text{Res}[f(z), -2] = \lim_{z \to -2} [(z+2)f(z)] = \lim_{z \to -2} \left(\frac{z^2 - 5z + 7}{z + 1} \right) = -21. \qquad \diamond$$

Example 3.7.2 Residues at Higher Order Poles
Given

$$f(z) = \frac{1}{(z+1)^2(z-3)^3},$$

find the residue at each of its singularities.

SOLUTION
The zeros of the denominator are $z = -1$ with multiplicity 2 and $z = 3$ with multiplicity 3. From Lemma 3.7.3,

$$\text{Res}[f(z), -1] = \frac{1}{(2-1)!} \lim_{z\to -1} \left\{ \frac{d}{dz}[(z+1)^2 f(z)] \right\} = \lim_{z\to -1} \left\{ \frac{d}{dz}\left(\frac{1}{(z-3)^3} \right) \right\} = -\frac{3}{256},$$

and

$$\text{Res}[f(z), 3] = \frac{1}{(3-1)!} \lim_{z\to 3} \left\{ \frac{d^2}{dz^2}[(z-3)^3 f(z)] \right\} = \frac{1}{2!} \lim_{z\to -3} \left\{ \frac{d^2}{dz^2}\left(\frac{1}{(z+1)^2} \right) \right\} = \frac{3}{256}.$$

◇

Example 3.7.3 Residues of a Meromorphic Function
Given

$$f(z) = \frac{e^{imz}}{z^2 + a^2}, \qquad (m, a \text{ real}),$$

find the residues of $f(z)$ at its singularities.

SOLUTION
The zeros of the denominator are $z = \pm ia$. From Lemma 3.7.4 with $g(z) = e^{ima}$ and $h(z) = z^2 + a^2$ we have

$$\text{Res}\,[f(z), ia] = \frac{g(ia)}{f'(ia)} = \left(\frac{e^{imz}}{2z} \right)_{z=ia} = -\frac{ie^{-ma}}{2a},$$

and

$$\text{Res}\,[f(z), -ia] = \frac{g(-ia)}{f'(-ia)} = \left(\frac{e^{imz}}{2z} \right)_{z=-ia} = \frac{ie^{-ma}}{2a}.$$

◇

Example 3.7.4 Residues of $\cot \alpha z$
Given

$$f(z) = \cot \alpha z, \qquad (\alpha \text{ real}),$$

find the residues at its singularities.

SOLUTION
In terms of sine and cosine functions

$$f(z) = \frac{\cos \alpha z}{\sin \alpha z},$$

so it is a meromorphic function for which the zeros of the denominator occur at $z_n = n/\alpha$, with $n = 0, \pm 1, \pm 2, \ldots$. Then from Lemma 3.7.4 with $g(z) = \cos \alpha z$ and $h(z) = \sin \alpha z$ we have

$$\operatorname{Res}[f(z), z_n] = \frac{g(z_n)}{h'(z_n)} = \left[\frac{\cos \alpha z}{(\sin \alpha z)'}\right]_{z=z_n} = \frac{1}{\alpha}, \quad \text{for } n = 0, \pm 1, \pm 2, \ldots. \quad \diamond$$

Example 3.7.5 Deducing a Residue from a Laurent Series Expansion
Given

$$f(z) = \frac{\cosh 2z - 1}{z^7},$$

find the residue at its singularity.

SOLUTION
Although this is a meromorphic function, Lemma 3.7.4 does not apply because when $z = 0$ both the numerator and the derivative of the denominator vanish, contradicting the requirements of the lemma that the numerator should not equal zero. To find the Laurent series expansion about the origin we write out the Maclaurin series expansion of $\cosh 2z - 1$ and divide the result by z^7 to obtain

$$f(z) = \frac{\cosh 2z - 1}{z^7} = \frac{2}{z^5} + \frac{2}{3}\frac{1}{z^3} + \frac{4}{45}\frac{1}{z} + \frac{2}{315}z + \cdots, \quad \text{for } |z| > 0,$$

from which the coefficient of the term in $1/z$ is seen to be $4/45$, so that

$$\operatorname{Res}[f(z), 0] = \frac{4}{45}.$$

Because $z = 0$ is the only singularity of $f(z)$, this is the required residue. $\diamond$

Example 3.7.6 A Residue at Infinity
Given

$$f(z) = \left(\frac{z^4}{2z^2 - 1}\right)\sin(1/z),$$

t the point at infinity.

ction for large z, we first rewrite it as

$$f(z) = \tfrac{1}{2}z^2\left(1 - \frac{1}{2z^2}\right)^{-1}\sin(1/z).$$

Expanding $(1 - 1/2z^2)^{-1}$ by the binomial theorem and replacing z by $1/z$ in the Maclaurin series expansion of $\sin z$, this becomes

$$f(z) = \tfrac{1}{2}z^2\left(1 + \frac{1}{2z^2} + \frac{1}{4z^4} + \cdots\right)\left(\frac{1}{z} - \frac{1}{3!\,z^3} + \frac{1}{5!\,z^5} - \cdots\right), \quad \text{for } |z| > 1/\sqrt{2}.$$

Inspection of this result shows that when expanded, the coefficient a_{-1} of $1/z$ is

$$a_{-1} = -\frac{1}{2\cdot 3!} + \frac{1}{4} = \frac{1}{6},$$

so from Equation (3.47) we see that the residue at infinity is

$$\mathrm{Res}\,[f(z), \infty] = -a_{-1} = -\tfrac{1}{6}. \qquad \diamond$$

Example 3.7.7 Residue of a Function with Branches
Given

$$f(z) = \frac{z^{\alpha-1}}{z+3}, \quad (\alpha \text{ real}),$$

and working with the principal branch of the function $z^{\alpha-1}$, find the residue at the singularity $z = -3$.

SOLUTION
The function $z^{\alpha-1}$ has branches (it is many-valued), and to make the function $f(z)$ single-valued, it is necessary to cut the z-plane along the negative real axis up to, and including, the origin. When this is done, in terms of the principal branch, the singularity at $z = -3$ becomes the point at the *top* of the cut at a radial distance 3 to the left of the origin, so in polar form it is necessary to set $z = 3e^{i\pi}$. The function $f(z)$ is a meromorphic function, its denominator only has a simple zero at $z = -3$, and it satisfies the conditions of Lemma 3.7.4, so setting $g(z) = z^{\alpha-1}$ and $h(z) = 3 + z$, it follows directly from this lemma that

$$\mathrm{Res}[f(z), 3e^{i\pi}] = \left(\frac{g(z)}{h'(z)}\right)_{z=3e^{i\pi}} = (z^{\alpha-1})_{z=3e^{i\pi}} = -3^{\alpha-1}e^{\alpha\pi i},$$

and so

$$\mathrm{Res}[f(z), -3] = -3^{\alpha-1}e^{\alpha\pi i}. \qquad \diamond$$

The connection between residues and complex integration is established in the fundamental theorem that follows. This theorem enables all of the complex integrals evaluated in Chapter 2 by means of residues, and it extends the evaluation of complex integrals to functions that are not amenable to the direct approach used in Chapter 2.

THEOREM 3.7.1 The Residue Theorem
Let f(z) be analytic inside and on a simple, closed piecewise smooth curve C, with the exception of a finite number of isolated singularities $z_1, z_2, \ldots, z_N$ inside C. Then

$$\int_C f(z)dz = 2\pi i \sum_{r=1}^{N} \text{Res}[f(z), z_r].$$

PROOF

Because the singularities of $f(z)$ inside C are isolated, it follows that each singularity may be enclosed within a circle C_r, centered on z_r in such a manner that none of the circles have points in common, while each lies entirely within C, as shown in Figure 3.13.

Integrating around C and using the generalized Cauchy–Goursat theorem (Theorem 2.2.3) gives

$$\int_C f(z)dz = \sum_{r=1}^{N} f(z)dz. \tag{3.48}$$

Expanding $f(z)$ as a Laurent series about each singular point z_r leads to expansions of the form

$$f(z) = \sum_{n=-\infty}^{\infty} a_n^{(r)}(z - z_r)^n, \quad \text{for } r = 1, 2, \ldots, N,$$

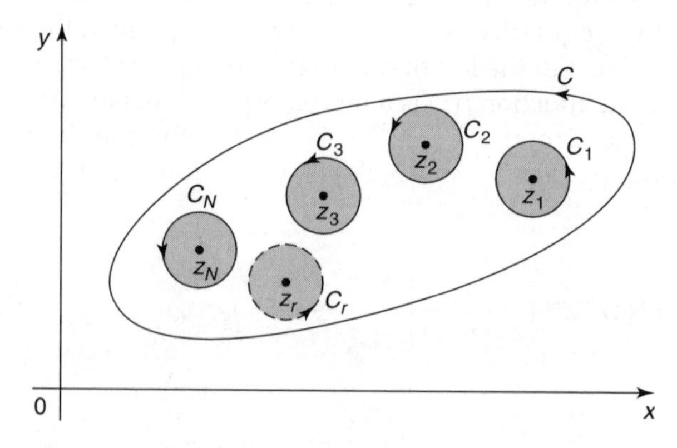

FIGURE 3.13
Singularities of $f(z)$ at $z_1, z_2, \ldots, z_N$ enclosed in circles $C_1, C_2, \ldots, C_N$.

where $a_n^{(r)}$ is the nth Laurent series coefficient of the expansion of $f(z)$ around z_r. From Equations (3.44) and (3.45) we have

$$\text{Res}[f(z), z_r] = a_{-1}^{(r)} = \frac{1}{2\pi i}\int_{C_r} f(z)dz, \tag{3.49}$$

and the statement of the theorem follows by substituting Equation (3.49) into Equation (3.48). ◆

Example 3.7.8 A Simple Application of the Residue Theorem
Given

$$f(z) = \frac{1}{z^3(z-2)},$$

find the residues at its poles and hence determine the value of

$$\int_{C_r} f(z)dz, \quad \text{for } r = 1, 2, 3, \dots,$$

when

(i) C_1 is the circle $|z - \frac{1}{2}| = 1$
(ii) C_2 is the rectangle ABCD with corners at A ($z = 3 + i$), B ($z = 1 + i$), C ($z = 1 - 2i$), and D ($z = 3 - 2i$)
(iii) C_3 is the circle $|z| = 3$.

SOLUTION
Inspection shows that $f(z)$ has a pole of order 3 at $z = 0$ and a simple pole at $z = 2$. Thus

$$\text{Res}[f(z), 0] = \frac{1}{2!}\left\{\frac{d^2}{dz^2}[z^3 f(z)]\right\} = \frac{1}{2}\left[\frac{d^2}{dz^2}\left(\frac{1}{z-2}\right)\right]_{z=0} = -\frac{1}{8},$$

and

$$\text{Res}[f(z), 2] = \lim_{z\to 2}[(z-2)f(z)] = \lim_{z\to 2}(1/z^3) = \frac{1}{8}.$$

As shown in Figure 3.14, consideration of the contours C_r shows that only the pole at the origin lies in C_1, only the pole at $z = 2$ lies in C_2, while both the

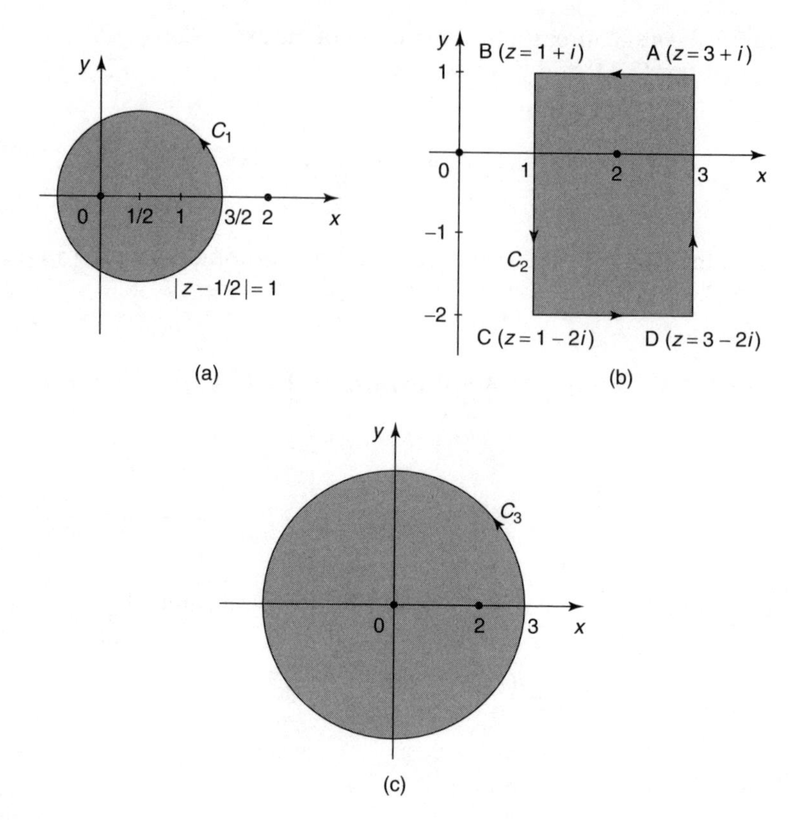

FIGURE 3.14
The contours C_1, C_2, and C_3.

pole at the origin and the pole at $z = 2$ lie inside C_3. Thus from the residue theorem we have

$$\int_{C_1} f(z)dz = 2\pi i\,\text{Res}[f(z), 0] = 2\pi i\left(-\tfrac{1}{8}\right) = -\tfrac{1}{4}\pi i,$$

$$\int_{C_2} f(z)dz = 2\pi i\,\text{Res}[f(z), 2] = 2\pi i\left(\tfrac{1}{8}\right) = \tfrac{1}{4}\pi i,$$

$$\int_{C_3} f(z)dz = 2\pi i\,\text{Res}[f(z), 0] + 2\pi i\,\text{Res}[f(z), 2] = 2\pi i\left(-\tfrac{1}{8} + \tfrac{1}{8}\right) = 0$$

◇

THEOREM 3.7.2 Contribution of a Residue at an Indented Simple Pole
Let an analytic function f(z) have a simple pole at z_0, and let C_ε be the arc AB of the circle $|z - z_0| = \varepsilon$ centered on z_0 with radius $\varepsilon > 0$ such that $\alpha \leqslant \text{Arg}[z - z_0] \leqslant \beta$, as shown in Figure 3.15. Then the limit of the integral along C_ε in the positive sense from A to B as $\varepsilon \to 0$ is given by

$$\lim_{\varepsilon \to 0} \int_{C_\varepsilon} f(z)dz = i(\beta - \alpha)\text{Res}[f(z), z_0].$$

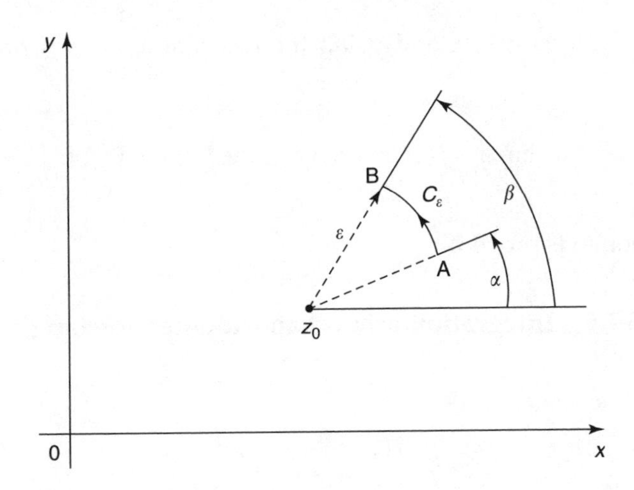

FIGURE 3.15
The arc C_ε centered on z_0.

PROOF

If $f(z)$ has a simple pole at z_0, it must have a Laurent series expansion of the form

$$f(z) = \frac{a_{-1}}{z - z_0} + \sum_{n=0}^{\infty} a_n (z - z_0)^n,$$

in a neighborhood of z_0.

As the Laurent series is uniformly convergent in its annulus of convergence, it follows that for a suitably small $\varepsilon > 0$

$$\int_{C_\varepsilon} f(z)dz = a_{-1}\int_{C_\varepsilon} \frac{dz}{z - z_0} + \sum_{n=0}^{\infty}\left[a_n \int_{C_\varepsilon} (z - z_0)^n dz\right].$$

Setting $z - z_0 = \varepsilon e^{i\theta}$, so that $dz = \varepsilon i e^{i\theta}\, d\theta$, this becomes

$$\int_{C_\varepsilon} f(z)dz = ia_{-1}\int_{\alpha}^{\beta} d\theta + \sum_{n=0}^{\infty}\left(ia_n\varepsilon^{n+1}\int_{\alpha}^{\beta} \exp[i(n+1)\theta]d\theta\right),$$

and so

$$\int_{C_\varepsilon} f(z)dz = ia_{-1}(\beta - \alpha) + \sum_{n=0}^{\infty} \frac{a_n\varepsilon^{n+1}}{n+1}\left\{\exp[i(n+1)\beta] - \exp[i(n+1)\alpha]\right\}.$$

Taking the limit as $\varepsilon \to 0$, and using the fact that $a_{-1} = \text{Res}[f(z), z_0]$, this becomes

$$\lim_{\varepsilon\to 0}\int_{C_\varepsilon} f(z)dz = i(\beta-\alpha)\text{Res}[f(z), z_0],$$

and the theorem is proved. ◆

Example 3.7.9 Integration around an Indented Rectangle
Given that

$$f(z) = \frac{e^{iz}}{z^2+4},$$

evaluate $\int_C f(z)dz$, where C is the rectangle in the complex plane with its corners at $z = 0, z = 2i, z = -3+2i$ and $z = -3$, where integration is performed in the positive sense (counterclockwise).

SOLUTION
The function $f(z)$ has simple poles at $z = -2i$ and $z = 2i$, but only the pole at $z = 2i$ is involved in the integration because it lies at a corner of rectangle C, while the pole at $z = -2i$ lies outside C. Consider Figure 3.16 where the pole at point P ($z = 2i$) is excluded by the small quarter circular indentation of radius ε. Then by the Cauchy–Goursat theorem, as the indented contour ABCODA contains no poles, it follows that

$$\int_{\text{ABCODA}} f(z)dz = 0.$$

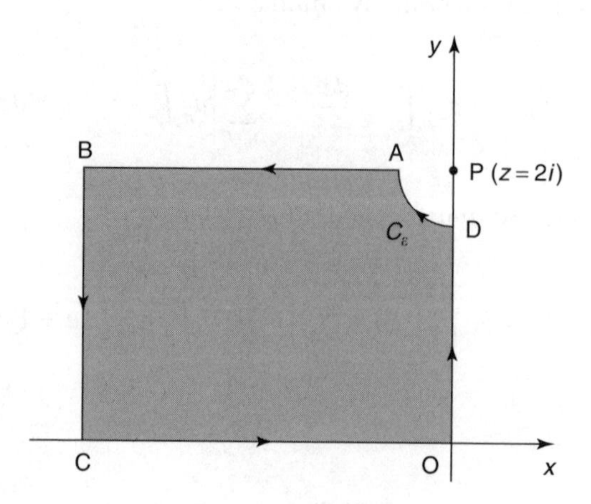

FIGURE 3.16
The indented rectangle.

When re-expressed in terms of integrals along parts of the indented rectangle this becomes

$$\int_{\mathrm{ABCOD}} f(z)dz + \lim_{\varepsilon\to 0}\int_{\mathrm{C}_\varepsilon} f(z)dz = 0.$$

Now

$$\mathrm{Res}[f(z), 2i] = [(z-2i)f(z)]_{z=2i} = -\tfrac{1}{4}e^{-2},$$

so to evaluate this integral we must make use of Theorem 3.7.6.

Using the notation of the theorem we have $\alpha = \mathrm{Arg[PD]} = -\pi/2$, and when z moves around C_ε from D to A it follows that $\beta = \mathrm{Arg[PA]} = -\pi$. Thus from Theorem 3.7.6 we find that

$$\lim_{\varepsilon\to 0}\int_{\mathrm{C}_\varepsilon} f(z)dz = i[-\pi - (-\pi/2)]\left(-\frac{i}{4e^2}\right) = -\frac{\pi}{8e^2}.$$

Proceeding to the limit at $\varepsilon \to 0$, the contour ABCODA reduces to the rectangle C, and so

$$\int_{\mathrm{C}} f(z)dz = \frac{\pi}{8e^2}.$$ ◇

Exercises 3.7

Find the residues of the following functions at their poles in the complex plane.

1. $f(z) = \dfrac{1}{z(1-z^2)}$
2. $f(z) = \dfrac{1}{\sin z}$
3. $f(z) = \dfrac{7z^4}{(1+z)^2}$
4. $f(z) = \dfrac{3\sin 2z}{(z+1)^3}$
5. $f(z) = \dfrac{e^{iz}}{z^2+a^2}$ (a real)

6. $f(z) = \dfrac{\cosh az}{\cosh z}$ $\quad (a \text{ real})$

Find the residue at infinity for the following functions.

7. $f(z) = \dfrac{z^2}{(z^2+1)^2}$

8. $f(z) = \dfrac{3\sin 2z}{(z+1)^3}$

Find the residues at the simple poles of the following functions.

9. $f(z) = \dfrac{1}{z \sin z}$

10. $f(z) = \dfrac{\operatorname{Ln} z}{(z+i)\cosh z}$

11. Find the residue at $z = \exp[i\pi/2n]$ of

$$f(z) = \frac{1}{1+z^{2n}}$$

12. Find the residue at $z = 2i$ of

$$f(z) = \frac{1}{(4+z^2)^4}$$

13. Find the residue at the simple pole of

$$f(z) = \frac{z^{\alpha-1}}{1+z}$$ for $0 < \alpha < 1$, using the principal branch of $z^{\alpha-1}$.

14. Given that

$$f(z) = \frac{g(z)}{(z-a)^n},$$

where $g(z)$ is analytic at $z = a$ and such that $g(a) \neq 0$, use Lemma 3.7.3 to find the form of $\text{Res}[f(z), a]$.

3.8 Applications of the Residue Theorem

The residue theorem may be used in a variety of different ways, a few of which are presented in this section. The most obvious application of the theorem is to the evaluation of definite integrals, which is a topic that has already been discussed in Chapter 2, where a direct appeal was made to the Cauchy integral theorems. The advantage obtained by using the residue theorem for this purpose is that the computational task is simplified, and a far wider class of integrals can be considered.

However, it still remains necessary to estimate the integral along any circular arc or part of a rectangle used to close a contour, and to show that the integral along these parts vanishes in the limit as the contour of interest is extended to infinity.

3.8.1 The Evaluation of Definite Integrals

Example 3.8.1 An Integrand with Simple Poles

Evaluate

$$\int_{-\infty}^{\infty} \frac{dx}{x^4 + 1}.$$

SOLUTION

Consider the function

$$f(z) = \frac{1}{z^4 + 1},$$

and the contour Γ_R shown in Figure 3.17, comprising the semicircle C_R with equation $|z| = R$ in the upper half plane, and the segment of the real axis $-R \leqslant x \leqslant R$. An application of the residue theorem gives

$$\int_{\Gamma_R} \frac{dz}{z^4 + 1} = 2\pi i \times (\text{sum of residues inside } \Gamma_R).$$

Now $z^4 = -1 = e^{(2k+1)\pi i}$ for integral k, so the zeros of the denominator of $f(z)$ are all simple and occur at the points

$$z_k = \exp[(2k+1)\pi i/4], \quad \text{for } k = 0, 1, 2, 3, \ldots,$$

showing that $f(z)$ has simple poles at these points. Only the poles at $z_0 = e^{\pi i/4}$ and $z_1 = e^{3\pi i/4}$ lie in the upper half plane, and hence will lie inside Γ_R when

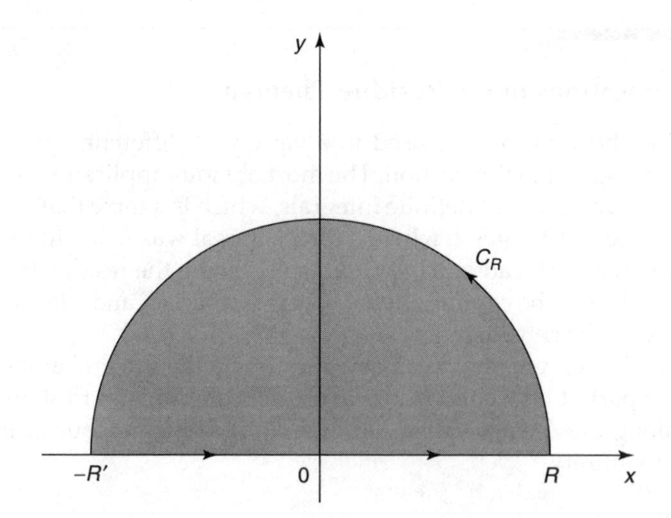

FIGURE 3.17
The contour Γ_R.

$R > 0$ is sufficiently large. So to apply the residue theorem only the residues at z_0 and z_1 are necessary. An application of Lemma 3.7.4 gives

$$\text{Res}[f(z), z_0] = \frac{1}{4z_0^3} = \tfrac{1}{4}e^{-3\pi i/4}, \text{ and } \text{Res}[f(z), z_1] = \frac{1}{4z_1^3} = \tfrac{1}{4}e^{-9\pi i/4} = \tfrac{1}{4}e^{-\pi i/4}.$$

Thus

$$\text{Res}[f(z), z_0] + \text{Res}[f(z), z_1] = -i\sqrt{2}/4,$$

and so

$$\lim_{R\to\infty} \int_{\Gamma_R} f(z)dz = 2\pi i(-i\sqrt{2}/4) = \frac{\pi}{\sqrt{2}}.$$

To complete the evaluation of the integral it is now necessary to consider the contributions made by integrating around the semicircle C_R and along the real axis. To do this we write

$$\int_{\Gamma_R} f(z)dz = \int_{C_R} \frac{dz}{z^4+1} + \int_{-R}^{R} \frac{dx}{x^4+1},$$

so proceeding to the limit as $R \to \infty$ and using the previous result this becomes

$$\lim_{R\to\infty} \int_{C_R} \frac{dz}{z^4+1} + \int_{-\infty}^{\infty} \frac{dx}{x^4+1} = \frac{\pi}{\sqrt{2}}$$

The contour Γ_R is shown in Figure 3.18, where upward semicircular indentations have been inserted around the simple poles at $z = 0$ and $z = \pm\pi$. Contour Γ_R contains no poles, so by Cauchy's theorem

$$\int_{\Gamma_R} f(z)dz = 0.$$

Replacing the integral around Γ_R by the sum of integrals around different parts of the contour gives

$$\int_{-R}^{-(\pi-\varepsilon_1)} f(z)dz + \int_{C_{\varepsilon_1}-} f(z)dz + \int_{-\pi+\varepsilon_1}^{-\varepsilon_2} f(z)dz + \int_{C_{\varepsilon_2}-} f(z)dz$$
$$+ \int_{\varepsilon_2}^{\pi-\varepsilon_2} f(z)dz\sqrt{2} + \int_{C_{\varepsilon_3}-} f(z)dz + \int_{\pi+\varepsilon_3}^{R} f(z)dz + \int_{C_R} f(z)dz = 0,$$

where $C_{\varepsilon_1}-$, $C_{\varepsilon_2}-$, and $C_{\varepsilon_3}-$, indicates that integration around these semicircles is in the *negative* sense (clockwise).

After rearranging terms and reversing the direction of integration around the semicircles with compensating changes of sign this becomes

$$\int_{-R}^{-(\pi-\varepsilon_1)} f(z)dz + \int_{-\pi+\varepsilon_1}^{-\varepsilon_2} f(z)dz + \int_{\varepsilon_2}^{\pi-\varepsilon_3} f(z)dz + \int_{\pi+\varepsilon_3}^{R} f(z)dz$$
$$= \int_{C_{\varepsilon_1}} f(z)dz + \int_{C_{\varepsilon_2}} f(z)dz + \int_{C_{\varepsilon_3}} f(z)dz - \int_{C_R} f(z)dz,$$

where now integration around all semicircular contours is in the positive sense.

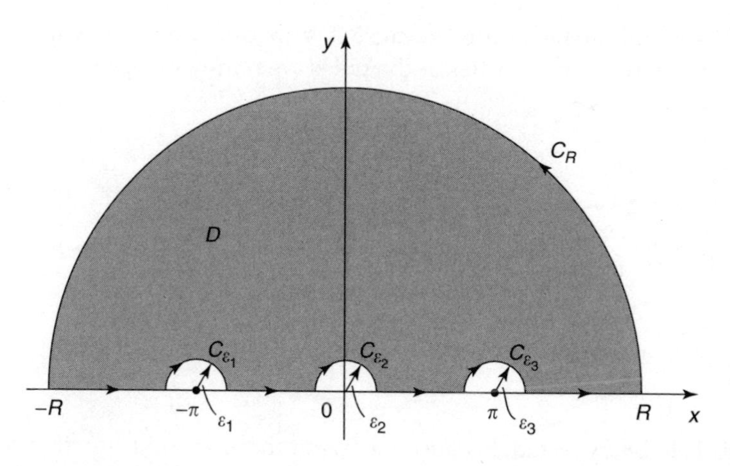

FIGURE 3.18
The indented contour Γ_R.

To investigate the contribution made by the integral around the semicircular contour C_R we set $f(z) = e^{iz}g(z)$, where

$$g(z) = \frac{1}{z(\pi^2 - z^2)}.$$

Then $|g(z)| = 1/|z||\pi^2 - z^2|$, so as $|\pi^2 - z^2| \geqslant |\pi^2 - |z|^2|$, $|g(z)| \leqslant 1/|z||\pi^2 - |z|^2|$, but $|z| = R$ on C_R, showing that $\lim_{|z|\to\infty} g(z) = 0$. As e^{iz} is of the form e^{imz} with $m > 0$, and $\lim_{|z|\to\infty} g(z) = 0$, it follows from Jordan's lemma that

$$\lim_{R\to\infty} \int_{C_R} f(z)dz = 0.$$

When integrating around each semicircular indentation, the contribution made by the each residue must be multiplied by $\frac{1}{2}$ because of Theorem 3.7.6, so proceeding to the limit as $R \to \infty$ and the radius of each indentation tends to zero, we find that because Cauchy values are involved we must write

$$\text{P.V.} \int_{-\infty}^{\infty} \frac{e^{ix}}{x(\pi^2 - x^2)} dx = \left(\frac{1}{2}\right) 2\pi i \left(\frac{1}{2\pi^2} + \frac{1}{\pi^2} + \frac{1}{2\pi^2}\right) = \frac{2i}{\pi^2}.$$

Equating the imaginary part on each side of this equation gives

$$\text{P.V.} \int_{-\infty}^{\infty} \frac{\sin x}{x(\pi^2 - x^2)} dx = \frac{2}{\pi^2}.$$

The P.V. symbol can be omitted because it is not difficult to show this integral is convergent (the singularities at the poles are removable), so

$$\int_{-\infty}^{\infty} \frac{\sin x}{x(\pi^2 - x^2)} dx = \frac{2}{\pi^2}.$$

Had the real parts been equated, we would have obtained

$$\text{P.V.} \int_{-\infty}^{\infty} \frac{\cos x}{x(\pi^2 - x^2)} dx = 0,$$

but this is to be expected, because the integrand is an odd function of x and so the Cauchy principal value of integral must vanish (the integral itself is divergent). ◇

Example 3.8.4 Integrating a Function with a Branch
Show that

$$\int_0^\infty \frac{x^{-\alpha}}{(x+1)^2}\,dx = \frac{\alpha\pi}{\sin\alpha\pi}, \qquad \text{for } |\alpha| < 1.$$

SOLUTION
We consider the function

$$f(z) = \frac{z^{-\alpha}}{(z+1)^2} \qquad \text{for } |\alpha| < 1,$$

and work with the principal branch of $z^{-\alpha}$. Because $f(z)$ has a branch point at the origin, to make the function single-valued, it is necessary to cut the complex plane along the positive real axis up to and including the origin, which must be indented. The contour Γ_R that will be used is shown in Figure 3.19, where in the limit we will let $\varepsilon \to 0$ and $R \to \infty$. Integrating around Γ_R, which contains the pole at $z = -1$, it follows from the residue theorem that

$$\int_{\Gamma_R} f(z)dz = 2\pi i \times \text{Res}[f(z), -1].$$

In terms of integrals around segments of Γ_R this becomes

$$\int_{AB} f(z)dz + \int_{C_R} f(z)dz + \int_{CD} f(z)dz + \int_{C_{\varepsilon-}} f(z)dz = 2\pi i \times \text{Res}[f(z), -1],$$

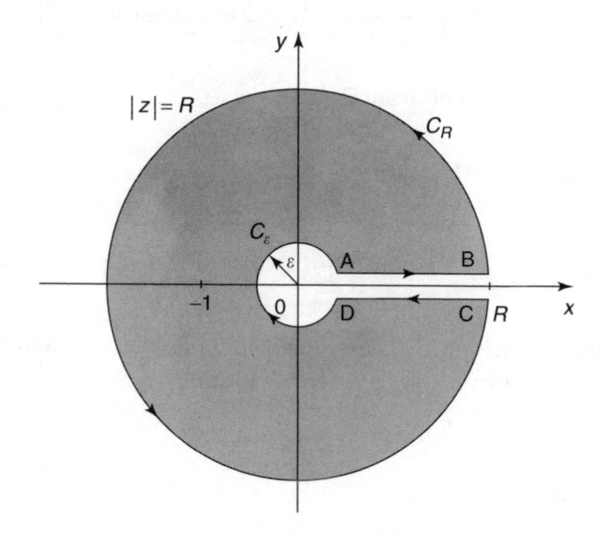

FIGURE 3.19
The indented contour Γ_R.

where $C_{\varepsilon-}$ indicates that integration around this indentation is in the negative sense. On C_R we have $z = Re^{i\theta}$, so that $dz = iRe^{i\theta}\,d\theta$ with the result that

$$\left|\int_{C_R} f(z)dz\right| = \left|\int_0^{2\pi} \frac{iRe^{i\alpha\theta}}{(R^2e^{2i\theta}+1)^2}\,d\theta\right| \leqslant \frac{R}{(R^2-1)}\int_0^{2\pi} d\theta = \frac{2\pi R}{(R^2-1)^2},$$

showing that

$$\lim_{R\to\infty}\left|\int_{C_R} f(z)dz\right| = \lim_{R\to\infty}\frac{2\pi R}{(R^2-1)^2} = 0, \quad \text{and so } \lim_{R\to\infty}\int_{C_R} f(z)dz = 0.$$

On C_ε we have $z = \varepsilon e^{i\theta}$, so $dz = i\varepsilon e^{i\theta}\,d\theta$, from which it follows that

$$\left|\int_0^{2\pi} \frac{i\varepsilon e^{i\alpha\theta}}{(\varepsilon^2 e^{2i\theta}+1)^2}\,d\theta\right| \leqslant \varepsilon\int_0^{2\pi} d\theta = 2\pi\varepsilon,$$

showing that the integral around $C_{\varepsilon-}$ and hence around C_ε vanishes as $\varepsilon \to 0$.

On the upper edge AB of the cut $z = x$, while on the lower edge CD we have $z = xe^{2\pi i}$. So the limiting form of the integral around Γ_R as $R \to \infty$ becomes

$$\int_0^\infty \frac{x^{-\alpha}}{(x+1)^2}\,dx + \int_\infty^0 \frac{(xe^{2\pi i})^{-\alpha}}{(x+1)^2}\,dx = 2\pi i \times \mathrm{Res}[f(z), -1],$$

and so

$$(1-e^{-2\alpha\pi i})\int_0^\infty \frac{x^{-\alpha}}{(x+1)^2}\,dx = 2\pi i \times \mathrm{Res}[f(z), -1].$$

Because $z = -1$ is a pole of order 2

$$\mathrm{Res}[f(z), -1] = \lim_{z\to-1}\left(\frac{d}{dz}[(z+1)^2 f(z)]\right)$$

$$= \lim_{z\to-1}\left(\frac{d}{dz}(z^{-\alpha})\right) = \lim_{z\to-1}\left(-\alpha z^{-(1+\alpha)}\right).$$

The principal branch of $z^{-\alpha}$ is involved, so when determining this limit we must replace $z \to -1$ by $z \to e^{\pi i}$, when we obtain

$$\mathrm{Res}[f(z), -1] = \alpha e^{-\alpha\pi i}.$$

Combining results gives

$$(1-e^{-2\alpha\pi i})\int_0^\infty \frac{x^{-\alpha}}{(x+1)^2}\,dx = 2\alpha\pi i e^{-\alpha\pi i},$$

so as the factor $(1 - e^{-2\alpha\pi i})$ will be non-zero for $|\alpha| < 1$, this becomes

$$\int_0^\infty \frac{x^{-\alpha}}{(x+1)^2}\,dx = \frac{2\alpha\pi i e^{-\alpha\pi i}}{1 - e^{-2\alpha\pi i}} = \frac{\alpha\pi}{\sin\alpha\pi}, \qquad \text{for } |\alpha| < 1.$$

◇

3.8.2 The Summation of Series

Convergence tests determine if a numerical series is convergent, but they give no information about the sum when it is convergent. The application of the residue theorem that follows shows how it can be used to find a closed form expression for the sum of certain types of series. We now state and prove the relevant theorem.

THEOREM 3.8.1 Summation of Series
Let $f(z)$ be a meromorphic function with a finite number of poles located at $z_1, z_2, \ldots, z_M$, none of which coincide with the points $z = 0, \pm 1, \pm 2, \ldots$, and take C_N to be the circle

$$|z| = N + \tfrac{1}{2},$$

with N a positive integer. Then

(i) $\displaystyle\sum_{n=-\infty}^{\infty} f(n) = -\pi \sum_{k=1}^{M} \operatorname{Res}[f(z)\cot\pi z, z_k],$

provided

$$\lim_{N\to\infty} \int_{C_N} f(z)\cot\pi z\,dz = 0,$$

and

(ii) $\displaystyle\sum_{n=-\infty}^{\infty} (-1)^n f(n) = -\pi \sum_{k=1}^{M} \operatorname{Res}[f(z)\csc\pi z, z_k],$

provided

$$\lim_{N\to\infty} \int_{C_N} f(z)\csc\pi z\,dz = 0.$$

PROOF

(i) By hypothesis, the only singularities of $f(z)$ occur at $z_1, z_2, \ldots, z_M$, so $f(z)$ is analytic at $z = \pm 1, \pm 2, \ldots$. Thus the function

$$F(z) = f(z)\cot\pi z = \frac{f(z)\pi\cos\pi z}{\sin\pi z}$$

is a meromorphic function with simple poles at $z = 0, \pm 1, \pm 2, \ldots$. Then it follows from Lemma 3.7.4 that

$$\begin{aligned} \operatorname{Res}[F(z), n] &= \operatorname{Res}[f(z)\pi \cot \pi z, n] = \left(\frac{f(z)\pi \cos \pi z}{\pi \cos \pi z} \right)_{z=n} \\ &= f(n), \quad \text{for } n = 0, \pm 1, \pm 2, \ldots . \end{aligned}$$

Applying the residue theorem to $\int_{C_N} F(z)dz$ yields

$$\begin{aligned} \int_{C_N} f(z)dz &= 2\pi i \underbrace{\sum_{n=-N}^{N} \operatorname{Res}[F(z), n]}_{\substack{\text{Residues at } z=n \\ \text{inside } C_N}} + 2\pi i \underbrace{\sum \operatorname{Res}[F(z), z_k]}_{\substack{\text{Residues at } z=z_k \\ \text{inside } C_N}} \\ &= 2\pi i \sum_{n=-N}^{N} f(n) + 2\pi i \underbrace{\sum \operatorname{Res}[F(z), z_k]}_{\substack{\text{Residues at } z=z_k \\ \text{inside } C_N}}. \end{aligned}$$

Proceeding to the limit as $N \to \infty$, it follows that eventually the circle C_N becomes large enough to contain all of the points $z_1, z_2, \ldots, z_M$ in the second summation, so that its summation index will run from $k = 1$ to M. Thus the last result becomes

$$\lim_{N\to\infty} \int_{C_N} F(z)dz = 2\pi i \left\{ \sum_{n=-\infty}^{\infty} f(n) + \sum_{k=1}^{M} \operatorname{Res}[F(z), z_k] \right\}.$$

Finally, if $|F(z)| \to 0$ sufficiently rapidly as $|z| \to \infty$ in order that

$$\lim_{N\to\infty} \int_{C_n} F(z)dz \lim_{N\to\infty} \int_{C_N} f(z) \cot \pi z \, dz = 0,$$

The preceding result reduces to

$$\sum_{n=-\infty}^{\infty} f(n) = -\pi \sum_{k=1}^{M} \operatorname{Res}[f(z) \cot \pi z, z_k],$$

and the proof of (i) is complete.

(ii) The proof of the second result follows in similar fashion, so the details are omitted, apart from noticing that $f(z)\csc \pi z$ has simple poles at $z = 0, \pm 1, \pm 2, \ldots$, so that from Lemma 3.7.4

$$\begin{aligned} \operatorname{Res}[f(z) \csc \pi z, n] &= \operatorname{Res}\left[\frac{\pi f(z)}{\sin \pi z}, n \right] = \left(\frac{\pi f(z)}{\pi \cos \pi z} \right)_{z=n} \\ &= (-1)^n f(n), \quad \text{for } n = 0, \pm 1, \pm 2, \ldots , \end{aligned}$$

subject to the obvious requirement that

$$\lim_{N\to\infty} \int_{C_n} f(z) \csc \pi z \, dz = 0.$$

◆

COROLLARY 3.8.1 Deformation of C_N
The family of circles C_N in Theorem 3.8.1 may be replaced by any other family of simple closed curves, like squares with corners at $(N + \frac{1}{2})(\pm 1 + i)$ and $(N + \frac{1}{2})(1 \pm i)$, that in the limit as $N \to \infty$ cover the entire complex plane.

PROOF

The result follows directly from the residue theorem by means of contour deformation. ◆

Verification of the conditions

$$\lim_{N\to\infty} \int_{C_N} f(z) \cot \pi z \, dz = 0 \quad \text{and} \quad \lim_{N\to\infty} \int_{C_N} f(z) \csc \pi z \, dz = 0$$

in Theorem 3.8.1 is simplified by the fact that $|\cot \pi z|$ and $|\csc \pi z|$ are uniformly bounded throughout the complex plane, so that for some constant $M > 0$

$$|\cot \pi z| < M \quad \text{and} \quad |\csc \pi z| < M \quad \text{for all } z.$$

This means that whatever the shape of the contour C_N, $|\cot \pi z|$ and $|\csc \pi z|$ may always be overestimated by some absolute constant $M > 0$. These results form the content of the following lemma.

LEMMA 3.8.1 $|\cot \pi z|$ and $|\csc \pi z|$ Are Uniformly Bounded
The functions $|\cot \pi z|$ and $|\csc \pi z|$ are uniformly bounded in the complex plane by a finite number $M > 0$, with the result that

$$|\cot \pi z| < M \quad \text{and} \quad |\csc \pi z| < M, \quad \text{for all } z.$$

PROOF

Consider the behavior of $|\cot \pi z|$ on the sides of a square with corners located at $z = \pm(N + \frac{1}{2}) \pm i(N + \frac{1}{2})$. Then on the side parallel to the real axis on which $z = x + (N + \frac{1}{2})i$, with $-(N + \frac{1}{2}) \leqslant x \leqslant (N + \frac{1}{2})$,

$$|\cot \pi z| = \left|\frac{e^{2\pi i z} + 1}{e^{2\pi i z} - 1}\right| = \left|\frac{1 + e^{-2(N+1/2)\pi} e^{2\pi i x}}{1 - e^{-2(N+1/2)\pi} e^{2\pi i x}}\right| \leqslant \frac{1 + e^{-2(N+1/2)\pi}}{1 - e^{-2(N+1/2)\pi}},$$

and so

$$|\cot \pi x| \leqslant \frac{1+e^{-\pi}}{1+e^{-\pi}} = \coth \tfrac{1}{2}\pi,$$

showing that $|\cot \pi z|$ is uniformly bounded on this side of C_N for $N = 1, 2, \ldots,$ and a similar result holds for the lower side of the square.

On the sides parallel to the imaginary axis on which $z = \pm(N + \frac{1}{2}) + iy$, because of periodicity in x we have

$$|\cot \pi z| = \left|\cot\left(\pi\left[\pm\left(n+\tfrac{1}{2}\right)+iy\right]\right)\right| = \left|\cot\left[\pi\left(\pm\tfrac{1}{2}+iy\right)\right]\right| = |\tanh \pi y| \leqslant 1.$$

So $|\cot \pi z|$ is uniformly bounded on all sides of the square C_N for $N = 1, 2, \ldots,$ and hence it is uniformly bounded in the complex plane. A similar argument establishes the uniform boundedness of $|\csc \pi z|$ throughout the complex plane, so the proof is complete. ◆

Example 3.8.5 Summation of Series

Find a closed form expression for

$$\sum_{n=1}^{\infty} \frac{1}{(n^2+a^2)^2}, \qquad (a > 0).$$

SOLUTION

The terms in the summation are all positive, so the result of Theorem 3.8.1(i) must be used. Inspection shows that to make use of the theorem we must consider the function

$$f(z) = \frac{1}{(z^2+a^2)^2} = \frac{1}{(z-ia)^2(z+ia)^2},$$

which becomes the general term of the series when $z = n$. This shows that $f(z)$ has poles of order 2 at $z = \pm ia$, so by Lemma 3.7.3,

$$\begin{aligned}\operatorname{Res}\left[\frac{\cot \pi z}{(z^2+a^2)^2}, ia\right] &= \left[\frac{d}{dz}\left(\frac{(z-ia)^2 \cot \pi z}{(z^2+a^2)^2}\right)\right]_{z=ia} \\ &= -\frac{1}{4a^2}\left(\pi \operatorname{csch}^2 \pi a + \frac{1}{a}\coth \pi a\right),\end{aligned}$$

and similarly,

$$\operatorname{Res}\left[\frac{\cot \pi z}{(z^2+a^2)^2}, -ia\right] = -\frac{1}{4a^2}\left(\pi \operatorname{csch}^2 \pi a + \frac{1}{a}\coth \pi a\right).$$

Substituting these results into Theorem 3.8.1(i) gives

$$\sum_{n=-\infty}^{\infty} \frac{1}{(n^2 + a^2)^2} = \frac{\pi}{2a^2}\left(\pi \operatorname{csch}^2 \pi a + \frac{1}{a} \coth \pi a\right).$$

Because $(n^2 + a^2)^2$ is an even function of n we can write

$$\sum_{n=-\infty}^{\infty} \frac{1}{(n^2 + a^2)^2} = \frac{1}{a^4} + 2\sum_{n=1}^{\infty} \frac{1}{(n^2 + a^2)^2},$$

and so arrive at the result

$$\sum_{n=1}^{\infty} \frac{1}{(n^2 + a^2)^2} = \frac{\pi}{4a^2}\left(\pi \operatorname{csch}^2 \pi a + \frac{1}{a} \coth \pi a\right) - \frac{1}{2a^4}.$$

The final justification of this result is provided by showing that the remaining condition in Theorem 3.8.1(i) is satisfied, namely that

$$\lim_{N\to\infty} \int_{C_N} \frac{\cot \pi z}{(z^2 + a^2)^2}\, dz = 0.$$

Using the usual integral inequality and Lemma 3.8.1(i) we have

$$\left|\int_{C_N} \frac{\cot \pi z}{(z^2 + a^2)^2}\, dz\right| \leqslant \int_{C_N} \frac{|\cot \pi z|}{|(z^2 + a^2)^2|}\, |dz| \leqslant M \int_{C_N} \frac{|dz|}{|(z^2 + a^2)^2|},$$

for some M bounding $|\cot \pi z|$. However, $|(z^2 + a^2)^2| > |z|^4$, and on C_N we have $\min\{|z|\} = N + \frac{1}{2}$, so that on C_N we have $|z|^4 > (N + \frac{1}{2})^2$, and hence

$$\left|\int_{C_N} \frac{\cot \pi z}{(z^2 + a^2)^2}\, dz\right| < \frac{M}{\left(N + \frac{1}{2}\right)^4} \int_{C_N} |dz|.$$

The integral $\int_{C_N} |dz|$ is the length of the perimeter of C_N, namely $4(N + 1)$, so that

$$\lim_{N\to\infty} \left|\int_{C_N} \frac{\cot \pi z}{(z^2 + a^2)^2}\, dz\right| < \lim_{N\to\infty} \frac{4M(2N + 1)}{\left(N + \frac{1}{2}\right)^2} = 0,$$

confirming that the condition in Theorem 3.8.1(i) is satisfied. So, in summary, we have proved that

$$\sum_{n=1}^{\infty}\frac{1}{(n^2+a^2)^2}=\frac{\pi}{4a^2}\left(\pi\operatorname{csch}^2\pi a+\frac{1}{a}\coth\pi a\right)-\frac{1}{2a^4}.$$

◇

3.8.3 Principle of the Argument

The theorem proved in this subsection has many applications, although the only application considered here will be to determine the number of zeros of a polynomial in a gives region of the complex plane.

THEOREM 3.8.2 Principle of the Argument
Let $f(z)$ be meromorphic in a domain D bounded by a simple closed curve C and let $f(z)$ have at most a finite number of poles in D, but neither zeros nor poles on C. If Z is the number of zeros and P is the number of poles of $f(z)$ in D, with each counted according to its multiplicity, then

$$\frac{1}{2\pi i}\int_C\frac{f'(z)}{f(z)}dz=Z-P,$$

or equivalently,

$$\frac{1}{2\pi}\Delta_C\arg[f(z)]=Z-P,$$

where $\Delta_C\arg[f(z)]$ is the change in $\arg[f(z)]$ as z moves once around the contour C.

PROOF

The singularities of the function $f'(z)/f(z)$ can only occur at the zeros and poles of $f(z)$, so by the residue theorem

$$\int_C\frac{f'(z)}{f(z)}dz=2\pi i\times(\text{the sum of residues inside } C \text{ at zeros and poles of } f(z)),$$

If $f(z)$ has a zero of order m at $z=z_1$, then in a neighborhood of z_1 it must have the representation

$$f(z)=(z-z_1)^m h(z),$$

where $h(z)$ is analytic at z_1 and such that $h(z_1)\neq 0$. Thus

$$\frac{f'(z)}{f(z)}=\frac{m}{(z-z_1)}+\frac{h'(z)}{h(z)},$$

and so $f'(z)/f(z)$ has a simple pole at z_1 with $\operatorname{Res}[f'(z)/f(z),z_1]=m$.

When this reasoning is applied to each zero of $f(z)$ inside C, the sum of all of the residues of $f'(z)/f(z)$ at the zeros of $f(z)$ inside C equals the sum of the order of each zero inside C, with each counted according to its multiplicity.

Next, suppose that $f(z)$ has a pole of order n at $z = z_2$. Then in a neighborhood of z_2 the function $f(z)$ will have the Laurent series representation

$$f(z) = \frac{k(z)}{(z - z_2)^n},$$

where $k(z)$ is analytic at z_2 and such that $k(z_2) \neq 0$. Consequently,

$$\frac{f'(z)}{f(z)} = -\frac{n}{(z - z_2)} + \frac{k'(z)}{k(z)},$$

showing that $f'(z)/f(z)$ has a simple pole at z_2 with $\mathrm{Res}[f'(z)/f(z), z_2] = -n$. Therefore the sum of all of the residues of $f'(z)/f(z)$ due to these poles inside C will equal the negative of the sum of the order of each such pole, with each counted according to its multiplicity.

Combining these results shows that if Z is the sum of all the zeros of $f(z)$ inside C, and P is the sum of all the poles in C, counted according to their multiplicity, then

$$\int_C \frac{f'(z)}{f(z)} dz = Z - P,$$

so the first part of the theorem is proved.

To prove the second statement we make use of the fact that as $f(z)$ is analytic in a neighborhood of C, and it has neither zeros nor poles on C, then $\{\ln[f(z)]\}' = f'(z)/f(z)$, so we may write

$$\frac{1}{2\pi i}\int_C \frac{f'(z)}{f(z)} dz = \frac{1}{2\pi i}\int_C \{\ln[f(z)]\}' dz = \frac{1}{2\pi i}\Delta_C \ln[f(z)],$$

where $\Delta_C \ln[f(z)]$ denotes the change in $\ln[f(z)]$ as z moves once around C in the positive sense. However,

$$\ln[f(z)] = \ln_e|f(z)| + i\arg[f(z)],$$

but $\ln_e|z|$ is a real valued function of z that returns to its original value as z moves once around C, so the change in $\arg[f(z)]$ is given by

$$\Delta_C \ln[f(z)] = \Delta_C \arg[f(z)].$$

Thus

$$\frac{1}{2\pi i}\int_C \frac{f'(z)}{f(z)} dz = \frac{1}{2\pi}\Delta_C \arg[f(z)],$$

and so

$$\frac{1}{2\pi}\Delta_C \arg[f(z)] = Z - P. \qquad \blacklozenge$$

It is the second statement in this theorem that gives rise to its name, and it is this same result that proves to be most useful in applications. To understand the geometrical meaning of the principle of the argument, it is necessary to consider the relationship between the variables z and w, where $w = f(z)$. The full significance of this is explored in the next chapter, but for now it is sufficient to recognize that to any given point z_0 in the z-plane, there will correspond a point $w_0 = f(z_0)$ in the w-plane. So if z moves around a simple closed curve. Although the nature of $f(z)$ may be such that as z moves once around C, the point w moves around a more complicated closed-curve C^* in the w-plane that may comprise a number of loops, and so may *not* be a simple closed curve.

If the origin in the w-plane lies *outside* C^*, then as z moves once around C in the positive sense, starting from a point z_0, it follows that $\arg[f(z)]$ changes from its initial value $\arg[f(z_0)]$, and then return to this same value when z returns to z_0, so the change in the argument is zero, corresponding to the fact that inside C the sum $Z - P = 0$. However, if the origin in the w-plane lies inside C^*, then as w moves it may encircle the origin in the w-plane more than once. When this happens, $(1/2\pi)\Delta_C \arg|f(z)|$ is a count of the number of times w circles around the origin $w = 0$, and its value equals $Z - P$. It is because the number $(1/2\pi)\Delta_C \arg|f(z)|$ counts the number of loops made by w around the origin in the w-plane that it is called the *winding number* with respect to $w = 0$.

These ideas become clearer if we consider an application of the principle of the argument to a function $f(z)$ that is a polynomial because a polynomial only has zeros in the complex z-plane, in which case $P = 0$, and then as z moves around a curve C in the z-plane, the winding number with respect to $w = 0$ provides a count the number of zeros of $f(z)$ inside C. For our example we take the cubic polynomial

$$f(z) = z^3 - z^2 + 2 = (z - 1 + i)(z - 1 - i)(z + 1),$$

whose factors show that $f(z)$ has zeros at $z_1 = 1 - i$, $z_2 = 1 + i$, and $z_3 = -1$. We now set $f(z) = u + iv$, and consider the curves C^* in the w-plane corresponding to z moving around each of four different circles in the z-plane. To display the results graphically we need to set $z = x + iy$ in $f(z)$, and then to use the result that $u = x^3 - 3xy^2 - x^2 + y^2 + 2$, and $v = 3x^2y - y^3 - 2xy$.

The diagram to the left of Figure 3.20(a) shows the zeros of $f(z)$ as the dots located at points P_1, P_2, and P_3, and a circle C_1 in the z-plane given by $|z + 1 - i| = 3/4$ that contains *no* zeros of $f(z)$. The diagram to the right of

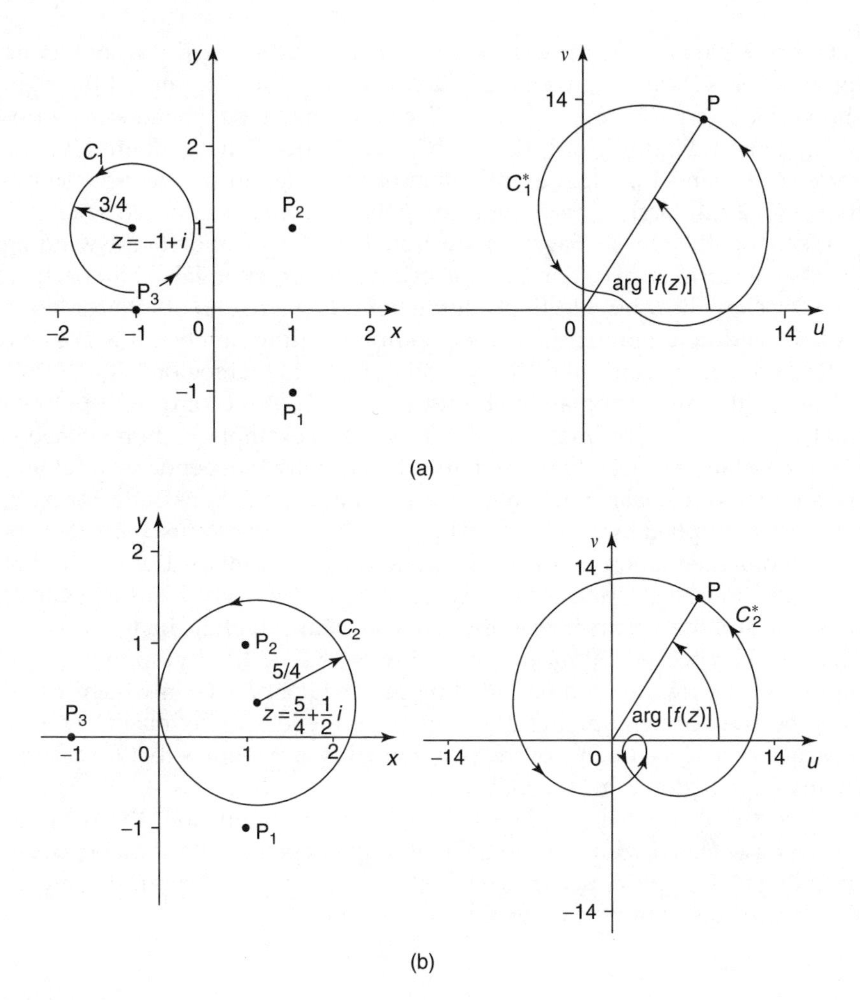

FIGURE 3.20
The curves C_i and C_i^* related by $w = f(z)$, and the zeros at P_1, P_2, and P_3.

Figure 3.20(a) shows the curve C_1^* corresponding to C_1, with a representative point P on C_1^*, together with its argument arg[$f(z)$], from which it can be seen that when z makes one complete revolution around C_1, the corresponding curve C_1^* is a *simple* closed curve that does *not* contain the origin in the w-plane, and $Z = (1/2\pi)\Delta_{C_1}\arg|f(z)| = 0$, confirming that $f(z)$ has no zeros in C_1.

Figure 3.20(b) shows the corresponding situation when C_2 is the circle $|z - \frac{5}{4} - \frac{1}{2}i| = \frac{5}{4}$ containing only the single zero at P_2, while the curve C_2^* in the diagram on the right is seen to contain the origin in the w-plane, and to encircle it once. In this case, examination of the behavior of arg[$f(z)$] shows that $Z = (1/2\pi)\Delta_{C_2}\arg[f(z)] = 1$, confirming that C_2 contains *one* zero of $f(z)$.

Figure 3.20(c) shows the case when C_3 is the circle $|z-1| = \frac{3}{2}$ containing the two zeros P_1 and P_2, when an examination of the diagram on the right shows that $Z = (1/2\pi)\Delta_{C3}\arg[f(z)] = 2$, confirming that C_3 contains *two* zeros of $f(z)$. Finally, Figure 3.20(d) shows the circle $C_4|z| = 2$ that contains all three zeros P_1, P_2, and P_3, when examination of the diagram on the right shows that $= (1/2\pi)\Delta_{C_4}\arg[f(z)] = 3$, confirming that C_4 contains *three* zeros of $f(z)$.

Taken together, these diagrams illustrate the interpretation of the winding number $(1/2\pi)\Delta_C \arg[f(z)]$ in the case of a typical polynomial $f(z)$. In particular, they show how by selecting a curve bounding a region of interest in the z-plane, and determining the corresponding winding number, it is possible to find how many zeros of the polynomial $f(z)$ lie in that region.

This qualitative information about the location of the zeros of a polynomial is often of importance in applications. For example, in homogeneous linear constant coefficient differential equations with independent variable x, the solutions are linear combinations of exponential functions of the form $e^{\lambda x}$, possibly multiplied by a polynomial in x, where the constants λ are zeros of a polynomial equation in λ (the characteristic polynomial). The solution of the equation is *stable*, and decay to zero as $x \to \infty$, if every λ has a negative real part, and this occurs if all of the zeros lie to the left of the imaginary axis. This can be checked by means of Theorem 3.8.2, either by applying it to regions to the right of and bounded by the imaginary axis, in which case Z must be zero, or by applying it to regions to the left of the imaginary axis, in which case if the degree of the polynomial is n, it follows that for a sufficiently large region Z must equal n.

When the theorem is applied to a meromorphic function, the winding number associated with a curve C^* bounding a region in the w-plane, corresponding to a simple closed-curve C in the z-plane, will determine the integer $Z - P$ associated with the curve C in the z-plane.

3.8.4 Rouché's Theorem

A result that is less powerful than the principle of the argument when identifying regions in which zeros of an analytic function occur, but is easier to use, is Rouché's theorem, the derivation of which from Theorem 3.8.2 is straightforward.

THEOREM 3.8.3 Rouché's Theorem
Let $f(z)$ and $g(z)$ be analytic inside a simply connected bounded domain D and also on its boundary C, and let them be such that

$$|f(z)| > |g(z)|$$

on the boundary of C. Then the functions $f(z)$ and $f(z) + g(z)$ have the same number of zeros in D, where each zero is counted according to its multiplicity.

PROOF

Because $|f(z)| > |g(z)|$ on C it follows that $|f(z)| > 0$ on C, and hence that $f(z) \neq 0$ on C. Let N_1 and N_2 be, respectively, the number of zeros of $f(z)$ and $f(z) + g(z)$ in D. Then by Theorem 3.8.2,

$$2\pi N_1 = \Delta_C \arg[f(z)],$$

and

$$2\pi N_2 = \Delta_C \arg[f(z) + g(z)] = \Delta_C \arg\left[f(z)\left[1 + \frac{g(z)}{f(z)}\right]\right]$$

$$= \Delta_C \arg[f(z)] + \Delta_C \arg\left[1 + \frac{g(z)}{f(z)}\right].$$

Now as $|f(z)| > |g(z)|$ on C, the point $w = 1 + g(z)/f(z)$ must always be a point inside the unit circle with its center at $w = 1$. Consequently, as z moves once around C, so w must move around a curve C^* that does *not* enclose the origin $w = 0$. Thus

$$\Delta_C \arg[w(z)] = \Delta_C \arg[1 + g(z)/f(z)] = 0,$$

showing that $N_1 = N_2$, and the proof is complete. ◆

This theorem is useful when seeking the number of zeros of a complicated analytic function $f(z)$ inside a circle $|z| = R$, and particularly so when $f(z)$ is such that the approximate location of its zeros is not obvious.

Example 3.8.6 Some Applications of Rouché's Theorem

(i) Show that the polynomial $P(z) = z^5 - 6z^2 + 13$ has no zeros inside the circle $|z| = 1$.

(ii) Find the number of zeros of $P(z) = 2z^5 + 9z^3 + 3z + 15i$ inside the unit circle $|z| = 1$.

(iii) Show the equation $e^z - \lambda z^5 = 0$ with $|\lambda| > e^R/R^5$ has five zeros inside the circle $|z| = R$.

SOLUTION

(i) Set $f(z) = 13$ and $g(z) = z^5 - 6z^2$. Then on $|z| = 1$, $|g(z)| \leqslant |z^5| + |-6z^2| = |z|^5 + 6|z|^2 = 7 < |f$ from Rouché's theorem, $f(z)$ and $P(z) = f(z) + g($ number of zeros inside $|z| = 1$, but $f(z)$ has no zeros that $P(z)$ has no zeros inside $|z| = 1$.

(ii) Set $f(z) = 15i$ and $g(z) = 2z^5 + 9z^3 + 3z$. Then on $|f(z)| = 15$, and $g(z) \leqslant 2|z|^5 + 9|z|^3 + 3|z| = 14$. Th $|f(z)| > |g(z)|$, so Rouché's theorem may be applied.

zeros inside the circle $|z| = 1$, it follows immediately that $P(z)$ can have no zeros inside this circle.

(iii) Let $f(z) = -\lambda z^5$ and $g(z) = e^z$. Then on $|z| = R$ it is given that $|\lambda|R^5 > e^R$, but $|g(z)| = |e^{R(\cos\theta + i\sin\theta)}| \leqslant e^R$, so $|f(z)| > |g(z)|$ on $|z| = R$, permitting the use of Rouché's theorem. As $f(z)$ has a zero of order 5 (at $z = 0$) inside $|z| = R$, it follows that $f(z) + g(z) = e^z - \lambda z^5$ also has five zeros inside $|z| = R$. ◇

3.8.5 The Fundamental Theorem of Algebra

The fundamental theorem of algebra was stated and proved in Theorem 2.6.5, but such is its importance that we take this opportunity to give a different and very simple proof based on Rouché's theorem.

THEOREM 3.8.4 The Fundamental Theorem of Algebra
Every polynomial of degree n with coefficients that may be either real or complex has precisely n zeros when their multiplicity is counted.

PROOF

Let

$$P(z) = a_0 z^n + a_1 z^{n-1} + a_2 z^{n-2} + \cdots + a_n,$$

be an arbitrary polynomial of degree n, where the coefficients $a_0, a_1, \ldots, a_n$ may be real or complex numbers, with $a_0 \neq 0$. Set $f(z) = a_0 z^n$ and $g(z) = a_1 z^{n-1} + a_2 z^{n-2} + \cdots + a_n$, then as $\lim_{|z|\to\infty} [g(z)/f(z)] = 0$, it follows a number $R > 0$ exists such that

$$\left|\frac{g(z)}{f(z)}\right| < 1 \quad \text{for } |z| > R.$$

Thus on the circle $|z| = R$ we have $|f(z)| > |g(z)| > 0$, showing that Rouché's theorem may be applied. The function $f(z)$ has n zeros inside $|z| = R$, so it follows that the polynomial $P(z) = f(z) + g(z)$ must also have precisely n zeros inside $|z| = R$. The theorem is proved. ◆

3.9 The Laplace Inversion Integral

A widely used technique for the solution of linear ordinary and partial differential equations is the *Laplace transformation*

$$F(s) = \int_0^\infty e^{-st} f(t)dt. \tag{3.50}$$

This integral transformation converts a function $f(t)$ of t (usually the time in a differential equation) into a function $F(s)$ of the Laplace transform variable s, where in general $s = \xi + i\eta$ is a complex variable. The relationship between $F(s)$ and $f(t)$ in Equation (3.50) is denoted symbolically by writing $F(s) = \mathcal{L}\{f(t)\}$, with the understanding that

$$\mathcal{L}\{f(t)\} = F(s) = \int_0^\infty e^{-st} f(t)dt. \tag{3.51}$$

The function $F(s)$ is called the *Laplace transform* of $f(t)$, and for it to exist it necessary to impose conditions on $f(t)$ that guarantee the convergence of the integral in Equation (3.50). A convenient condition that guarantees the existence of a Laplace transform is as follows.

3.9.1 Condition for the Existence of a Laplace Transform

If $f(t)$ is such that $|f(t)| \leqslant Me^{\alpha t}$ for some positive constant M and some real α with $\mathrm{Re}\{s\} > \alpha$, then $\mathcal{L}\{f(t)\}$ exists. When this condition is satisfied, the function $f(t)$ is said to be of *exponential order* α. The condition that $f(t)$ is of exponential order ensures that the factor e^{-st} in integral Equation (3.50) decreases sufficiently rapidly as t increases for the integral to converge. Although this condition ensures the existence of many Laplace transforms, functions exist which do not satisfy this condition, and yet still have a Laplace transform, showing this to be a *sufficient* but *not* a *necessary* condition for the existence of the transform (see Exercises 2 and 7).

When the transformation in Equation (3.50) exists, it establishes a relationship between a function $f(t)$ and its Laplace transform $F(s)$, and the pair of functions $\{f(t), F(s)\}$ related by Equation (3.50) is called a *Laplace transform pair*. Once a table of Laplace transform pairs has been established, knowledge of a function $f(t)$ in the table will determine its Laplace transform $F(s)$ and, conversely, given $F(s)$ it will determine the function $f(t)$ that gave rise to $F(s)$. Using such a table in this reverse order is called *inverting* a Laplace transform, and when $F(s) = \mathcal{L}\{f(t)\}$, it is convenient denote the inverse result by using the notation $f(t) = \mathcal{L}^{-1}\{F(s)\}$, in which case $\mathcal{L}^{-1}\{F(s)\}$ is called the *inverse Laplace transform* of $F(s)$. By way of example a short table of elementary Laplace transforms is given in Table 3.1.

Table 3.1

Some Laplace Transform Pairs

$f(t)$	$F(s)$
1	$1/s$
t	$1/s^2$
e^{at}	$1/(s-a)$
$\sin at$	$a/(s^2+a^2)$
$\cos at$	$s/(s^2+a^2)$
$e^{bt}\sin at$	$a/[(s-b)^2+a^2]$
$e^{bt}\cos at$	$(s-b)/[(s-b)^2+a^2]$

Table 3.1 shows, for example, that $\mathcal{L}\{e^{at}\} = 1/(s - a)$, and that $\mathcal{L}^{-1}\{(s - a)\} = e^{at}$. The *linearity* of the Laplace transform follows from the linearity of the operation of integration, so when $f(t)$ and $g(t)$ have the respective Laplace transforms $F(s)$ and $G(s)$, it follows that $\mathcal{L}\{f(t) + g(t)\} = F(s) + G(s)$. This simple result enables Laplace transform pairs to be combined to yield more complicated transform pairs.

A typical example of the use of this linearity property is provided by the calculation of $\mathcal{L}\{\cos at\}$, by making use of $\mathcal{L}\{e^{iat}\}$ and the complex definition of the cosine function $\cos at = \frac{1}{2}(e^{iat} + e^{-iat})$. By definition [Equation (3.50)]

$$\mathcal{L}\{e^{iat}\} = \int_0^\infty e^{-st} e^{iat}\, dt = \lim_{u\to\infty} \left(\frac{e^{(ia-s)t}}{ia - s} \right)\Bigg|_{t=0}^{u},$$

so if $s = \xi + i\eta$, the limit as $t \to \infty$ will exist and equal $1/(s - ia)$ provided $\xi > 0$, which is equivalent to requiring that $\text{Re}\{s\} > 0$. So we have established that $\mathcal{L}\{e^{iat}\} = 1/(s - ia)$ for $\text{Re}\{s\} > 0$. Using this result with the definition $\cos at = \frac{1}{2}(e^{iat} + e^{-iat})$ shows that

$$\mathcal{L}\{\cos at\} = \frac{1}{2}\left(\frac{1}{s - ia} + \frac{1}{s + ia} \right) = \frac{s}{s^2 + a^2}, \qquad \text{for Re}\{s\} > 0.$$

A different example of the use of the linearity property is provided by the transform

$$\mathcal{L}\{2 + 3\sin at\} = \frac{2}{s} + \frac{3a}{s^2 + a^2} = \frac{2s^2 + 3as + 2a^2}{s(s^2 + a^2)},$$

where the result $\mathcal{L}\{2 + 3\sin at\} = 2\mathcal{L}\{1\} + 3\mathcal{L}\{\sin at\}$ that has been used is a special case of the general result $\mathcal{L}\{\alpha f(t) + \beta g(t)\} = \alpha\mathcal{L}\{f(t)\} + \beta\mathcal{L}\{g(t)\}$, with α and β arbitrary constants.

Inverting this Laplace transform is a little harder because it requires the expression on the right, that is typical of the way more complicated Laplace transforms requiring inversion arise, to be simplified by means of partial fractions, as

$$\begin{aligned}\mathcal{L}^{-1}\left\{ \frac{2s^2 + 3as + 2a^2}{s(s^2 + a^2)} \right\} &= \mathcal{L}^{-1}\left\{ \frac{2}{s} + \frac{3a}{s^2 + a^2} \right\} \\ &= 2\mathcal{L}^{-1}\left\{ \frac{1}{s} \right\} + 3\mathcal{L}^{-1}\left\{ \frac{s}{s^2 + a^2} \right\} \\ &= 2 + 3\sin at,\end{aligned}$$

where now the linearity of the inverse Laplace transform has been used in the form $\mathcal{L}^{-1}\{\alpha F(s) + \beta G(s)\} = \alpha\mathcal{L}^{-1}\{F(s)\} + \beta\mathcal{L}\{G(s)\}$.

For many purposes a table of Laplace transform pairs, such as Table 3.1, suffices when solving simple differential equations, and also in other applications. However, it is often the case that the table does not contain an inverse Laplace transform that is required, nor can it be obtained by manipulating entries in the Table 3.1. When this occurs it is necessary to make use of *the complex Laplace inversion integral*

$$f(t) = \frac{1}{2\pi i}\int_{c-i\infty}^{c+i\infty} e^{st}F(s)ds, \tag{3.52}$$

where the meaning of this integral must be explained. In the notation of Equation (3.52), when working with the Laplace transform in the complex s-plane, the general complex variable z that has been used previously when working with analytic functions is replaced by the complex variable s, while the argument of the real function f is denoted by the real variable t, to represent *time*.

To interpret Equation (3.52) it is necessary to consider the contour Γ_R in Figure 3.21 comprising the path ABCDEA called the *Bromwich contour* in the s-plane, where the chord ED is part of the line $\xi = c$ parallel to the imaginary s-axis, and $F(s)$ is defined for $\text{Re}\{s\} > a$, say, with c any real number such that $c > a$.

The evaluation of the contour integral in Equation (3.52) is performed in the usual way, though it must be shown that the integral around the arc $C_R^{(+)}$ above the imaginary s-axis, and the integral around the arc $C_R^{(-)}$ below the

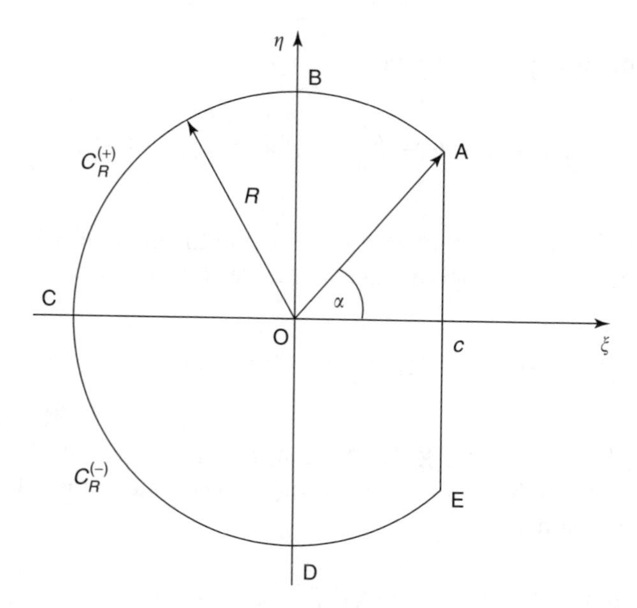

FIGURE 3.21
The contour Γ_R for the complex Laplace inversion integral.

imaginary axis, both vanish in the limit as $R \to \infty$. When this is so, in the limit at $R \to \infty$, the inverse Laplace transform becomes

$$\mathcal{L}^{-1}\{F(s)\} = \sum_{k=1}^{N} \text{Res}[e^{st}F(s), s_k], \tag{3.53}$$

where the complex numbers $s_1, s_2, \ldots, s_N$ are the locations of the finite number of poles of $F(s)$ inside the limiting form of the contour Γ_R. If $F(s)$ has a branch point, the contour Γ_R must be modified to include a suitable branch cut, such as the one used in Example 3.9.1.

Before proceeding to an example we first prove he following lemma. This is often useful when working with result Equation (3.52), as it gives conditions that establish the vanishing of integrals around the arcs $C_R^{(+)}$ and $C_R^{(-)}$ of the Bromwich contour in the limit as $R \to \infty$.

LEMMA 3.9.1 A Useful Lemma when Inverting Laplace Transforms
Let $C_R^{(+)}$ *and* $C_R^{(-)}$ *denote the circular arcs* ABC *and* CDE *of radius R shown, respectively, in Figure 3.21. If* $|F(s)| < MR^{-k}$ *on the arc* $s = Re^{i\theta}$, *for* $-\pi \leqslant \theta \leqslant \pi$, *with M and k positive constants and* $t > 0$, *then*

$$\lim_{R\to\infty} \int_{C_R^{(+)}} e^{st}F(s)ds = 0 \quad \text{and} \quad \lim_{R\to\infty} \int_{C_R^{(-)}} e^{st}F(s)ds = 0.$$

PROOF

Starting with the arc $C_R^{(+)}$ we have

$$\int_{C_R^{(+)}} e^{st}F(s)ds = \int_{\text{AB}} e^{st}F(s)ds + \int_{\text{BC}} e^{st}F(s)ds,$$

so as $s = Re^{i\theta}$ and $ds = iRe^{i\theta}\,d\theta$ on this arc, using the inequality $|F(s)| < MR^{-k}$ and proceeding in the usual manner we arrive at the estimate

$$\left|\int_{C_R^{(+)}} e^{st}F(s)ds\right| \leqslant MR^{-k}\int_{\text{AB}} |e^{st}||ds| + MR^{-k}\int_{\text{BC}} |e^{st}||ds|.$$

Now on $C_R^{(+)}$, $|e^{st}| = |\exp[Rt(\cos\theta + i\sin\theta)| = \exp(Rt\cos\theta)$, but on AB, $0 \leqslant \cos\theta \leqslant c/R$, so that $|e^{st}| < |e^{ct}|$, causing our estimate of the integral along the arc AB to become

$$\begin{aligned} MR^{-k}\int_{\text{AB}} |e^{st}||ds| &\leqslant MR^{-k}e^{ct}\int_{\alpha}^{\pi/2} d\theta = MR^{1-k}e^{ct}\left(\tfrac{1}{2}\pi - \alpha\right) \\ &= MR^{1-k}\text{Arcsin}(c/R). \end{aligned}$$

Proceeding to the limit as $R \to \infty$ this gives

$$\lim_{R\to\infty} \int_{C_R^{(+)}} e^{st} F(s) ds = 0.$$

The next stage of the proof involves estimating the integral around the arc BC. A similar argument shows that

$$MR^{-k} \int_{\mathrm{BC}} |e^{st}||ds| \leqslant MR^{1-k} \int_{\pi/2}^{\pi} \exp(Rt\cos\theta) d\theta,$$

and after setting $u = \theta - \frac{1}{2}\pi$ this becomes

$$MR^{-k} \int_{\mathrm{BC}} |e^{st}||ds| \leqslant MR^{1-k} \int_{0}^{\pi/2} \exp(-Rt\sin u) du.$$

After applying the Jordan inequality $(\sin u)/u > 2/\pi$ to this result becomes

$$\begin{aligned} MR^{-k} \int_{\mathrm{BC}} |e^{-st}||ds| &\leqslant MR^{1-k} \int_{0}^{\pi/2} \exp[-(2Rt/\pi)u] du \\ &= MR^{1-k}(1 - e^{-Rt}). \end{aligned}$$

As $k > 0$, the limit of this last expression as $R \to \infty$ is zero. So, taken together with the previous result for the integral around AB, this has proved that

$$\lim_{R\to\infty} \int_{C_R^{(+)}} e^{st} F(s) ds = 0.$$

The corresponding result for the integral along the arc $C_R^{(-)}$ follows in similar fashion, so the lemma is proved. ◆

Instead of inverting a simple Laplace transform with the aid of the complex Laplace inversion integral Equation (3.53), we use the integral to invert a more complicated Laplace transform with a branch point at the origin.

Example 3.9.1 Inverting a Laplace Transform when $F(s)$ Has a Branch Point

Use the Laplace inversion integral to find

$$\mathcal{L}^{-1}\left\{\frac{e^{-a\sqrt{s}}}{s}\right\}, \quad (a > 0).$$

SOLUTION

We start by using the inversion integral Equation (3.53) to write

$$f(t) = \mathcal{L}^{-1}\{F(s)\} = \frac{1}{2\pi i}\int_{c-i\infty}^{c+i\infty} \frac{e^{st}e^{-a\sqrt{s}}}{s}\,ds.$$

This shows that the integrand we will denote by $\phi(s) = e^{st}e^{-a\sqrt{s}}/s$ has both a branch point and a simple pole at the origin, where

$$\text{Res}[\phi(s), 0] = [s\phi(s)]_{s=0} = 1.$$

Because of the branch point the Bromwich contour must be modified to include a branch cut and an indentation around the origin, as shown in Figure 3.22.

Because $\phi(s)$ has no singularities inside the modified Bromwich contour Γ_R we have

$$\frac{1}{2\pi i}\int_{\Gamma_R}\phi(s)ds = \frac{1}{2\pi i}\left\{\int_{\text{FA}}\phi(s)ds + \int_{C_R^{(+)}}\phi(s)ds + \int_{\text{BC}}\phi(s)ds + \int_{C_{\varepsilon-}}\phi(s)ds \right.$$
$$\left. + \int_{\text{DE}}\phi(s)ds + \int_{C_R^{(-)}}\phi(s)ds\right\} = 0.$$

An application of Lemma 3.9.1 shows that in the limit as $R \to \infty$ the integrals around $C_R^{(+)}$ and $C_R^{(-)}$ vanish. On BC $s = re^{i\pi}$ and $ds = ire^{i\pi}\,dr = -ir\,dr$, so that

$$\int_{\text{BC}}\phi(s)ds = \int_R^{\varepsilon}\frac{e^{-rt}e^{-ia\sqrt{r}}}{-r}(-dr) = -\int_{\varepsilon}^{R}\frac{e^{-rt}e^{ia\sqrt{t}}}{r}\,dr,$$

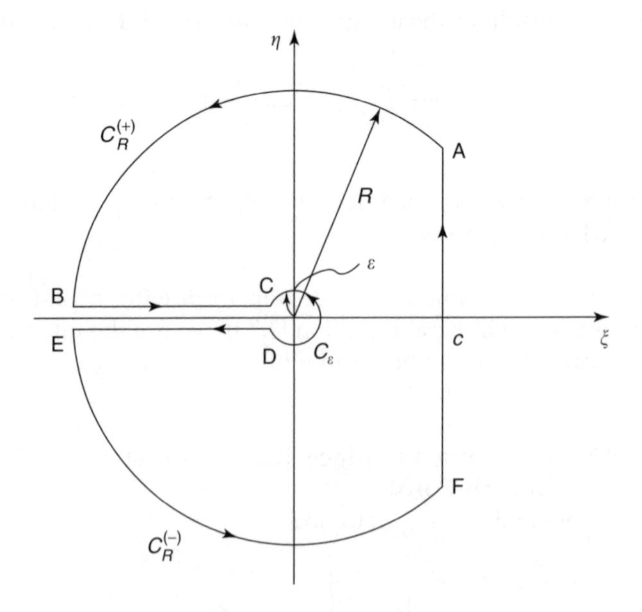

FIGURE 3.22
The modified Bromwich contour Γ_R with a branch cut.

while on DE $s = re^{-i\pi}$, leading to the result

$$\int_{DE} \phi(s)ds = \int_{\varepsilon}^{R} \frac{e^{-rt}e^{-ia\sqrt{r}}}{-r}(-dr) = \int_{\varepsilon}^{R} \frac{e^{-rt}e^{ia\sqrt{r}}}{r}dr.$$

Proceeding to the limit the integral along FA becomes

$$\lim_{\substack{R\to\infty,\\ \varepsilon\to 0}} \int_{FA} \phi(s)ds = \int_{c-i\infty}^{c+i\infty} \phi(s)ds,$$

so combining results and proceeding to the limit as $R \to \infty$ and $\varepsilon \to 0$ we obtain

$$\int_{c-i\infty}^{c+i\infty} \phi(s)ds + \int_{0}^{\infty} \frac{e^{-rt}}{r}\left(e^{ia\sqrt{r}} - e^{-ia\sqrt{r}}\right)dr - 2\pi i = 0,$$

and after division by $2\pi i$, this becomes

$$f(t) = \frac{1}{2\pi i}\int_{c-i\infty}^{c+i\infty} \phi(s)ds = 1 - \frac{1}{\pi}\int_{0}^{\infty} \frac{e^{-rt}\sin a\sqrt{r}}{r}dr.$$

This is the required inverse transform, though it is not in a convenient form. The integral can be simplified by first making the change of variable $r = u^2$, when it becomes

$$f(t) = 1 - \frac{2}{\pi}\int_{0}^{\infty} \frac{\exp[-(tu^2)]\sin au}{u}du.$$

The last term on the right can be further simplified by using the result

$$\frac{2}{\pi}\int_{0}^{\infty} \frac{\exp[-(tu^2)]\sin au}{u}du = \frac{2}{\sqrt{\pi}}\int_{0}^{a/(2\sqrt{t})} \exp(-v^2)dv.$$

For the sake of brevity, we only outline how this result can be derived. It is found by regarding the expression on the left as a function $G(a, t)$, with t fixed, differentiating it twice under the integral sign with respect to a, and integrating by parts, to obtain the differential equation $G''(a, t) = -aG'(a, t)/2t$. Solving this linear differential equation subject to the initial condition $G'(0, t) = 1/\sqrt{\pi t}$ with respect to a, found from the form of $G'(a, t)$, gives

$$G(a, t) = \frac{2}{\sqrt{\pi}}\int_{0}^{a/2\sqrt{t}} \exp(-v^2)dv.$$

To simplify matters still further we now need to introduce the special function erf(x), defined as the integral

$$\mathrm{erf}(x) = \frac{2}{\sqrt{\pi}} \int_0^x \exp(-u^2)du,$$

and called the *error function*. This function is used throughout statistics, in the study of diffusion processes, and elsewhere. The function $\mathrm{erfc}(x) = 1 - \mathrm{erf}(x)$ is called the *complementary error function*, and as $\mathrm{erf}(0) = 0$ and $\mathrm{erf}(\infty) = 1$, it follows that $\mathrm{erfc}(0) = 1$ and $\mathrm{erfc}(\infty) = 0$. Thus we have shown that $G(a, t) = \mathrm{erf}(a/2\sqrt{t})$, so that

$$\mathcal{L}^{-1}\left\{\frac{e^{-a\sqrt{s}}}{s}\right\} = 1 - \mathrm{erf}\left(\frac{a}{2\sqrt{t}}\right) = \mathrm{erfc}\left(\frac{a}{2\sqrt{t}}\right).$$

The error function and the complementary error function are examples of what are called *higher transcendental functions* which are functions not expressible in terms of elementary functions. Other examples of such functions that arise in applications of mathematics are the *Bessel functions* $J_v(x)$ and $Y_v(x)$, the *Legendre polynomials* $P_n(x)$ and the *Sine integral* $\mathrm{Si}(x) = \int_0^x (\sin t/t)dt$ and the *Fresnel integral* $\mathrm{S}(x) = \int_0^x \sin^2 t\, dt$, to mention only a few.

Exercises 3.9

1. (a) Use definition in Equation (3.50) of the Laplace transform to show that if $f(t) = e^{at}\cos bt$, with a and b real, then

$$F(s) = \mathcal{L}\{f(t)\} = \frac{s-a}{(s-a)^2 + b^2}, \qquad \text{for } \mathrm{Re}\{s\} > a.$$

   (b) Apply the complex Laplace inversion integral Equation (3.53) to recover $f(t)$ from $F(s)$ by showing that $\mathcal{L}^{-1}\{F(s)\} = e^{at}\cos bt$.

2. Use integration by parts and the change of variable $u = (s-1)t$, with the definition of a Laplace transform and the standard result $\int_0^\infty u^{1/2}e^{-u}\,du = \frac{1}{2}\sqrt{\pi}$, to show that

$$\mathcal{L}\left\{e^t/\sqrt{t}\right\} = \sqrt{\frac{\pi}{s-1}}, \qquad \text{for } s > 1.$$

   Confirm that although $f(t) = e^t/\sqrt{t}$ has a Laplace transform, this function is *not* of exponential order. This example illustrates that although the condition that a function $f(t)$ be of exponential order is *sufficient* to ensure it has a Laplace transform, the condition is *not* a *necessary* one.

3. Use the inversion integral Equation (3.53) to find

$$\mathcal{L}^{-1}\left\{\frac{2}{(s-b)^3}-\frac{2a}{(s-b)^2}+\frac{a^2}{(s-b)}\right\}, \quad \text{for } \mathrm{Re}\{s\} > b.$$

4. Use the inversion integral Equation (3.53) to find $\mathcal{L}^{-1}\{F(s)\}$, given that

$$F(s) = \frac{1}{(s-a)^3}+\frac{2(s-a)}{(s-a)^2+b^2}, \quad \text{for } \mathrm{Re}\{s\} > a.$$

5. Use the inversion integral Equation (3.53) to find $\mathcal{L}^{-1}\{F(s)\}$, given that

$$F(s) = \frac{1}{(s^2+a^2)^2}, \quad \text{for } \mathrm{Re}\{s\} > a.$$

6. Use the inversion integral Equation (3.53) to find $\mathcal{L}^{-1}\{F(s)\}$, given that

$$F(s) = \frac{1}{(s^2-a^2)^2}, \quad \text{for } \mathrm{Re}\{s\} > a.$$

7. Use the inversion integral Equation (3.53) with the Bromwich contour containing the branch cut shown in Figure 3.22 to find $f(t) = \mathcal{L}^{-1}\{e^{-a\sqrt{s}}\}$ in the form of an integral in which a and t appear as parameters, given that the principal branch of the function $s^{-1/2}$ is to be used.

4

Conformal Mapping

4.1 Geometrical Aspects of Analytic Functions: Mapping

An important part of the study of analytic functions is concerned with their geometrical properties and it is this aspect of the subject that finds wide ranging applications to physical problems whose solutions depend on the Laplace equation. The geometrical properties of analytic functions form a subject called *conformal mapping* and some of the most important of these mappings by analytic functions will be considered in this chapter. Applications of conformal mapping to heat conduction, electric fields and fluid mechanics will form the subject matter of Chapter 5.

The complex function

$$w = f(z) = u(x, y) + iv(x, y), \quad \text{for } z \in D$$

relates points $z = x + iy$ in the *domain of definition* D of f to points $w = u + iv$ in the *range* R of f, that is, R is the set of all points w that correspond to points z in D. This suggests a natural approach to adopt when considering the relationship between the complex variables z and w, which has already been used in Section 3.9 when considering the principle of the argument. The idea is to introduce two different complex planes, one the z-plane and the other the w-plane, where points in the w-plane are determined by $w = f(z)$ for $z \in D$. This idea, due to Riemann, is used here to examine the geometrical connection imposed between these two planes when $f(z)$ is required to be *analytic*.

Our concern is how an analytic function $w = f(z)$ brings regions in the z-plane into correspondence with regions in the w-plane, and conversely. The approach adopted will be to find how conveniently chosen arcs in a region of the z-plane are transformed into corresponding arcs in the w-plane.

To clarify matters, let us agree to denote by $\mathbb{C}_z$ the set of point comprising the entire z-plane (previously denoted by $\mathbb{C}$), and by $\mathbb{C}_w$ the set of points comprising the entire w-plane. Then if the function f has the domain of definition D, the correspondence between points $z \in D$ in $\mathbb{C}_z$ and the points $w = f(z)$ that belong to part of $\mathbb{C}_w$ is called a *mapping* from the z-plane to the w-plane (also a *transformation*) of the domain D by f *into* $\mathbb{C}_w$. However, if R is the range of f, that is the set of points $w = f(z)$ in $\mathbb{C}_w$ with $z \in D$, the correspondence is called a mapping of D by f *onto* R.

The point $w_0 = f(z_0)$ corresponding to $z \in D$ is called the *image point* in the w-plane of the point z_0 in the z-plane under the mapping f. More briefly, when it is understood that the mapping function f is specified, the point w_0 is called the image of z_0. When referring back from the w-plane to the z-plane the point z_0 is called the *preimage* of w_0.

The difference between *into* and *onto* mappings can be summarized as follows. In a mapping by f of D *into* a set S, not every point of S is the image of a point in D, whereas in the case of a mapping of D *onto* a set R, every point of R is the image of at least one point in D. These ideas are illustrated symbolically in Figure 4.1 in which γ is the boundary of the domain of definition D of f, and its image Γ is the boundary of the range R of f, with S a set in $\mathbb{C}_w$, and R contained in S though not every point in S belongs to R.

Of special interest here will be the class of mappings f for which the relationship $f(z_1) = f(z_2)$ implies $z_1 = z_2$. These are called *one-one* mappings because they have the property that distinct points z_1 and z_2 have distinct image points w_1 and w_2. In particular, when a mapping by f of D is both *onto* R and *one-one*, this means that one z corresponds to one w, and conversely, so then an inverse function exists that maps any point w back to a point z in a unique manner. The function inverse to f is usually denoted by f^{-1} (not to be confused with $(1/f)$, with the understanding that if $w = f(z)$, then $z = f^{-1}(z)$. Alternatively expressed, this says that if domain D is mapped *onto* the range R of f one-one, then the inverse function f^{-1} maps R one-one *onto* D.

4.1.1 The Linear Transformation $w = az + b$

The simplest transformation involving an analytic function is the *linear transformation*

$$w = az + b \tag{4.1}$$

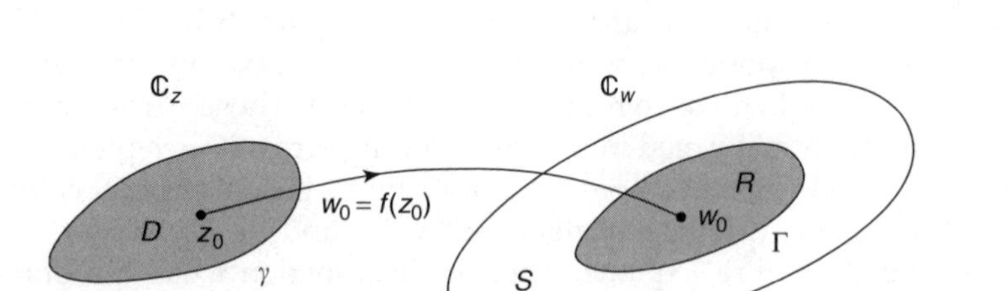

FIGURE 4.1
Function f maps D *into* S, but *onto* R.

where a and b are arbitrary complex numbers. To see the effect of this transformation, imagine the w- and z-planes to be superimposed, with their origins and real and complex axes coincident.

Consider first the case when $a = 1$, so that $w = z + b$. Then all of the points in the w-plane are obtained from corresponding points in the z-plane by being shifted by the same constant amount b. A transformation, of this type is called *translation*. As b is constant, the translation is uniform throughout the entire w-plane. Such a mapping will reproduce in the w-plane any curve in the z-plane without change of shape, orientation, or scale. Thus the w-plane can be thought of as having been obtained from the z-plane as a result of the entire z-plane having been translated without change of scale by an amount b. The origin in the w-plane is thus relocated at the point b in the z-plane.

Now consider the case when $b = 0$ so that $w = az$. We have the obvious results that

$$|w| = |az| = |a||z| \quad \text{and} \quad \arg w = \text{Arg } a + \text{Arg } z.$$

The first result shows that the modulus of any point w is obtained from the modulus of a corresponding point z by multiplying $|z|$ by $|a|$. If $|a| > 1$ this represents a *magnification* (or *dilation*), and if $|z| < 1$ it represents a *contraction*. Because a is constant, magnification is uniform throughout the entire w-plane. The second result shows that $\arg w$ is obtained from $\text{Arg } z$ by the addition of the constant angle $\text{Arg } a$. This corresponds to a *rotation* counterclockwise about the origin of the point z (or the vector representing it) through an angle $\text{Arg } a$. Thus the mapping $w = az + b$ is obtained from the z-plane by making a uniform magnification (contraction) and a rotation counterclockwise about the origin through an angle $\text{Arg } a$.

The combination of these effects shows that the linear transformation $w = az + b$ follows from the z-plane by first making a counterclockwise rotation about the origin in the z-plane through an angle $\text{Arg } a$, and then making a translation by an amount b. It is important to notice the order in which these operations are performed, with the rotation first and then the translation. This is because the operations do not commute, so that interchanging their order will change the nature of the transformation. This can be understood more clearly if the effect of the transformation is obtained by combining the two separate transformations $\zeta = az$ and $w = \zeta + b$ to get $w = az + b$. Had the order been reversed and started with the translation $\zeta = z + b$ and then the rotation introduced was introduced to give $w = a\zeta$, the result would have been $w = az + ab$. Returning to the transformation $w = az + b$, as the magnification and translation are uniform, the effect of this transformation on any curve or shape in the z-plane will be to reproduce the shape exactly in the w-plane, but with its orientation and overall scale changed.

Notice that this transformation can be reverse with these same effects preserved because $z = (w - b)/a$, so that to each point z_0 in the z-plane there corresponds a unique point $w_0 = az_0 + b$ in the w-plane and, conversely, to each point w_0 in the w-plane there corresponds a unique point $z_0 = (w_0 - b)/a$ *in the* z-plane.

In the special case $a = 1$ and $b = 0$ the linear transformation becomes an *identity mapping*, because then the w-plane is simply an identical copy of the z-plane.

Example 4.1.1 A Typical Linear Transformation
Figure 4.2 shows the effect on an L-shaped region in the z-plane when it is mapped to the w-plane by the linear transformation $w = (1/\sqrt{2})(1 + i)z + 1 + 2i$. In terms of the previous notation, $a = (1 + i)/\sqrt{2}$ and $b = 1 + 2i$, so $|a| = 1$ and Arg $a = \frac{1}{4}\pi$. Thus the effect of this transformation is first to produce a counterclockwise rotation through an angle $\frac{1}{4}\pi$ in which the magnification factor, $|a| = 1$ and then to perform a uniform translation by an amount $b = 1 + 2i$. To help see what has happened, the images of points at the corners P and Q in the z-plane are shown as the points P* and Q* in the w-plane, while the region D interior to the shape in the z-plane is shown as the corresponding interior region D^* in the w-plane. Notice that this transformation has preserved angles between lines, like the right angles at P and Q, and the corresponding right angles at P* and Q*. ◇

4.1.2 Composition of Mappings

When working with complicated mappings, it is often helpful to map from the z- to the w-plane by mapping via one or more intermediate complex planes. For example, in the simplest case it is frequently useful to consider a given mapping from the z-plane to the w-plane to be the result of a mapping via the ζ-plane by means of the two successive mappings $\zeta = f(z)$ and $w = g(\zeta)$. Eliminating ζ shows the z- and w-planes are connected by the result $w = g[f(z)]$.

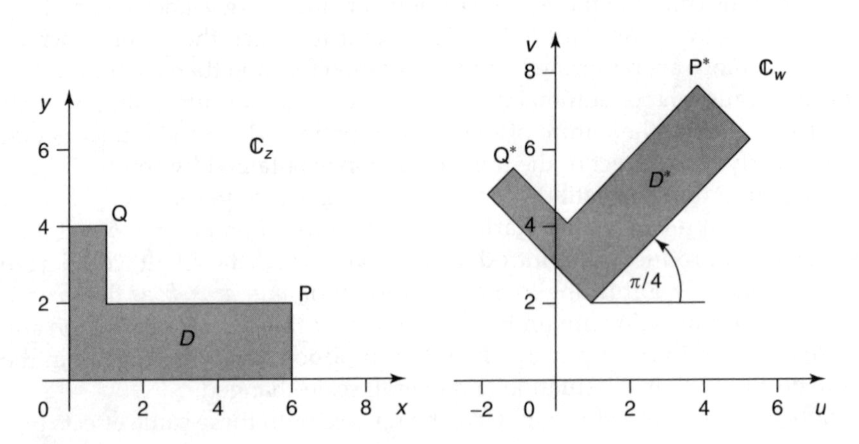

FIGURE 4.2
The linear transformation $w = (1/\sqrt{2})(1 + i)z + 1 + 2i$.

To make matters more precise, let f be a mapping from D into R and g be a mapping of R into S. Then the mapping from D to S that assigns to any point $z \in D$ the point $g[f(z)]$ of S is called the *composition* of f and g, and often denoted by $g \circ f$. Symbolically, the composition of f and g which maps z to w is written

$$w = (g \circ f)(z) = g[f(z)].$$

The mapping from the z- to the w-plane is then said to be the composition of the mappings f and g. On occasion, a composite function such as $g[f(z)]$ is referred to somewhat imprecisely as a *function of a function*.

The simplest example of the composition of two mappings has already been encountered in connection with the linear transformation in Equation (4.1). In that equation, it was convenient to regard $w = az + b$ as the composition of $\zeta = az$ and $w = \zeta + b$ when examining the geometrical implications of the mapping. More complicated examples of composite mappings are encountered later. The next example illustrates how the linear transformation can be used to map a given region in the z-plane onto a region in the w-plane in a particular way.

Example 4.1.2 Mapping a Quadrant onto a Quadrant

Find the linear transformation that maps the region $x \geqslant 1$, $y \leqslant -2$ in the z-plane onto the first quadrant of the w-plane in such a way that points A_1 at $1 - 2i$ and B_1 at $3 - 2i$ in the z-plane map to the points $w = 0$ and $w = i$ in the w-plane, respectively.

SOLUTION

We will solve this problem in two different ways, first by a simple algebraic method and then step by step via some intermediate transformations to illustrate their geometrical significance.

METHOD 1

The general linear transformation is $w = az + b$ where the two constants a and b are arbitrary, so to solve the problem we must find a and b to make points A_1 and B_1 in the z-plane map, respectively, to $w = 0$ and $w = i$ in the w-plane. The linear transformation preserves shapes, but can introduce a uniform magnification, or contraction, a rotation, and a translation. With only two constants in the transformation, the mapping of A_1 and B_1 to the indicated points will be sufficient to determine a and b. The mapping of A_1 to $w = 0$ implies that $0 = a(1 - 2i) + b$, while the mapping of B_1 to $w = i$ implies that $i = a(3 - 2i) + b$. Solving for a and b gives $a = \frac{1}{2}i$ and $b = -1 - \frac{1}{2}i$, so the required transformation is $w = \frac{1}{2}iz - 1 - \frac{1}{2}i$.

METHOD 2

The region to be mapped is shown as D_1 inside the fourth quadrant in the first diagram in Figure 4.3, and the translation $\zeta = z - 1 + 2i$ maps D_2 onto the fourth quadrant of the ζ-plane, as shown in Figure 4.3(b). The transformation

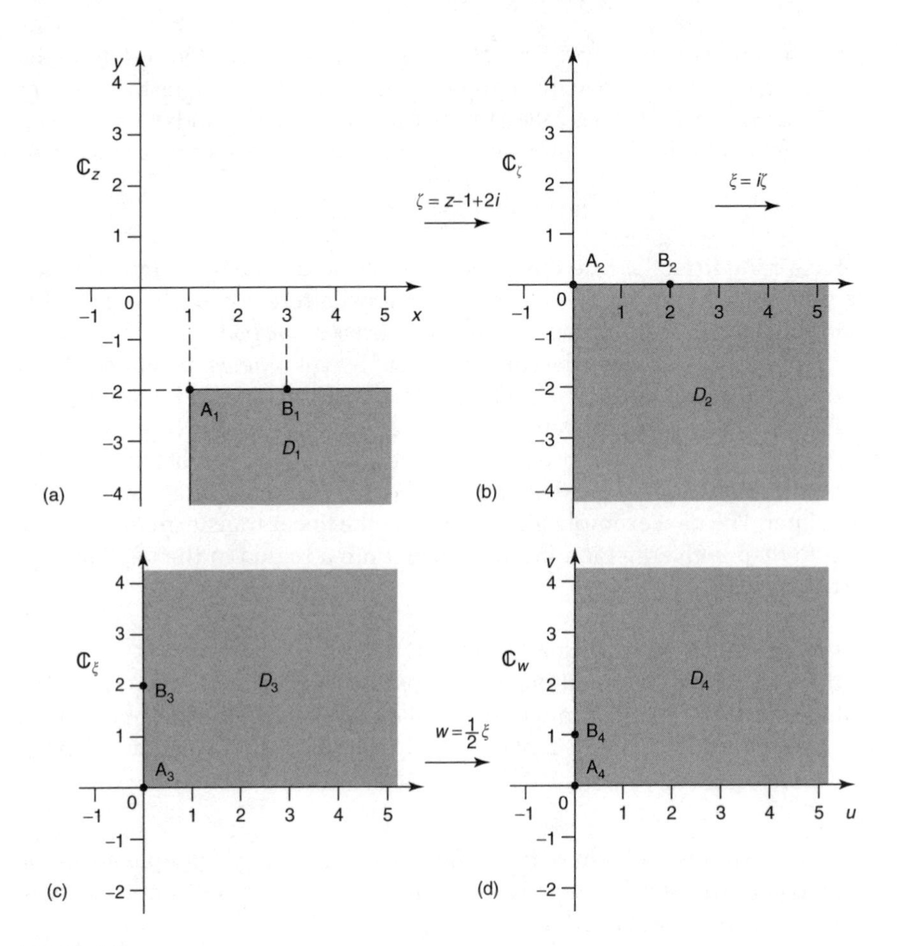

FIGURE 4.3
The mapping $w = \frac{1}{2}iz - 1 - \frac{1}{2}i$ obtained via intermediate transformations.

$\zeta = i\xi$ rotates D_2 counterclockwise through an angle $\pi/2$ and maps it onto region D_3 in the ξ-plane, as shown in Figure 4.3(c). Finally, region D_3 must be scaling to make the point B_3 at $\xi = 2i$ to become the point $v = i$ in the w-plane, when D_3 will map to D_4 as shown in Figure 4.3(d). The necessary scale factor is $\frac{1}{2}$, so the final transformation becomes $w = \frac{1}{2}\xi$. The composition of these mappings obtained by elimination of ξ and ζ becomes $w = \frac{1}{2}iz - 1 - \frac{1}{2}i$, in agreement with the result found by Method 1. ◇

4.1.3 The Mapping $w = z^c$

Setting $w = \rho e^{i\phi}$ and $z = re^{i\theta}$ in the mapping $w = z^c$ $(c > 0)$ gives for $r > 0$,

$$\rho = |w| = r^c \tag{4.2}$$

and

$$\phi = \arg w = c\operatorname{Arg} z + 2k\pi = c\theta + 2k\pi, \quad k = 0, \pm 1, \pm 2, \ldots. \tag{4.3}$$

This shows that when $c \neq 1$ a nonuniform scale factor r^c is introduced, with a *magnification* occurring when $r > 1$, that is for points *outside* the unit circle $|z| = 1$, and a *contraction* when $r < 1$, that is for points inside the unit circle. The radial line through the origin with $r > 0$ and polar angle θ_0, called a *ray*, maps onto the ray $\rho > 0$, $\phi_0 = c\theta_0$ in the w-plane. The function $w = z^c$ will thus map rays into rays, but the function is many-valued and so will provide a many-valued mapping unless θ is suitably restricted. This happens because as θ increases by 2π causing the ray $r > 0$ with polar angle θ to move once round the origin in the z-plane, the corresponding ray in the w-plane will move around the origin in the w-plane through an angle $\phi = c\theta$. This same reasoning shows that when $0 < c < 1$, $w = z^c$ will provide a mapping of the z-plane *into* the w-plane. The mapping becomes the identity mapping when $c = 1$, and when $c > 1$ it becomes a many valued mapping *onto* the w-plane.

Let us consider first the case $c = n$ (an integer), when $w = z^n$ will give an n-valued mapping of the z-plane onto the w-plane. However, if the region D in the z-plane to be mapped by $w = z^n$ is restricted to

$$D = \left(r > 0, -\frac{\pi}{b} < \theta \leqslant \frac{\pi}{n} \right),$$

this wedge-shaped region will be mapped just once onto the z-plane, cut along the negative real axis up to and including the origin.

Recalling the nth root of a complex number, we see that each of the n-inverse mappings $z = f^{-1}(w)$ given by

$$z = w^{1/n} = f_k^{-1}(w) = \rho^{1/n} \exp\left[i\left(\frac{\phi + 2k\pi}{n} \right) \right], \quad \text{with } \rho > 0, -\pi < \phi \leqslant \pi, \tag{4.4}$$

and $k = 0, 1, 2, \ldots, n-1$ provides a mapping of this same cut plane just once onto the wedge-shaped region D_k in the z-plane given by

$$D_k = \left(r > 0,\ (2k-1)\frac{\pi}{n} < \theta \leqslant (2k+1)\frac{\pi}{n} \right), \tag{4.5}$$

for $k = 0, 1, 2, \ldots, n-1$. Each of the functions $z = f_k^{-1}(w)$ is a *branch* of the nth root function that maps the cut w-plane just once onto a different wedge-shaped region D_k in the z-plane. The origin is a *branch point* of the mapping, and taken together, the regions $D_0, D_1, \ldots, D_{n-1}$ make up the entire z-plane.

The mapping corresponding to $k = 0$ in Equation (4.4) is the *principal branch* of the nth-root function, and together with $w = z^n$, provides a one-one mapping between D_0 and the cut w-plane. Regions D_0 and D_2 are shown together with the cut w-plane in Figure 4.4.

Note should be taken of the fact that the branch cut need not necessarily be taken along the negative real axis in the w-plane. The branch point will remain located at the origin, but the cut may be rotated counterclockwise through an angle α to lie along the ray $\rho > 0$, $-(\pi + \alpha) < \phi \leqslant \pi - \alpha$ if in Equation (4.4) the condition $-\pi < \phi \leqslant \pi$ is replaced by $-(\pi + \alpha) < \phi \leqslant \pi - \alpha$. The regions D_k are then all rotated through an angle α/n to become the regions $D_k^{(\alpha)}$ shown in Figure 4.5, in which is also shown the cut w-plane. However, the

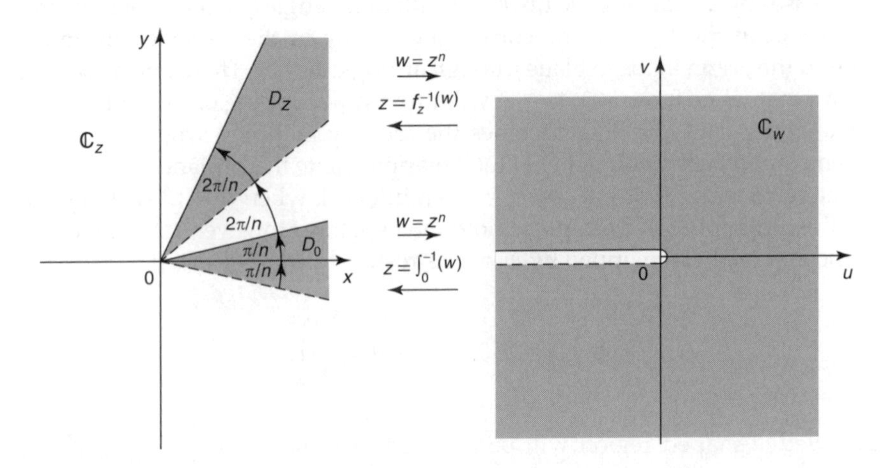

FIGURE 4.4
Mapping by $w = z^n$ and $z = f_k^{-1}(w)$ between the cut w-plane and regions D_k.

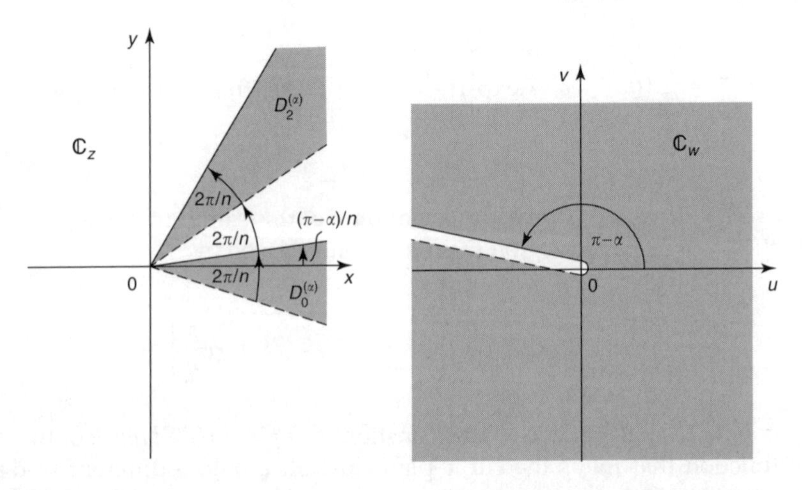

FIGURE 4.5
Mapping by $w = z^n$ with a branch cut along $\rho \geqslant 0$, $\phi = \pi - \alpha$.

convention that the cut is taken along the negative real axis, in agreement with the convention that the principal value of the argument ϕ lies in the interval $-\pi < \phi \leqslant \pi$.

4.1.4 The Mapping z^2

To understand the general effect of the nonuniform magnification occurring in the mapping $w = z^c$ when $c > 1$, it will suffice to consider the case $w = z^2$. Although first we must be clear about how the z-plane is mapped onto the w-plane. As the argument θ increases through the interval $-\frac{1}{2}\pi < \theta \leqslant \frac{1}{2}\pi$, so the argument ϕ increases through the interval $-\pi < \phi \leqslant \pi$. This shows that the right half of the z-plane is mapped onto the w-plane cut along the negative real axis up to and including the origin. Thus the fourth quadrant in the lower half of the z-plane maps onto the lower half of the w-plane, and the first quadrant in the upper half of the z-plane maps onto the upper half of the w-plane.

The left half of the z-plane corresponds to the argument θ increasing through the interval $\frac{1}{2}\pi < \theta \leqslant \frac{3}{2}\pi$, so the argument ϕ increases from π to 3π, which because of periodicity is equivalent to ϕ increasing through the interval $\frac{1}{2}\pi < \theta \leqslant \frac{3}{2}\pi$, showing that the left half of the z-plane is also mapped onto the same cut w-plane, though now the order in which upper and lower half planes are mapped is reversed. This situation is illustrated in Figure 4.6 where the four octants in each half plane are numbered 1 to 4 in the order in which they are mapped onto the numbered quadrants in the cut w-plane.

To show the effect of mapping. We find out how the coordinate lines x = constant and y = constant in the z-plane are mapped onto the w-plane. Any other lines (curves) in the z-plane could have been used, but the mapping of the coordinate lines is the simplest and it suffices to illustrate the geometrical properties of the mapping $w = z^c$.

Setting $w = u + iv$ and $z = x + iy$ in

$$w = z^2, \tag{4.6}$$

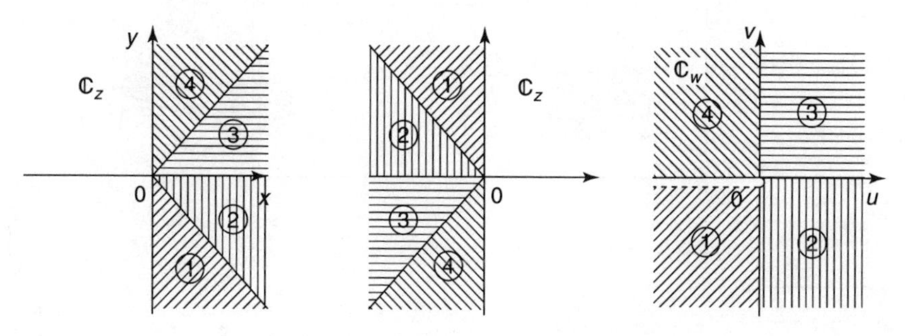

FIGURE 4.6
Mapping by $w = z^2$ of the right and left half z-planes onto the w-plane.

we find that

$$u = x^2 - y^2 \quad \text{and} \quad v = 2xy. \tag{4.7}$$

This shows that the line $x = a$ $(a > 0)$ in the z-plane maps onto the w-plane as the parabola

$$u = a^2 - \frac{v^2}{4a^2}, \tag{4.8}$$

while the line $y = b$ $(b > 0)$ in the z-plane maps onto the w-plane as the parabola

$$u = \frac{v^2 4}{4b^2} - b^2. \tag{4.9}$$

Each of the parabolas in Equation (4.8) lie to the left of $x = a^2$, while each of the parabolas in Equation (4.9) lie to the right of $x = -b^2$.

Figure 4.7 shows how a rectangular region D in the first quadrant of the z-plane is mapped onto the upper half of the w-plane by $w = z^2$ to form the region D^* with parabolic boundaries. Points P*, Q*, R*, and S^* are the images of P, Q, R, and S. The unit semicircle $|z| = 1$, $\text{Re}\{z\} > 0$ in the z-plane and its image the unit circle $|w| = 1$ in the w-plane are shown as dashed curves. The effect on the mapping of the z-plane by the nonuniform magnification factor r^2, together with the doubling of the argument, can be seen by comparing the shaded regions inside and outside $|z| = 1$ and $|w| = 1$. Notice that each of the internal angles at the corners P, Q, R, and S in the z-plane which are equal to $\frac{1}{2}\pi$, has been preserved at the image points P*, Q*, R*, and S*. This important feature will be discussed later, but notice also how the mapping has *contracted* a region inside the unit circle, and *magnified* one outside it. The image of the rectangular region D' in the third quadrant of the z-plane is again the

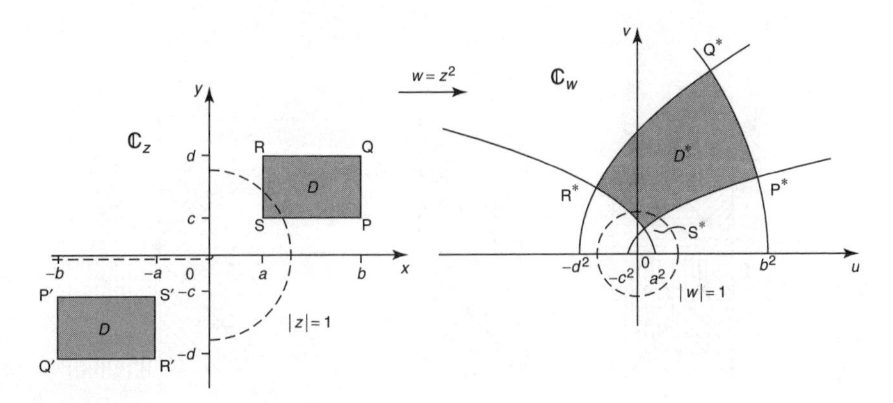

FIGURE 4.7
Mapping by $w = z^2$.

same region D^* in the w-plane. This shows the precise way in which each half of the z-plane is mapped onto the cut w-plane.

4.1.5 The Mapping $w = z^{1/2}$

The final step when studying the mapping $w = z^c$ involves consideration of the case $0 < c < 1$. We take $c = 1/2$ as typical, and consider the mapping

$$w = z^{1/2}. \tag{4.10}$$

Notice that the principal branch of this function maps the z-plane onto the cut w-plane, with the cut extending along the negative real axis up to and including the origin.

Squaring Equation (4.10) and setting $w = u + iv$ and $z = x + iy$ gives

$$x = u^2 - v^2 \quad \text{and} \quad y = 2uv. \tag{4.11}$$

This shows that the line $x = a$ $(a > 0)$ in the z-plane maps onto the w-plane as the hyperbola

$$a = u^2 - v^2, \tag{4.12}$$

while the line $y = b$ $(b > 0)$ in the z-plane maps onto the hyperbola

$$b = 2uv. \tag{4.13}$$

Figure 4.8 shows how the rectangular region D in the first quadrant of the z-plane maps onto the upper half of the w-plane. The unit circle $|z| = 1$ in the

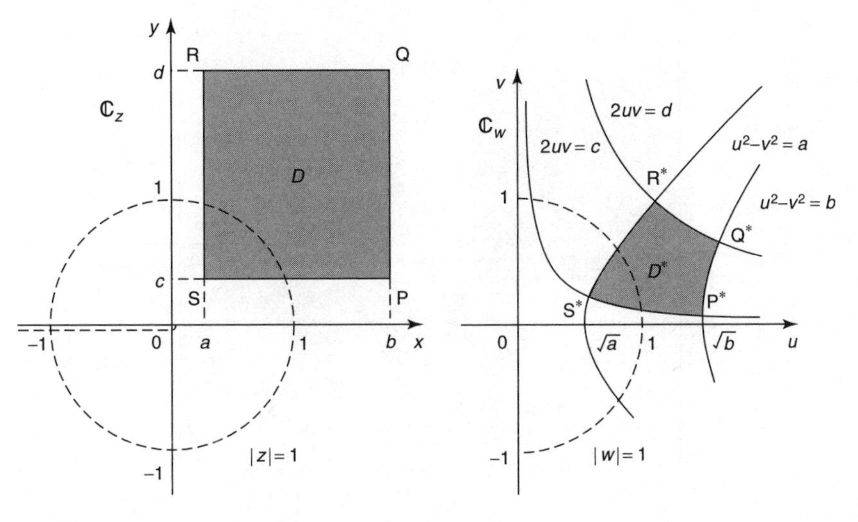

FIGURE 4.8
Mapping by the principal branch of $w = z^{1/2}$.

z-plane and its image the unit semicircle in w-plane $|w| = 1$, $\text{Re}\{w\} > 0$ have again been shown as dashed lines. Here the points P^*, Q^*, R^*, and S^* are the images of P, Q, R, and S in the z-plane, where again angles between intersecting curve have been preserved.

If, instead of the principal branch of $w = z^{1/2}$ the other branch of the square root function had been used to map D, the image D' would lie in the left half of the w-plane.

The mapping of this same region D by the other branch is shown in Figure 4.9. This other branch has no special name, but follows from Equation (4.4) as $w = f_1^{-1}(z)$, where w and z have been interchanged, ϕ has been replaced by θ, $n = 2$ and $k = 1$. It is seen that D' is obtained from D^* by a reflection first in the imaginary axis, and then in the real axis (or in the reverse order), while P′, Q′, R′, and S′ are the respective images of P, Q, R, and S.

4.1.6 The Inversion Mapping $w = 1/z$

The inversion mapping

$$w = 1/z \tag{4.14}$$

provides a one-one mapping of the z-plane onto the w-plane, with the exception of the points $z = 0$ and $w = 0$, neither of which has an image in the other plane. Setting $z = re^{i\theta}$ and $w = \rho e^{i\phi}$, the inversion mapping in Equation (4.14) becomes

$$\rho e^{i\phi} = \frac{1}{r} r^{-i\theta} \quad \text{for } r \neq 0, \tag{4.15}$$

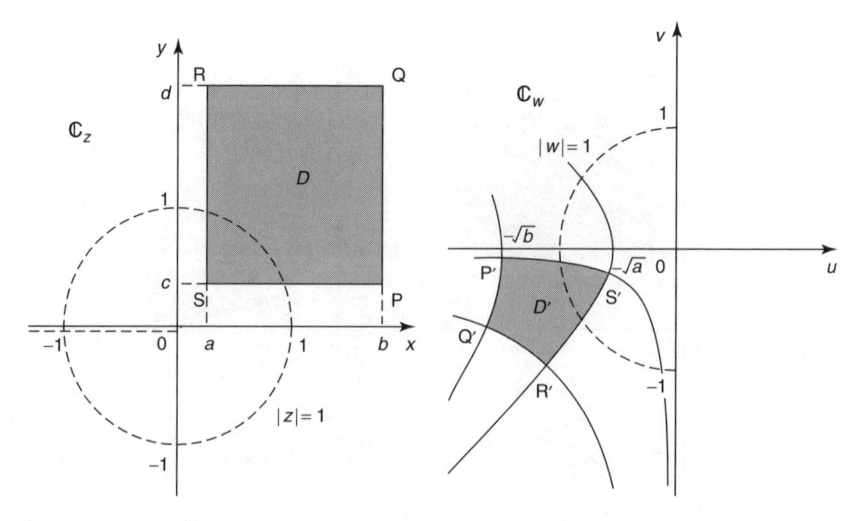

FIGURE 4.9
Mapping by the branch $w = f_1^{-1}(z)$ of the square root function.

from which we find that

$$\rho = 1/r \quad \text{and} \quad \phi = -\theta. \tag{4.16}$$

This shows the effect of the mapping from the w-plane to the z-plane is first to replace $|z| = r$ by $|w| = 1/r$, and then to change the sign of the argument from Arg $z = \theta$ to Arg $w = -\theta$.

It is helpful to relate the inversion mapping to the geometrical operation of *inversion in the unit circle*. This involves replacing each point P on any radial line drawn from the origin O to the point Q on the same line, where OQ $= 1/(\text{OP})$, where OP and OQ are the respective radial distances from O to P and Q. It is seen from Equation (4.16) that in geometrical terms the inversion mapping involves first an inversion in the unit circle, and then a reflection in the real axis. Points *exterior* to $|z| = 1$ map onto the *interior* of $|w| = 1$, and conversely, while points on $|z| = 1$ map to $|w| = 1$, but with a *reflection* in the real axis.

This geometrical interpretation is shown in Figure 4.10, in which both z at P and its image $w = 1/z$ at T are shown on the same diagram.

The inversion mapping has the property that it maps straight lines and circles in one plane onto straight lines and circles in the other plane, though not necessarily in this order. The most direct proof of this property starts from the general equation

$$A(x^2 + y^2) + Bx + Cy + D = 0, \tag{4.17}$$

with A, B, C, and D real with $B^2 + C^2 > 4AD$, which represents a circle in the z-plane if $A \neq 0$, and a straight line if $A = 0$.

Then, as $x^2 + y^2 = z\bar{z}$, $x = \frac{1}{2}(z + \bar{z})$ and $y = 1/(2i)(z - \bar{z})$, Equation (4.17) can be rewritten as

$$Az\bar{z} + \tfrac{1}{2}B(z + \bar{z}) + 1/(2i)C(z - \bar{z}) = 0.$$

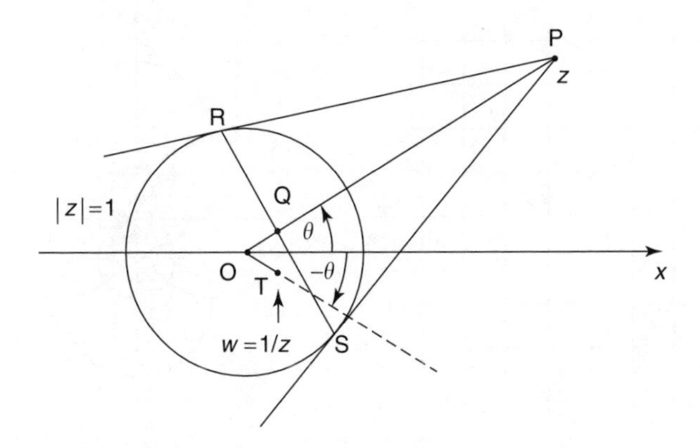

FIGURE 4.10
The inversion mapping $w = 1/z$.

However, as $z = 1/w$, $u = \frac{1}{2}(w + \overline{w})$, $v = 1/(2i)(w - \overline{w})$ this equation is easily transformed into

$$A + Bu - Cv + D(u^2 + v^2) = 0, \tag{4.18}$$

which represents a circle in the w-plane if $D \neq 0$, or a straight line if $D = 0$.

Figure 4.11 shows how the coordinate lines x = constant and y = constant in the z-plane map onto families of coaxial circles in the w-plane that are tangent at the origin to either the real or the imaginary axis.

4.1.7 Fixed Points

A *fixed point* of the mapping $w = f(z)$ is a point z_0 that maps onto itself, and so is a solution of the equation

$$z_0 = f(z_0). \tag{4.19}$$

In the case of the linear transformation

$$w = az + b, \tag{4.20}$$

the fixed point z_0 is determined by the equation

$$z_0 = az_0 + b, \tag{4.21}$$

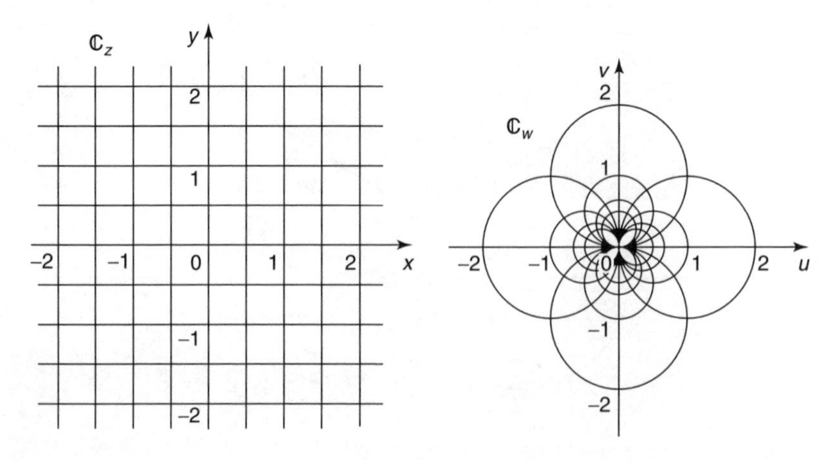

FIGURE 4.11
The mapping $w = 1/z$.

which has as its solution

$$z_0 = b/(1-a) \quad \text{for } a \neq 1. \tag{4.22}$$

when $a \neq 1$, the linear transformation in Equation (4.20) has the finite fixed point $b/(1-a)$. However, when $a = 1$ the linear transformation reduces to a pure translation, and no fixed point exists for $z \in \mathbb{C}$. If transformation in Equation (4.20) is considered in the extended complex plane the point at infinity may be considered to be a fixed point. The translation corresponding to $a = 1$ in Equation (4.20) then has ∞ as its fixed point, while for the identity mapping corresponding to $a = 1$, $b = 0$ in Equation (4.20), every point is a fixed point. The inversion mapping $w = 1/z$ is seen to have the two real fixed points ± 1.

The fixed point of $w = az + b$ $(a \neq 0)$ has an interesting and useful geometrical interpretation. Subtracting Equation (4.21) from Equation (4.20) and recognizing that $w_0 = z_0$, gives the result

$$w - w_0 = a(z - z_0). \tag{4.23}$$

This shows that in a linear transformation the w-plane is obtained from the z-plane by a rotation about the fixed point $z_0 = b/(1-a)$ through an angle $\operatorname{Arg} a$, together with a uniform magnification (contraction) by a factor $|a|$.

An immediate consequence of this result is that circles centered on the fixed point will retain their center under a linear transformation, and any pencil of straight lines through z_0 as focus will remain a pencil of lines through that same point (a *pencil of line* is defined as a family of straight lines drawn through a common point called its *focus*).

Exercises 4.1

1. Sketch the line $y = x$ in the z-plane. Find and sketch its image in the w-plane if $w = (1-2i)z + 1$, shade the region in the w-plane corresponding to the half of the z-plane which lies above the line $y = x$ in the z-plane,
2. Find the magnification factor, angle of rotation and the point to which the origin in the z-plane is translated by the linear transformation $w = az + b$ when:
   (i) $a = 2, b = 1 + i$
   (ii) $a = (1-i)/\sqrt{2}, b = i$
   (iii) $a = 4 + i4\sqrt{3}, b = -i$
   (iv) $-i, b = 3 + i$
3. Sketch the parabola $y = x^2$. By interpreting the transformation $w = \frac{1}{2}(1 - i\sqrt{3})z - 1$ in terms of a magnification, a rotation and a translation, find and sketch the image of the parabola in the w-plane.
4. Given that $w = (3+4i)z + 3$, find and sketch the image in the w-plane of the interior of the circle $|z + 2 - i| < 2$ in the z-plane.

5. Find the linear transformation from the z-plane to the w-plane which produces a magnification of 5, a rotation of $-\pi/3$, and maps $z = 2$ onto $w = 1 - 2i$.
6. A linear transformation maps a triangle in the z-plane with vertices at $z_1 = 1 - i$, $z_2 = 3 - i$ and $z_3 = 2 - 4i$ onto a triangle in the w-plane with corresponding vertices located at $w_1 = \frac{1}{2}(13 + 3i)$, $w_2 = \frac{1}{2}(13 + 21i)$, and $w_3 = 20 + 6i$. Find the transformation, its magnification factor, the rotation it produces and the translation that is involved.
7. Given that $\alpha < \beta$ and δ are real, find the linear transformation that maps the strip $\alpha \leqslant x \leqslant \beta$ in the z-plane onto the strip $0 \leqslant u \leqslant \delta$ in the w-plane such that (i) $w(\alpha) = 0$, and (ii) $w(\alpha) = \delta$. In each case, state how the points in the strip in the z-plane which lie above the real axis map onto the strip in the w-plane.
8. Find the family of curves in the w-plane that are produced when $w = 1/z$ maps the family of straight lines $y = x + k$ from the z-plane to the w-plane.
9. Find the linear transformation with the fixed point $1 + i$ that produces a contraction of $1/3$ and a rotation of $-\pi/3$.
10. Find the linear transformation that maps the interior of the circle $|z| = 1$ onto the interior of the circle $|w - w_0| = R$ in such a way that their centers correspond, and the diameter of the circle in the z-plane along the real axis is mapped to a diameter of the circle in the w-plane that is rotated counterclockwise to make an angle α relative to the real axis in the w-plane.

4.2 Conformal Mapping

To develop the geometrical aspects of complex functions beyond the material in Section 4.1, it becomes necessary to define a conformal mapping. This concept is fundamental to all that follows and is, indeed, the cornerstone of the geometrical approach to analytic functions introduced by Riemann (1826–1866).

Let γ_1 and γ_2 be any two simple arcs in the z-plane, which may always be described in terms of the parameter t by writing

$$\begin{aligned} \gamma_1&: z = Z_1(t) = r_1(t) + is_1(t), \\ \gamma_2&: z = Z_2(t) = r_2(t) + is_2(t), \end{aligned} \tag{4.24}$$

where $r_i(t)$ and $s_i(t)$ for $i = 1, 2$ are real functions of t that are defined in some interval $a \leqslant t \leqslant b$. Let us further suppose that these arcs intersect at an arbitrary point P located at z_0 in the z-plane corresponding to $t = t_0$, so that $z_0 = Z_1(t_0) = Z_2(t_0)$. Then, as t increases, the respective points $Z_1(t)$ and $Z_2(t)$ move along the arcs γ_1 and γ_2 in the manner indicated by the arrows in Figure 4.12(a). This defines a natural sense of direction or *orientation* along each arc called the *positive sense* along the arc. In general, curves with such an associated sense of direction are called *oriented (directed) curves*. For example, $z = R(\cos t + i\sin t)$ for $R > 0$, $0 \leqslant t \leqslant \pi/2$ is a directed curve in the form of an arc of a circle of radius R centered on the origin and lying in the first quadrant, with the positive sense in the counterclockwise direction.

The angle α between the two oriented arcs γ_1 and γ_2 at P in Figure 4.12(b) is defined as the angle between the two corresponding oriented tangents T_1 and T_2 at P, with α chosen so that it lies in the interval $0 \leqslant \alpha \leqslant \pi$.

Let $w = f(z)$ be an analytic function that maps the oriented arcs γ_1 and γ_2 intersecting at the point P located at z_0 in the z-plane onto their images the oriented arcs Γ_1 and Γ_2 in the w-plane intersecting at the point P′ located at $w_0 = f(z_0)$. The function $w = f(z)$ is said to give a *conformal mapping* of the z-plane onto the w-plane if the angle α between the oriented arcs γ_1 and γ_2 at an arbitrary point P in the z-plane is the same as the angle between the oriented image arcs Γ_1 and Γ_2 at the corresponding point P′ in the w-plane, and the sense of the angle between these arcs is also preserved. Here, by the preservation of the *sense* of an angle under a mapping, we mean that if at P the tangent T_2 is obtained from the tangent T_1 by a counterclockwise rotation through the angle α, then at P′ the tangent T_2' is obtained from T_1' in precisely the same manner. Thus a conformal mapping is a mapping that preserves angles between intersecting curves together with the sense in which the angle is measured. We prove that an analytic function $w = f(z)$ has the property that it maps one complex plane conformally onto another at all points other than those where $f'(z) = 0$.

To accomplish this task, we first assume the functions $Z_1(t)$ and $Z_2(t)$ to be continuously differentiable so that the arcs γ_1 and γ_2 along which we will need to compute derivatives are smooth curves with smoothly changing tangents. This will be the case if the functions $r_i(t)$ and $s_i(t)$ in Equation (4.24) are themselves continuously differentiable. Using a vector representation of a complex quantity, consideration of Figure 4.13 shows that the vector

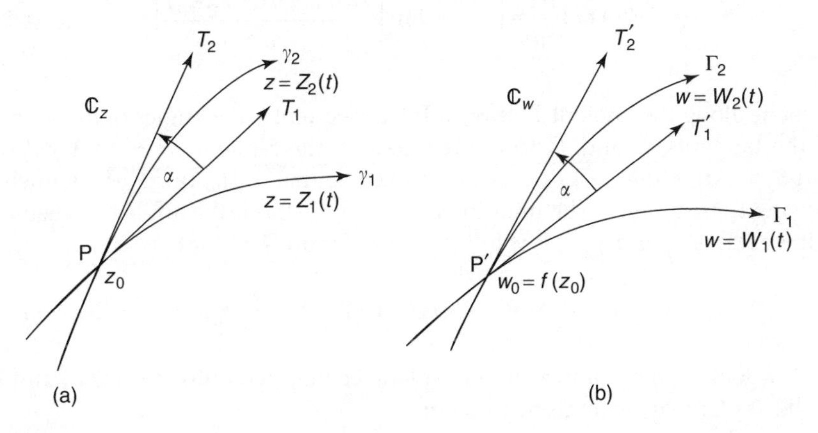

FIGURE 4.12
Conformal mapping of γ_1 and γ_2 onto Γ_1 and Γ_2 by $w = f(z)$.

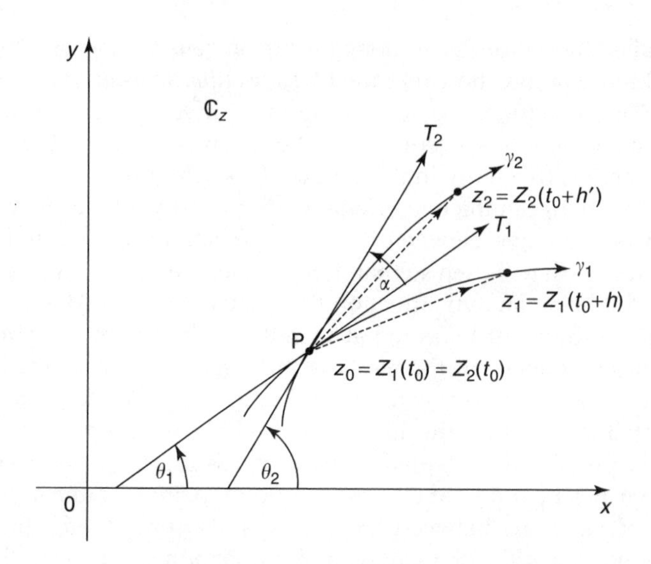

FIGURE 4.13 Chords drawn from P to points on γ_1 and γ_2.

$Z_1(t_0 + h) - Z_1(t_0)$ is a chord drawn from $z_0 = Z_1(t_0)$ to $Z_1(t_0 + h)$ on γ_1. Because $h \to 0$, so the chord approaches the direction of the tangent T_1 at P, and thus the vector derivative given by

$$\dot{Z}_1(t_0) = \left(\frac{dZ_1}{dt}\right)_{t=t_0} = \lim_{h\to 0}\left(\frac{Z_1(t_0 + h) - Z_1(t_0)}{h}\right) \tag{4.25}$$

must lie along the tangent T_1 to γ_1 at P. The same form of argument shows that the vector derivative

$$\dot{Z}_2(t_0) = \left(\frac{dZ_2}{dt}\right)_{t=t_0} = \lim_{h\to 0}\left(\frac{Z_2(t_0 + h) - Z_2(t_0)}{h}\right) \tag{4.26}$$

must lie along the tangent T_2 to γ_2 at P. Consequently the angles of inclination of the tangents T_1 and T_2 to the real axis in the z-plane are $\theta_1 = \operatorname{Arg}\dot{Z}_1(t_0)$ and $\theta_2 = \operatorname{Arg}\dot{Z}_2(t_0)$, respectively, provided neither $\dot{Z}_1(t_0)$ nor $\dot{Z}_2(t_0)$ vanishes causing these angles to be undefined. Therefore, if α is the angle between the oriented arcs γ_1 and γ_2 at P, it follows from Figure 4.13 that

$$\alpha = \theta_2 - \theta_1 = \operatorname{Arg}\dot{Z}_2(t_0) - \operatorname{Arg}\dot{Z}_1(t_0). \tag{4.27}$$

Now let the arcs γ_1 and γ_2 in the z-plane be mapped onto the arcs Γ_1 and Γ_2 in the w-plane by an analytic function

$$w = f(z). \tag{4.28}$$

Then since $z = Z_1(t)$ on γ_1, the image Γ_1 of γ_1 in the w-plane is given by the composite function $w = W_1(t) = f[Z_1(t)]$, while the image Γ_2 of γ_2 in the w-plane is given by the composite function $w = W_2(t) = f[Z_2(t)]$. Applying the chain rule for differentiation of these composite functions we arrive at the results

$$\dot{W}_1(t_0) = \left(\frac{df}{dZ_1}\frac{dZ_1}{dt}\right)_{t=t_0} = \left(\frac{df}{dZ_1}\right)_{t=t_0} \dot{Z}_1(t_0), \tag{4.29}$$

and

$$\dot{W}_2(t_0) = \left(\frac{df}{dZ_2}\frac{dZ_2}{dt}\right)_{t=t_0} = \left(\frac{df}{dZ_2}\right)_{t=t_0} \dot{Z}_2(t_0). \tag{4.30}$$

In Equation (4.29) $(df/dZ_1)_{t=t_0}$ is the derivative of $f(z)$ at $t = t_0$ computed by taking the limit along the smooth arc γ_1, while in Equation (4.30) $(df/dZ_2)_{t=t_0}$ is the derivative of $f(z)$ at $t = t_0$ computed by taking the limit along the smooth arc γ_2. However, as $f(z)$ is assumed to be analytic, the derivative is independent of the path along which the limit defining the derivative is taken, and depends only on the point z_0 in question. Thus we have $(df/dZ_1)_{t=t_0} = (df/dZ_2)_{t=t_0} = f'(z_0)$, and therefore Equations (4.29) and (4.30) may be written

$$\dot{W}_1(t_0) = f'(z_0)\dot{Z}_1(t_0), \tag{4.31}$$

and

$$\dot{W}_2(t_0) = f'(z_0)\dot{Z}_2(t_0). \tag{4.32}$$

The form of argument leading to Equations (4.24) and (4.25) also shows that the vector $\dot{W}_1(t_0)$ lies along the tangent T_1' to Γ_1 at P′ and that the vector $\dot{W}_2(t_0)$ lies along the tangent T_2' to Γ_2 at P′. Consequently the angle between Γ_1 and Γ_2 at P′ is the angle between the complex vectors representing these derivatives.

Taking the arguments of Equations (4.31) and (4.32), setting $\phi_1 = \operatorname{Arg} \dot{W}_1(t_0)$, $\phi_2 = \operatorname{Arg} \dot{W}_2(t_0)$, and using the fact that the argument of a product equals the sum of the respective arguments gives

$$\phi_1 = \operatorname{Arg} f'(z_0) + \theta_1 \quad \text{and} \quad \phi_2 = \operatorname{Arg} f'(z_0) + \theta_2, \tag{4.33}$$

provided $f'(z_0) \neq 0$, because then $\operatorname{Arg} f'(z_0)$ becomes indeterminate. Subtraction of these equations, coupled with the use of Equation (4.27) brings us to the fundamental result

$$\phi_2 - \phi_1 = \theta_2 - \theta_1 = \alpha. \tag{4.34}$$

This establishes our claim that an analytic function gives rise to a conformal mapping because the sense in which the corresponding angles θ and ϕ are measured is the same. To justify this last remark it is only necessary to notice that just as $\theta_2 - \theta_1$ is the angle between γ_2 and γ_1 at P, so also is $\phi_2 - \phi_1$ the angle between Γ_2 and Γ_1 at P′ while the point z_0 is arbitrary apart from the requirement that $f'(z_0) \neq 0$. The geometrical interpretation of Equation (4.34) is shown in Figure 4.14 and we have thus proved the conformal mapping theorem.

THEOREM 4.2.1 Conformal Mapping Theorem
The analytic function $w = f(z)$ maps the z-plane conformally onto the w-plane at all points other than those where the derivative $f'(z)$ is zero. ◆

Example 4.2.1 The Linear Transformation
So far simple geometrical arguments have been used to show that the linear transformation $w = az + b$ $(a \neq 0)$ preserves geometrical similarity of shape when mapping from the z-plane to the w-plane. However, as w is analytic and $dw/dz = a \neq 0$, the same result is established for all $z \in \mathbb{C}$ by virtue of Theorem 4.2.1. ◇

Example 4.2.2 The Mapping $w = z^2$
Applying Theorem 4.2.1 to the mapping $w = z^2$ studied in Section 4.1 establishes that this transformation maps the z-plane conformally onto the w-plane for all z other than $z = 0$ at which point $dw/dz = 2z = 0$. The preservation of angles between intersecting lines in the z-plane and their images in the w-plane has already been remarked upon, and a comparison of the orientation of the points P, Q, R, and S and their images P*, Q*, R*, and S* in Figure 4.7 confirms that the *sense* of the angles at these corners has also been preserved. ◇

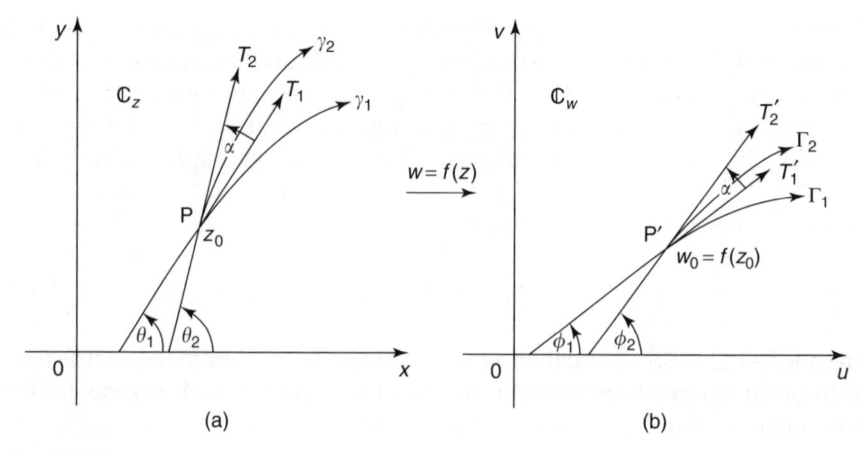

FIGURE 4.14
Geometrical interpretation of Equation (4.34).

Take z and z_0 to be neighboring points in the z-plane, with $f(z)$ and $f(z_0)$ their images in the w-plane under the mapping $w = f(z)$, where $f(z)$ is an analytic function. Then the *linear scale factor* involved in the mapping $w = f(z)$ involved when mapping distinct points is $|f(z) - f(z_0)|/|z - z_0|$. In the limit as $z \to z_0$ it follows from the definition of a derivative that the *linear scale factor* $\rho(z_0)$ for the mapping at z_0 is

$$\rho(z_0) = \lim_{z \to z_0} \left| \frac{f(z) - f(z_0)}{z - z_0} \right| = |f'(z_0)| . \tag{4.35}$$

This depends only on the function $f(z)$ and the point z_0, and so shows that in a suitably small neighborhood of z_0 the linear scale factor is constant and equal to $\rho(z_0) = |f'(z_0)|$. The *area scale factor* in a suitably small neighborhood of z_0 is $\rho^2(z_0)$.

The rotation produced by a conformal mapping $w = f(z)$ at z_0 follows directly from Equation (4.33), which shows that an arc in the z-plane is rotated through the angle $\operatorname{Arg} f'(z_0)$ when it is mapped onto the w-plane. In summary, in some suitably small neighborhood of z_0, a conformal mapping onto the w-plane involves a uniform scale change by the factor $\rho(z_0) = |f'(z_0)|$ and a rotation through the angle $\operatorname{Arg} f'(z_0)$.

Later, when we come to consider solving what are called boundary value problems for the Laplace equation, we see the role of conformal mapping. In brief, it replaces solving the equation in a region of complicated shape by solving an equivalent problem is a much simpler shape. Once this has been accomplished, the solution of the simpler problem can then be transformed back to provide the solution of the original more complicated problem. For this to be successful and to yield a unique solution, as is usually required of solutions to physical problems, the mapping between the planes must be one-one. In view of this, we now look more closely at this aspect of mappings.

The conformal mapping produced by the analytic function $w = f(z)$ implies the real variable transformation from the variables (x, y) to the variables (u, v) defined by

$$u = u(x, y) \quad \text{and} \quad v(x, y). \tag{4.36}$$

In general, although we shall consider transformations such as those in Equation (4.36) that map a point (x_0, y_0) onto a *unique* point (u_0, v_0), where $u_0 = u(x_0, y_0)$ and $v_0 = v(x_0, y_0)$, it is not always true that the inverse mapping from (u, v) to (x, y) is unique. However, the *inverse function theorem* from calculus tells us that if the Jacobian of the transformation $\partial(u, v)/\partial(x, y)$ is continuous and nonvanishing at (x_0, y_0), then in a suitably small neighborhood of (u_0, v_0) a unique inverse transformation occurs

$$x = x(u, v) \quad \text{and} \quad y = y(u, v), \tag{4.37}$$

such that $x_0 = x(u_0, v_0)$ and $y_0 = y(u_0, v_0)$. This in turn defines in this neighborhood of (u_0, v_0) a unique complex function $z = g(w)$ that is the *local inverse* of the function $w = f(z)$. Here the term *local* is used in the sense that the inverse function $g(z)$ is valid only in some neighborhood of the point (u_0, v_0) in the w-plane. This is in contrast to the term *global* which, if it were to be used in this context, would mean that the inverse function holds for all points in the w-plane.

By definition, the Jacobian of transformation in Equation (4.36) is

$$\frac{\partial(u,v)}{\partial(x,y)} = \begin{vmatrix} \frac{\partial u}{\partial x} & \frac{\partial u}{\partial y} \\ \frac{\partial v}{\partial x} & \frac{\partial v}{\partial y} \end{vmatrix} = \frac{\partial u}{\partial x}\frac{\partial v}{\partial y} - \frac{\partial u}{\partial y}\frac{\partial v}{\partial x}. \tag{4.38}$$

Because $f(z)$ is analytic, it follows from the Cauchy–Riemann equations and from Equation (4.26) that

$$\frac{\partial(u,v)}{\partial(x,y)} = \left(\frac{\partial u}{\partial x}\right)^2 + \left(\frac{\partial v}{\partial x}\right)^2 = |f'(z)|^2. \tag{4.39}$$

Thus the condition for the existence of a local inverse in some neighborhood of $w_0 = f(z_0)$ is seen to be that $f'(z_0) \neq 0$.

The derivative of the inverse $z = g(w)$ is most easily found by differentiation of the composite expression $z = g[f(z)]$, which merely states that if $z = g(w)$, then $w = f(z)$. Differentiation gives

$$1 = g'[f(z)]f'(z) = g'(w)f'(z),$$

so at $z = z_0$,

$$g'(w_0) = 1/f'(z_0). \tag{4.40}$$

This has established our next theorem.

THEOREM 4.2.2 A Condition for the Local Inverse to Exist
Let z_0 be any point at which $w = f(z)$ maps the z-plane conformally onto the w-plane. Then, provided $f'(z_0) \neq 0$, $w = f(z)$ has a local inverse $z = g(w)$ in a neighborhood of $w_0 = f(z_0)$, at which point $g'(w_0) = 1/f'(z_0)$. ◆

It must again be stressed that this is a local and not a global result. The fact that $f'(z)$ does not vanish in a region does *not* ensure the existence of a unique inverse in the region is well illustrated by the following example.

Example 4.2.3 A Local Inverse Does Not Imply a Global Inverse
The function $f(z) = e^z$ is an entire function with $f'(z) = e^z$. The Jacobian is given by $|f'(z)|^2 = |e^x e^{iy}| = e^{2x} \neq 0$ for any $z \in \mathbb{C}_z$, so by Theorem 4.2.2 a local inverse exists at each point $w \in \mathbb{C}_w$, where $w = f(z)$. However a global inverse does not exist because for any z_0, all of the points $z = z_0 + 2n\pi i$ for $n = 0, \pm 1, \pm 2, \ldots$ have the same image $w_0 = e^{z_0}$. ◇

The conformality of a mapping $w = f(z)$, where $f(z)$ is a nonconstant analytic function, has been seen to fail at a point where $f'(z_0) = 0$. Such points are called *critical points* of the mapping and, for example, $z = 0$ is the only critical point of the mapping $w = z^2$.

We have already seen that this function maps the ray $\theta = \theta_0$, $r > 0$ drawn through the critical point (the origin) in the z-plane onto the ray $\phi = \phi_0$, $\rho > 0$ drawn through the origin in the w-plane. Thus a wedge shaped region with interior angle α in the z-plane with its vertex at the critical point $z = 0$, is mapped by $w = z^2$ onto a similarly located wedge shaped region in the w-plane, but with interior angle 2α. The conformality is seen to be lost at the critical point $z = 0$, where in this case a doubling of the angle between rays takes place, though elsewhere conformality is preserved (see Figure 4.7).

Similar reasoning applies to the mapping $w = z^n$, which also has $z = 0$ as its only critical point, though in this case a wedge shaped region with internal angle α is mapped onto a similar wedge shaped region with internal angle $n\alpha$.

The reason for the difference in the way the same angle α between rays drawn through a critical point is mapped in these two cases forms the result of the next theorem. To examine the anomalous behavior of a mapping at a critical point of an analytic function, it will be necessary use the result of Theorem 2.5.2, to the effect that an analytic function has derivatives of all orders. This means that in a neighborhood of a point z_0, an analytic function has derivatives $f^{(n)}(z_0)$ of all orders, and so may be represented there for suitably small $|z - z_0|$ by a complex Taylor series expanded about z_0.

Thus, it follows that if $f(z)$ is analytic at z_0, but is such that its first $m - 1$ derivatives vanish at z_0, so that z_0 is a critical point because $f'(z_0) = 0$, in a neighborhood of z_0 the function $f(z)$ must have the Taylor series representation

$$f(z) = f(z_0) + (z - z_0)^m \frac{f^{(m)}(z_0)}{m!} + (z - z_0)^{m+1} \frac{f^{(m+1)}(z_0)}{(m+1)!} + \cdots. \quad (4.41)$$

So in a neighborhood of the critical point z_0 we may set

$$f(z) - f(z_0) = (z - z_0)^m h(z),$$

where the function

$$h(z) = (z - z_0)^m \frac{f^{(m)}(z_0)}{m!} + (z - z_0)^{m+1} \frac{f^{(m+1)}(z_0)}{(m+1)!} + \cdots$$

is analytic at z_0, with $h(z_0) \neq 0$. Writing $w = f(z)$, and taking the argument of Equation (4.41) gives

$$\arg(w - w_0) = \mathrm{Arg}[f(z) - f(z_0)] = m\mathrm{Arg}(z - z_0) = \mathrm{Arg}h(z). \quad (4.42)$$

Now consider the oriented arcs γ_1 and γ_2 intersecting at an angle α at a point P located at the critical point z_0, as shown in Figure 4.15(a), and their images Γ_1 and Γ_2 intersecting at an angle Φ at the point P′ located at $w_0 = f(z_0)$, as shown in Figure 4.15(b). Then as $z \to z_0$ along the arc γ_1 we have $\lim_{z\to z_0} \mathrm{Arg}(z - z_0) = \theta_1$, while as $z \to z_0$ along the arc γ_2 we have $\lim_{z\to z_0} \mathrm{Arg}(z - z_0) = \theta_2$. In each case we have the result

$$\lim_{z\to z_0} \mathrm{Arg}h(z) = \mathrm{Arg}[f^{(m)}(z_0)/m!] = \mathrm{Arg}\, f^{(m)}(z_0).$$

Correspondingly, as $z \to z_0$ along γ_1, so $w \to w_0$ along Γ_1 with $\lim_{w\to w_0} \mathrm{Arg}(w - w_0) = \phi_1$, and similarly, as $z \to z_0$ along γ_2, so $w \to w_0$ along Γ_2 with $\lim_{w\to w_0} \mathrm{Arg}(w - w_0) = \phi_2$.

Subtracting the limit of Equation (4.42) as $z \to z_0$ along γ_1 from the limit of Equation (4.42) as $z \to z_0$ along γ_2 we find that

$$\phi_2 - \phi_1 = \Phi = m(\theta_2 - \theta_1) = m\alpha, \quad (4.43)$$

showing that $\Phi = m\alpha$.

This result, that explains the nature of the anomalous mapping at a critical point, will now be stated as a theorem.

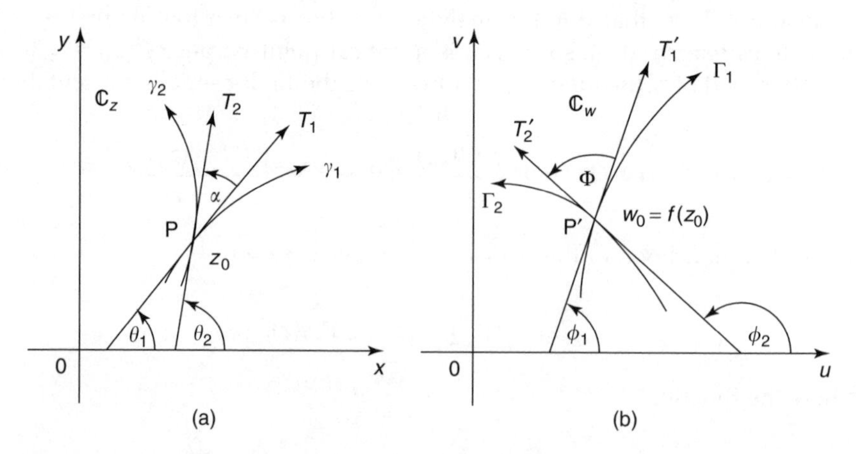

FIGURE 4.15
Anomalous mapping at a critical point by $w = f(z)$.

THEOREM 4.2.3 Mapping at a Critical Point
Let the analytic function $f(z)$ have a critical point at z_0, and suppose also that $f^{(1)}(z_0) = f^{(2)}(z_0) = f^{(3)}(z_0) = \cdots = f^{(m-1)}(z_0) = 0$, for $m > 2$, but that $f^{(m)}(z_0) \neq 0$. Then the mapping $w = f(z)$ will increase the angle between any two oriented arcs that intersect at the critical point z_0 by the factor m, while preserving the sense of the angle. ◆

Example 4.2.4 Mapping at the Critical Point of $w = z^n$
The function $f(z) = z^n$ is an analytic function for positive integral n, and has the single critical point $z = 0$. As $0 = f^{(1)}(0) = f^{(2)}(0) = \cdots = f^{(n-1)}(0)$, but $f^{(n)}(0) = n!$, the conditions of Theorem 4.2.3 are satisfied. Thus any two arcs intersecting at an angle α at $z = 0$ will be mapped onto corresponding arcs through the origin in the w-plane that intersect at an angle $n\alpha$. At all other points the mapping will be conformal. ◇

The fact that the sense of an angle is also preserved at a critical point has an important consequence, as will now be explained. Let P be a point on a simple closed curve γ in the z-plane, which may pass through critical points of $f(z)$. Suppose γ is mapped onto the simple closed-image curve Γ in the w-plane by the analytic function $w = f(z)$, with point P′ on Γ the image of P on γ. Then because of the properties of a conformal mapping, it follows that as P moves around γ in the positive sense, points to the left of γ when moving in the direction of P will map to points to the left of Γ when moving in the direction of P′.

This property of conformal mapping is always used when deciding how regions (areas) in one plane map onto regions (areas) in another plane. A different though equivalent way of determining how mapped regions correspond involves choosing a convenient test point in one region of interest, and then seeing to which region it is mapped by $w = f(z)$.

These ideas are well illustrated by considering the effect of the inversion mapping $w = 1/z$. Consider Figure 4.16, where the diagram on the left

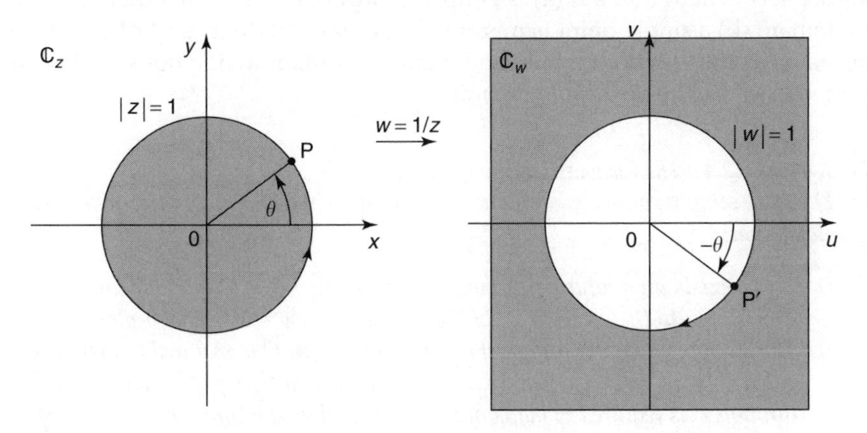

FIGURE 4.16
Mapping of areas by $w = 1/z$.

represents a point P moving in the *counterclockwise* sense around the unit circle $|z| = 1$, in which case points to the left of P as it moves around the unit circle lie *inside* the unit circle. The effect of the inversion mapping $w = 1/z$ on the points inside this unit circle is shown in the diagram on the right, where although the unit circle $|z| = 1$ maps to the unit circle $|w| = 1$, the effect of the inversion mapping is to make point P′ in the diagram on the right move in the opposite sense to that of P, namely in the *clockwise* sense around $|w| = 1$. Thus, in the w-plane, the points that lie to the left of P′ as it moves around $|w| = 1$ in the *clockwise* sense all lie *outside* the unit circle $|w| = 1$.

A question of fundamental theoretical importance is when one region can be mapped onto another region by means of a *unique* conformal mapping. However, when conformal mapping is used to solve boundary value problems for the Laplace equation, as described in Chapter 5, there are often real advantages when the required mapping can be performed in more than one way. Basically, the idea involved is to map an awkwardly shaped region into a more conveniently shaped one in which the original problem is easier to solve. The solution of the initial problem is then obtained by mapping back the simpler solution to the original region.

When more than one mapping is possible it can happen that mapping back in some cases is simpler than in others, in which case the most convenient mapping is the one that is usually used.

The existence of a transformation that maps a given domain onto another one, subject to certain condition, is guaranteed by the *Riemann mapping theorem* that will be stated but not proved, as this would take us beyond the scope of this first account. Although before stating this fundamental theorem we must first make clear what is meant by a *simply connected domain*.

A domain D is said to be simply connected if every simple closed curve in D can be contracted to a single point in D. When expressed intuitively, this definition means that a simply connected domain can contain no holes, cuts or missing points within its interior. This intuitive definition is illustrated in Figure 4.17 where domain (a) is simply connected, while the others are not; in domain (b) a single point is missing; in domain (c) there is a hole; while in domain (d) there is both a hole and a cut. A domain that is not simply connected is said to be *nonsimply connected.*

THEOREM 4.2.4 The Riemann Mapping Theorem
Let D be any simply connected domain that does not comprise all of the z-plane. Then it follows that:

(i) There exists an analytic function $w = f(z)$ that maps D one-one and conformally onto the interior D^* of the unit circle $|w| = 1$ in the w-plane.
(ii) The mapping becomes unique if a given point z_0 in D is required to map to a given point w_0 in D^* (so that $f(z_0) = w_0$) and, in addition, a specified direction through z_0 is required to map onto a specified direction through w_0. ◆

This theorem answers the question of when a unique mapping function exists between specific domains because if each is mapped to a unit circle,

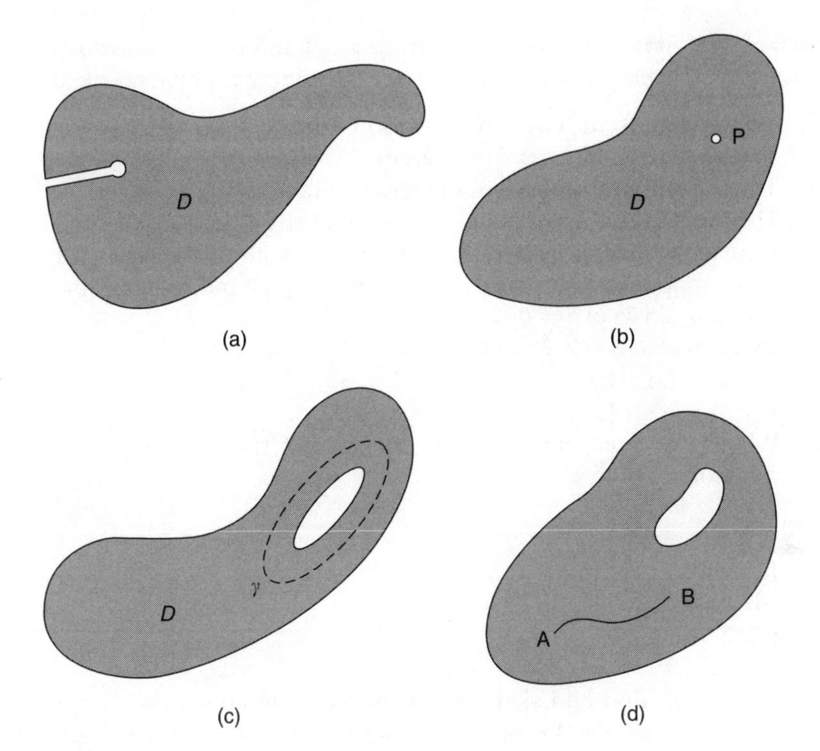

FIGURE 4.17
Domain (a) is simply connected, but domains (b) through (d) are nonsimply connected.

then one can be mapped to the other. However it is only an existence theorem, because it does not say how such a mapping function between given domains can be found. A simple illustration of the need for the conditions in part (ii) of the theorem is provided by considering the mapping of $|z| < 2$ onto $|w| = 1$. For if all that is required is that $z = 0$ is mapped to $w = 0$, then $w = \frac{1}{2}e^{i\alpha}z$ is such a mapping for *any* real α. This mapping introduces a uniform contraction of $\frac{1}{2}$, and a rotation through the angle α. If in addition it is also required that the positive real axis in the z-plane maps to the positive imaginary axis in the w-plane it is necessary to set $\alpha = \pi/2$, when the mapping becomes unique.

Exercises 4.2

In the following cases, sketch the indicated curve in the z-plane and show its orientation by the addition of an arrowhead.

1. $z = t + 4it^2$, for $-2 \leqslant t \leqslant 3$.
2. $z = 1 + 3i + 2e^{it}$, for $-\pi/2 \leqslant t \leqslant \pi$.
3. $z = t + i/t$, for $0 < t < \infty$.
4. $z = a\cosh t + ib\sinh t$, $a > 0$, $b > 0$, $-\infty < t < 0$.
5. $z = a(t + i - ie^{-it})$, $a > 0$ and $0 \leqslant t \leqslant 2\pi$.
6. $z = a(\cos^3 t + i\sin^3 t)$, $a > 0$ and $0 \leqslant t \leqslant 2\pi$.

In each of the following cases, by computing $\dot{z}(t)$ find the angle relative to the positive real axis made by the oriented tangent to the curve at the specified point.

7. The tangent to the curve in Exercise 1 when $t = 1$.
8. The tangent to the curve in Exercise 2 when $t = 0$.
9. The tangent to the curve in Exercise 5 when $t = \pi/4$.
10. The tangent to the curve in Exercise 6 when $t = \pi/4$.
11. For each of the following functions find the scale factor $\rho(z)$ involved in the mapping, and hence its magnitude at the indicated point.
    (a) $w = \sin 2z$ at $z = 0$.
    (b) $w = e^z$ at $z = 2 + i$.
    (c) $w = \operatorname{Ln} z$ at $z = 1 + i\sqrt{3}$.
    (d) $w = \cosh 4z$ at $z = i$.
12. When they exist, find the critical points of the following analytic functions.
    (a) e^z
    (b) $z^2 + 3z + 1$
    (c) $\sin z$
    (d) $\cosh z$
    (e) $z + 1/z$
    (f) ze^z
13. If $w = 1/z$, find and sketch the way the region $0 < r_1 \leqslant r \leqslant r_2 < 1$, $\pi/4 \leqslant \theta \leqslant \pi/3$ in the z-plane is mapped onto the w-plane. Indicate how the corners of the region in the z-plane map to the corners of the image in the w-plane.
14. Prove that the inversion mapping $w = 1/z$ maps the disk $|z - 1| < 1$ in the z-plane onto the region $u > 1/2$ in the extended w-plane. Indicate how any three distinct points on the circle map to the w-plane.

4.3 The Linear Fractional Transformation

This section deals with one of the most useful of the elementary mappings called *the linear fractional transformation* (also the *bilinear transformation* or the *Mobius transformation*). Its usefulness comes from the fact that it can map circles and straight lines onto circles and straight lines, although not necessarily in this order, and in the process it maps the interior of a circle onto either the interior or exterior of another circle or onto a half plane. These and related problems frequently arise in applications, and the flexibility of the linear fractional transformation allows it to be used to full advantage in such circumstances.

Let a, b, c and d be real or complex constants. Then a linear fractional transformation $w = T(z)$ is defined to be an analytic function of the form

$$w = T(z) = \frac{az + b}{cz + d}, \qquad \text{with } ad - bc \neq 0. \tag{4.44}$$

Transformation $T(z)$ defines w uniquely in terms of z, except when $z = -d/c$, or z is the point at infinity. Conversely, the inverse of the transform $z = T^{-1}(w)$ when derived from Equation (4.44) is of the form

$$z = T^{-1}(w) = \frac{-dw + b}{cw - a}, \tag{4.45}$$

is also a linear fractional transformation. It defines z uniquely in terms of w, except when $w = a/c$ or w is the point at infinity. These restrictions on z and w may be removed by working in the extended complex plane and defining the images of $T(-d/c)$ and $T^{-1}(a/c)$ in a suitable manner. To see this, write Equation (4.44) as

$$T(z) = \frac{a + (b/z)}{c + (d/z)},$$

and then define

$$T(\infty) = \lim_{z\to\infty}\left(\frac{a + (b/z)}{c + (d/z)}\right) = \frac{a}{c}, \tag{4.46}$$

so it follows that $T^{-1}(a/c) = \infty$. The same form of argument applied to Equation (4.45) leads to the definition

$$T^{-1}(\infty) = \lim_{w\to\infty}\left(\frac{-d + (b/w)}{c - (a/w)}\right) = -\frac{d}{c}, \tag{4.47}$$

from which it follows that $T(-d/c) = \infty$.

In terms of definitions in Equations (4.46) and (4.47), the transformation $w = T(z)$ then provides a one-one mapping between every point in the extended w-plane and every point in the extended z-plane.

The need for the condition $ad - bc \neq 0$ attached to Equation (4.44) may be seen either by writing $T(z)$ as

$$T(z) = \frac{a}{c} + \left(\frac{bc - ad}{c}\right)\left(\frac{1}{cz + d}\right), \quad \text{with } c \neq 0, \tag{4.48}$$

or by noticing that

$$\frac{dw}{dz} = \frac{ad - bc}{(cz + d)^2}. \tag{4.49}$$

Equation (4.48) shows the necessity of the condition $ad - bc \neq 0$ in order that $T(z)$ does not degenerate and map every point z onto the single point $w = a/c$. If $c = 0$ the transformation reduces to the linear transformation studied earlier. The result in Equation (4.49) shows the condition $bc - ad \neq 0$ to be necessary that the mapping is conformal.

A simple calculation establishes that if S and T are two linear fractional transformations, the composite transformation $w = S[T(z)]$ is also a linear fractional transformation. Of course, the same is true of the composition of any number of linear fractional transformations.

The most important geometrical mapping properties of linear fractional transformations relate to straight lines and circles, and they follow directly from Equation (4.48) when it is written as the sequence of mappings

$$w_1 = cz + d, \quad w_2 = 1/w_1, \quad w = \frac{a}{c} + \left(\frac{bc - ad}{c}\right) w_2. \tag{4.50}$$

Both the first and last of these three mappings is a linear transformation that will map any geometrical shape into a similar shape, while the second mapping is the inversion mapping that has already been shown to map straight lines and circles onto straight lines and circles. Consequently, the composite mapping in Equation (4.44) will map all possible straight lines and circles in the z-plane onto all possible straight lines and circles in the w-plane, though not necessarily in this order. This result forms the substance of the next theorem.

THEOREM 4.3.1 Mapping by T(z) of Straight Lines and Circles
All straight lines and circles in the z-plane will be mapped by any linear fractional transformation onto all straight lines and circles in the w-plane, although not necessarily in this order. ◆

The *fixed points* of Equation (4.44) follow by solving

$$z = \frac{az + b}{cz + d},$$

leading to the equation

$$cz^2 - (a - d)z - b = 0. \tag{4.51}$$

Thus, it follows that a nontrivial linear fractional transformation will have at most two distinct fixed points in the extended complex plane. One of these may be finite and the other coincide with the point at infinity, as in the case when $T(z)$ degenerates to the linear transformation $w = az + b$, with $a \neq 1$. No finite fixed points exist in this last case for $a = 1$ because the transformation then reduces to a uniform translation $w = z + b$. In such a transformation both fixed points may be considered to have coalesced at the point at infinity. Should more than two fixed points be known to exist for a linear fractional transformation, then it must be the identity mapping in which all points are fixed points.

To use the transformation in applications, it is necessary to construct a linear fractional transformation that will map given points in a prescribed manner. Since a circle is completely defined by specifying three distinct points on its circumference, this information should suffice to specify $T(z)$. This might appear to be contradicted by their being four constants a, b, c, and d in the transformation $T(z)$ which are to be determined from these three conditions. The difficulty is resolved once it is observed that the transformation is fully determined by the *three* ratios a/c, b/c and d/c because it can be written as

$$T(z) = \frac{(a/c)z + (b/c)}{z + (d/c)}. \tag{4.52}$$

This is possible as $c \neq 0$, for was this not to be the case $T(z)$ would reduce to a linear transformation.

Rather than use Equation (4.44) directly when seeking particular mappings, we now derive a result that is both simple and completely general in its application. Let z_1, z_2, z_3 and z_4 be any four distinct points in the extended z-plane with the distinct images w_1, w_2, w_3 and w_4 under the transformation $w = T(z)$. When these points are finite we have for $m \neq n = 1, 2, 3, 4$ that

$$w_m - w_n = K(z_m - z_n), \tag{4.53}$$

with $K = (ad - bc)/[(cz_m + d)/(cz_n + d)]$. From this it is easily shown that

$$\frac{(z_1 - z_4)(z_3 - z_2)}{(z_1 - z_2)(z_3 - z_4)} = \frac{(w_1 - w_4)(w_3 - w_2)}{(w_1 - w_2)(w_3 - w_4)} \tag{4.54}$$

which, since it is independent of the coefficients of $T(z)$, is true for all linear fractional transformations. The expression on the left of Equation (4.54) is called the *cross ratio* of the four points z_1, z_2, z_3 and z_4 and is abbreviated by (z_1, z_2, z_3, z_4). The result in Equation (4.54) shows that the cross ratio of the points in the z-plane equals the cross ratio of the corresponding image points in the w-plane, so being independent of the coefficients of $T(z)$ it is *invariant* under any linear fractional transformation.

If one of the points in the z- or w-plane is the point at infinity, the quotient of the factors in which it appears can be replaced by unity. We show this by supposing that $z_3 = \infty$, and then noticing that

$$\lim_{z_3 \to \infty} \frac{(z_1 - z_4)(z_3 - z_2)}{(z_1 - z_2)(z_3 - z_4)} = \lim_{z_3 \to \infty} \frac{(z_1 - z_4)(1 - z_2/z_3)}{(z_1 - z_2)(1 - z_4/z_3)} = \frac{z_1 - z_4}{z_1 - z_2},$$

which can be obtained from the original expression by replacing the quotient $(z_3 - z_2)/(z_3 - z_4) = 1$.

Replacing z_4 and w_4 in Equation (4.54) by the variables z and w, respectively, defines w in terms of z, and hence $w = T(z)$, by means of the given points z_1, z_2, and z_3 by their prescribed images w_1, w_2, and w_3. This forms the next result that is used when deriving specific linear fractional transforms.

THEOREM 4.3.2 Mappings Obtained by Using the Cross Ratio
Let the three distinct points z_1, z_2, and z_3 in the z-plane be assigned the distinct images w_1, w_2, and w_3 in the w-plane. Then the unique linear fractional transformation $w = T(z)$ which accomplishes this mapping is given by solving for w the equation

$$\left(\frac{w - w_1}{w - w_3}\right)\left(\frac{w_2 - w_3}{w_2 - w_1}\right) = \left(\frac{z - z_1}{z - z_3}\right)\left(\frac{z_2 - z_3}{z_2 - z_1}\right).$$

When one of the points in either plane is the point at infinity, the quotient of the factors involving that point is 0 be replaced by 1. ◆

Example 4.3.1 Mapping the Interior of a Circle onto the Exterior of a Circle

Find the linear fractional transformation that maps the points $-i$, 1, and on a circle i in the z-plane onto the respective points 0, $1 + i$, and 2 on a circle in the w-plane. Determine the domain in the w-plane that is the image of the interior of the circle in the z-plane.

SOLUTION
Setting $z_1 = -i$, $z_2 = 1$, $z_3 = i$, and $w_1 = 0$, $w_2 = 1 + i$, $w_3 = 2$ in the result of Theorem 4.3.2 gives

$$\left(\frac{w - 0}{w - 2}\right)\left(\frac{1 + i - 2}{1 + i - 0}\right) = \left(\frac{z + i}{z - i}\right)\left(\frac{1 - i}{1 + i}\right).$$

After simplification, this becomes

$$w = \frac{z + i}{z},$$

which is the required transformation $T(z)$. The points in the z-plane define the circle $|z| = 1$, while the points in the w-plane define the circle $|w - 1| = 1$, both of which are shown in Figure 4.18. The points z_1, z_2, and z_3 lie in a counterclockwise sense around the circle $|z| = 1$, while the image points w_1, w_2, and w_3 lie around the circle $|w - 1| = 1$ in the opposite (clockwise) sense. So as the *interior* of the circle $|z| = 1$ lies to the left when moving from z_1 to z_2 to z_3, so the area to the left when moving from w_1, w_2, and w_3 in the w-plane must lie *outside* the circle $|w - 1| = 1$. These areas are shaded in the diagrams in Figure 4.18. Another way to confirm this is by noticing that the point $z = 1/2$

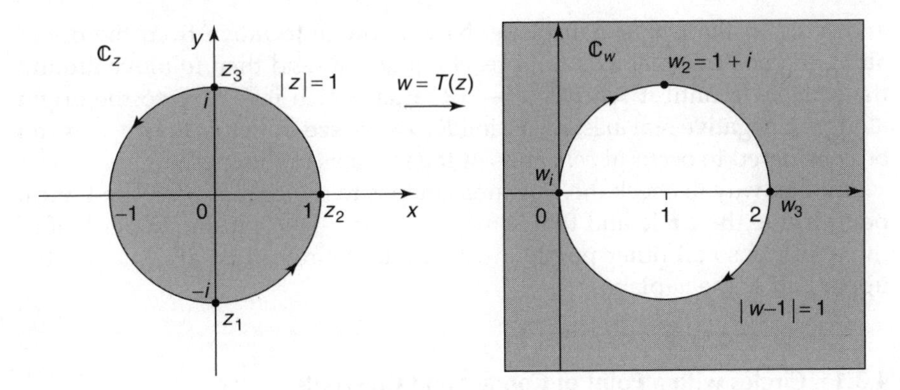

FIGURE 4.18
Mapping the interior of a circle onto the exterior of a circle by the transformation $w = T(z) = (z + i)/z$.

inside the circle in the z-plane maps to the point $w = 1 + 2i$ *outside* the circle in the w-plane. ◇

Example 4.3.2 Mapping the Interior of a Circle onto a Half Plane

Find the linear fractional transformation that maps the points i, $-i$, and 1 on the unit circle in the z-plane onto the respective points 0, 1, and ∞ on the real axis in the w-plane. Determine the domain in the w-plane that is the image of the interior of the circle in the z-plane.

SOLUTION
Setting $z_1 = i$, $z_2 = -i$, $z_3 = 1$, and $w_1 = 0$, $w_2 = 1$, $w_3 = \infty$ in the cross-ratio expression in Theorem 4.3.1 and replacing the factor involving a quotient of the point at infinity by unity, we find that

$$\left(\frac{w-0}{1-0}\right) \cdot 1 = \left(\frac{z-i}{z-1}\right)\left(\frac{-i-1}{-i-1}\right),$$

which simplifies to

$$w = \left(\frac{1-i}{2}\right)\left(\frac{z-i}{z+i}\right).$$

This is the required transformation $T(z)$. Moving counterclockwise around the circle in the z-plane from z_1 to z_2 and then to z_3, the interior of the circle lies to the left. Moving from the image points w_1 to w_2 and then to w_3, the area to the left is seen to be the upper-half of the w-plane. Notice that the point at infinity in the w-plane occurs at each end of the real axis.

The explanation for this is as follows. Consider the points $w = \pm R$ $(R > 0)$ on the real axis to be joined by a semicircle of radius R centered on the origin and lying in the upper half-plane. Now allow w to move from the origin along the positive real axis until reaching $w = R$, and then to move around the semicircle until it reaches $w = -R$, after which it returns to the origin along the negative real axis. By letting $R \to \infty$ we see from this that $w_3 = \infty$ can be considered to occur at *both* ends of the real axis in the w-plane.

Another way to check the way areas map is to notice that $z = 0$ is a typical point inside the circle and that it maps to $w = -\frac{1}{2} + \frac{1}{2}i$ in the upper-half of the w-plane, so all other points inside the unit circle must also map to the upper half of the w-plane. ◇

4.3.1 Circles with a Point of Contact and Crescents

The last example showed that if one of the three given points defining a circle in the z-plane is mapped to the point at infinity by a linear fractional transformation, the circle is mapped onto a straight line in the w-plane. This simple observation can be used to advantage in applications necessitating the mapping of circles with a point of contact, and overlapping circles defining crescent-shaped regions called *lunes*. Examples of each of these types of domain are shown in Figure 4.19(a) to (d).

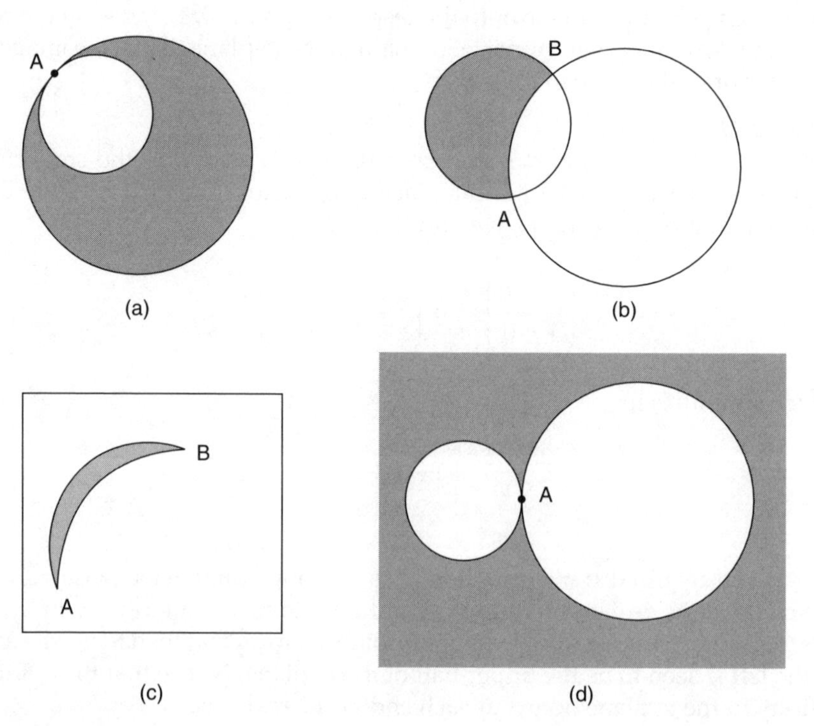

FIGURE 4.19
Circles with a point of contact and lunes (crescents).

In Figure 4.19(a) to (d), the circles have a common point of contact at A, and in (a) and (d) their tangents at the point of contact actually coincide. The contact is said to be *internal* in case (a) and *external* in case (d). If one of these circles is mapped by a linear fractional transformation in such a way that the image of point A on the circle is mapped to infinity, it will be mapped onto a straight line. When the two tangents coincide at A, the conformal nature of the mapping will cause the other circle to be mapped onto a parallel straight line. Thus in each of the two cases (a) and (d), the shaded regions will be mapped onto the infinite strip between the two parallel straight lines.

The situation is different in cases (b) and (c), and if one of the circles is mapped by a linear fractional transformation in such a way that a cusp, say at A, is mapped to the point at infinity, the circle will be mapped onto a straight line passing through the image of B. The other circle is mapped onto a different straight line which will also pass through the image of B. Therefore, in these two cases the shaded domains will map onto a wedge-shaped domain with its vertex at the image of B.

It will be seen later that the mapping $w = e^z$ maps an infinite strip onto a half plane. Consequently, as Example 4.3.2 has shown how the interior of a circle can be mapped onto a half plane, it follows that each shaded domain in Figure 4.19 may be mapped onto the interior of a circle.

Example 4.3.3 Mapping Circles with Exterior Contact onto a Strip
Construct a linear fractional transformation that will map the exterior of the two circles $|z - 1| = 1$ and $|z + 1| = 1$ onto a strip in the w-plane in such a way that the points $z_1 = 2$, $z_2 = 1 + i$, and $z_3 = 0$ on the first circle map onto the respective points $w_1 = 0$, $w_2 = 1$, and $w_3 = \infty$ in the w-plane.

SOLUTION
Substituting into the cross-ratio formula in Theorem 4.3.1 and replacing the quotient involving w_3 by 1 gives

$$\left(\frac{w-0}{1-0}\right)\cdot 1 = \left(\frac{z-2}{z-0}\right)\left(\frac{1+i-0}{1+i-2}\right),$$

and after simplification this becomes

$$w = i\left(\frac{2-z}{z}\right).$$

The representative points $z_4 = -1 + i$, $z_5 = -2$, and $z_6 = -1 - i$ on the circle on the left in Figure 4.20 are mapped onto $w_4 = 1 - 2i$, $w_5 = -2i$, and $w_6 = -1 - 2i$, respectively, all of which lie in the line $\text{Im}\{w\} = -2$ in the w-plane. The usual form of argument shows that the shaded domain *outside* the two circles maps onto the *interior* of the strip shown in the w-plane, thus, the exterior of the

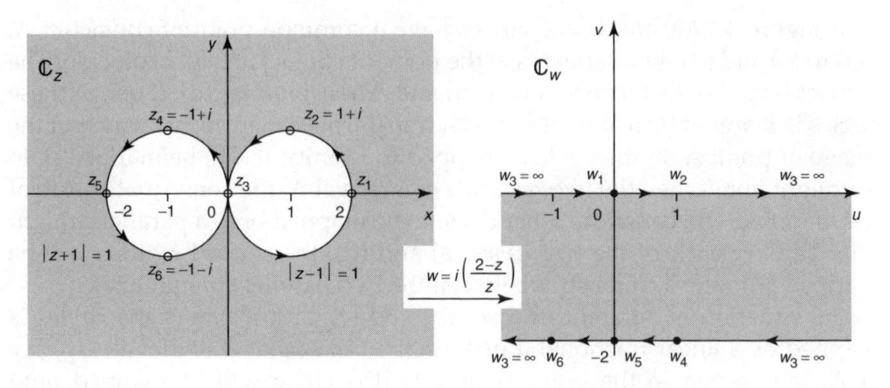

FIGURE 4.20
Mapping the exterior of circles with an exterior point of contact onto a strip by $w = i(2 - z)/z$.

two circles makes exterior contact at the origin in the z-plane map onto the interior of the infinite strip $-2 < v < 0$ in the w-plane. The fact that the point $z = i$ in the z-plane also shows that point *outside* the circles maps to the point $w = 2 - i$ that is *inside* the infinite strip in the w-plane. Remember that the conformal nature of a mapping means that if one representative point in a domain D of interest in the z-plane maps to a domain D^* in the w-plane, then all other points in D will follow it by mapping to points in D^*.

Here again the point w_3 at infinity is shown at the ends of the strip in the w-plane. The ends may be considered to be joined at infinity, making it possible to encompass the entire strip by moving along one boundary, and then back along the other one. ◇

Example 4.3.4 Mapping a Crescent onto a Wedge

Construct a linear fractional transformation that will map the crescent-shaped domain (lune) contained between the circles $|z| = 1$ and $|z - 1| = \sqrt{2}$ onto the w-plane in such a way that the points $z_1 = 1$, $z_2 = i$, and $z_3 = -i$ on the unit circle map onto the respective points $w_1 = -1$, $w_2 = 0$, and $w_3 = \infty$, respectively.

SOLUTION

Substituting into the cross-ratio formula in Theorem 4.3.1, and replacing the quotient of factors involving the point $w_3 = \infty$ by 1, gives

$$\left(\frac{w+1}{0+1}\right) = \left(\frac{z-1}{z+i}\right)\left(\frac{i+i}{i-1}\right),$$

and after simplification this becomes

$$w = i\left(\frac{i-z}{i+z}\right).$$

The crescent shaped domain (lune) shown as the shaded area on the left in Figure 4.21 is mapped to the wedge-shaped region in the w-plane with its vertex at $w = 0$, where the usual arguments show how the areas correspond. This can also be confirmed by noticing that the point $z = -3/4$ inside the lune maps to the point $w = \frac{1}{25}(24 + 7i)$ inside the wedge-shaped domain.

That the angle of the wedge is $\pi/4$ can be seen from purely geometrical considerations, because the internal angle between the two tangents at a cusp in the z-plane is $\pi/4$ and this angle and its sense is preserved by the conformal transformation. ◇

4.3.2 Symmetry and Linear Fractional Transformations

In some applications mapping problems arise involving a combination of straight lines and circles in which a certain symmetry exists. In such cases it is natural that a linear fractional transformation $w = T(z)$ should be used to map the boundaries onto circles; however, it is also desirable that the mapping preserves some aspects of the symmetry. Typical examples of such situations are shown in Figure 4.22(a) and (b); in each of which the straight line L passing through A and B is a line of symmetry.

In order to make use of this symmetry when constructing a linear fractional transformation is necessary to be clear what is meant by symmetry with respect to a straight line L, and then to generalize the notion of to symmetry with respect to a circle. Points A and B are said to be *symmetric with respect to the straight line L* in Figure 4.23(a) when AB is the perpendicular bisector of L so that AP = PB. The points A and B are said to be *symmetric with respect to the circle* γ, or to be *inverse points*, if they are derived by means of the construction shown in Figure 4.23(b). It then follows from similar triangles that A and B are such that

$$\text{OA} \times \text{OB} = \text{OC}^2,$$

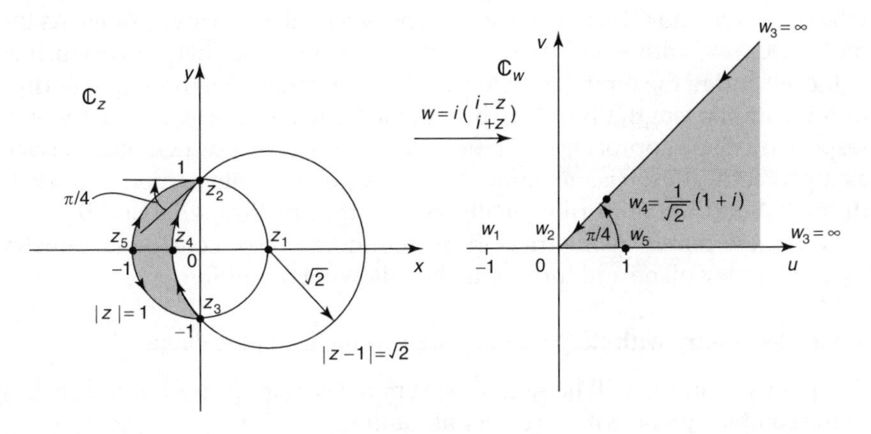

FIGURE 4.21
Mapping a crescent onto a wedge by $w = i(i - z)/(i + z)$.

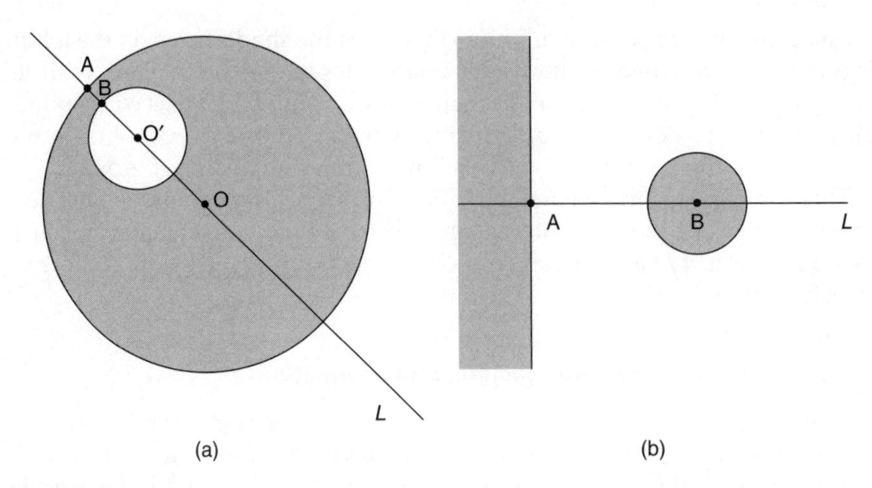

FIGURE 4.22
Domains possessing a line of symmetry.

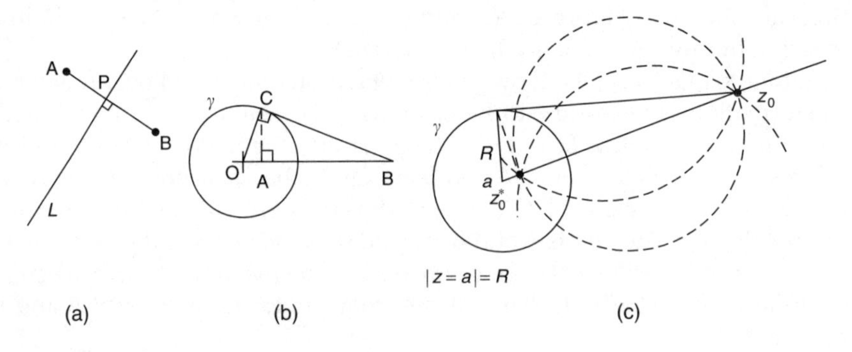

FIGURE 4.23
Symmetry with respect to a straight line L and a circle γ.

where OA, OB and OC are the distances between the respective points. As the radius $\text{OC} \to \infty$, with A and B fixed, so the center O of the circle moves further to the left and in the limit the circle γ tends to a straight line through P normal to AB, with the length $\text{PB} \to \text{AP}$. Thus, in the limit as $\text{OC} \to \infty$, symmetry with respect to a circle approaches symmetry with respect to a straight line. It is left as a geometrical exercise to show that at its point of intersection, any circle through A and B cuts γ orthogonally, as does the line through A and B.

With these purely geometrical ideas in mind, we now consider symmetry in the complex plane and formulate the following definition.

4.3.3 Symmetry with Respect to a Circle in the Complex Plane

The points z_0 and z_0^* will be said to be symmetric with respect to the circle γ in the complex plane with its center at a and radius R if

$$(z_0 - a)(\overline{z}_0^* - \overline{a}) = R^2.$$

To motivate this definition, notice that the equation of γ, usually written $|z - a| = R$, may also be written as

$$(z - a)(\overline{z} - \overline{a}) = R^2. \tag{4.55}$$

A direct comparison of the definition of symmetry with respect to circle γ with Equation (4.55) then shows that z_0 and z_0^* must lie on a straight line through the center O of γ. So from our previous geometrical arguments, if z_0 and z_0^* are symmetric with respect to the circle γ, the line joining them will be orthogonal to γ, as will every circle passing through the points z_0 and z_0^* [Figure 4.23(c)].

The concept of symmetry is given meaning in the extended complex plane if we define the points a (the center of the circle) and ∞ to be symmetric with respect to γ. The main result concerning symmetry in connection with the general linear fractional transform $w = T(z)$ is stated in the next theorem.

THEOREM 4.3.3 Preservation of Symmetry by $W = T(z)$
Let z_0 and z_0^* be any two points in the z-plane that are symmetric with respect to a given straight line L or circle γ and let the z-plane be mapped onto the w-plane by a linear fractional transformation $w = T(z)$. Then if Λ and Γ are the respective images of L and γ, under $w = T(z)$, the points $w_0 = T(z_0)$ and $w_0^* = T(z_0^*)$ are symmetric with respect to Λ and Γ.

PROOF

To establish this result it is necessary to show that the straight line through w_0 and w_0^*, and any circle C_w that passes through this pair of points is orthogonal to Γ. First we notice that the inverse transformation $z = T^{-1}(w)$ is itself a linear fractional transformation with $T^{-1}(w_0) = z_0, T^{-1}(w_0^*) = z_0^*$, and γ is the pre-image of Γ. Because $z = T^{-1}(w)$ is a linear fractional transformation, the pre-image C_z of C_w is automatically a straight line or circle passing through the pair of points z_0 and z_0^*. However, these are symmetric with respect to γ, and therefore, C_z is orthogonal to γ. The conformal nature of the mapping then ensures that C_w is orthogonal to Γ, and the result is proved. ◆

Theorem 4.3.3 is used in the following text to prove two results, each of which is simple to use, while each has many applications.

THEOREM 4.3.4 Mapping an Upper Half-Plane onto a Disk
The most general linear fractional transformation that maps the upper-half z-plane onto the unit disk $|w| < 1$, with the arbitrary point z_0 mapping to the center of the disk, is of the form

$$w = k\left(\frac{z - z_0}{z - \overline{z}_0}\right),$$

where $|k| = 1$ and $\text{Im}\{z_0\} > 0$.

PROOF

Let $T(z)$ be a general linear fractional transformation, which may always be written as

$$w = T(z) = k\left(\frac{z-a}{z-b}\right).$$

Since z_0 lies in the upper-half z-plane, we require $\text{Im}\{z_0\} > 0$, and if z_0 is to map to the origin in the w-plane, it follows that $a = z_0$. Now the point $\bar{z}_0$ which is symmetric with respect to z_0 relative to the real axis in the z-plane must, by virtue of Theorem 4.3.3, be mapped onto the point symmetric to the center ($w = 0$) of the disk with respect to the circle $|w| = 1$. Since the linear fractional transformation is defined throughout the extended complex plane, this is the point at infinity in the w-plane: therefore $w = \infty$ must be the image of $z = \bar{z}_0$, which shows that $b = \bar{z}_0$. Thus $T(z)$ must be of the form

$$w = k\left(\frac{z-z_0}{z-\bar{z}_0}\right).$$

Now if the real axis is to be mapped onto $|w| = 1$, it follows that for $z = x$ (real), $x - z_0$ is the complex conjugate of $x - \bar{z}_0$, so that $|x - z_0| = |x - \bar{z}_0|$. Consequently, taking the modulus of w, we have

$$|w| = 1 = \left|k\left(\frac{x-z_0}{x-\bar{z}_0}\right)\right| = |k|\frac{|x-z_0|}{|x-\bar{z}_0|} = |k|.$$

This has established that $|k| = 1$, and as x was an arbitrary real number this also proves that the real axis is mapped onto the boundary $|z| = 1$, so the theorem is proved. ◆

The indeterminacy concerning k is not surprising because $|k| = 1$ implies that $k = e^{i\theta}$, with θ arbitrary, so k simply represents a pure rotation in the w-plane. This rotation cannot change the symmetry of the origin of the disk with respect to the boundary $|w| = 1$, so the amount of the rotation is immaterial from the point of view of such symmetry.

Example 4.3.5 Mapping a Straight Line and Circle onto an Annulus
Map the real z-axis and an arbitrary circle of radius ρ in the z-plane that lies above it, onto an annulus of unit outside radius centered on the origin in the w-plane.

SOLUTION
Any configuration of this type may always be transformed into the situation found in Figure 4.24(a) by means of a translation. Thus we shall consider

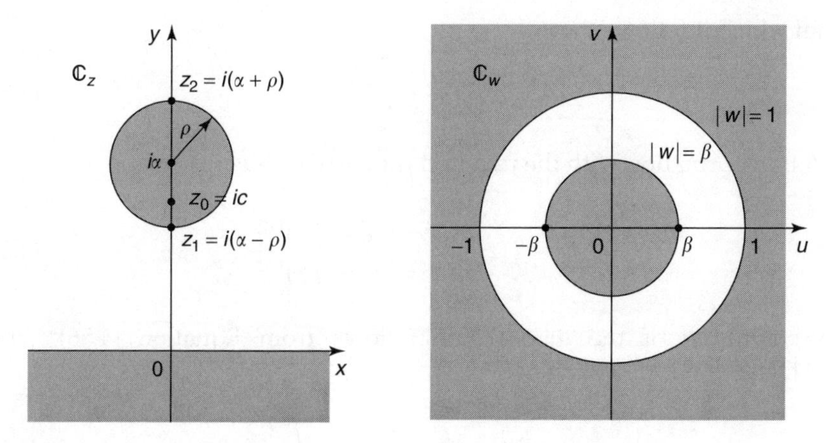

FIGURE 4.24
Mapping a straight line and circle onto an annulus.

only this case in which the imaginary axis in the w-plane becomes a line of symmetry. If, in Theorem 4.3.4, we set $k = 1$ (which is always permissible), and take z_0 to be purely imaginary, with $z_0 = ic$ (c real), it follows that the imaginary axis in the z-plane will map onto the real axis in the w-plane.

It is known from Theorem 4.3.3 that the transformation preserves symmetry with respect to straight lines and circles, so the symmetry present in Figure 4.24(a) is preserved when it is mapped onto the w-plane. The symmetry about the imaginary axis in the z-plane in Figure 4.24(a) is preserved and mapped to produce symmetry about the real axis in the w-plane in Figure 4.24(b).

To choose the value of c that is to cause the circle $|z - i\alpha| = \rho$ to map onto the inner circle of radius β centered on the origin in Figure 4.24(b) we proceed as follows. Altering c will not affect the symmetry about the real axis, but it will move the center of the image circle along the real axis in the w-plane. Accordingly, c must be chosen so that the ends $z_1 = i(\alpha - \rho)$ and $z_2 = i(\alpha + \rho)$ of the diameter located on the imaginary axis in the z-plane are mapped onto points on the real axis in the w-plane that are symmetrically located relative to the origin, and distant β from it. Thus if $w = T(z)$ is the required transformation in Theorem 4.3.4, we require c to be such that

$$T(z_1) = -T(z_2). \tag{4.56}$$

Because $k = 1$ and $z_0 = ic$, $T(z)$ has the form

$$T(z) = \frac{z - ic}{z + ic},$$

and Equation (4.56) implies that

$$\frac{i(\alpha - \rho) - ic}{i(\alpha - \rho) + ic} = -\left(\frac{i(\alpha + \rho) - ic}{i(\alpha + \rho) + ic}\right),$$

from which it follows that

$$c = (\alpha^2 - \rho^2)^{1/2}. \tag{4.57}$$

A transformation with the required property is thus

$$w = T(z) = \frac{z - i(\alpha^2 - \rho^2)^{1/2}}{z + i(\alpha^2 - \rho^2)^{1/2}}. \tag{4.58}$$

The radius β of the inner circle follows from Equation (4.56), since $\beta = |T(z_1)| = |T(z_2)|$, giving

$$\beta = \frac{\rho}{\alpha + (\alpha^2 - \rho^2)^{1/2}}. \tag{4.59}$$

◇

THEOREM 4.3.5 Mapping a Disk onto a Disk
The most general linear fractional transformation that maps the disk $|z| < 1$ onto the disk $|w| < 1$, with the arbitrary point z_0 in the disk $|z| < 1$ mapping to the center of the disk $|w| < 1$ is of the form

$$w = k\left(\frac{z - z_0}{\overline{z}_0 z - 1}\right),$$

where $|k| = 1$ and $|z_0| < 1$.

PROOF

Let $T(z)$ be the general fractional linear transformation, and again express it in the form

$$w = T(z) = K\left(\frac{z - a}{z - b}\right).$$

If z_0 is interior to the disk $|z| < 1$, then $|z_0| < 1$, so if z_0 is to map to the origin in the w-plane we must have $a = z_0$. The point z_0^* symmetric relative to z_0 with respect to the circle $|z| = 1$ must, by Theorem 4.3.3, be mapped into a point that is symmetric relative to the center $w = 0$ of the circle $|w| = 1$. Thus z_0^* is mapped to the point at infinity, showing that $b = z_0^*$.

This has established that

$$w = T(z) = K\left(\frac{z - z_0}{z - z_0^*}\right).$$

Setting $a = 0$, $R = 1$ in Equation (4.55) shows that for the unit circle centered on the origin in the z-plane $z_0^* = 1/\bar{z}_0$, so that

$$w = T(z) = k\left(\frac{z - z_0}{\bar{z}_0 z - 1}\right), \qquad \text{with } k = \bar{z}_0 K.$$

It now only remains to find k and to show that the circle $|z| = 1$ maps onto the circle $|w| = 1$.

If the transformation $w = T(z)$ has this property then $z = 1$ will map onto a point on $|w| = 1$. Therefore, setting $z = 1$ in $T(z)$ and taking the modulus shows that

$$|w| = 1 = \left|k\left(\frac{1 - z_0}{\bar{z}_0 - 1}\right)\right| = |k|\,\frac{|1 - z_0|}{|1 - \bar{z}_0|} = |k|,$$

because $1 - z_0$ and $1 - \bar{z}_0$ are complex conjugates and so have the same modulus, and we may set $k = e^{i\theta}$, with θ real and arbitrary.

$T(z)$ will have been shown to map $|z| = 1$ onto $|w| = 1$ if we can demonstrate that all z and w satisfying these two conditions also satisfy $w = T(z)$. This is easily accomplished by forming the expression $w\bar{w} = T(z)[\overline{T(z)}]$, and then showing that the resulting expression is true when $w\bar{w} = |w|^2$ and $z\bar{z} = |z|^2$ are both replaced by unity, for if $|w|^2 = |z|^2 = 1$, then $|w| = 1$ and $|z| = 1$. We have

$$w\bar{w} = k\bar{k}\left(\frac{z - z_0}{\bar{z}_0 z - 1}\right)\left(\frac{\bar{z} - \bar{z}_0}{\bar{z}_0 z - 1}\right) = |k|^2\left(\frac{z - z_0}{\bar{z}_0 z - 1}\right).$$

This expression is found to be an identity when $|k|^2 = 1$ and $w\bar{w} = z\bar{z} = 1$ thereby completing the proof of the theorem. Here again $k = e^{i\theta}$, with θ real and arbitrary, represents a rotation that does not influence the symmetry of $w = 0$ relative to $|w| = 1$. ◆

Example 4.3.6 Mapping the Region between Eccentric Circles onto an Annulus

Find a linear fractional transformation that maps the region D between two circles lying one within the other, as shown on the left in Figure 4.25, onto the annulus shown on the right of Figure 4.25.

SOLUTION

A rotation, change of scale and translation will reduce any such configuration to the one shown on the left of Figure 4.25, so we shall only consider such a case in which $\alpha > 0$ and the real axis is a line of symmetry.

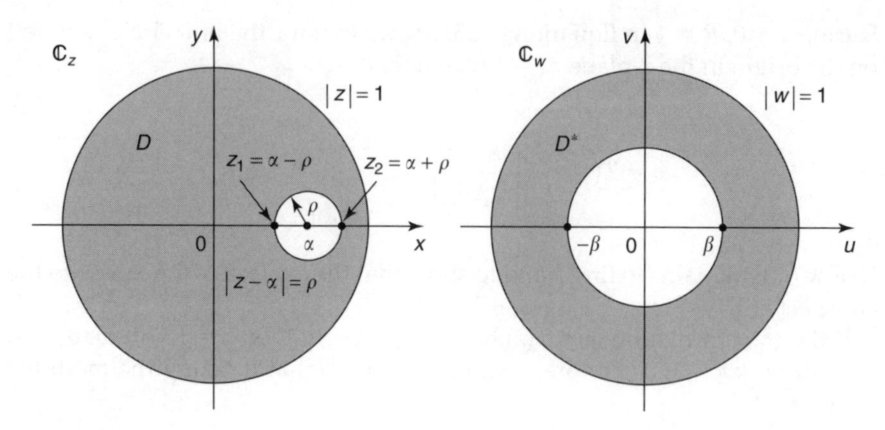

FIGURE 4.25
Mapping a domain between eccentric circles onto an annulus.

Since the line of symmetry lies along the real axis in the z-plane, and the transformation in Theorem 4.3.5 preserves symmetry with respect to straight lines and circles, we need to take the point z_0 in Theorem 4.3.5 on the real axis, so we set $z_0 = c$ (real). With this choice for z_0 we see from the theorem that the transformation will map real points onto real points if k is real, so for convenience we set $k = 1$, which satisfies the condition $|k| = 1$. The transformation preserves symmetry about the real axis in the z-plane, which is mapped onto the real axis in the w-plane. The choice of c will simply move the center of the image circle in the w-plane along the real axis in that plane. The choice of c must be such that the points $z_1 = \alpha - \rho$ and $z_2 = \alpha + \rho$ at opposite ends of the diameter of the inner circle on the left of Figure 4.25 map onto the real axis in the w-plane in a symmetrical manner about the origin, and distant β from it as shown in the diagram on the right of Figure 4.25.

By symmetry this mapping will then ensure that the image of the inner circle on the left of Figure 4.25, when mapped to the w-plane will become a circle a radius β with its center at $w = 0$.

Thus the transformation $w = T(z)$ must be such that $T(z_1) = -T(z_2)$. Because we are setting $k = 1$ and $z_0 = c$ is real, the transformation becomes

$$T(z) = \frac{z - c}{cz - 1}.$$

Because $T(z_1) = -T(z_2)$, this last result leads to c being determined by the equation

$$\frac{\alpha - \rho - c}{c(\alpha - \rho) - 1} = -\left(\frac{\alpha + \rho - c}{c(\alpha + \rho) - 1}\right),$$

and so by the quadratic equation for c

$$\alpha c^2 - (1 + \alpha^2 - \rho^2)c + \alpha = 0.$$

The roots of this equation are real and distinct, because the discriminant $(1 + \alpha^2 - \rho^2)^2 > 4\alpha^2 > 0$, because we must have $\rho < 1 - \alpha$ to ensure that the smaller circle lies within the larger one. Inspection of the quadratic equation for c shows the product of the roots to be unity, so as the roots are real only the smaller root can lie on the real axis inside the circle $|z| = 1$. Taking this choice for c in

$$w = \frac{z - c}{cz - 1}$$

then gives the required mapping onto the annulus. The radius β of the inner circle in the annulus is given by

$$\beta = \left|\frac{\alpha - \rho - c}{c(\alpha - \rho) - 1}\right| = \left|\frac{\alpha + \rho - c}{c(\alpha + \rho) - 1}\right|,$$

because $\beta = |T(z_1)| = |T(z_2)|$. ◇

Exercises 4.3

Express the following mappings as a sequence of mappings and describe the effect of each mapping in the sequence.

1. $w = 2i(1 - z)/(1 + z)$
2. $w = (2z + 3i)/(iz + 4)$
3. $w = 2i(1 - z)/(1 + z)$
4. $w = (1 + iz)/(1 - iz)$
5. Find the fixed points of the transformations
   (i) $w = (2z - 1)/z$
   (ii) $w = (1 + z)/(1 - iz)$
   (iii) $w = (3z - 1)/(4z + 2)$
6. Find the image in the w-plane of the boundary and interior of the circle $|z| = 1$ under the transformation $w = (1 + z)/(1 - z)$.
7. Find the image in the w-plane of the boundary and interior of the circle $|z| = 1$ under the transformation $w = (iz + 1)/z$.
8. Find the image in the w-plane of the upper-half of the z-plane from which is removed the unit disk $|z| \leqslant 1$, $\text{Im}\{z\} > 0$, when mapped by the function $w = (z + 1)/z$.
9. Find the image in the w-plane of the boundary and interior of the annular domain $\frac{1}{2} \leqslant |z| \leqslant 1$ under the transformation $w = \frac{1}{2}i(1 - z)/(1 + z)$.

10. Find the image in the w-plane of the boundary and interior of $|z| = 1$ under the transformation $w = i(2z + i)/(iz - 2)$.

In each of Exercises 11 through 14 find the linear fractional transformation that maps the z-plane onto the w-plane in the manner indicated.

11. $z = -1, 0, 1$ onto the respective points $w = 0, i, 3i$.
12. $z = 1, i, -i$ onto the respective points $w = -1, 0, 1$.
13. $z = 0, 1, 2$ onto the respective points $w = 0, 1, \infty$.
14. $z = -3, 0, 3$ onto the respective points $w = \infty, 1, 2$.
15. Find a linear fractional transformation that will map the region exterior to the two circles $|z - 2| = 2$ and $|z + 2| = 2$ onto a strip in the w-plane in such a way that the points $z_1 = 0, z_2 = 2 - 2i$, and $z_3 = 4$ map to the respective points $w_1 = \infty$, $w_2 = 0$, and $w_3 = 2$.
16. Construct a linear fractional transformation that will map the crescent-shaped region contained between the circles $|z| = 3$ and $|z - 4| = 5$ onto the w-plane in such a way that the points $w_1 = 0$, $w_2 = 1$, and $w_3 = \infty$ in the w-plane. How is the crescent mapped?
17. Prove that the transformation $w = i(1 + z)/(1 - z)$ maps the unit disk $|z| < 1$ onto the upper half of the w-plane. For what values of α (real) will it map the crescent-shaped domain between $|z| = 1$ and $|z - i\alpha| = 3$ onto a wedge shaped domain in the w-plane, and where will these wedges be located?
18. Find the linear fractional transformation $w = T(z)$ that will map the right half of the z-plane, from which has been removed the disk $|z - 3| < 1$, onto the annulus $\beta < |w| < 1$ in the w-plane in such a way that the imaginary axis in the z-plane maps onto $|w| = 1$. What is the value of β?
19. Map the upper half of the z-plane onto the unit disk $|w| < 1$ by a linear fractional transformation $w = T(z)$ in such a way that $T(2i) = 0$ and $\text{Arg}\,\{T'(2i)\} = 0$.
20. Map the disk $|z| < 1$ onto the disk $|w| < 1$ by a linear fractional transformation $w = T(z)$ in such a way that $T(\frac{1}{2}i) = 0$ and $\{\text{Arg}\, T'(\frac{1}{2}i)\} = \pi/2$.

4.4 Mappings by Elementary Functions

It has been shown that the linear fractional transformation is simple in form and it permits considerable flexibility when mapping circles and straight lines because freedom of choice is allowed when mapping three points in the extended z-plane onto three points in the extended w-plane. With the exception of the Schwarz–Christoffel transformation to be considered in the next section, mappings by elementary functions offer very little flexibility of behavior when mapping domains. The main importance of mappings by elementary functions derives in part from their ability to map specially shaped domains into more convenient ones, and in part because their composition with other mappings makes it possible to transform domains with even more complicated shapes into ones with simpler shapes. The full significance of this becomes apparent in Chapter 5 once the solutions of boundary value problems for the Laplace equation are considered.

4.4.1 The Exponential Function

The exponential function $f(z) = e^z$ is an entire function, its derivative $f'(z) = e^z$ does not vanish anywhere in the complex plane, so it has no critical points. Thus it follows that

$$w = e^z \tag{4.60}$$

defines a conformal mapping throughout the entire complex plane. The useful property of this mapping is its ability to map rectangular domains in the z-plane onto sectors of an annulus in the w-plane.

Setting $w = Re^{i\phi}$ and $z = x + iy$ in Equation (4.60) gives

$$Re^{i\phi} = e^x e^{iy},$$

so that

$$R = e^x \quad \text{and} \quad \phi = y \text{ (modulo } 2\pi). \tag{4.61}$$

The image of the line $x = a$ in the z-plane is the circle of radius $R = e^a$ centered on the origin in the w-plane, while the image of the line $y = k$ in the z-plane is the ray $\phi = k$ radiating out from the origin in the w-plane. Thus the image of the rectangular domain D in the z-plane defined by $a \leqslant x \leqslant b$, $c \leqslant y \leqslant d$ and shown in Figure 4.26, is the sector D^* of an annulus in the w-plane defined by

$$e^a \leqslant R \leqslant e^b, \quad c \leqslant \phi \leqslant d.$$

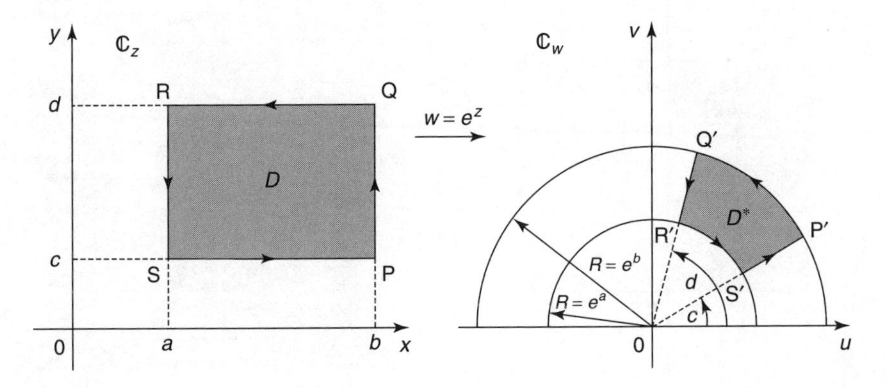

FIGURE 4.26
The mapping $w = e^z$.

Both of these domains are shown in Figure 4.26, with P′, Q′, R′, and S′ the images of P, Q, R, and S. Because of the periodicity of e^z with respect to y, the mapping will be one-one when $d - c \leqslant 2\pi$. It should be noticed that because the exponential function never vanishes for any z, the origin in the w-plane is *not* the image of any point in the z-plane.

Previously, when the exponential function was introduced, it was shown that e^z has a period of $2\pi i$, and that the *fundamental strip* $-\pi \leqslant y \leqslant \pi$ in the z-plane is mapped by $w = e^z$ onto the entire w-plane, with a cut along the negative real axis. Because of periodicity this is also true for any strip of width 2π with its sides parallel to x-axis.

The linear scale factor of a mapping $w = f(z)$ is $\rho(z) = |f'(z)|$, so for the exponential function $\rho(z) = |e^z|$. Setting $e^z = Re^{i\phi}$ shows that $\rho(z) = R$. From this equation it can be seen that the mapping produces a contraction when points inside $|z| = 1$ are mapped onto the w-plane, and a magnification when points outside $|z| = 1$ are mapped onto the w-plane. The general effect of the linear scale factor can be seen from Figure 4.27, in which points D, E, A, B, C on the fundamental strip in the z-plane are seen to be mapped to the points D′, E′, A′, B′, and C′ on the unit circle in the w-plane, with a cut along the negative real axis. This also shows that points in the fundamental strip to the left of the imaginary axis in the z-plane map to points inside the unit circle in the w-plane, while points in the fundamental strip to the right of the imaginary axis map to points outside the unit circle in the w-plane. In general, a curve on which the linear scale factor of a mapping $\rho(z) = 1$ is called the *isometric curve* for the mapping, so in the case of $w = e^z$ the isometric curve of the mapping is the circle $|w| = 1$.

4.4.2 The Logarithmic Function

The logarithmic function $w = \ln z = u + iv$ is the inverse of the exponential function, and so maps the z-plane cut along the negative real axis, one-one

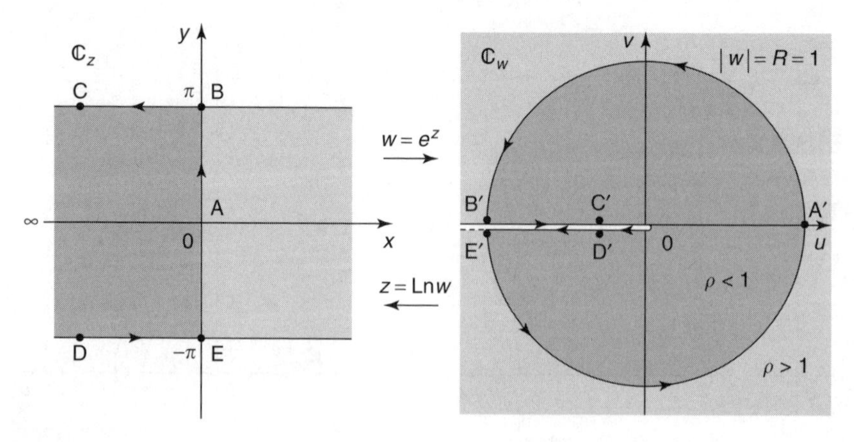

FIGURE 4.27
The magnification and contraction produced by $w = e^z$.

onto any strip in the w-plane of the form $(2k-1)\pi \leqslant v \leqslant (2k+1)\pi$, for $k = 0$, $\pm 1, \pm 2, \ldots$. Because the exponential function defines a conformal mapping, so also does the logarithmic function, and the principal branch $w = \operatorname{Ln} z$ maps the cut z-plane onto the strip $-\pi < v \leqslant \pi$ in the w-plane. The mapping properties of the logarithmic function can be derived from those of the exponential function by reversing the direction of the mapping. This amounts to interchanging the w- and z-planes while still considering the mapping to be from the z-plane to the w-plane.

4.4.3 The Mapping $w = \sin z$

The function $f(z) = \sin z$ is an entire function with $f'(z) = \cos z$ vanishing only at $z = (k + \frac{1}{2})\pi$, $k = 0, \pm 1, \pm 2, \ldots$. Thus $w = \sin z$ will provide a conformal mapping of the z-plane onto the w-plane everywhere except at these critical points. Setting $z = x + iy$ and $w = u + iv$ in $w = \sin z$ gives

$$u + iv = \sin x \cosh y + i \cos x \sinh y. \tag{4.62}$$

The periodicity of u, v with respect to x makes it necessary to restrict z if $w = \sin z$ is to provide a one-one mapping. A possible restriction involves requiring z to lie in the semi-infinite strip D given by $-\pi/2 \leqslant x \leqslant \pi/2$, $y \geqslant 0$. To see the nature of the one-one mapping involved, notice from Equation (4.62) that this strip is mapped onto the domain D^* comprising the upper half of the w-plane. This follows because the line $x = -\pi/2$, $y \geqslant 0$ maps onto the line $u \leqslant -1$, $v = 0$, the line $x = \pi/2$, $y \geqslant 0$ maps onto the line $u \geqslant 1$, $v = 0$, and the line $-\pi/2 \leqslant x \leqslant \pi/2$, $y = 0$ maps onto the line segment $-1 \leqslant u \leqslant 1$, $v = 0$. When moving around D in the positive sense the interior of D lies to the left, and this corresponds to moving along the real axis in the w-plane in the direction of increasing u, when the area to the left is the upper half of the w-plane. The mapping will not be conformal at $w = \pm 1$ because these points are the images of the critical points $z = \pm\pi/2$. Because $d^2w/dz^2 = -\sin z \neq 0$ at $z = \pm\pi/2$ it follows from Theorem 4.2.3 that the angles at these points will be doubled under the mapping.

The image of the line $x = a$ in the semi-infinite strip D follows from Equation (4.62) as $u = \sin a \cosh y$, and $v = \cos a \sinh y$, so as $\cosh^2 y - \sinh^2 y = 1$, elimination of y gives the equation of an hyperbola in the w-plane

$$\frac{u^2}{\sin^2 a} - \frac{v^2}{\cos^2 a} = 1. \tag{4.63}$$

The image of the line $y = b$ in the semi-infinite strip D follows from Equation (4.62) as $u = \sin x \cosh b$, and $v = \cos x \sinh b$, so eliminating x gives the equation of an ellipse in the w-plane

$$\frac{u^2}{\cosh^2 b} + \frac{v^2}{\sinh^2 b} = 1. \tag{4.64}$$

As a and b were arbitrary, and the lines x = constant and y = constant are mutually orthogonal, the conformal nature of the transformation means that the families of hyperbolas and ellipses are mutually orthogonal. Notice that the foci of both the family of hyperbolas and the family of ellipses lie at $w = \pm 1$, the images of the critical points $z = \pm\pi/2$. Thus the mapping is onto *confocal* hyperbolas and ellipses.

If, now the semi-infinite strip in the z-plane is extended to become the infinite strip $-\pi/2 \leqslant x \leqslant \pi/2$ with y unrestricted, an application of the previous argument shows that the semi-infinite strip $-\pi/2 \leqslant x \leqslant \pi/2$, $y \leqslant 0$ maps to the lower half of the w-plane. However, as the semi-infinite lines $x = -\pi/2$, $y > 0$ and $x = -\pi/2$, $y < 0$ both map onto $u < -1$, $v = 0$, it is necessary to cut the w-plane to make the mapping one-one. The cut is made so that all of the real axis in the w-plane is removed, with the exception of the segment $-1 < u < 1$, $v = 0$. Figure 4.28 shows how lines parallel to the real and imaginary axes in the w-plane map to the cut w-plane.

The conformal mapping in the reverse direction from the cut w-plane to the infinite strip in z-plane is provided by the principal branch of the inverse sine function $z = \text{Arcsin}\, w$. When considered as one of the elementary functions, the properties of the mapping $w = \text{Arcsin}\, z$ follow from the mapping $w = \sin z$ by interchanging the roles of the w- and z-planes.

The linear scale factor for the conformal mapping $w = \sin z$ is

$$\rho(z) = |d\sin(z)/dz| = |\cos z| = (\cosh^2 y - \sin^2 x)^{1/2}.$$

Representative curves of $\rho(z)$ are shown in Figure 4.29 for $-\pi \leqslant x \leqslant \pi, y > 0$, with the curve corresponding to $\rho = 1$ being the isometric curve for the mapping $w = \sin z$.

The arguments already used show that the horizontal strip D defined as $-\pi < x < \pi, 0 < a \leqslant y \leqslant b$ is mapped by $w = \sin z$ onto an elliptic annulus

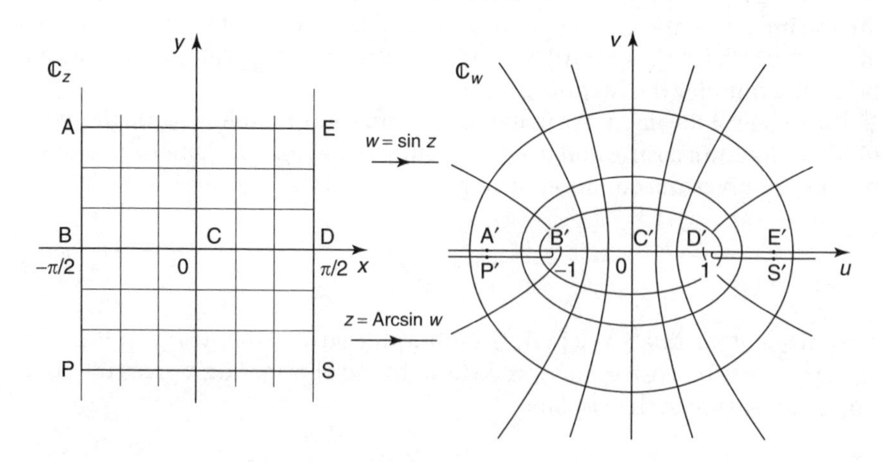

FIGURE 4.28
The mapping by $w = \sin z$ of $|z| \leqslant \pi/2$ onto the cut w-plane.

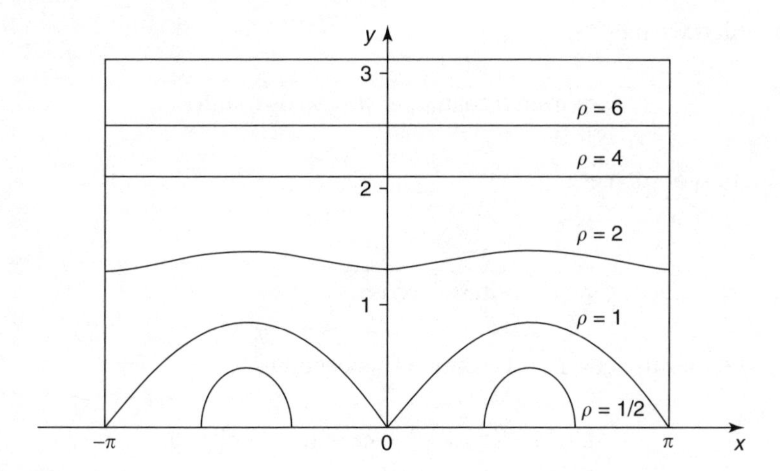

FIGURE 4.29
The linear scale factor ρ for $w = \sin z$, with $\rho = 1$ the isometric curve.

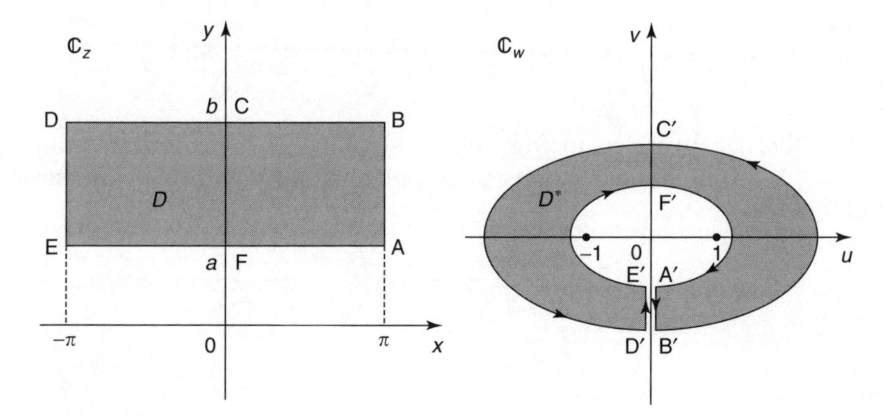

FIGURE 4.30
The mapping by $w = \sin z$ of the strip $-\pi < x < \pi, 0 < a < y < b$.

D^* cut along the negative imaginary axis to make the mapping one-one. This is shown in Figure 4.30 and as before the coordinate lines $x =$ const. and $y =$ const. within the strip map, respectively, onto confocal hyperbolas and ellipses with their foci at $w = \pm 1$, the images of the critical points at $z = \pm\pi/2$.

4.4.4 The Mapping w = Arcsin z/a

Let us determine explicitly in terms of x and y the real and imaginary parts of the principal branch of the inverse function $w = \text{Arcsin}\, z/a$, with $a > 0$ real. This function is equivalent to $z = a \sin w$, or by setting $w = u + iv$, to

$$x + iy = a \sin u \cosh v + ia \cos u \sinh v,$$

from which we have

$$x = a \sin u \cosh v, \qquad y = a \cos u \sinh v.$$

Elimination of v gives

$$\frac{x^2}{\sin^2 u} - \frac{y^2}{\cos^2 u} = a^2.$$

Using the identity $\cos^2 u = 1 - \sin^2 u$ this becomes

$$a^2 \sin^4 u - (x^2 + y^2 + a^2)\sin^2 u + x^2 = 0.$$

Solving this biquadratic equation for $\sin^2 u$ gives

$$\sin^2 u = \frac{1}{2a^2}\left(x^2 + y^2 + a^2 - \sqrt{(x^2 + y^2 + a^2)^2 - 4a^2x^2}\right),$$

where the negative sign in front of the square root is necessary because $\sin^2 u \leqslant 1$. After some tedious manipulation that we omit, it turns out that

$$\sin u = \pm\frac{1}{2a}\left(\sqrt{(x+a)^2 + y^2} - \sqrt{(x-a)^2 + y^2}\right),$$

so

$$u = \pm\arcsin\left(\frac{\sqrt{(x+a)^2 + y^2} - \sqrt{(x-a)^2 + y^2}}{2a}\right),$$

where the real function $\arcsin t$ lies in the interval $-\pi/2 < \arcsin t < \pi/2$.

To decide the choice of sign for u, notice from Figure 4.28 (with the w- and z-planes interchanged) that points in the first quadrant of the z-plane are mapped by the function $w = \text{Arcsin}\, z/a$ onto points in the first quadrant of the w-plane lying within the infinite strip $-\pi/2 < u < \pi/2$. This is only possible if the positive sign is selected, so we have established that

$$u = \text{Re}\{\text{Arcsin}\, z/a\} = \arcsin\left(\frac{\sqrt{(x+a)^2 + y^2} - \sqrt{(x-a)^2 + y^2}}{2a}\right). \tag{4.65}$$

A corresponding argument shows that

$$v = \text{Im}\{\text{Arcsin}\, z/a\} = [\text{sgn}(y)]\,\text{arccosh}\left(\frac{\sqrt{(x+a)^2+y^2}+\sqrt{(x-a)^2+y^2}}{2a}\right), \tag{4.66}$$

where the real function $\text{arccosh}\, t = \ln_e(t + \sqrt{t^2 - 1})$ for $t \geqslant 1$. The factor $\text{sgn}(y)$ in Equation (4.66) called the *signum function* and defined as

$$\text{sgn}(y) = \begin{cases} 1 & \text{for } y \geqslant 0 \\ -1 & \text{for } y < 0, \end{cases}$$

is necessary because the mapping $w = \text{Arcsin}\, z/a$ requires y and v to have the same sign for $-\pi/2 < u < \pi/2$.

The results in Equations (4.65) and (4.66) show explicitly how the z-plane, which is cut so the points $|x/a| \geqslant 1$, $y = 0$ are deleted, is mapped by $w = \text{Arcsin}\, z/a$ onto the infinite vertical strip $-\pi/2 \leqslant u \leqslant \pi/2$ in the w-plane (i.e., see Figure 4.28).

4.4.5 The Mapping $w = \cos z$

The properties of the mapping

$$w = \cos z \tag{4.67}$$

follow directly from those of $w = \sin z$ by using the result

$$w = \cos z = \sin(z + \pi/2). \tag{4.68}$$

This shows that the same mapping as before is involved, namely the mapping $w = \sin z$, once a translation to the right by $\pi/2$ has been made.

4.4.6 The Hyperbolic Functions $w = \sinh z$ and $w = \cosh z$

The mappings provided by these entire functions can either be established directly, or by appeal to the two previous mappings using the results

$$w = \sinh z = -i \sin(iz), \tag{4.69}$$

and

$$w = \cosh z = \cos(iz). \tag{4.70}$$

To see how to interpret the mapping $w = \sinh z$, we can make use of the composition of the mappings

$$\zeta = iz,\ \eta = \sin\zeta \text{ and } \quad w = -i\eta.$$

This sequence of mappings shows that the mapping $w = \sinh z$ may be regarded as a combination of a positive rotation by $\pi/2$ ($\zeta = iz$), followed by the sine function mapping ($\eta = \sin\zeta$), and then a negative rotation through an angle $\pi/2$ ($w = -i\eta$).

The mapping $w = \cosh z$ is even simpler, because writing is as the composition of $\zeta = iz$ and $w = \cos\zeta$ shows that it comprises a positive rotation through an angle $\pi/2$ corresponding to $\zeta = iz$, followed by the cosine mapping ($w = \cos z$).

The critical points of the mapping $w = \sinh z$ are the zeros of $\cosh z$, namely $z = (k + \frac{1}{2})\pi i$, $k = 0, \pm 1, \pm 2, \ldots$. Correspondingly, the critical points of $w = \cosh z$ occur at $z = k\pi i$, $k = 0, \pm 1, \pm 2, \ldots$.

The restriction of z to parts of the z-plane to make the sine and cosine functions give one-one mappings onto suitably cut w-planes implies corresponding restrictions on z if the hyperbolic functions are to give one-one mappings.

4.4.7 The Mapping $w = z + 1/z$: The Joukowski Transformation

The transformation

$$w = z + \frac{1}{z}, \tag{4.71}$$

is called the *Joukowski transformation* (or mapping). It will be seen that when z is suitably restricted this transformation provides a one-one conformal mapping of part of the z-plane onto the cut w-plane. In particular, the Joukowski transformation has a number of useful mapping properties involving straight lines and circles.

We have

$$\frac{dw}{dz} = 1 - \frac{1}{z^2},$$

so the critical points occur at $z = \pm 1$. Their images are at $w = \pm 2$, and the mapping will not be conformal at either of these points.

Setting $z = re^{i\theta}$, $-\pi < \theta \leqslant \pi$ in Equation (4.71) gives

$$w = u + iv = \left(1 + \frac{1}{r}\right)\cos\theta + i\left(1 - \frac{1}{r}\right)\sin\theta, \tag{4.72}$$

so

$$u = \left(r + \frac{1}{r}\right)\cos\theta, \quad v = \left(r - \frac{1}{r}\right)\sin\theta. \tag{4.73}$$

These results show that the unit circle $|z| = r = 1$ maps onto the line segment $-2 < u < 2$, $v = 0$, while the points in the z-plane outside the unit circle map onto the rest of the w-plane. However, the points inside the unit circle also map onto the rest of the w-plane, so when mapping the entire z-plane it follows that the w-plane is mapped *twice*. To make the mapping one-one it is necessary to restrict z so that it either lies *inside* the circle $|z| = 1$, or *outside* it. In each case points on the segment $-2 < u < 2$, $v = 0$ must be excluded from the w-plane by a cut along the real axis extending from $u = -2$ to $u = 2$.

This double covering of the w-plane also becomes apparent if we set

$$w = T(z) = z + \frac{1}{z}, \tag{4.74}$$

for it then follows that

$$T(z) = T(1/z). \tag{4.75}$$

This shows that every point in the w-plane, with the exception of the points $w = \pm 2$, has two distinct pre-images in the z-plane. Denoting these distinct pre-images by z_1 and z_2 it must follow that

$$T(z_1) = T(z_2), \tag{4.76}$$

which in turn implies that

$$z_1 z_2 = 1. \tag{4.77}$$

The form of v in Equation (4.72) shows that for $r < 1$ (points in the disk $|z| < 1$), the upper half of the disk maps onto the deleted lower half of the w-plane, that is the lower half plane without the points on the cut. When $r > 1$ (points outside the disk $|z| < 1$) the lower half of the disk map onto the deleted upper half of the w-plane. This behavior is shown in Figure 4.31, where the cut extends from $w = -2$ to $w = 2$, shown as points B′ and D′.

A routine calculation establishes that the circle $|z| = r > 1$ maps onto the ellipse

$$\frac{u^2}{a^2} + \frac{v^2}{b^2} = 1 \tag{4.78}$$

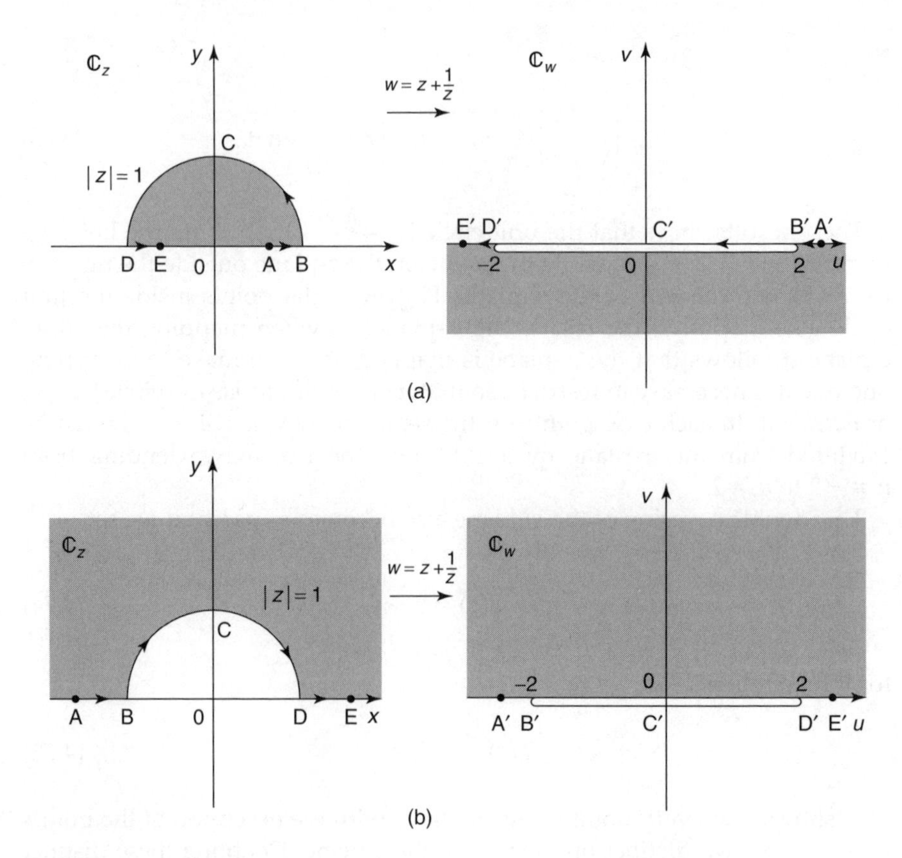

FIGURE 4.31
The mapping $w = z + 1/z$.

with semimajor and minor axes $a = r + 1/r$ and $b = |r - 1/r|$, with its foci at $w = \pm 2$. Using the fact that $a^2 - b^2 = 4$ to eliminate r from Equation (4.72) shows that the ray $\theta = \theta_0$ maps onto the hyperbola

$$\frac{u^2}{c^2} - \frac{v^2}{d^2} = 1, \tag{4.79}$$

because mapping is conformal, the orthogonality of the coaxial circles and rays in the z-plane means that the family of confocal ellipses and hyperbolas in the w-plane are also mutually orthogonal.

The double covering of the w-plane by the transformation can be seen in yet another way by solving Equation (4.71) to obtain

$$z = \frac{1}{2}\left(w + \sqrt{w^2 - 4}\right). \tag{4.80}$$

When using this result it is important that the correct branch of the square root function is assigned, according to the restriction on z. Thus if $|z| > 1$ the

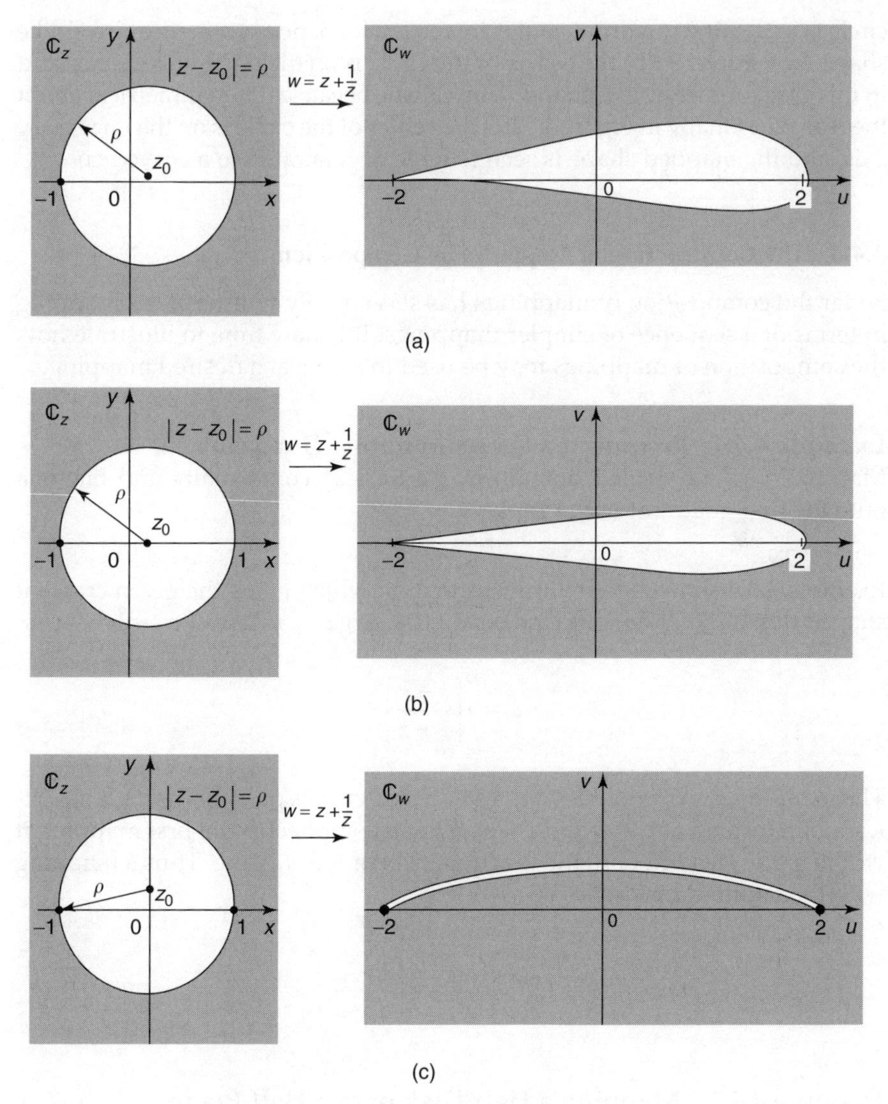

FIGURE 4.32
Air foil-like shapes mapped by the Joukowski transformation.

branch to be taken must be the one for which $|w + \sqrt{w^2 - 4}| > 2$, because only then will the upper half of the cut w-plane map to the exterior of $|z| = 1$ in the upper half of the z-plane.

The Joukowski transformation is of historical importance because it was the first transformation to map the boundary and exterior of a circle onto the boundary and exterior of an airfoil-like shape. Depending on the location of the center of the circle, so will depend the shape to which it is mapped. Figure 4.32 shows some typical mappings. In Figure 4.32(a) the center of the circle is in the first quadrant, and is image is a realistic asymmetric airfoil-like shape. In Figure 4.32(b) the center of the circle is on the positive real axis, and in this case the airfoil-like shape is unrealistic because it is symmetrical about the real axis. Finally in Figure 4.32(c) the center of the circle is on the imaginary axis, and the mapped shape is seen to have degenerated to a curved cut.

4.4.8 The Construction of Mappings by Composition

So far the composition of mappings has served only to interpret a mapping in terms of a sequence of simpler mappings. It is now time to illustrate how the composition of mappings may be used to arrive at a desired mapping.

Example 4.4.1 Mapping a Crescent onto a Quadrant
Map the crescent shaped domain in Figure 4.21 conformally and one-one onto the first quadrant of the w-plane.

SOLUTION
Example 4.3.4 showed that a transformation which maps the given crescent shaped domain (a lune) onto the octant $0 \leqslant \operatorname{Arg}\zeta \leqslant \pi/4$ in the ζ-plane was

$$\zeta = i\left(\frac{i-z}{i+z}\right).$$

The mapping $w = \zeta^2$ doubles angles between curves passing through its critical point located at the origin $\zeta = 0$. So when applied to the first mapping it will map the crescent onto the first quadrant of the w-plane. Thus a mapping with the required properties is

$$w = i^2\left(\frac{i-z}{i+z}\right)^2 = \left(\frac{1+iz}{i+z}\right)^2.$$

◇

Example 4.4.2 Mapping a Half Disk onto a Half Plane
Map the upper half of the disk $|z| < 1$ conformally and one-one onto the upper half of the w-plane.

SOLUTION
The linear fractional transformation

$$\zeta = \frac{1+z}{1-z}$$

maps the points $z_1 = -1, z_2 = 1$, and $z_3 = i$ on the diameter and top of the unit circle in the z-plane onto the points $\zeta_1 = 0, \zeta_2 = \infty$, and $\zeta_3 = i$ in the first quadrant of the ζ-plane. So again using the mapping $w = \zeta^2$, the first quadrant of the ζ-plane will be mapped onto the upper half of the w-plane, so a mapping with the required properties is

$$w = \left(\frac{1+z}{1-z}\right)^2.$$

◇

Notice that neither of the composite mappings in these two examples is the only one with the required property. The main difference between other mappings with the same general property is that they are likely alter the scaling and pattern of the lines u = const. and v = const. in the w-plane.

Exercises 4.4

1. Find the image in the w-plane under the mapping $w = e^{az}$ of (i) the infinite strip $0 \leqslant y \leqslant \pi/a$, and (ii) the infinite strip $-\pi/a \leqslant y \leqslant 0$, where a is real and positive. How must the w-plane be cut in order that the infinite strip $-\pi/a \leqslant y \leqslant \pi/a$ is mapped one-one onto the w-plane?
2. Find the image in the w-plane of the rectangle $a \leqslant x \leqslant b$, $-\pi/2 \leqslant y \leqslant \pi$ in the z-plane under the mapping $w = 2 + 3ie^z$, and indicate how the corners of the rectangle map onto the w-plane.
3. Find the image in the w-plane of (i) the upper half and (ii) the lower half of the infinite strip $0 \leqslant x \leqslant \pi/4$ under the mapping $w = e^{2iz}$.
4. Find the image in the w-plane of the sector $0 < a < r < b$, $-\pi/2 \leqslant \text{Arg}\{z\} \leqslant \pi/2$ under the mapping $w = (1 + 2i) + 2\text{Ln}\, z$.
5. Find the image in the w-plane of the rectangle $\pi/6 \leqslant x \leqslant \pi/3$, $-1 \leqslant y \leqslant 2$ under the mapping $w = \sin z$.
6. Show that if a is real, $w = \cos az$ maps the infinite strip $\pi/a \leqslant x \leqslant 2\pi/a$ in the z-plane onto the w-plane cut along the negative real axis.
7. By finding the real and imaginary parts of $w = \sinh az$, show that if a is real it maps the infinite strip $-\pi/(2a) \leqslant y \leqslant \pi/(2a)$ one-one onto the w-plane cut along the imaginary axis from $y = 1$ to $y = \infty$ and from $y = -1$ to $y = -\infty$.
8. By finding the real and imaginary parts of $w = \cosh az$, show that if a is real it maps the infinite strip $0 \leqslant y \leqslant \pi/a$ one-one onto the w-plane cut along the real axis from $u = 1$ to $u = \infty$, and from $u = -1$ to $u = -\infty$.
9. Show $w = \tanh z$ maps the lines $x = a$ and $y = b$ onto circles and circular arcs, respectively, provided $a \neq 0$ and $b \neq k\pi/2$, with k an integer.
10. Show by finding the real and imaginary parts of $w = \tanh az$ that when a is real, the infinite strip $-\pi/(2a) \leqslant y \leqslant \pi/(2a)$ is mapped onto the w-plane cut along the real axis from $u = 1$ to $u = \infty$ and from $u = -1$ to $u = -\infty$.

11. Show that when a is real, $w = \tanh az$ maps the infinite strip $0 \leqslant y \leqslant \pi/a$ onto the w-plane cut along the real axis from $u = -1$ to $u = 1$.
12. Show that $w = \tan z$ maps the infinite strip $-\pi/4 < x < \pi/4$ onto the interior of the unit disk $|w| < 1$. Determine how the quarters of the strip in each quadrant map onto the disk.
13. Find how the unit disk $|z| < 1$ from which is cut the line segment extending from $z = a$ to $z = 1$ is mapped onto the w-plane by $w = \frac{1}{2}(z + 1/z)$ when (i) $0 < a < 1$ and (ii) $-1 < a < 0$.
14. Find a mapping that maps the sector of the unit disk $|z| < 1$, $0 \leqslant \text{Arg}\{z\} \leqslant 2\pi/n$ $(n \geqslant 1)$ onto the upper half of the w-plane.

4.5 The Schwarz–Christoffel Transformation

The Schwarz–Christoffel transformation shows how a domain in the z-plane bounded by a polygon whose sides never intersect can be mapped onto the upper half of the complex plane. Polygons of this type, with nonintersecting sides, are called *simple polygons,* and they arise in many applications when conformal mapping is used to solve boundary value problems for the Laplace (see Chapter 5). For a general polygon the mapping function is not usually expressible in terms of elementary functions and in such cases to make use of the mapping numerical methods must be used. However, included among this class of polygons are what are called *degenerate polygons* for which a vertex is located at infinity, or where two sides are folded back on each other so they coincide. Degenerate polygons can lead to much simpler mapping functions that can often be expressed in terms of elementary functions, so the examples given in this section will be restricted to these simpler problems.

We must preface our discussion of the Schwarz–Christoffel transformation by introducing the notation that will be used when mapping a specified polygon and by adding some general comments about *geometrical symmetry.* Consider Figure 4.33, in which five finite points x_1 to x_5 together with the point at infinity ($z = \infty$) along the real z-axis are to be mapped onto the vertices w_1 to w_6, respectively, of a simple *closed polygon* in the w-plane by the mapping $w = f(z)$ with $w_i = f(x_i)$ for $i = 1$ to 5, with $w_6 = f(\infty)$.

Let the vertices of the polygon be numbered in the positive sense (counterclockwise), and let τ_i be the positively oriented tangent at w_i as shown in Figure 4.33. Then the *exterior* angle of the polygon at w_i, denoted by θ_i, is the angle measured from the tangent at that vertex to the next side, with the angle when measured counterclockwise will be considered to be positive. Thus, the angle θ_2 at w_2 is positive, whereas the angle θ_3 at w_3 is negative.

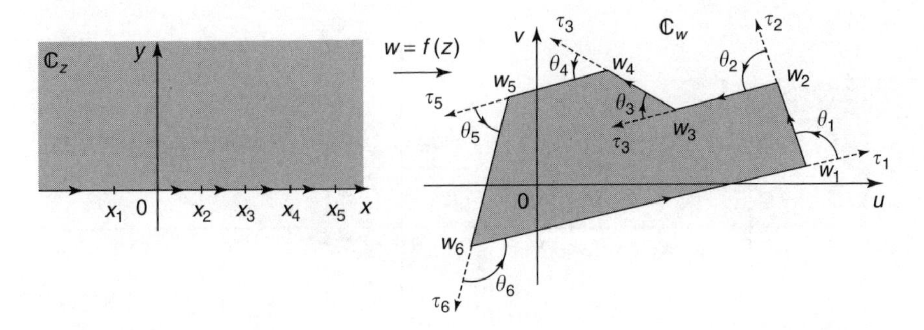

FIGURE 4.33
The geometry of the Schwarz–Christoffel transformation of a simple closed six-sided polygon with $\theta_1 + \theta_2 + \theta_3 + \theta_4 + \theta_5 > \pi$.

For a general simple closed polygon with n vertices the sum of the n exterior angles must equal 2π, so that

$$\theta_1 + \theta_2 + \cdots + \theta_n = 2\pi.$$

It follows from this simple result that only $n - 1$ of the exterior angles are independent because

$$\theta_n = 2\pi - (\theta_1 + \theta_2 + \cdots + \theta_{n-1}), \tag{4.81}$$

where at vertex w_i the angle θ_i must be such that $-\pi < \theta_i < \pi$ for all i. So, for θ_n to satisfy this constraint, it is necessary that we require that in a simple closed polygon with n vertices

$$\theta_1 + \theta_2 + \cdots + \theta_{n-1} > \pi. \tag{4.82}$$

Suppose, if possible, that the condition in Equation (4.82) is not true for some polygon and that

$$\theta_1 + \theta_2 + \cdots + \theta_{n-1} < \pi. \tag{4.83}$$

Then Equations (4.81) and (4.83) imply that the remaining angle $\theta_n > \pi$, which is impossible because then the polygon cannot close. Such an *open polygon* has only $n - 1$ sides of finite length, and the ends of the two open sides of the polygon must be taken to coincide with the point at infinity ($w = \infty$). An example of an open polygon with five finite vertices and the sixth at $w = \infty$ is shown in Figure 4.34.

Before introducing the Schwarz–Christoffel transformation, it is necessary to say something about the *similarity of polygons*. Two closed polygons are

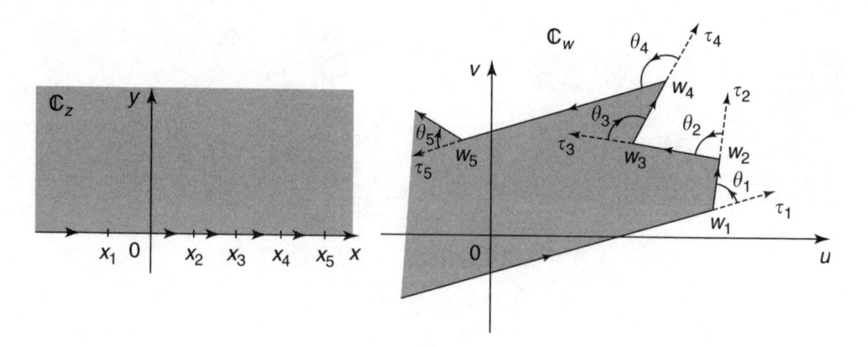

FIGURE 4.34
The Schwarz–Christoffel transformation of an open polygon with five finite vertices and $\theta_1 + \theta_2 + \theta_3 + \theta_4 + \theta_5 < \pi$.

geometrically similar if their corresponding exterior angles are equal and if the lengths of the corresponding sides are proportional. Only for a triangle is similarity guaranteed by the specification of three exterior angles. For a quadrilateral, the exterior angles must be given together with the additional condition that two pairs of corresponding sides are proportional. In the case of a pentagon, the exterior angles must be given together with two conditions concerning the proportionality of sides. So in general, when the exterior angles of two polygons are specified, for them to be similar it is necessary that $n - 3$ additional proportionality conditions ($3 - 3 = 0$ for a triangle, $4 - 3 = 1$ for a quadrilateral and $5 - 3 = 2$ for a pentagon, etc.).

This has important implications when considering the conformal one-one mapping of a polygon onto a half plane. Consider, for example, a closed polygon with n vertices $w_1, w_2, \ldots, w_n$, and let them be arranged sequentially in counterclockwise order. Then a conformal mapping of the boundary of the polygon onto the real axis with vertices mapping to the points $x_1, x_2, \ldots, x_n$, respectively, requires that $x_1 < x_2 < \cdots < x_n$. The assignment of a vertex w_i to a point x_i is a proportionality condition, so from the previous discussion, we see that only three of the points $x_1, x_2, \ldots, x_n$ can be assigned arbitrarily, while the remainder must be determined in the course of the transformation in such a way that the polygon has the correct shape.

If one of the points $x_1, x_2, \ldots, x_n$ is taken to be the point at infinity ($z = \infty$), one degree of freedom is lost, and only two of the remaining $n - 1$ points may be assigned arbitrarily. For example, both in Figure 4.33 for a simple closed polygon and Figure 4.34 for an open polygon only two of the points x_1 to x_5 may be assigned arbitrarily.

THEOREM 4.5.1 The Schwarz–Christoffel Transformation
Let it be required to map conformally and one-one by means of a transformation $w = f(z)$ the upper half of the z-plane onto the interior of a simple polygon in the w-plane, with its vertices arranged in counterclockwise order $w_1, w_2, \ldots, w_n$. Let it also be required that the first $n - 1$ vertices are, respectively, the images of the finite

points $x_1, x_2, \ldots, x_{n-1}$ on the real axis of the z-plane, and that w_n is the image of $z = \infty$, where

$$x_1 < x_2 < \cdots < x_{n-1}. \tag{4.84}$$

Take the exterior angles at the vertices of the polygon to be $\theta_1, \theta_2, \ldots, \theta_n$, respectively. Then the derivative $f'(z)$ of the required transformation is

$$f'(z) = A(z - x_1)^{-\theta_1/\pi}(z - x_2)^{-\theta_2/\pi} \cdots (z - x_{n-1})^{-\theta_{n-1}/\pi}. \tag{4.85}$$

(i) *The mapping function $f(z)$ is given by the indefinite integral*

$$f(z) = A\int (z - x_1)^{-\theta_1/\pi}(z - x_2)^{-\theta_2/\pi} \cdots (z - x_{n-1})^{-\theta_{n-1}/\pi} dz + B, \tag{4.86}$$

where the complex constants A and B are determined by the way the vertices are required to map onto the real axis in the z-plane, with the principal branches of the functions $(z - x_i)^{-\theta_i/\pi}$ being used for $i = 1, 2, \ldots, n - 1$.

(ii) *Any two of the points $x_1, x_2, \ldots, x_{n-1}$ may be assigned arbitrarily while the others, subject to the constraint in Equation (4.84), must be determined by the transformation and the shape of the polygon.*

PROOF

A direct proof of the Schwarz–Christoffel transformation is complicated and would be inappropriate here, so instead we will simply demonstrate that it has the required properties. It will suffice to examine Equation (4.85), since this determines how the positively oriented tangent at any point in the z-plane is rotated by the transformation. Tasking the argument of Equation (4.85) with $z = x$ (on the real axis) gives

$$\arg\{f'(x)\} = \text{Arg}\{A\} - \frac{1}{\pi}\left[\theta_1 \text{Arg}(x - x_1) + \theta_2 \text{Arg}(x - x_2) + \cdots + \theta_{n-1} \text{Arg}(x - x_{n-1})\right]. \tag{4.87}$$

The mapping is already arranges so that $x_n = \infty$, so let us agree to set $x_0 = -\infty$. When x lies in the interval $x_{j-1} < x < x_j$ we have $\text{Arg}(x - x_i) = 0$ for $i < j$ and $\text{Arg}(x - x_i) = \pi$ for $i \geqslant j$. This can be seen from Figure 4.35 by allowing point P at z to move to a point on the real axis in the interval $x_{j-1} < x < x_j$.

Now allow x to increase from $-\infty$ to ∞ and consider how $\arg\{f'(x)\}$ changes. For x in the interval $x_{j-1} < x < x_j$ no change is made in $\arg\{f'(x)\}$, corresponding to the image $w = f(x)$ moving along the real axis from $w_{j-1} = f(x_{j-1})$ to $w_j = f(x_j)$. Similarly, for x in the next interval $x_j < x < x_{j+1}$ $\arg\{f'(x)\}$ is again constant (though different) corresponding to the image $w = f(x)$ moving along the real axis from $w_j = f(x_j)$ to $w_{j+1} = f(x_{j+1})$. However, when x crosses the point x_j, corresponding to w passing the point $w_j = f(x_j)$, the value of

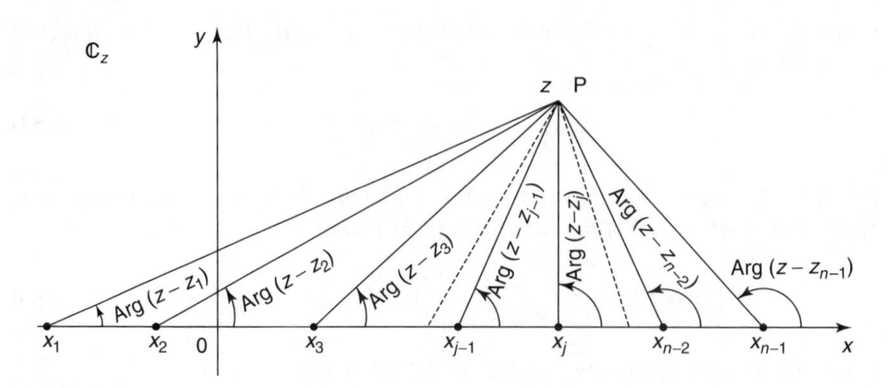

FIGURE 4.35
Arguments of the vectors $(x - x_i)$.

$\arg(x - x_j)$ decreases discontinuously by the angle π, with the result that $\arg\{f'(x)\}$ increases by the angle θ_j.

This shows that both the segment $x_{j-1} < x < x_j$ and the segment $x_j < x < x_{j+1}$ map onto a straight line segment in the w-plane with a common point at $w_j = f(x_j)$, with the second segment being rotated counterclockwise through an angle θ_j relative to the first segment. This has established that two contiguous segments on the real axis of the z-plane are mapped onto two contiguous straight-line segments in the w-plane meeting at the point w_i, where the exterior angle is θ_j. This argument applies to any two contiguous segments determined by the points $x_1, x_2, \ldots, x_n$, so we have succeeded in showing that the real z-axis maps onto a polygon in the w-plane with the correct exterior angles.

Writing constant A in Equation (4.85) as $A = Ce^{i\lambda}$, with C real, shows that the effect of A is to scale the mapping by a factor C and to rotate it through the angle λ. Neither of these operations affect the geometrical symmetry of the transformation. When Equation (4.85) is integrated to obtain Equation (4.86) the constant B simply introduces a translation, and so once again the geometrical symmetry is not changed. The function $f(z)$ defined in this way is not analytic on all of the real axis in the z-plane, because the magnitudes of angles between tangents to the left and right of the points corresponding to $x_1, x_2, \ldots, x_n$ are not preserved, though $f(z)$ is analytic in the upper half of the z-plane.

The principal branches of the functions $(z - x_i)^{-\theta_i/\pi}$ must be used because these functions must agree with the corresponding real functions when z is on the real axis. The choice of independent points from the points $x_1, x_2, \ldots, x_{n-1}$ has already been discussed, and it has been established that only two can be assigned arbitrarily. Finally, as the mapping is analytic for $\text{Im}\{z\} \geqslant 0$ (except for the points $x_1, x_2, \ldots, x_{n-1}$), the manner in which z moves along the x-axis coupled with the sense in which w moves around the polygon ensures that the upper half of the z-plane is mapped onto the interior of the polygon. ◆

It should be clearly understood that in general the integral in Equation (4.86) that determines $f(z)$ defines a function that is not expressible in terms of elementary functions. However, when degenerate polygons are involved, the integrand simplifies to the extent that the integral can be evaluated in terms of familiar functions.

REMARKS

1. It is often convenient to express the indefinite integral in Equation (4.86) of Theorem 4.5.1 in the form

$$f(z) = w_0 + A\int_{z_0}^{z} (\zeta - x_1)^{-\theta_1/\pi}(\zeta - x_2)^{-\theta_2/\pi}\cdots(\zeta - x_{n-1})^{-\theta_{n-1}/\pi}\,d\zeta, \quad (4.88)$$

   where z_0 is a point at which $f(z)$ is analytic and $w_0 = f(z_0)$ is its image in the w-plane. The point z_0 is usually taken as one of the two real values that may be assigned arbitrarily from amongst the $x_1, x_2, \ldots, x_{n-1}$, because then w_0 becomes the corresponding vertex of the polygon in the w-plane.

2. If, in Theorem 4.5.1, instead of taking w_0 to be the image of $z = \infty$ it is taken to be the image of a finite number x_n (real) with $-\infty < x_1 < x_2 < \cdots < x_n < \infty$, result Equation (4.85) must be replaced by

$$f'(z) = A(z - x_1)^{-\theta_1/\pi}(z - x_2)^{-\theta_2/\pi}\cdots(z - x_{n-1})^{-\theta_{n-1}/\pi} \times (z - x_n)^{-\theta_n/\pi}, \quad (4.89)$$

   and result from Equation (4.86) by

$$f(z) = A\int (z - x_1)^{-\theta_1/\pi}(z - x_2)^{-\theta_2/\pi}\cdots(z - x_{n-1})^{-\theta_{n-1}/\pi} \times (z - x_n)^{-\theta_n/\pi}\,dz + B. \quad (4.90)$$

   Three of the points $x_1, x_2, \ldots, x_n$ may then be assigned arbitrarily.

3. If in the modified Theorem 4.5.1 represented by Equations (4.89) and (4.90) the real points $x_1, x_2, \ldots, x_n$ are replaced by any n distinct points $z_1, z_2, \ldots, z_n$ on the unit circle $|z| = 1$, the Schwarz–Christoffel transformation maps the interior of the disk $|z| < 1$ conformally and one-one onto the interior of the polygon. In this form, three of the points $z_1, z_2, \ldots, z_n$ may be assigned arbitrarily.

The Approach Used in the Illustrative Examples

Each of the following examples illustrating the application of the Schwarz–Christoffel transformation involves the mapping of the upper half of the z-plane onto a degenerate polygon. To emphasize how the final degenerate polygon can be obtained as the limiting form of a finite polygon, in each case the mapping function is derived in two steps. In the first step the integral in Equation (4.86) determining the mapping onto a finite polygon is derived, and only then in the second step is the limiting form of the integrand in Equation (4.86) found by allowing the finite polygon to tend to its degenerate form, after which the integral Equation (4.86) is evaluated.

This two step approach, although not strictly necessary, has been adopted to illustrate some important features of the transformation. The purpose of the first step is to demonstrate the general ideas involved when the mapping of a half plane onto an arbitrary finite polygon is required. In particular, this step shows how an integral of the type in Equation (4.86) is to be derived for the mapping of a half plane onto a finite polygon, and in the process it also shows that the resulting algebraic integrand is sufficiently complicated that Equation (4.86) cannot be evaluated in terms of elementary functions. The other reason for this two step approach is to show how, when the polygon is allowed to becomes degenerate, the integrand in Equation (4.86) can become sufficiently simple for the integral itself to be evaluated, thereby allowing the required transformation to be found. It will be seen from the examples that, even when mapping onto a fairly simple degenerate polygon, the algebraic integrand in Equation (4.86) can lead to surprisingly complicated mapping function.

Once the assignment of values to the points $z_i = x_i$, the corresponding images w_i, and the angles θ_i has become familiar, the intermediate step of first mapping onto a finite polygon before proceeding to the mapping onto its degenerate form can be omitted because the points, their images and the angles θ_i involved can be deduced directly from the shape of the degenerate polygon itself.

This approach will be assumed in the exercises at the end of this section, because in each case only the selected points on the real axis of the z-plane, and the way they are required to map onto a degenerate polygon in the w-plane are shown in the diagrams, together with the form of the mapping that is to be derived from Equation (4.86).

Example 4.5.1 Mapping a Semi-Infinite Strip

Map the upper half of the z-plane onto the semi-infinite strip $u > 0, 0 < v < a$ in the w-plane in such a way that $z = -1$ and $z = 1$ have the respective images $w = ia$ and $w = 0$ (a real).

SOLUTION

Consider the semi-infinite strip as the limit of the triangle in Figure 4.35 as $\alpha \to \pi/2$ and let points with primed letters denote the images of points with unprimed letters under the required transformation $w = f(z)$. Two points may be assigned arbitrarily on the x-axis, so we set $x_1 = -1$ and $x_2 = 1$. These have the images $w_1 = ia$ and $w_2 = 0$, and inspection of Figure 4.36 shows that $\theta_1 = \pi/2$ and $\theta_2 = \pi - \alpha$. Thus from Equation (4.86) in Theorem 4.5.1 we see that the transformation is given by

$$w = f(z) = \mathrm{A}\int (z+1)^{-(\pi/2)/\pi}(z-1)^{-(\pi-\alpha)/\pi}\,dz + \mathrm{B}.$$

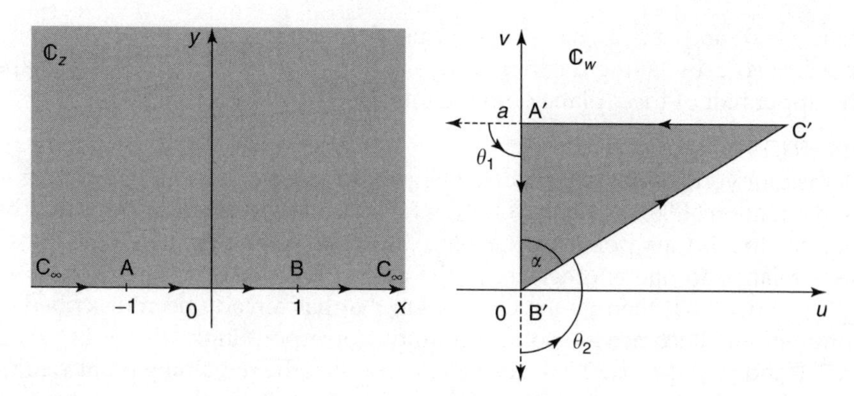

FIGURE 4.36
Mapping a semi-infinite strip.

This integral is complicated and is not expressible in terms of elementary functions. However consider the limit as $\alpha \to \pi/2$, when the triangle tends to the semi-infinite strip $u > 0$, $0 < v < a$ (a degenerate polygon), in which case w is given by the simpler integral

$$w = \mathrm{A} \int \frac{dz}{(z^2 - 1)^{1/2}} + \mathrm{B},$$

and so

$$w = \mathrm{A}\ \mathrm{Arccosh}\, z + \mathrm{B}.$$

As $w = 0$ when $z = 1$ we conclude that $\mathrm{B} = 0$, and as $w = ia$ when $z = -1$, it follows that

$$ia = \mathrm{A}\ \mathrm{Arccosh}(-1).$$

Now $\cosh \pi i = \cos \pi = -1$, so as the principal branch must be used which shows that $\mathrm{Arccosh}(-1) = \pi i$, so that $ia = \mathrm{A} i\pi$, or $\mathrm{A} = a/\pi$. Thus the required transformation is $w = \frac{a}{\pi}\,\mathrm{Arccosh}\, z$. Equivalently, the transformation from the semi-infinite strip in the w-plane to the upper half of the z-plane is given by

$$z = \cosh(\pi w/a).$$

It is recalled that this transformation was considered, although not derived, in the previous section. ◇

Example 4.5.2 Mapping an Infinite Strip

Find the integral determining the function $f(z)$ that maps the upper half of the z-plane onto the rhombus in the w-plane shown in Figure 4.37 in such a way that $z = 0$ and $z = 1$ have the respective images $w = -L + ia/2$ and $w = L + ia/2$. By taking the limit as $L \to \infty$, find the transformation that maps the upper half of the z-plane onto the infinite strip $0 < v < a$ (a real).

SOLUTION

Of the four vertices of the rhombus in Figure 4.37, one will be mapped to $z = \infty$ as in Equation (4.86) in Theorem 4.5.1, while two of the remaining three may be mapped to arbitrary points on the real axis on the z-plane, provided their positions relative to one another are in the correct order. Accordingly, let us set $x_2 = 0$ and $x_3 = 1$; then by the convention that has already been described x_4 automatically becomes the point at infinity. Correspondingly $w_2 = -L + ia/2$, $w_3 = 0$ and $w_4 = L + ia/2$. We are *not* free to assign the remaining point x_1 arbitrarily, but from Equation (4.84) in Theorem 4.5.1 it follows that it must be such that $x_1 < x_2 < x_3$. Consequently, we set $x_1 = -\lambda$ ($\lambda > 0$), where the parameter λ is to be determined. Thus, in terms of the angles in Figure 4.37, the transformation corresponding to Equation (4.86) becomes

$$w = f(z) = \mathrm{A}\int (z+\lambda)^{-\theta_1/\pi}(z-0)^{-\theta_2/\pi}(z-1)^{-\theta_1/\pi}\,dz + \mathrm{B}.$$

This integral is not expressible in terms of elementary functions, but our concern will be with the case when $L \to \infty$ when the rhombus will tend in the limit to the infinite strip $0 < v < a$, so that $\theta_1 \to 0$ and $\theta_2 \to \pi$. So in the limit the transformation simplifies to

$$w = \mathrm{A}\int \frac{dz}{z} + \mathrm{B},$$

so that

$$w = \mathrm{A}\,\mathrm{Ln}\,z + \mathrm{B}.$$

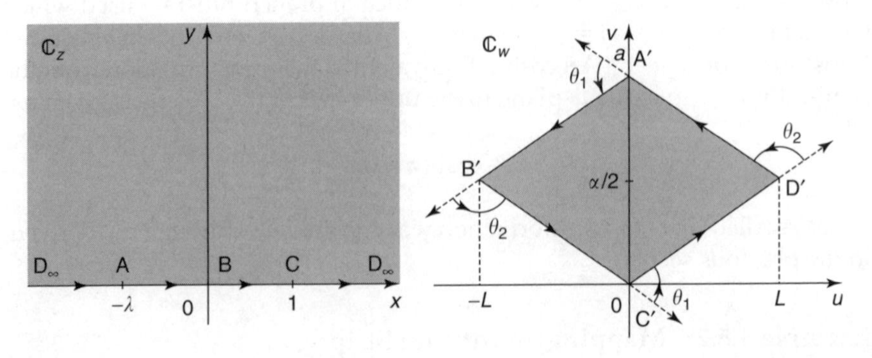

FIGURE 4.37
Mapping an infinite strip.

Now $w = 0$ when $z = 1$, so $B = 0$, while $w = ia$ when $z = -\lambda$, so that $ia = A\,\mathrm{Ln}(-1)$, showing that

$$ia = A(\ln_e \lambda + i\pi),$$

which is only possible if $\lambda = 1$, when $A = a/\pi$, so the value of the parameter λ has been determined. In actual fact, symmetry also dictates this value for λ.

Thus the required transformation is given by

$$w = \frac{a}{\pi}\,\mathrm{Ln}\,z.$$

Equivalently, the transformation from the infinite strip in the w-plane to the upper half of the z-plane is

$$z = \exp(\pi w/a).$$

This mapping was also considered, though not derived, in the previous section. ◇

Example 4.5.3 Mapping a Finite Slit

Find the integral determining the function $f(z)$ that maps the upper half of the z-plane onto the domain of the w-plane shown in Figure 4.38 in such a way that $z = 0$ and $z = 1$ have the respective images $w = ia$ and $w = h + ik$ with $h > 0$ and $k > 0$. By taking the limits as $h \to 0$ and $k \to 0$, find the transformation that maps the upper half of the z-plane onto the w-plane cut along the imaginary axis from $v = 0$ to $v = a$ $(a > 0)$.

SOLUTION

This example illustrates how symmetry arguments may sometimes be used to simplify a mapping problem. The polygon in the w-plane has the five

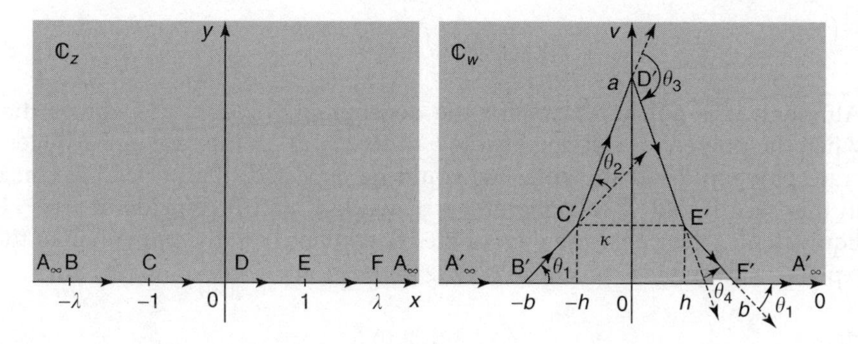

FIGURE 4.38
Mapping a slit.

finite vertices B′, C′, D′, E′, and F′, and one infinite vertex A'_∞ at $w = \infty$. Two of the five may be assigned arbitrarily, and the others must then be arranged in the order necessary to preserve the shape of the domain that is to be mapped. For the two arbitrarily chosen points on the x-axis we will take $x_3 = 0$ and $x_4 = 1$, corresponding to the vertices at $w_3 = ia$ and $w_4 = ih + ik$. From this and Equation (4.84) of Theorem 4.5.1 it follows we must set $x_5 = \lambda$ ($\lambda > 0$) which is to be determined and $x_6 = \infty$, and so we must also have $w_5 = b$ ($b > h$) and $w_6 = \infty$.

The symmetry of the domain to be mapped, together with the choice made for x_3 and x_4, then dictates that C is at $x_2 = -1$ and B is at $x_1 = -\lambda$, while A_∞ is again the point at infinity in the z-plane. Thus transformation in Equation (4.86) becomes

$$w = f(z) = A\int (z+\lambda)^{-\theta_1/\pi}(z+1)^{-\theta_2/\pi}(z-0)^{-\theta_3/\pi} \times (z-1)^{-\theta_2/\pi}(z-\lambda)^{-\theta_1/\pi}\,dz + B.$$

In the limit as $h \to 0$, $k \to 0$ we have $\theta_1 \to 0$, $\theta_2 \to \pi/2$, and $\theta_3 \to -\pi$, which makes the sides C′D′ and D′E′ fold back on themselves and the shape in the w-plane reduce to the upper half of the w-plane cut (slit) along the imaginary axis from $v = 0$ to $v = a$. The preceding transformation then reduces to

$$w = A\int \frac{z}{\sqrt{z^2-1}}\,dz + B,$$

and hence to

$$w = A\sqrt{z^2-1} + B.$$

The limiting form of the degenerate polygon in the w-plane is such that C′ and E′ move to the origin, causing $z = \pm 1$ to correspond to $w = 0$, so that $B = 0$. In addition, $z = 0$ corresponds to $w = ia$, which shows that $A = a$.

Thus the required transformation becomes

$$w = a\sqrt{z^2-1}.$$

Although it is not necessary for the degenerate polygon, this shows that when the polygon is not degenerate $b = a\sqrt{\lambda^2 - 1}$. In the case of the degenerate polygon $b = 0$ showing, as would be expected, that $\lambda = \pm 1$ because in this case B and C coincide at $x = 1$, while E and F coincide at $x = -1$. Equivalently, the transformation of the cut w-plane onto the upper half of the z-plane is given by

$$z = \left(1 + \frac{w^2}{a^2}\right)^{1/2}.$$

The mapping discussed in this example can be derived using a much simpler degenerate polygon than the one shown in Figure 4.38 (see Exercise 5). The polygon in Figure 4.38 was chosen to demonstrate the simplification brought about by symmetry, which in this case amounted to B being located at $z = -\lambda$ and F being located at $x = \lambda$. ◇

Example 4.5.4 Mapping a Semi-Infinite Slit

Map the upper half of the z-plane onto the upper half of the w-plane cut (slit) along the semi-infinite line $u < -1, v = \pi$.

SOLUTION

To find this mapping we first consider the polygon shown in Figure 4.39 in which the points $z = -1$ and $z = 0$ are required to map onto $w = -1 + i\pi$ and $w = -b$, respectively, with $b > 0$. The points A'_∞ and D'_∞ are at $w = \infty$, corresponding to the points A_∞ and D_∞ at $z = \infty$.

The required transformation corresponding to Equation (4.86) in Theorem 4.5.1 follows by setting $x_1 = -1, x_2 = 0$, to give

$$w = f(z) = A\int (z+1)^{-\theta_1/\pi}(z-0)^{-\theta_2/\pi}\,dz + B.$$

This integral cannot be evaluated in terms of elementary functions, but we are only interested in the degenerate case where $b \to \infty$, when the side B′C′ folds back onto the side $B'A'_\infty$ as in Figure 4.40, causing the degenerate polygon to reduce to a cut (slit) in the w-plane along the line $u = -1, v = \pi$. When this occurs $\theta_1 \to -\pi$ and $\theta_2 \to \pi$, causing the transformation to simplify to

$$w = A\int \left(1 + \frac{1}{z}\right) dz + B,$$

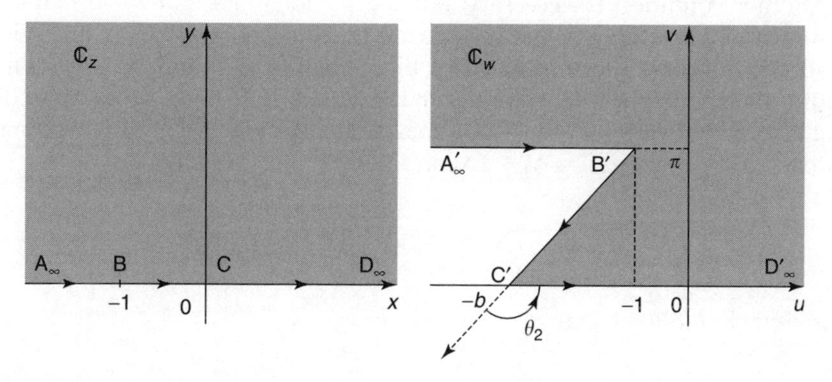

FIGURE 4.39
Mapping an infinite slit as $b \to \infty$.

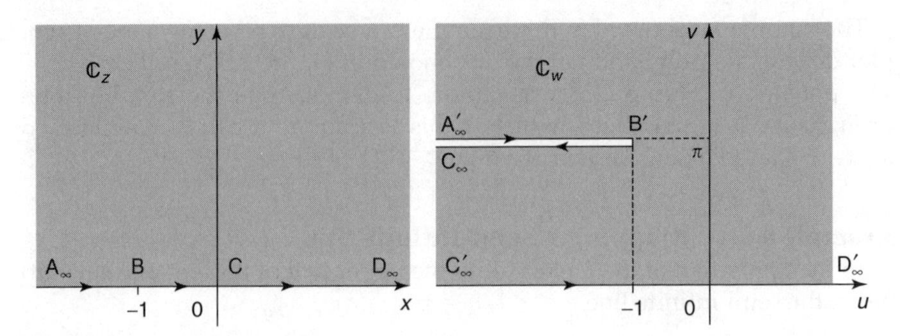

FIGURE 4.40
Cut w-plane.

with the solution

$$w = A(z + \text{Ln}\, z) + B.$$

To determine the complex constants A and B, it is necessary to find conditions that must be satisfied by this solution. An obvious condition is that $z = -1$ maps to $w = -1 + i\pi$, when the previous result becomes

$$-1 + i\pi = A(-1 + \text{Ln}(-1)) + B.$$

Using the principal branch of the logarithmic function gives $\text{Ln}(-1) = i\pi$, so the first condition to be satisfied by A and B is

$$-1 + i\pi = A(-1 + i\pi) + B.$$

Another condition is necessary, but the obvious one that $z = 0$ maps to $w = \infty$ cannot be used, because it leads to a meaningless expression involving A, B and infinity. The resolution of the difficulty is found by examining Figure 4.40 which shows that points on the real line CD_∞ map to points on the real line $C'_\infty D'_\infty$. Setting $w = u + iv$, we see that all points $z = x > 0, y = 0$ map onto $v = 0$. So setting $A = A_1 + iA_2$ and $B + B_1 + iB_2$ we have

$$w = u + iv = (A_1 + iA_2)(z + \text{Ln}\, z) + B_1 + iB_2,$$

from which it follows that

$$v = A_1 y + A_2 x + A_2 \ln_e |z| + A_1 \text{Arg}\{z\} + B_2.$$

Setting $v = 0$, $z = x > 0$ and $y = 0$, when $\text{Arg}\{z\} = \text{Arg}\{x\} = 0$, this reduces to

$$0 = A_2(x + \ln_e x) + B_2 \qquad (x > 0).$$

This result can only be true for all $x > 0$ if $A_2 = B_2 = 0$. In which case the condition $-1 + i\pi = A(-1 + i\pi) + B$ simplifies to

$$-1 + i\pi = A_1(-1 + i\pi) + B_1.$$

Equating real and imaginary parts gives $A_1 = 1$ and $B_1 = 0$, showing that the required transformation is

$$w = z + \text{Ln}\, z.$$

Notice that the cut in the w-plane acts as a barrier that separates values of w above the cut from those below it. Later, this fact will be seen to be important in applications involving boundary value problems for the Laplace equation because there such a cut will represent a physical barrier that separates the solution above the cut from the one below it. For example, in a steady-state heat flow problem, the cut could represent a barrier with different temperatures on either side of it, with heat unable to flow across the barrier. ◇

Exercises 4.5

In each of the following exercises the Schwarz–Christoffel transformation is to be used to derive the given transformation that maps the upper half of the z-plane onto the w-plane in the way shown in the accompanying diagram. Specific points in the z-plane that are to be used when deriving the transformation are indicated by unprimed letters, while the specific points to which they are to map in the w-plane are indicated by primed letters.

1.

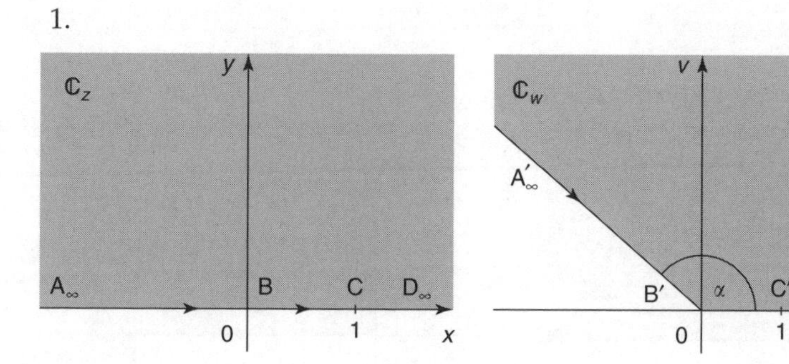

FIGURE 4.41

The transformation to be derived is $w = z^{\alpha/\pi}$.

2.

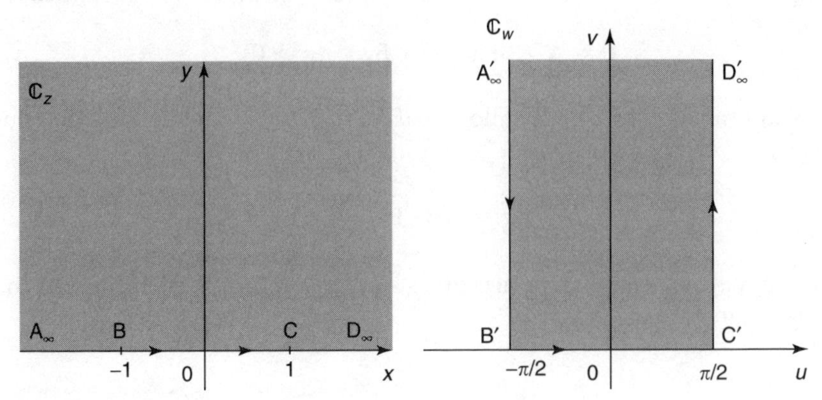

FIGURE 4.42

The transformation to be derived is $w = \text{Arcsin}\, z$

3.

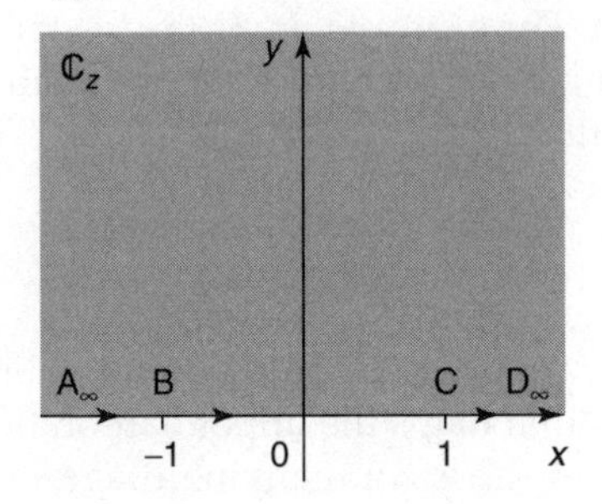

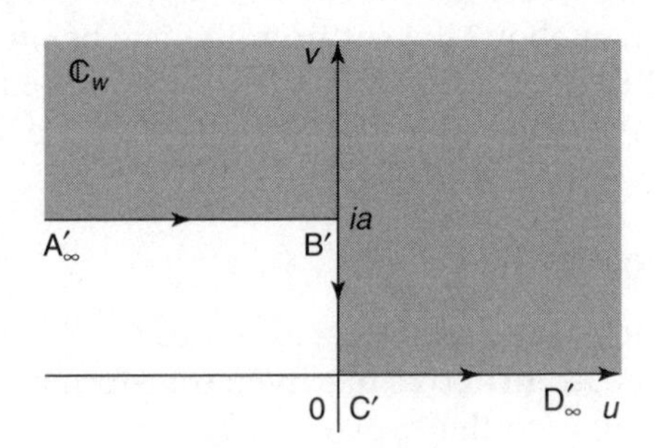

FIGURE 4.43

The transformation to be derived is $w = \dfrac{a}{\pi}\left(\sqrt{z^2-1} + \text{Ln}\left(z + \sqrt{z^2-1}\right)\right)$

4.

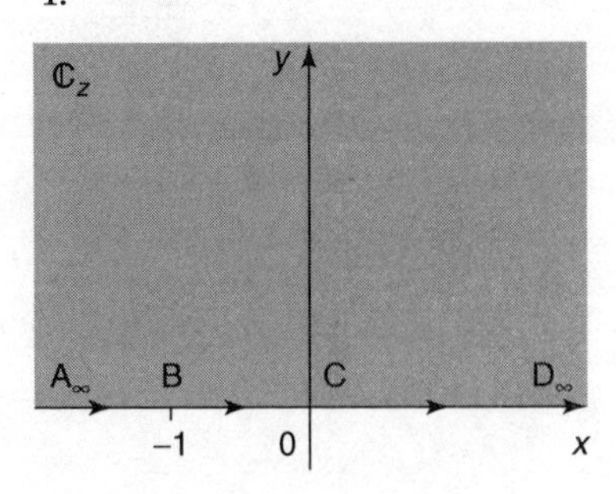

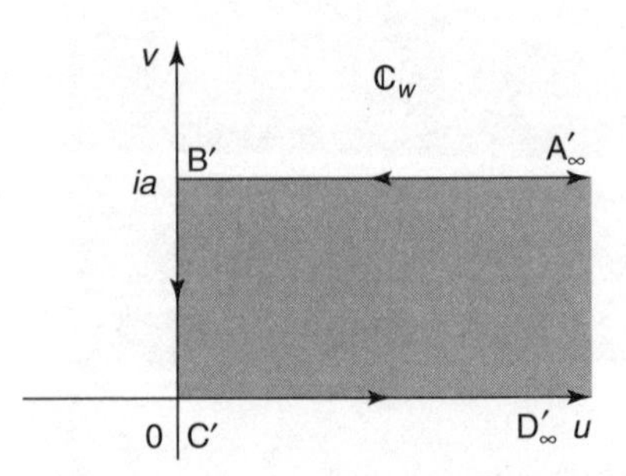

FIGURE 4.44

The transformation to be derived is $w = \dfrac{a}{\pi}\text{Ln}\left(1 + 2z + 2\sqrt{z^2+1}\right)$

5.

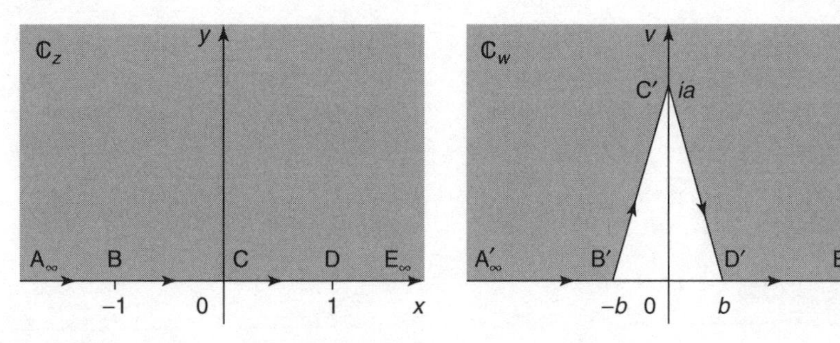

FIGURE 4.45

The transformation to be derived is $w = a\sqrt{z^2 - 1}$

6.

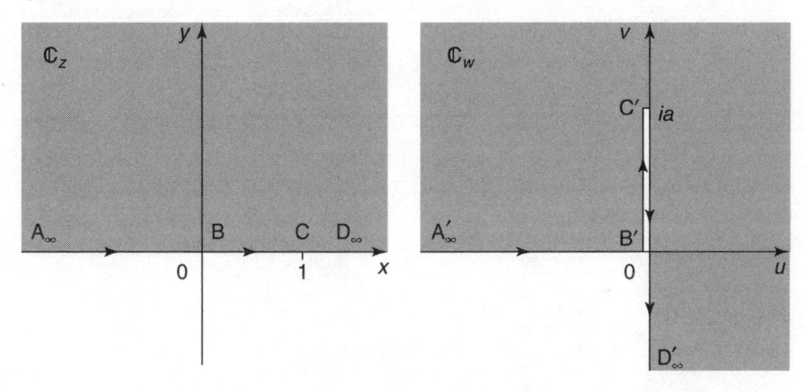

FIGURE 4.46

The transformation to be derived is $w = -\frac{1}{2} iaz^{1/2}(z - 3)$

7.

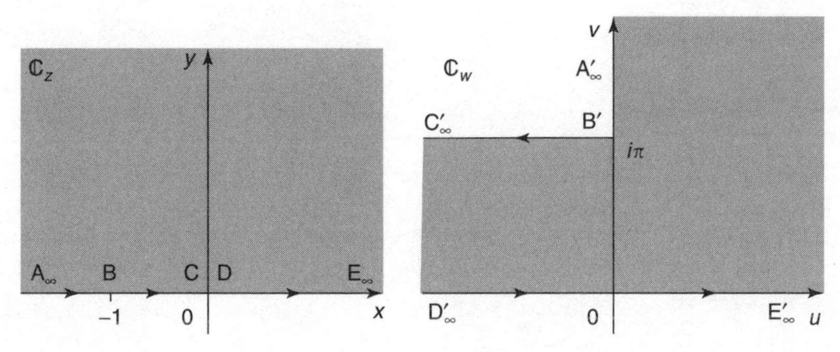

FIGURE 4.47

The transformation to be derived is $w = 2\sqrt{z+1} + \text{Ln}\left(\frac{\sqrt{z+1} - 1}{\sqrt{z+1} + 1}\right)$

5

Boundary Value Problems, Potential Theory, and Conformal Mapping

5.1 Laplace's Equation and Conformal Mapping: Boundary Value Problems

This section provides some physical motivation for the study of conformal mapping that follows. Its purpose is to establish the connection between a conformal mapping and the solution of what is called a two-dimensional boundary value problem for the Laplace equation, which is studied later in some detail. We start by recalling that a two-dimensional harmonic function $\phi(x, y)$ in a domain D of the (x, y)-plane satisfies the *Laplace equation*

$$\Delta\phi(x, y) = \frac{\partial^2 \phi}{\partial x^2} + \frac{\partial^2 \phi}{\partial y^2} = 0. \tag{5.1}$$

Let $w = f(z) = u + iv$ be an analytic function that maps a domain D in the z-plane one-one and conformally onto a domain D^* in the w-plane. Let us examine what happens to the Laplace Equation (5.1) if the Cartesian coordinates (x, y) are changed to the differentiable curvilinear coordinates variables $u = u(x, y)$ and $v = v(x, y)$ in the function $f(z) = u + iv$, when $\phi(x, y)$ becomes the function $\Phi(u, v) = \phi(u(x, y), v(x, y))$. From the chain rule for differentiation, we have

$$\frac{\partial \phi}{\partial x} = \frac{\partial \Phi}{\partial u}\frac{\partial u}{\partial x} + \frac{\partial \Phi}{\partial v}\frac{\partial v}{\partial x}, \tag{5.2}$$

and a further differentiation with respect to x gives

$$\frac{\partial^2 \phi}{\partial x^2} = \left[\frac{\partial}{\partial x}\left(\frac{\partial \Phi}{\partial u}\right)\right]\left(\frac{\partial u}{\partial x}\right) + \left(\frac{\partial \Phi}{\partial u}\right)\left(\frac{\partial^2 u}{\partial x^2}\right) + \left[\frac{\partial}{\partial x}\left(\frac{\partial \Phi}{\partial v}\right)\right]\left(\frac{\partial v}{\partial x}\right)$$
$$+ \left(\frac{\partial \Phi}{\partial v}\right)\left(\frac{\partial^2 v}{\partial x^2}\right). \tag{5.3}$$

Examination of Equation (5.2) shows that the operation of differentiation of ϕ with respect to x is related to the operation of differentiation of $\Phi(u, v)$ with respect to u and v by the linear operator

$$\frac{\partial(.)}{\partial x} = \frac{\partial u}{\partial x}\frac{\partial(.)}{\partial u} + \frac{\partial v}{\partial x}\frac{\partial(.)}{\partial v}. \tag{5.4}$$

In terms of this operator in Equation (5.3) becomes

$$\frac{\partial^2\phi}{\partial x^2} = \frac{\partial^2\Phi}{\partial u^2}\left(\frac{\partial u}{\partial x}\right)^2 + \frac{\partial^2\Phi}{\partial v^2} + \left(\frac{\partial v}{\partial x}\right)^2 + \left(\frac{\partial^2\Phi}{\partial u\,\partial v} + \frac{\partial^2\Phi}{\partial v\,\partial u}\right)\left(\frac{\partial u}{\partial x}\right)\left(\frac{\partial v}{\partial x}\right)$$
$$+ \frac{\partial\Phi}{\partial u}\left(\frac{\partial^2 v}{\partial x^2}\right) + \frac{\partial\Phi}{\partial v}\left(\frac{\partial^2 v}{\partial x^2}\right). \tag{5.5}$$

The corresponding expression for $\partial^2\phi/\partial y^2$ follows directly from Equation (5.5) by replacing x by y whenever it occurs in a partial derivative.

Combining the expressions for $\partial^2\phi/\partial x^2$ and $\partial^2\phi/\partial y^2$ gives

$$\Delta\phi = \frac{\partial^2\Phi}{\partial u^2}\left[\left(\frac{\partial u}{\partial x}\right)^2 + \left(\frac{\partial u}{\partial y}\right)^2\right] + \frac{\partial^2\Phi}{\partial v^2}\left[\left(\frac{\partial v}{\partial x}\right)^2 + \left(\frac{\partial v}{\partial y}\right)^2\right]$$
$$+ \left(\frac{\partial^2\Phi}{\partial u\,\partial v} + \frac{\partial^2\Phi}{\partial v\,\partial u}\right)\left(\frac{\partial u}{\partial x}\frac{\partial v}{\partial x} + \frac{\partial u}{\partial y}\frac{\partial v}{\partial y}\right) + \frac{\partial\Phi}{\partial u}\Delta u + \frac{\partial\Phi}{\partial v}\Delta v. \tag{5.6}$$

The last three terms in Equation (5.6) vanish; the first because of the Cauchy–Riemann equations and the remaining two because u and v are harmonic functions so that $\Delta u = \Delta v = 0$. Finally, after using the Cauchy–Riemann equations in the first two terms we find that Equation (5.6) can be written in the concise form

$$\Delta\phi = |f'(z)|^2\Delta\Phi. \tag{5.7}$$

If ϕ is harmonic

$$\Delta\phi = |f'(z)|^2\Delta\Phi = 0,$$

provided that $|f'(z)|^2 \neq 0$, it follows directly that Φ must also be harmonic. It is recalled that the transformation $w = f(z) = u + iv$ is conformal everywhere except at the critical points determined by those values of z for which $f'(z) = 0$. So we have proved that under a conformal transformation $w = u + iv$, the harmonic nature of ϕ is preserved when transformed to Φ. The only points in D^* where Φ fails to be harmonic is at the critical points of $f(z)$, which are of course the points where the conformality of the transformation breaks down. We now state this result as a theorem.

THEOREM 5.1.1 *Harmonic Functions under a Conformal Mapping*
Let the analytic function $w = f(z)$ provide a mapping that is one-one and conformal between a domain D in the z-plane and its image D^* in the w-plane. Set $w = u + iv$ and let the change of variable $u = u(x, y)$, $v = v(x, y)$ transform the function $\phi(x, y)$ into the function $\Phi(u, v)$. Then if $\phi(x, y)$ is harmonic in D, the function $\Phi(u, v)$ will be harmonic in D^*, with the exception of the critical points of $f(z)$ where $f'(z) = 0$. ◆

To understand the implication of this theorem, suppose the boundary of D has a complicated shape and that a conformal transformation $w = f(z)$ maps D onto a domain D^* whose boundary has a simple shape. Then the task of finding a harmonic function ϕ in D that satisfies certain conditions on the boundary of D can be replaced by the much simpler task of finding function Φ in D^* that satisfies corresponding conditions on its boundary. The fact that the transformation $w = f(z)$ is one-one and conformal means that once the solution Φ has been found in D^*, it can be transformed back to the corresponding solution ϕ in D.

Our next task will be to define the nature of a boundary value problem for the Laplace equation and to consider what conditions can be imposed on ϕ when it is on the boundary of D (the boundary conditions).

5.1.1 Boundary Value Problems of the First Kind: Dirichlet Problems

Consider the problem illustrated in Figure 5.1(a) in which the real valued function $\phi(x, y)$ is required to satisfy the Laplace equation in the two-dimensional domain D with boundary γ such that $\phi(x, y)$ varies in a prescribed way around γ. To be precise, suppose that

$$\Delta\phi = 0 \quad \text{in } D, \tag{5.8}$$

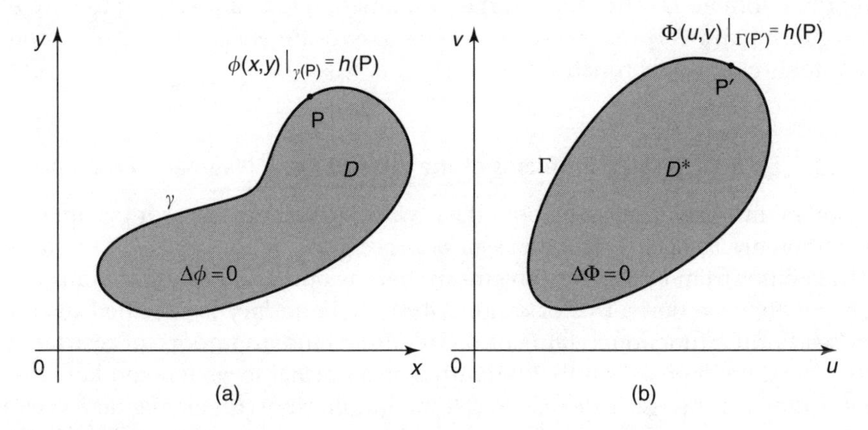

FIGURE 5.1
(a) Dirichlet problem for ϕ in D. (b) Dirichlet value problem for Φ in D^*.

while on the boundary γ of D

$$\phi(x, y)\big|_{\gamma(P)} = h(\text{P}), \tag{5.9}$$

where h is a given function that defines the value of ϕ at each point P on γ. Then the problem of finding a solution of Equation (5.8) subject to condition in Equation (5.9) is called a *boundary value problem* for the Laplace equation. The condition in Equation (5.9) assigns the *boundary value condition* to be satisfied by the harmonic function ϕ on γ. Boundary value problems of this type are of great practical importance and they are called *boundary value problems of the first kind,* or *Dirichlet problems.* To avoid interrupting this part of the text, the proof that such problems have a unique solution in a bounded domain will be given in Appendix 5.1.

Let us suppose that the analytic function $w = f(z)$ in Theorem 5.1.1 maps the domain D, which can be considered to be in the z-plane, conformally and one-one onto the domain D^* in the w-plane, as in Figure 5.1(b). Then the boundary value $h(\text{P})$ to be satisfied by ϕ at P on γ is transferred by the transformation to the image point P′ on the boundary Γ of D^* that is the image of γ. Then, from Theorem 5.1.1, the function $\Phi(u, v)$ will also be harmonic and such that

$$\phi(x, y)\big|_{\gamma(\text{P})} = \Phi(u, v)\big|_{\Gamma(\text{P}')} = h(\text{P}). \tag{5.10}$$

Thus Theorem 5.1.1 ensures that a Dirichlet boundary value problem for ϕ in D is transferred to an equivalent Dirichlet boundary value problem for Φ in D^*.

The one-one nature of the conformal transformation ensures that these two boundary value problems are equivalent. So if a transformation can be found that maps the complicated boundary γ onto an equivalent but much simpler boundary Γ, the task of finding a solution for Φ will be much simpler than that of finding a solution for ϕ. Once the solution Φ has been found, the solution ϕ follows by transforming the solution Φ back from domain D^* to the original domain D. This, then, is the fundamental idea underlying the use of a conformal transformation when solving a two-dimensional Dirichlet problem for an harmonic functions ϕ.

5.1.2 Boundary Value Problems of the Second Kind: Neumann Problems

Another important class of boundary value problems involving Laplace's equation are *boundary value problems of the second kind,* or *Neumann problems.* These differ from Dirichlet problems in that the boundary condition imposed on γ is that the derivative of ϕ normal to the boundary is specified over γ, instead of the functional value of ϕ. To understand an important restriction that must be imposed on the derivative of ϕ normal to γ, denoted hereafter by $\partial\phi/\partial n$, it is necessary to assume some familiarity with elementary vector calculus. First, though, a Neumann problem in which $\partial\phi/\partial n = 0$ over the entire boundary γ is called a *homogeneous Neumann problem*. Such a problem

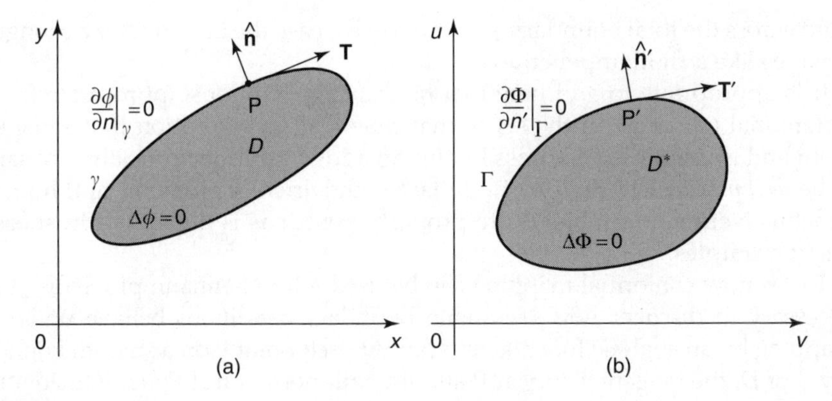

FIGURE 5.2
(a) A homogeneous Neumann problem for ϕ in D. (b) A homogeneous Neumann problem for Φ in D^*.

is illustrated in Figure 5.2(a) where the notation $\partial\phi/\partial n|_\gamma = 0$ is used to signify that $\partial\phi/\partial n$ vanishes at each point of γ that forms the boundary of D. To find a restriction that must be placed on $\partial\phi/\partial n$, we must consider the line integral $\int_\gamma(\partial\phi/\partial n)ds$, where ds is a line element of γ. Then if dS is the element of area in D, and $\hat{\mathbf{n}}$ is the outward drawn unit normal to γ at P, the directional derivative of ϕ in the direction of $\hat{\mathbf{n}}$ is $\partial\phi/\partial n = \hat{\mathbf{n}}\cdot\operatorname{grad}\phi$. An application of the two-dimensional form of the Gauss divergence theorem then shows that

$$\int_\gamma \frac{\partial\phi}{\partial n}ds = \int_\gamma \hat{\mathbf{n}}\cdot\operatorname{grad}\phi\, ds = \int_D \Delta\phi\, dS. \tag{5.11}$$

However ϕ is harmonic, so that $\Delta\phi = 0$, showing that in a Neumann problem $\partial\phi/\partial n$ must be such that

$$\int_\gamma \frac{\partial\phi}{\partial n}ds = 0. \tag{5.12}$$

It follows directly from Equation (5.12) that a homogeneous Neumann problem is always permissible because the homogeneous boundary condition automatically satisfies Equation (5.12).

To give a possible physical meaning to the condition in Equation (5.12) it is necessary to anticipate a later result. This is that a harmonic function ϕ can describe what is called the *velocity potential* of an incompressible fluid such as water. In such a case, the velocity potential has the property that the vector grad ϕ is the velocity $\mathbf{v}$ of the fluid. So because $\partial\phi/\partial n = \hat{\mathbf{n}}\cdot\operatorname{grad}\phi$, we see that $\partial\phi/\partial n$ is the velocity of the fluid *normal* to the boundary γ. In such circumstances, a homogeneous Neumann problem corresponds to a two-dimensional fluid flow in which fluid at the boundary γ can flow tangential to the boundary, but *not* across it, so that γ behaves like a rigid wall confining the fluid. If $\partial\phi/\partial n \neq 0$, the line integral $\int_\gamma(\partial\phi/\partial n)ds$ measures the net flow of fluid across the total boundary γ, and in such a case the boundary γ no longer behaves like a rigid impenetrable wall.

It is appropriate to add here that boundary conditions appropriate for a differential equation, in the sense that they lead to a solution that satisfies them and is unique except possibly for an added arbitrary constant, are said to be *well posed*, or *properly posed*. In fact both Dirichlet problems and homogeneous Neumann problems are properly posed, as is illustrated in subsequent examples.

To see how conformal mapping can be used with Neumann problems, it is necessary to discover how Neumann boundary conditions behave under a mapping by an analytic function $w = f(z)$. At each point P on a smooth boundary γ of D, the tangent $\mathbf{T}$ to γ at P and the unit normal $\hat{\mathbf{n}}$ at P are defined and mutually orthogonal. Thus a conformal transformation of the boundary γ in Figure 5.2(a) to the boundary Γ in Figure 5.2(b) ensures that vectors $\mathbf{T}$ and $\hat{\mathbf{n}}$ at P on γ transform to vectors $\mathbf{T}'$ and $\hat{\mathbf{n}}'$ at P′ on Γ so that the image of γ are again mutually orthogonal and such that $\mathbf{T}'$ is tangent to Γ at P′. Thus a homogeneous Neumann boundary condition on ϕ is transformed into an equivalent homogeneous Neumann condition for Φ on Γ. Thereafter, a conformal transformation plays the same role as with a Dirichlet problem, by enabling γ to be transformed to a simple boundary Γ, thereby permitting Φ to be found, and hence ϕ.

It is important to recognize that if the Neumann condition is nonhomogeneous, then although the tangent and normal at any point P on γ will transform the equivalent tangent and normal at P′ on Γ, where P′ is the image of P, the boundary condition itself alters its form as a result of the transformation. However, such problems are not considered here.

We mention that sometimes a homogeneous combination of Dirichlet and Neumann conditions are imposed on the boundary, leading to a special case of a *boundary value problem of the third kind*, in which the boundary condition takes the general form

$$\alpha \frac{\partial \phi}{\partial n} + \beta \phi = 0 \quad \text{on } \gamma. \tag{5.13}$$

If, as sometimes happens, Dirichlet conditions are imposed on part of the boundary and Neumann conditions on the remainder of it, the functions α and β will be discontinuous.

The relevance of the Laplace equation to the description of heat conduction, diffusion, the flow of an incompressible fluid and the distribution of an electrostatic field is established later and in general as such problems involve three space dimensions and time, they are *time-dependent*, or *unsteady* problems. However, in all of these physical problems it is possible that as time increases the solution settles down and becomes independent of time, leading to what is called a *steady-state* solution that in these cases, turns out to be the Laplace equation, although more will be said about this later. The general study of the Laplace equation in any number of space dimensions is called *potential theory*, so the steady-state applications just mentioned, together with others that do not concern us here, all form part of this branch of mathematics. Conformal transformations are two-dimensional, so it is necessary that the meaning of a two-dimensional problem is made clear and its relevance to the more general threedimensional is established.

Two-dimensional problems are problems in the three space dimensions, say x, y, and z, in which a coordinate axis like the z-axis can be chosen in such a way the solution in every plane z = constant depends only on x and y.

To illustrate matters, let us accept for the moment that the Laplace equation describes the following physical situation in which fluid flows in a steady manner past a cylindrical barrier of constant radius with its axis normal to the direction of flow. This is illustrated in Figure 5.3, where because the flow is steady, the cylinder has a constant radius and the walls of the cylinder are normal to the direction of flow, the lines followed by particles of fluid with velocity $\mathbf{q}$ (the *streamlines*) at similar positions P_1 and P_2 in the respective planes $z = z_1$ and $z = z_2$ will be the same. So in this case the streamlines in any plane z = constant are the same, and though z does not enter into the solution it is convenient to consider the streamlines to be in the plane $z = 0$. This example also illustrates the need for a Neumann boundary condition on the surface of the cylinder because fluid cannot penetrate the wall of the cylinder so the component of fluid velocity normal to the wall must be zero.

Another physical example concerns the steady-state temperature distribution in a very long bar of metal of semicircular cross-section when the cylindrical boundary is maintained at a constant temperature T_1, and the plane face is thermally insulated except for a centrally located narrow strip that is

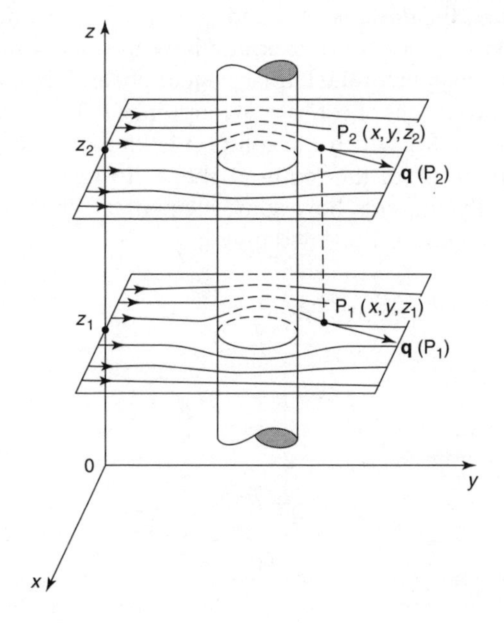

FIGURE 5.3
Steady flow past a cylinder.

maintained at a constant temperature T_0. The determination of the temperature at any point in the bar involves solving the Laplace equation with the Dirichlet condition on the curved surface $T = T_1$, the Dirichlet condition $T = T_0$ on the strip on the plane side, and the Neumann condition $\partial T/\partial n = 0$ on the thermally insulated surface. This is because if $\partial T/\partial n \neq 0$, a temperature gradient will cause heat to flow. Because the bar is very long and the temperatures of the surfaces are constant, when well away from the ends of the bar the variation of temperature with distance measured along the bar can be ignored, so the temperature distribution can again be represented by a two-dimensional situation in any plane normal to the length of the bar. Figure 5.4 shows a typical situation, where the curves $T = T_i$ represent lines along which the temperature T is constant. Such lines of constant temperature are called *isothermal* lines.

In general, the geometry of such two-dimensional problems dictates the choice of coordinates and in most cases these are likely to be either the rectangular Cartesian coordinates (x, y) or the plane polar coordinates (r, θ).

Before closing this introductory section, and prior to considering applications of conformal mapping, it is necessary to draw attention to the fact that two different methods will be considered when finding a solution of a boundary value problem for the two-dimensional Laplace equation in a nontrivially shaped domain.

The most obvious method starts with a real harmonic potential function ϕ known to be the solution of a boundary value problem in a domain D with a simply shaped boundary γ. A conformal mapping is then used to transform the solution into a corresponding solution with the same boundary conditions, but in a domain D^* of interest that has a boundary Γ with a more complicated shape. This is called the *direct method* and is first to be considered here. When using this method it is necessary to know how to solve a number of simple boundary value problems for the Laplace equation, and then to be sufficiently familiar with the properties of conformal mappings to know how to transform the simple domain D for which the solution is known, into a corresponding solution in a domain D^* of the required shape. To help with this approach, Table 5.1 graphically presents how some elementary analytic functions map simple shapes into more complicated ones.

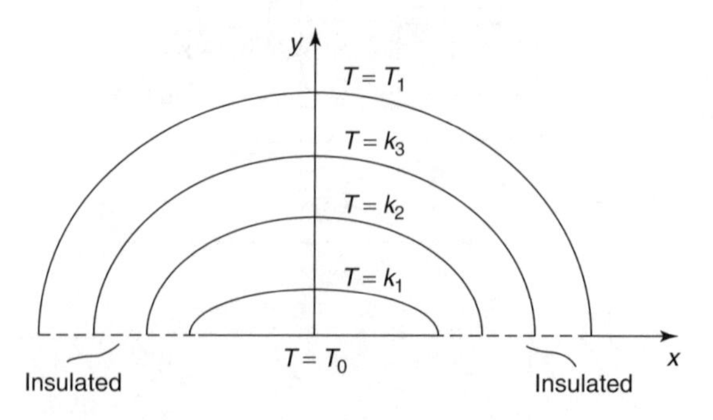

FIGURE 5.4
Isothermals in a semicircular metal bar with a thermally insulated strip.

Table 5.1

Some Useful Conformal Maps

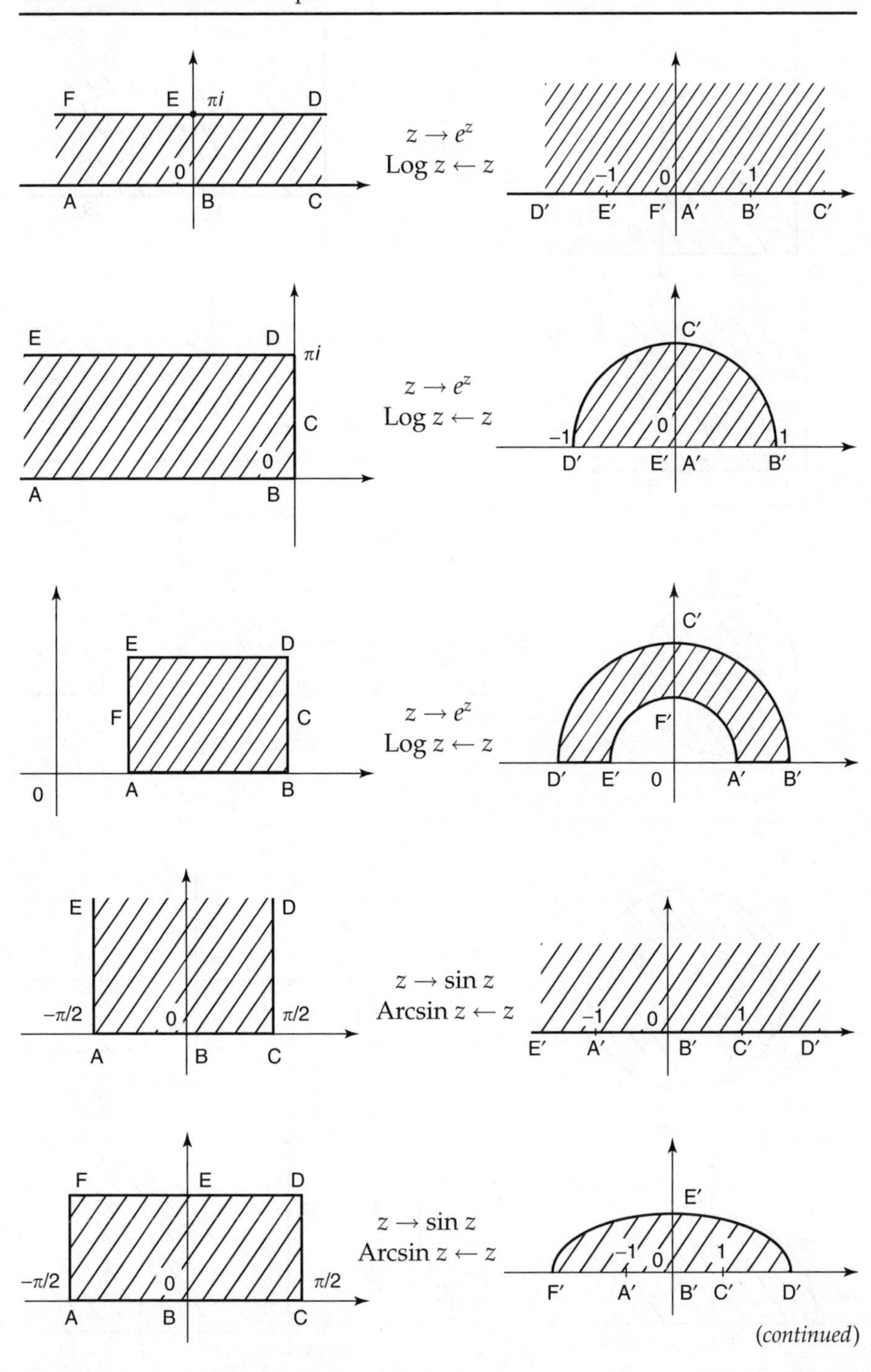

(*continued*)

Table 5.1
(*Continued*)

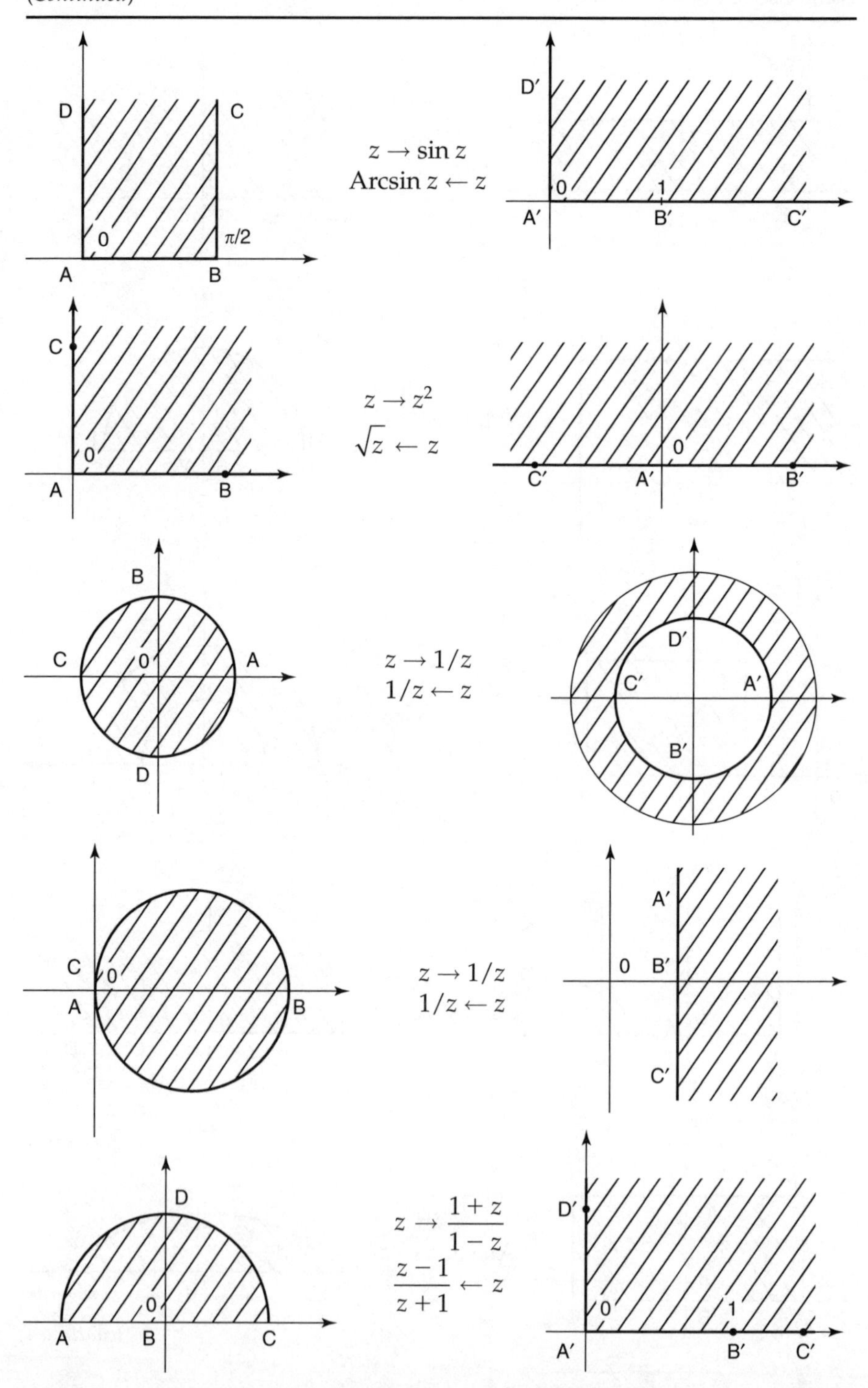

Table 5.1
(Continued)

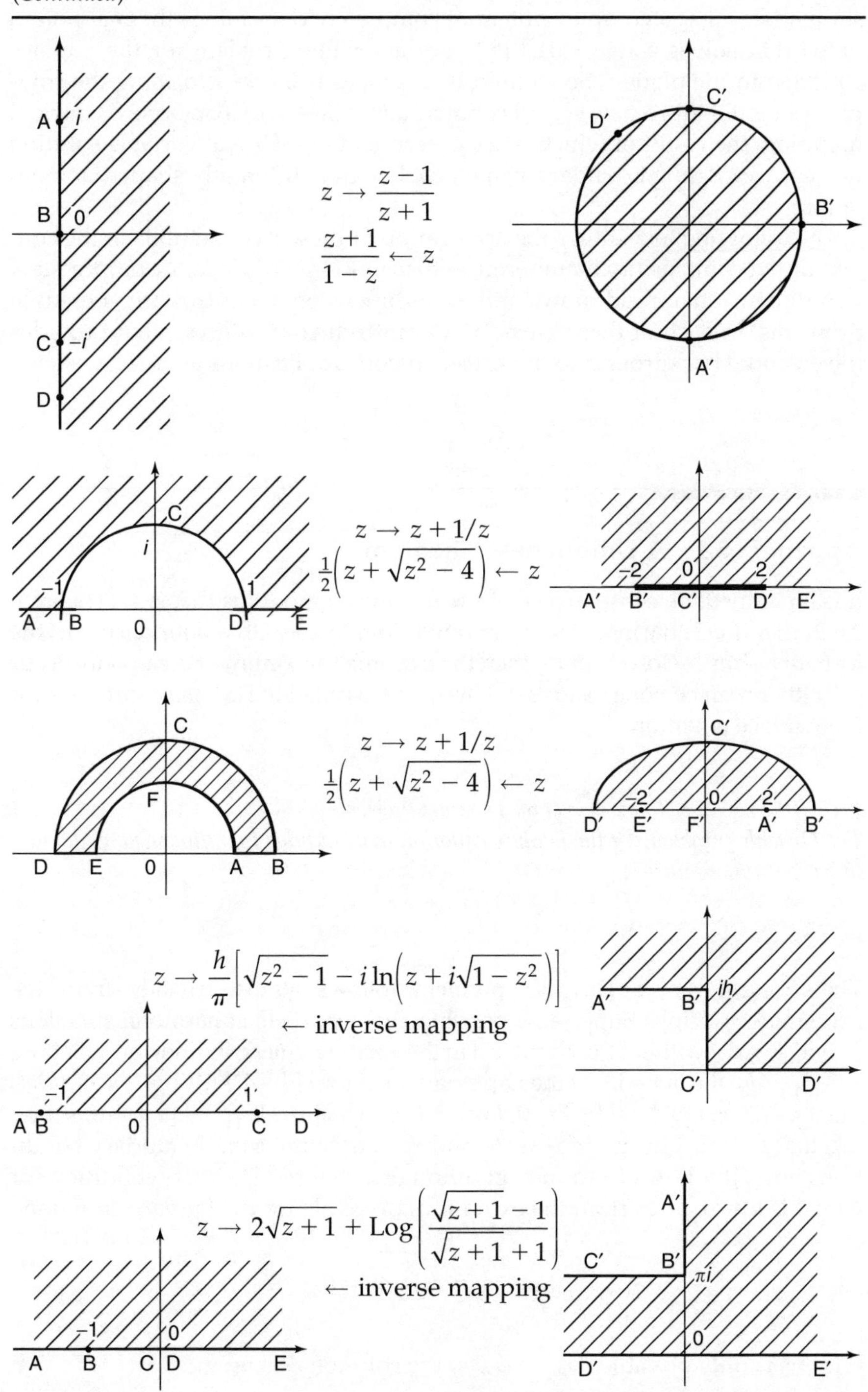

The second method to be considered, the *indirect method* begins with an analytic function of a complex variable called a *complex potential*. It will be shown later that a complex potential completely determines the real potential ϕ that solves a real variable boundary value problem for the Laplace equation in the plane. The method then proceeds by transforming the complex potential into a new complex potential by means of a one-one conformal mapping, the result of which is the determination of a real variable solution of a corresponding boundary value problem in a differently shaped domain of interest.

The following text shows the application of these two methods in the context of two-dimensional temperature distributions in solids, incompressible two-dimensional fluid flow, and two-dimensional electrostatic potential problems. In each of these areas, a brief introduction is given that provides the essential background to the subject before applications are made.

Appendix 5.1 A Uniqueness Theorem

It is an immediate consequence of the maximum modulus theorem (Theorem 2.6.2) that if ϕ is harmonic in a bounded domain D with a boundary γ in the form of a simple closed curve, then the extrema (max/min) of ϕ can only occur on γ (there can be none inside D). This result is called the *extremum principle* for the Laplace equation.

THEOREM 5.1.2 Uniqueness of the Dirichlet Problem
The Dirichlet problem for the Laplace equation in a bounded two-dimensional domain D has a unique solution.

PROOF

The *uniqueness* of a Dirichlet problem follows almost trivially from the extremum principle. Suppose, if possible, that two distinct harmonic functions ϕ_1 and ϕ_2 exist within D, each subject to the *same* Dirichlet condition on γ. Setting $u = \phi_1 - \phi_2$, the linearity of the Laplacian operator $L[.] = (\partial^2(.)/\partial x^2) + (\partial^2(.)/\partial y^2)$ means that as $L[u_1] = L[u_2] = 0$, $L[u] = L[u_1 + u_2] = L[u_1] + L[u_2] = 0$, showing that u is also harmonic. As ϕ_1 and ϕ_2 satisfy the same boundary conditions on γ, it follows from the definition of u that the Dirichlet condition for u on γ is $u = 0$. Thus from the extremum principle for the Laplace equation

$$\max u|_\gamma = \min u|_\lambda = 0,$$

but this is only possible if $\phi_1 \equiv \phi_2$, so the solution ϕ is unique. ◆

5.2 Standard Solutions of the Laplace Equation

Before solving physical problems using conformal mapping methods, we first need to obtain solutions for some standard mathematical boundary value problems for the Laplace equation. These standard problems arise frequently when conformal mapping methods are used in conjunction with the direct method, and also elsewhere. In this context it is necessary to introduce the notion of what is called a *level surface* in a space of three dimensions, and then for what is to follow to generalize this to its counterpart in a two-dimensional problem called a *level curve*.

In general, if $\phi(x, y, z)$ is a real valued function in the three-space dimensions x, y, and z, a surface on which $\phi(x, y, z) = \text{const.}$ is called a *level surface* for the function. So, for example, if $\phi(x, y, z)$ represents the temperature T_0 at a point $P(x, y, z)$ in a metal block, a surface $\phi(x, y, z) = T_0$ is a surface inside the block at each point of which the temperature $T = T_0$. In heat conduction such a surface is called an *isothermal surface* (a surface of constant temperature). Similarly if $\phi(x, y, z)$ represents a potential ϕ, as in an electrostatic problem, the surface $\phi(x, y, z) = \phi_0$ is called an *equipotential surface* (a surface of constant potential). Recalling the definition of a two-dimensional problem, the analog of *a level surface* is seen to be a *level curve*. So if, for example, $\phi(x, y)$ represents the temperature T in a long metal bar with constant cross-section, constant boundary conditions along its walls and such that the z-axis is directed along the bar, the curve $\phi(x, y) = T_0$ is a curve (on a cross-section $z = \text{const.}$ of the bar) on which the temperature has a constant value T_0. This is the case in Figure 5.2 where the curves $T = T_i$ are *isothermal curves* or *isothermals*. For example, if $\phi(x, y, z) = x^2/a^2 + y^2/b^2 + z^2/c^2$, then a level surface would be the ellipsoid $\phi(x, y, z) = k$ $(k > 0)$, while its analog in two-space dimensions would be the level surface in the form of the ellipse $\phi(x, y) = x^2/a^2 + y^2/b^2 = k$ $(k > 0)$.

5.2.1 A Dirichlet Problem in an Infinite Strip

Let us find a real function $\phi(x, y)$ that is harmonic in an arbitrarily oriented infinite strip D of width h, which assumes the constant boundary values k_1 and k_2 on its sides (Dirichlet conditions) as shown in Figure 5.5(a).

The solution will be found in two stages; the first of which involves solving the simpler boundary value problem for the harmonic function $\Phi(x, y)$ in the strip D_1 that occupies the region $0 \leqslant X \leqslant h$ shown in Figure 5.5(b) with Φ subject to the boundary conditions

$$\Phi(0, Y) = k_1 \qquad \text{and} \qquad \Phi(h, Y) = k_2, \qquad \text{for all } Y. \tag{5.14}$$

Because the strip is parallel to the Y-axis and the boundary conditions on $X = 0$ and $X = h$ are independent of Y, the required harmonic function

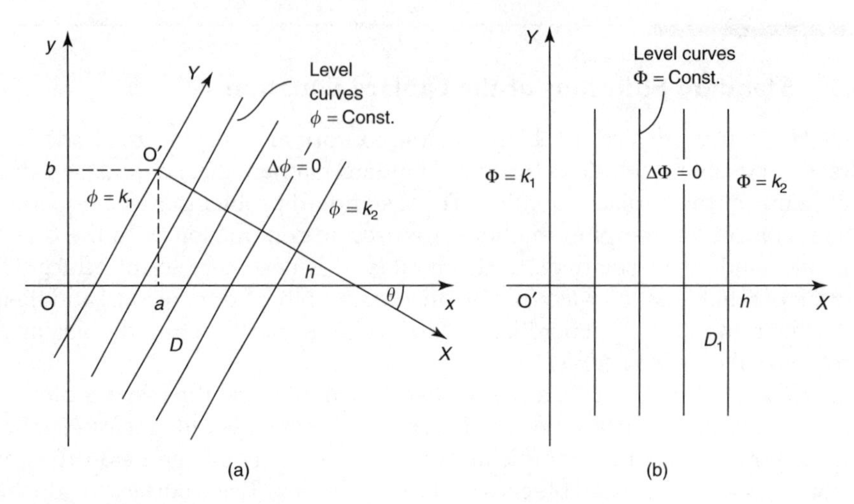

FIGURE 5.5
A Dirichlet problem in an arbitrarily oriented infinite strip of width h.

$\Phi(X, Y)$ can only be a function of X, say $U(X)$, so that

$$\Phi(X, Y) = U(X), \quad \text{for } 0 \leqslant X \leqslant h, \quad \text{for all } Y. \tag{5.15}$$

Substituting this form for $\Phi(X, Y)$ into the Laplace equation $\Phi_{XX} + \Phi_{YY} = 0$, where $\Phi_{XX} = \partial^2\Phi/\partial X^2$ and $\Phi_{YY} = \partial^2\Phi/\partial Y^2$, reduces it to $d^2U/dX^2 = 0$. Integrating this gives $U(X) = pX + q$ with p and q arbitrary integration constants. As $U(X)$ must satisfy boundary conditions in Equation (5.14) we have

$$(\text{at } X = 0) \;\; k_1 = q \quad \text{and} \quad (\text{at } X = h) \;\; k_2 = ph + q,$$

so after solving for p and q the required harmonic function $\Phi(X, Y)$ is found to be

$$\Phi(X, Y) = k_1 + \left(\frac{k_2 - k_1}{h}\right) X, \tag{5.16}$$

for $0 \leqslant X \leqslant h$, and all Y, and it is clear this solution is unique. The level curves $\Phi(X, Y) =$ const. are seen to be straight lines parallel to the sides of the strip, as shown in Figure 5.5(b).

The second stage of the solution leading to the required result could either be obtained by using a linear conformal transformation or, as will be done here, by an equivalent real variable transformation. Consider Figure 5.5(a) where the origin O′ for the orthogonal coordinates (X, Y) is located at the point $x = a$ and $y = b$ in the (x, y)-plane, and the normal to the strip is inclined at an angle θ to the x-axis. Then if an arbitrary point P has the coordinates (x, y) and (X, Y) in these two coordinate systems, elementary geometry shows these are related by the linear transformation

$$X = (x - a)\cos\theta - (y - b)\sin\theta, \quad Y = (x - a)\sin\theta + (y - b)\cos\theta. \tag{5.17}$$

A linear transformation of coordinates is equivalent to a linear fractional conformal transformation, so it will leave Laplace's equation invariant (Theorem 5.1.1), with the result that the harmonic solution Φ in D_1 will transform to the harmonic solution ϕ in D, while the boundary conditions will remain unchanged. Thus substituting Equation (5.17) into Equation (5.16) gives the required harmonic function $\phi(x, y)$ in D

$$\phi(x, y) = k_1 + \left(\frac{k_2 - k_1}{h}\right)[(x - a)\cos\theta - (y - b)\sin\theta], \tag{5.18}$$

for any (x, y) in strip D. The level curves are simply straight lines in D parallel to the sides of the strip. The solution ϕ in D is unique.

5.2.2 A Dirichlet Problem in a Sector

We are required to find a real harmonic function ϕ in a sector D on whose radial sides constant Dirichlet conditions are imposed. So the natural coordinates to use are the plane polar coordinates (r, θ). As shown in Figure 5.6, let the sector have its origin at O ($r = 0$), with its radial sides at $\theta = 0$ and $\theta = \alpha$, so that $0 \leqslant \theta \leqslant \alpha$, and let the boundary conditions be

$$\phi(r, \alpha) = k_1 \quad \text{and} \quad \phi(r, 0) = k_2 \qquad \text{for all } r > 0. \tag{5.19}$$

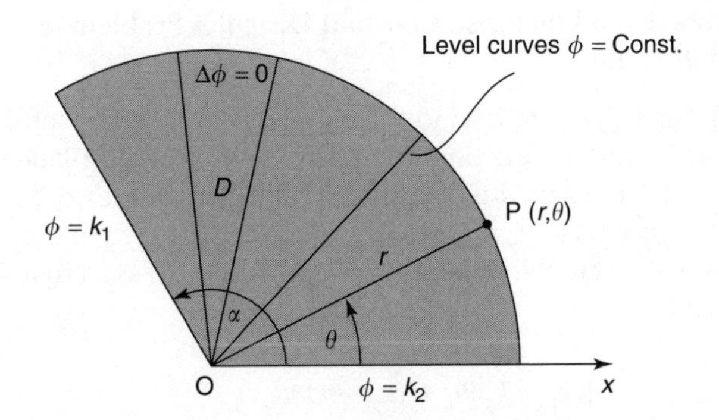

FIGURE 5.6
Dirichlet problem in the sector $0 \leqslant \theta \leqslant \alpha$ ($\alpha < \pi$).

In plane polar coordinates the Laplace equation takes the form

$$\frac{\partial^2\phi}{\partial r^2} + \frac{1}{r}\frac{\partial\phi}{\partial r} + \frac{1}{r^2}\frac{\partial^2\phi}{\partial\theta^2} = 0, \tag{5.20}$$

and ϕ is to satisfy the Dirichlet boundary conditions in Equation (5.19) for all $r > 0$. Because the Dirichlet boundary conditions are constant, this can only be possible if $\phi(r, \theta)$ is a function only of θ, so setting $\phi = U(\theta)$ the Laplace equation reduces to $d^2U/d\theta^2 = 0$, subject to the boundary conditions

$$U(0) = k_2 \quad \text{and} \quad U(\alpha) = k_1. \tag{5.21}$$

Thus the general solution for ϕ has the form $\phi(r, \theta) = U(\theta) = p\theta + q$, with p and q arbitrary integration constants. As this general solution must satisfy boundary conditions in Equation (5.21), solving for p and q and substituting their values into $U(\theta) = p\theta + q$ shows that

$$\phi(r,\theta) = k_2 + \left(\frac{k_1 - k_2}{\alpha}\right)\theta, \tag{5.22}$$

for $0 \leqslant \theta \leqslant \alpha$ and all $r > 0$.

By regarding point P in Figure 5.6 as a point z in the complex plane and using the polar representation for z in which $\theta = \operatorname{Arg} z$, the result in Equation (5.22) becomes

$$\phi(r,\theta) = k_2 + \left(\frac{k_1 - k_2}{\alpha}\right)\operatorname{Arg} z, \tag{5.23}$$

with $0 \leqslant \operatorname{Arg} z \leqslant \alpha$. This form of the solution is needed in subsequent applications. A special case arises when $\alpha = \pi$ because then region D becomes the upper half plane $y > 0$, with different constant Dirichlet conditions prescribed on the positive and negative parts of the real axis (the x-axis).

5.2.3 A Generalized Piecewise Constant Dirichlet Problem in the Half Plane

This last result can be generalized to solve the important and useful problem of finding the real harmonic function $\Phi(u, v)$ in the upper half plane $v > 0$ of the (u, v)-plane that assumes the $n + 1$ constant boundary values $k_1, k_2, \ldots, k_{n+1}$, on the respective segments $(-\infty, u_1), (u_1, u_2), (u_2, u_3), \ldots, (u_n, \infty)$ of the u-axis.

If $P(u, v)$ is any point in the upper half plane $v > 0$, we assert that the required harmonic function is

$$\Phi(u, v) = k_{n+1} + \frac{1}{\pi}[(k_1 - k_2)\theta_1 + (k_2 - k_3)\theta_2 + \cdots + (k_n - k_{n+1})\theta_n], \tag{5.24}$$

where the angles $\theta_1, \theta_2, \ldots, \theta_n$, with $0 \leqslant \theta_i \leqslant \pi$, are shown in Figure 5.7.

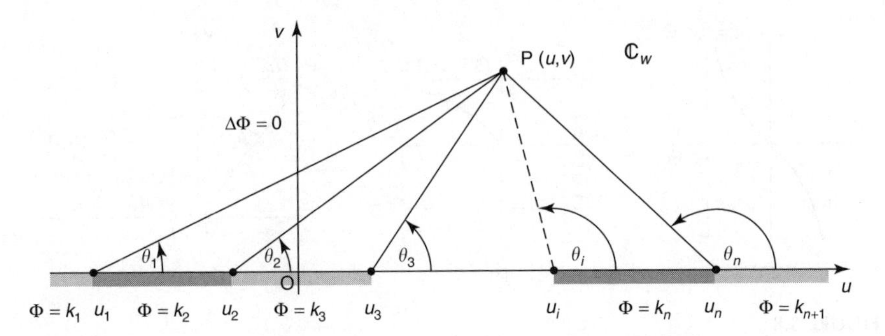

FIGURE 5.7
A generalized piecewise constant Dirichlet problem in $v > 0$.

By regarding P as the point $w = u + iv$ in the complex w-plane, the result from Equation (5.24) may be written

$$\Phi(u, v) = k_{n+1} + \frac{1}{\pi}[(k_1 - k_2)\mathrm{Arg}(w - u_1) + (k_2 - k_3)\mathrm{Arg}(w - u_2) + \cdots + (k_n - k_{n+1})\mathrm{Arg}(w - u_n)], \tag{5.25}$$

for $\mathrm{Im}\{w\} > 0$.

To demonstrate that this function provides the required solution we will consider the complex function

$$f(z) = ik_{n+1} + \frac{1}{\pi}[(k_1 - k_2)\mathrm{Ln}(w - u_1) + (k_2 - k_3)\mathrm{Ln}(w - u_2) + \cdots + (k_n - k_{n+1})\mathrm{Ln}(w - u_n)].$$

Each function in $f(z)$ is analytic for $v > 0$, so the sum of their imaginary parts, which is simply the function $\Phi(u, v)$, must be harmonic in the upper half plane. Next, allowing P to move along the u-axis from right to left, we see that each angle θ_i will be zero until P passes u_i, when it will jump to the value π, and thereafter remain at that value. To see this, think of P moving just above the u-axis along the line $v = \varepsilon$, with $\varepsilon > 0$ arbitrarily small, and then let $\varepsilon \to 0$. Then as P moves from right to left along the u-axis, so $\Phi(u, 0)$ will assume, successively, the constant values $k_{n+1}, k_n, \ldots, k_1$ in the respective segments $(\infty, u_n), (u_{n-1}, u_n), \ldots, (u_1, -\infty)$ of the u-axis.

This has shown that Φ is harmonic for $v > 0$, and that it assumes the required boundary values on the u-axis, so this is the required solution of the stated Dirichlet problem and it is unique.

To present the solution in Equation (5.24) in a more convenient form, it needs to be expressed in terms of u and v and to do this we need to make use

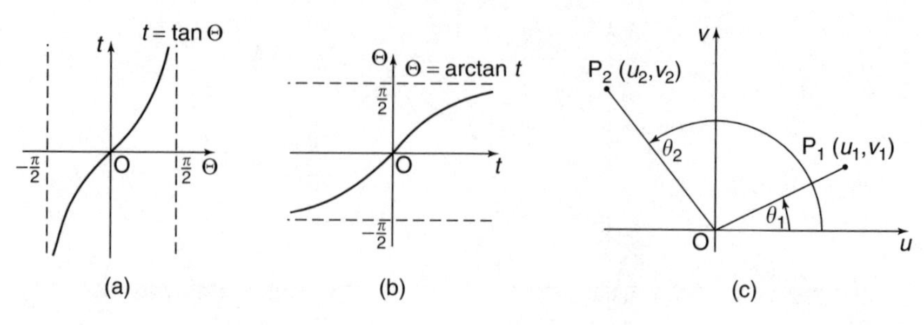

FIGURE 5.8
(a) $t = \tan\Theta$, (b) $\Theta = \arctan t$, (c) Point P with $0 < \theta < \pi$.

of the inverse tangent function. However, in Equation (5.24) the angles θ_i are all required to be such that $0 < \theta_i < \pi$ for any point P in $v > 0$, so the inverse tangent function to be used must be defined such that its range lies in this interval, with the function being *continuous* throughout the interval. To see how such a function is to be defined, it is necessary to consider the behavior of the tangent function $t = \tan\Theta$ defined in the interval $-\frac{1}{2}\pi < \Theta < \frac{1}{2}\pi$, together with the corresponding standard inverse tangent function $\Theta = \arctan t$, illustrated in Figure 5.8(a) and (b).

If $P(u, v)$ is a point in the upper half of the (u, v)-plane, setting $\theta = \arctan t$, with $t = v/u$, we see that θ will increase monotonically from 0 to $\pi/2$ as t increases from 0 to $+\infty$, corresponding to the line OP rotating counterclockwise from coincidence with the positive u-axis in the first quadrant, to coincidence with the positive v-axis in the same quadrant [see Figure 5.8(c)]. As the line OP continues to rotate counterclockwise into the second quadrant, the argument t of the function $\arctan t$ becomes negative and changes monotonically from an initial value of $-\infty$ until it reaches the value 0 when the line OP coincides with the negative u-axis in the second quadrant. This means that if θ is to continue to increase monotonically from $\pi/2$ to π, the angle θ in the second quadrant must be defined as $\pi + \arctan t$. Finally, in order to make θ change continuously, it is necessary to define θ to be $\pi/2$ for $t = \pm\infty$. We have thus arrived at the form of the inverse tangent function that must be used with the solution in Equation (5.24). To distinguish this function from the ordinary inverse tangent function it will be denoted by Arctan t, where

$$\textbf{Arctan}\, t = \begin{cases} \arctan t, & \text{for } t > 0 \\ \pi + \arctan t, & \text{for } t < 0 \\ \pi/2, & \text{for } t = \pm\infty. \end{cases} \tag{5.26}$$

For example, if $P(u, v)$ in Figure 5.8(c) is the point $(\sqrt{3}, 1)$, then $t = 1/\sqrt{3}$, in which case

$$\theta = \textbf{Arctan}\, t = \arctan(1/\sqrt{3}) = \pi/6.$$

However, if P(u, v) in Figure 5.8(c) is the point $(-1, \sqrt{3})$, then $t = -\sqrt{3}$, in which case

$$\theta = \textbf{Arctan}\, t = \textbf{Arctan}(-\sqrt{3}) = \pi + \arctan(-\sqrt{3}) = 2\pi/3.$$

Returning to Equation (5.25) and using this definition of the inverse tangent function, $\Phi(u, v)$ becomes

$$\Phi(u,v) = k_{n+1} + \frac{1}{\pi}\left[(k_1 - k_2)\,\textbf{Arctan}\left(\frac{v}{u-u_1}\right) + (k_2 - k_3)\,\textbf{Arctan}\left(\frac{v}{u-u_2}\right) + \cdots + (k_n - k_{n+1})\,\textbf{Arctan}\left(\frac{v}{u-u_n}\right)\right]. \tag{5.27}$$

The level curves $\Phi(u, v) = \text{const.}$ are no longer simple curves and in general they must be found by numerical computation.

In terms of this definition of the inverse tangent function, the solution in Equation (5.23) in the sector shown in Figure 5.6 becomes

$$\phi(x,y) = k_2 + \left(\frac{k_1 - k_2}{\alpha}\right)\textbf{Arctan}\left(\frac{y}{x}\right), \tag{5.28}$$

for $0 < \textbf{Arctan}(y/x) < \alpha$.

Example 5.2.1 A Simple Piecewise Constant Dirichlet Problem in the Half Plane

Find the function $\Phi(u, v)$ that is harmonic in the upper half plane $v > 0$ and on the u-axis assumes the values $\Phi(u, 0) = 3$ on the interval $(\infty < u < -2)$, $\Phi(u, 0) = 0$ on the interval $(-2 < u < 1)$, and $\Phi(u, 0) = -2$ on the interval $(1 < u < \infty)$, and construct some solution curves.

SOLUTION

The solution follows directly from Equation (5.27) by setting $k_1 = 3$, $k_2 = 0$, and $k_3 = -2$, with $u_1 = -2$ and $u_2 = 1$. The result is

$$\Phi(u,v) = -2 + \frac{1}{\pi}\left[3\textbf{Arctan}\left(\frac{v}{u+2}\right) + 2\textbf{Arctan}\left(\frac{v}{u-1}\right)\right], \quad v > 0.$$

Figure 5.9 shows solution curves for $v = 0, 0.25, 0.75, 1.5$, and 3.0, where the solution for $v = 0$ represents the piecewise constant Dirichlet conditions on the u-axis. ◇

The full significance of the solution in Equation (5.27) becomes apparent when a piecewise constant Dirichlet problem for the Laplace equation is considered for a simply connected infinite domain D with a boundary comprising

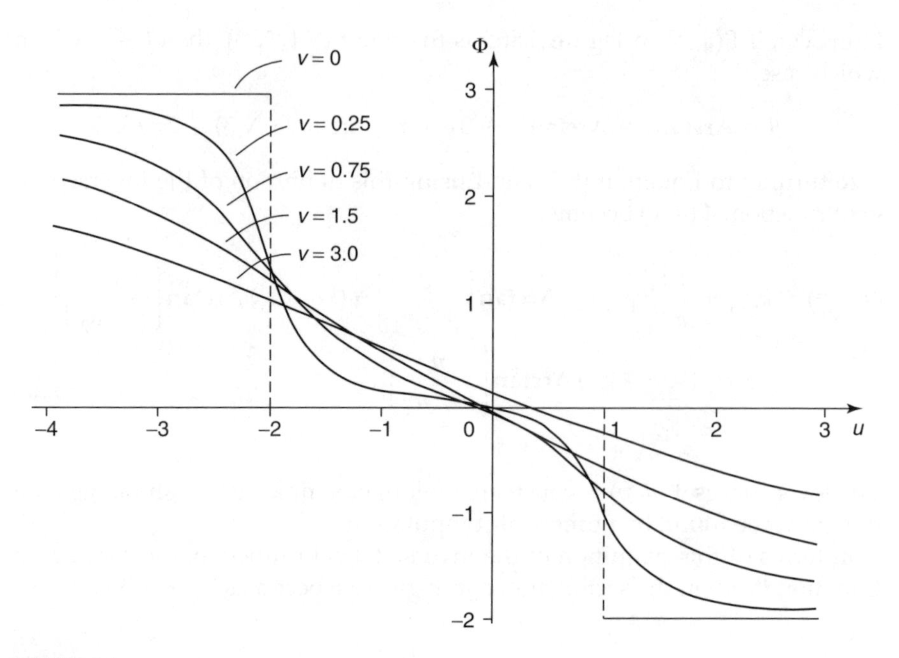

FIGURE 5.9
The solution Φ in the half plane $v > 0$.

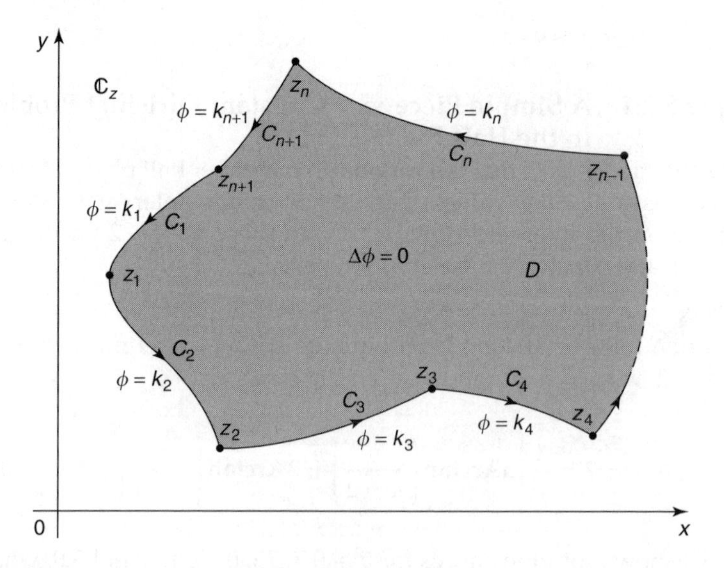

FIGURE 5.10
A piecewise constant Dirichlet problem in a simply connected domain D.

a set of $n + 1$ contiguous piecewise smooth arcs C_i, as shown in Figure 5.10. Let us see how to find a real function ϕ, harmonic in D, such that on arc C_i the function ϕ satisfies the constant Dirichlet condition $\phi = k_i$, for $i = 1, 2, \ldots, n + 1$, where the $n + 1$ arcs C_i are arranged around D in the positive sense.

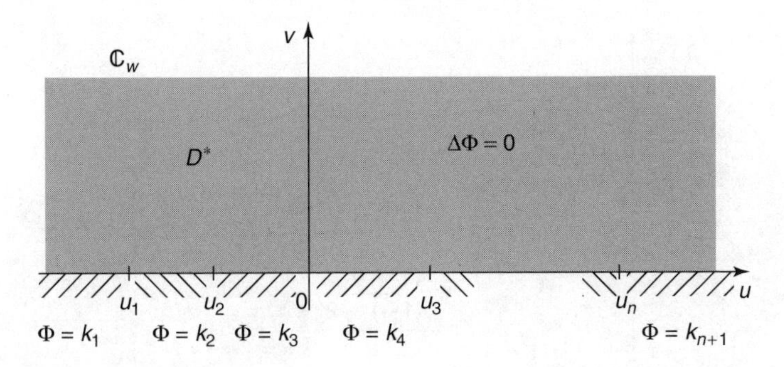

FIGURE 5.11
The mapping of D onto D^* comprising the half plane $v > 0$.

Consider the domain D in the z-plane shown in Figure 5.10 and let the analytic function

$$w = f(z) = u(x, y) + iv(x, y) \tag{5.29}$$

map domain D conformally and one-one onto a domain D^* occupying the upper half of the w-plane in such a manner that the points $z_1, z_2, \ldots, z_{n+1}$ have as their images the points $u_1 < u_2 < \cdots < u_n$ on the real u-axis, with z_{n+1} mapping to $u_{n+1} = +\infty$. Then the arc C_i will map onto the corresponding segment $u_i < u < u_{i+1}$ of the real axis in the w-plane, as in Figure 5.11, while the boundary value $\phi = k_i$ on C_i will become $\Phi = k_i$ on the segment $u_i < u < u_{i+1}$, where Φ is the form taken by ϕ when the change of variables is made from (x, y) to (u, v).

The result in Equation (5.27) now provides the solution Φ in D^*, so to find the solution ϕ in D we express u and v in terms of x and y as given in Equation (5.29) to obtain

$$\begin{aligned}\phi(x, y) = k_{n+1} + \frac{1}{\pi}\Bigg[&(k_1 - k_2)\mathbf{Arctan}\left(\frac{v(x, y)}{u(x, y) - u_1}\right)\\ &+ (k_2 - k_3)\mathbf{Arctan}\left(\frac{v(x, y)}{u(x, y) - u_2}\right) + \cdots + (k_n - k_{n+1})\\ &\times \mathbf{Arctan}\left(\frac{v(x, y)}{u(x, y) - u_n}\right)\Bigg] \quad \text{for } y > 0.\end{aligned} \tag{5.30}$$

Example 5.2.2 A Piecewise Constant Dirichlet Problem for a Disk
Find a function $\phi(x, y)$ that is harmonic in the disk $|z - i| < 1$ and on its boundary assumes the values

$$\begin{aligned}&\phi = 2 \quad \text{for } 0 < \operatorname{Arg}(z - i) < \pi/2,\\ &\phi = 3 \quad \text{for } \pi/2 < \operatorname{Arg}(z - i) < \pi,\\ &\phi = -1 \quad \text{for } -\pi/2 < \operatorname{Arg}(z - i) < 0,\\ &\phi = 3 \quad \text{for } -\pi < \operatorname{Arg}(z - i) < -\pi/2.\end{aligned}$$

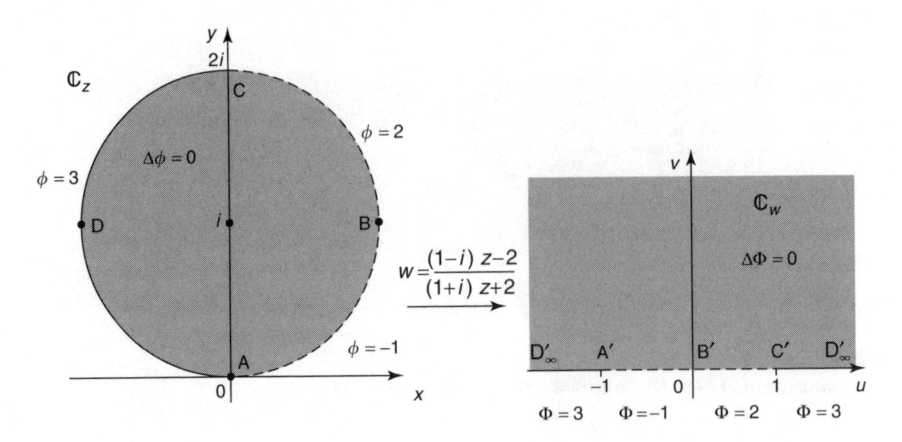

FIGURE 5.12
Equivalent boundary value problems for a disk and a half plane.

SOLUTION
The boundary value problem is illustrated in the diagram on the left in Figure 5.12, and we will solve it using the direct method. Any conformal one-one mapping $w = f(z)$ of this disk onto the upper half of the w-plane will enable us to obtain the required solution via the result in Equation (5.30). For simplicity we will use the linear fractional transformation to map the interior of the disk onto the upper half of the w-plane.

The points A ($z = 0$), B ($z = 1 + i$), and C ($z = 2i$) in the z-plane mark the ends of segments of the circular boundary on which ϕ is constant and so is taken as the three points in the z-plane that may always be assigned arbitrarily in a linear fractional conformal transformation when mapping the disk onto the upper half of the w-plane. To map the interior of the disk onto the upper half of the w-plane we need to assign the images A′, B′, and C′ of A, B, and C on the real axis of the w-plane, subject only to the condition that they occur in this order as we move from left to right along the u-axis. This last requirement is to ensure that the interior of the disk maps onto the *upper* half of the w-plane. Accordingly we make the arbitrary assignments A′ ($w = -1$), B′ ($w = 0$), and C′ ($w = 1$), as shown in Figure 5.12.

The form of the mapping follows directly from Theorem 4.3.1 is

$$w = \frac{(1-i)z - 2}{(1+i)z + 2}. \tag{5.31}$$

Setting $w = f(z) = u(x, y) + iv(x, y)$ it follows from Equation (5.31) that

$$u(x, y) = \frac{2(y-1)}{x^2 + y^2 + 2x - 2y + 2}, \qquad v(x, y) = \frac{2y - x^2 - y^2}{x^2 + y^2 + 2x - 2y + 2}. \tag{5.32}$$

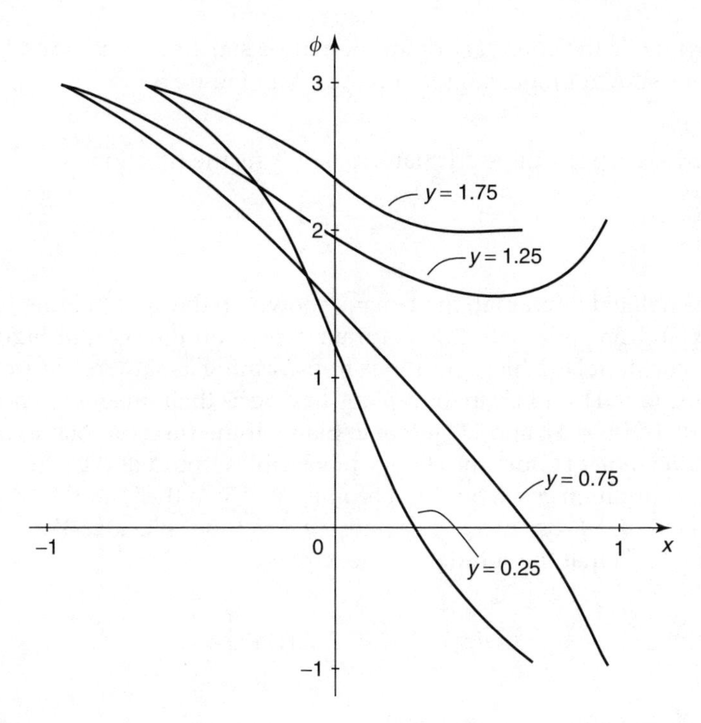

FIGURE 5.13
The solution ϕ in the disk.

Identifying the diagram on the right of Figure 5.12 with Figure 5.11 shows that $k_1 = 3$, $k_2 = -1$, $k_3 = 2$, and $k_4 = 3$, while $u_1 = -1$, $u_2 = 0$, and $u_3 = 1$. The required solution then follows from Equation (5.30) in the form

$$\phi(x, y) = 3 + \frac{1}{\pi}\left[4\mathbf{Arctan}\left(\frac{v}{u+1}\right) - 3\mathbf{Arctan}\left(\frac{v}{u}\right) - \mathbf{Arctan}\left(\frac{v}{u-1}\right)\right], \quad (5.33)$$

where u and v are defined in Equation (5.32) in terms of x and y, with z confined to the unit disk $|z - 1 < 1|$.

The solution of the Dirichlet problem is unique, so although a different assignment of the images A′, B′, and C′ would change the form of the mapping, the solution ϕ at a given point (x, y) would remain the same. The variation of ϕ along the chords y = const. in the disk are shown in Figure 5.13 for $y = 0.25$, 1.25, and 1.75. ◇

Example 5.2.3 A Piecewise Constant Dirichlet Problem in an Indented Quadrant

Find the function $\phi(x, y)$ that is harmonic in the first quadrant if the (x, y)-plane, from which has been removed the interior of the unit circle centered on the origin, while on the boundary of this domain ϕ satisfies the constant Dirichlet conditions shown in the diagram on the left in Figure 5.14.

SOLUTION
It is easily seen from Table 2.1 that if $w = u + iv$, the function

$$w = z + \frac{1}{z} \tag{5.34}$$

maps the deleted domain in the z-plane shown on the left of Figure 5.14 conformally and one-one onto the quadrant shown on the right of Figure 5.14, with the constant Dirichlet conditions transforming as shown. The points A_∞, B ($z = i$), C ($z = 1$), and D_∞ in the z-plane having as their images the points A'_∞, B' ($w = 0$), C' ($w = 1$), and D'_∞ in the w-plane. If the function $\phi(x, y)$ becomes the function $\Phi(u, v)$ under the change of variables from x and y to u and v, the boundary condition $\phi = 3$ on BA_∞ becomes $\Phi = 3$ on $B'A'_\infty$, and the boundary condition $\phi = -1$ on BCD_∞ becomes $\Phi = -1$ on $B'C'D'_\infty$. We see from Equation (5.27) that the solution in the w-plane is

$$\Phi(u, v) = -1 + \frac{4}{\pi}\,\mathbf{Arctan}\left|\frac{u}{v}\right|, \tag{5.35}$$

so the solution of the original problem in the (x, y)-plane is

$$\phi(x, y) = -1 + \frac{4}{\pi}\,\mathbf{Arctan}\left|\frac{y(x^2 + y^2 - 1)}{x(x^2 + y^2 + 1)}\right|, \tag{5.36}$$

because from Equation (5.34) with $z = x + iy$

$$u = \frac{x(x^2 + y^2 + 1)}{x^2 + y^2} \quad \text{and} \quad v = \frac{y(x^2 + y^2 - 1)}{x^2 + y^2}. \tag{5.37}$$

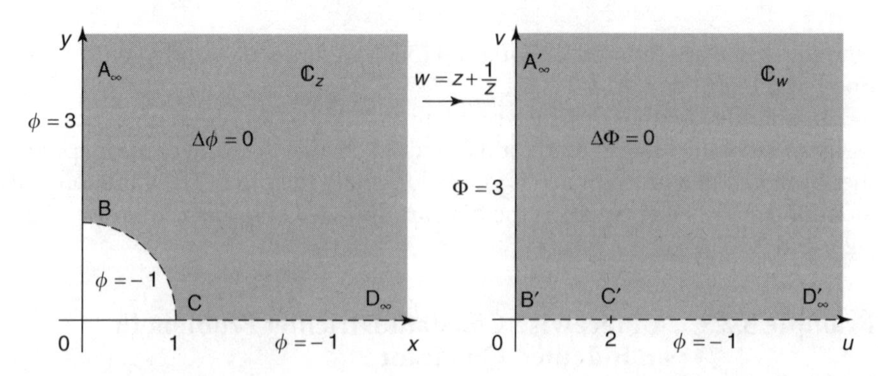

FIGURE 5.14
Mapping a deleted quadrant onto a quadrant.

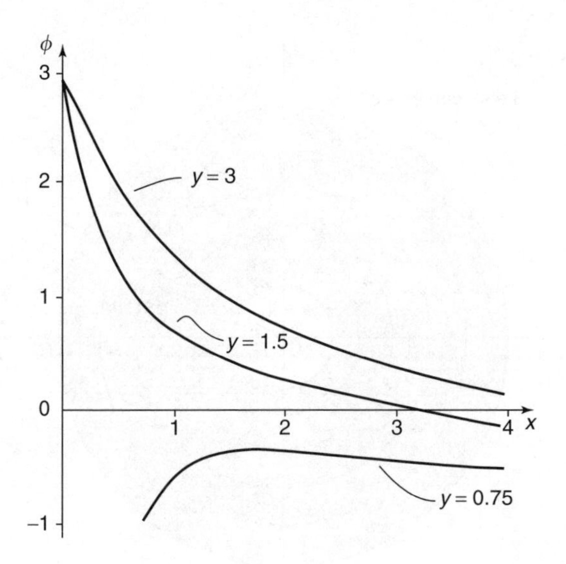

FIGURE 5.15
The solution ϕ in the deleted quadrant.

The variation of ϕ along the lines y = const. $x > 0$ in the deleted quadrant are shown in Figure 5.15 for $y = 0.75$, 1.5, and 3.0. ◇

5.2.4 A Dirichlet Problem in an Annulus

We now find a function $\phi(x, y)$ that is harmonic in the annulus $R_1 < r < R_2$ with $r = (x^2 + y^2)^{1/2}$, that assumes the boundary value $\phi = k_1$ or $r = R_1$, and $\phi = k_2$ or $r = R_2$. The problem is illustrated in Figure 5.16 and it should be remembered that an annulus is not simply connected.

The radial symmetry of the problem shows that the solution can only depend on r, so setting $\phi = U(r)$ and substituting into Equation (5.20) reduces it to the ordinary differential equation

$$\frac{d^2U}{dr^2} + \frac{1}{r}\frac{dU}{dr} = 0 \tag{5.38}$$

subject to the constant Dirichlet conditions

$$U(R_1) = k_1 \quad \text{and} \quad U(R_2) = k_2. \tag{5.39}$$

Solving this differential equation and using the boundary conditions gives

$$U(r) = \phi(r, \theta) = k_1 + \left(\frac{k_1 - k_2}{\ln_e(R_2/R_1)}\right)\ln_e R_1 + \left(\frac{k_2 - k_1}{\ln_e(R_2/R_1)}\right)\ln_e r, \tag{5.40}$$

for any θ. The solution is unique and the level curves are concentric circles.

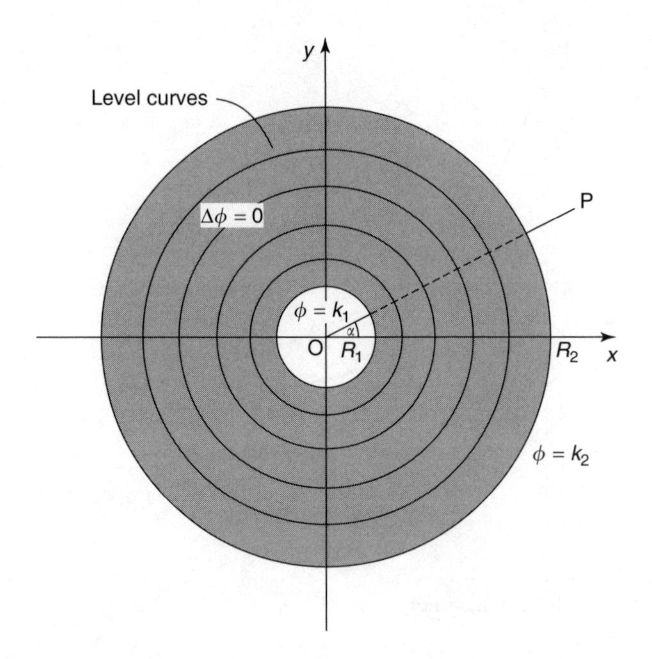

FIGURE 5.16
A constant Dirichlet problem for an annulus.

In terms of x and y, the solution takes the form

$$\phi(x, y) = k_1 + \left(\frac{k_1 - k_2}{\ln_e(R_2/R_1)}\right)\ln_e R_1 + \left(\frac{k_2 - k_1}{2\ln_e(R_2/R_1)}\right)\ln_e(x^2 + y^2), \qquad (5.41)$$

Suppose we consider any ray OP drawn through the origin at an angle α to the x-axis as in Figure 5.16. Then because the level curves are concentric circles, they all intersect the ray at right angles showing that the derivative of ϕ normal to the ray must be zero. In terms of the plane polar coordinates (r, θ) this is equivalent to the condition

$$\frac{\partial\phi}{\partial\theta} = 0 \quad \text{on } \theta = \alpha \qquad \text{for } R_1 < r < R_2. \qquad (5.42)$$

This is a homogeneous Neumann condition on the part of the ray that intersects the annulus and it is true for any angle α. With this idea in mind we see that Equation (5.40) provides the solution to the mixed boundary value problem illustrated in Figure 5.17. In this problem Dirichlet conditions are given on the circular arcs AD and BC, while Neumann conditions are given on its radial sides AB and DC.

Although this is one of the simplest nontrivial mixed boundary value problems, with the aid of conformal mapping it can lead to the solution of

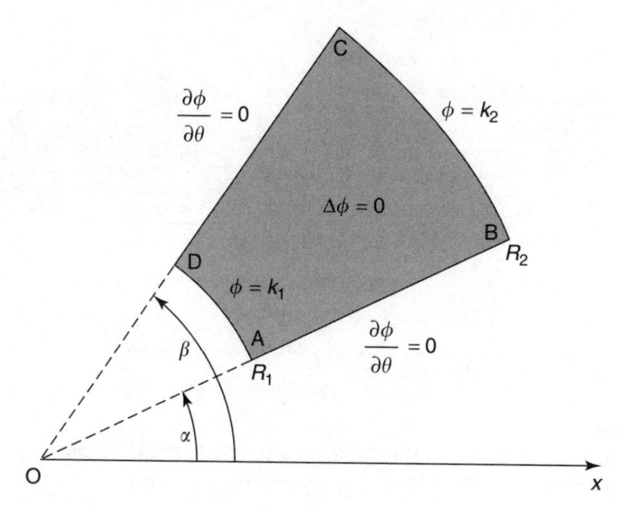

FIGURE 5.17
A mixed boundary value problem.

problems of similar type, although in far more complicated domains. This is illustrated by the next two examples.

Example 5.2.4 The Transformation of a Mixed Boundary Value Problem

Formulate and solve the mixed boundary value problem in the w-plane under which the conformal mapping $w = z + 1/z$ is equivalent to the following problem in the z-plane. The function ϕ is harmonic in the half-annular domain $a < r < b$ $(1 < a < b)$, $0 < \theta < \pi$, while on its boundary

(i) ϕ satisfies the Dirichlet conditions $\phi = k_1$ on $r = a$ and $\phi = k_2$ on $r = b$ for $0 < \theta < \pi$, and
(ii) ϕ satisfies the Neumann condition

$$\frac{\partial \phi}{\partial n} = 0 \quad \text{on } \theta = 0 \quad \text{and} \quad \theta = \pi, \quad \text{for } a < r < b.$$

SOLUTION

The mixed boundary value problem in the z-plane involves the half-annular domain illustrated on the left of Figure 5.18. This problem is equivalent to the one illustrated in Figure 5.16, but with $\alpha = 0, \beta = \pi, R_1 = a$, and $R_2 = b$. From Equation (5.40) the solution of the problem in the z-plane is seen to be

$$\phi(r, \theta) = k_1 + \left(\frac{k_2 - k_1}{\ln_e(b/a)}\right) \ln_e(r/a), \tag{5.43}$$

for $a < r < b$ and $0 < \theta < \pi$.

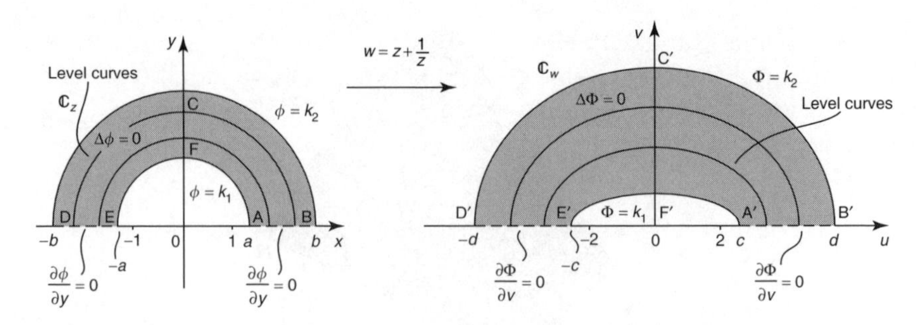

FIGURE 5.18
Equivalent mixed boundary value problems.

Now it can be seen from Table 5.1 that the conformal mapping $w = z + 1/z$ maps the semicircle $r = R, 0 < \theta < \pi$ in the z-plane onto the half-ellipse

$$\left(\frac{Ru}{R^2+1}\right)^2 + \left(\frac{Rv}{R^2-1}\right)^2 = 1, \quad v > 0 \tag{5.44}$$

in the w-plane. Thus the half-annular domain shown on the left of Figure 5.18 is mapped onto the half-elliptical annular domain shown on the right of Figure 5.18. The points A and F in the z-plane have as their images the points A′ and F′ in the w-plane, with A′ the point $(c, 0)$ and F′ the point $(0, d)$, where $c = a + 1/a$ and $d = a - 1/a$. Similarly, points B and C in the z-plane have as their images the points B′ and C′ in the w-plane, with B′ the point $(c', 0)$ and C′ the point $(0, d')$, with $c' = b + 1/b$ and $d' = b - 1/b$.

Let $\phi(r, \theta)$ become $\Phi(u, v)$ under the transformation from the z- to the w-plane, in which we set $w = u + iv$. Then we know that the boundary conditions for Φ are the same as those for ϕ, so they transform as shown on the right in Figure 5.18.

To solve the problem in the w-plane, we observe first that on the level curve $r = R$ in the z-plane it follows from Equation (5.43) that $\phi = k$, where

$$k = k_1 + \left(\frac{k_2 - k_1}{\ln_e(b/a)}\right) \ln_e(R/a). \tag{5.45}$$

Consequently $\Phi = k$ on every point of the ellipse in the w-plane corresponding to the semicircle $r = R$ in the z-plane with $a < R < b$ and $0 < \theta < \pi$.

The problem is now solved because the level curves in the w-plane are the upper half of confocal ellipses with their foci at the points $(2, 0)$ and $(-2, 0)$ in the w-plane.

Although $\Phi(u, v)$ cannot be expressed explicitly in terms of u and v, its value is determined in terms of R as a parameter. Once R is specified, as Φ is determined in terms of R by Equation (5.45), it follows that this is the value of Φ at all points of the semi-ellipse in Equation (5.44). If required, using this same value of R with a suitable choice for u, the corresponding value for v follows from Equation (5.44) in the form

$$v = \left(\frac{R^2-1}{R}\right)\left[1-\left(\frac{Ru}{R^2+1}\right)^2\right]^{1/2}, \quad \text{with } v > 0. \qquad \diamond$$

Example 5.2.5 A Dirichlet Problem in the Plane and a Related Mixed Boundary Value Problem

Find the function $\phi(x, y)$ that is harmonic in the z-plane and satisfies the Dirichlet conditions $\phi = k_1$ on the semi-infinite line segment $x < -1, y = 0$ and $\phi = k_2$ on the semi-infinite line segment $x > 0, y = 0$.

SOLUTION

The boundary value problem is illustrated on the left of Figure 5.19 and examination of Table 5.1 shows that the function $w = \text{Arcsin}\, z$ will map the required domain in the z-plane, cut along the two specified semi-infinite line segments, one-one and conformally onto the strip $-\pi/2 < u < \pi/2$ in the w-plane.

The boundary conditions imposed on each side of the cuts ABP and EDS are the same and they map onto the straight line boundaries P′B′A′and S′D′E′ of the strip in the w-plane as the respective constant Dirichlet conditions $\Phi = k_1$ and $\Phi = k_2$. If $\phi(x, y)$ becomes $\Phi(u, v)$ under the transformation, the solution $\Phi(u, v)$ in the w-plane follows directly from the Standard Case 1, Equation (5.18), with $\theta = 0$, $a = -\pi/2$, and $h = \pi$, in the form

$$\Phi(u, v) = \tfrac{1}{2}(k_1 + k_2) + \left(\frac{k_2 - k_1}{\pi}\right)u. \tag{5.46}$$

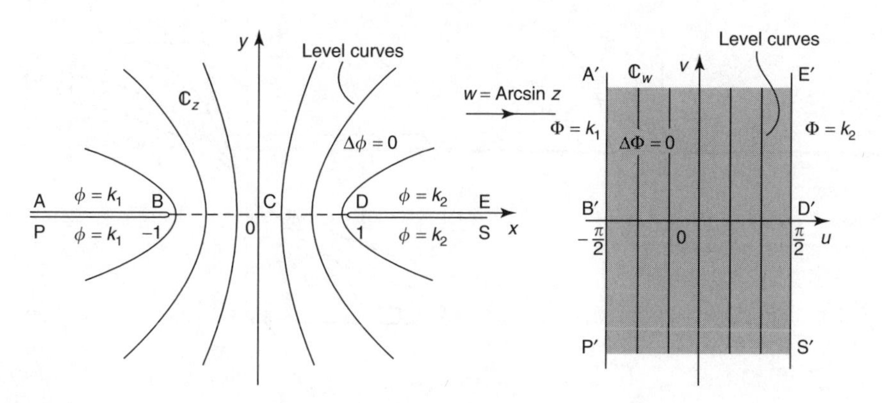

FIGURE 5.19
Mapping a cut z-plane onto a strip in the w-plane.

As $u = \text{Re}\{\text{Arcsin}\, z\}$, the corresponding solution in the (x, y) plane is

$$\phi(x, y) = \tfrac{1}{2}(k_1 + k_2) + \left(\frac{k_2 - k_1}{\pi}\right)\text{Re}\{\text{Arcsin}\, z\}. \tag{5.47}$$

Using the result of Equation (4.64) this becomes

$$\phi(x, y) = \tfrac{1}{2}(k_1 + k_2) + \left(\frac{k_2 - k_1}{\pi}\right)\text{Arcsin}\left(\frac{\sqrt{(x+1)^2 + y^2} - \sqrt{(x-1)^2 + y^2}}{2}\right), \tag{5.48}$$

where the range of the branch of the inverse sine functions to be used is $(-\pi/2, \pi/2)$.

The level curves in the strip are straight lines parallel to the sides of the strip and they map into the family of confocal hyperbolas shown on the left of Figure 5.19. Because the x-axis is a line of symmetry for the solution in the (x, y)-plane, it follows that across the line segment $-1 < x < 1$, $y = 0$, the solution satisfies the condition $\partial\phi/\partial n = 0$, where n is normal to the x-axis. This is simply a homogeneous Neumann condition. Thus by considering the solution in Equation (5.48) in the upper half plane, the solution of the mixed boundary value problem illustrated on the left of Figure 5.20 is obtained. Inspection of Equation (5.48) shows that $\phi = \frac{1}{2}(k_1 + k_2)$ along the level curve passing through the origin in the (x, y)-plane. The diagram on the right of Figure 5.20 shows the solution corresponding to $k_1 = -2$ and $k_1 = 1$, with the variation given for the lines $y = 0, 0.5, 1.5$, and 3.0. ◇

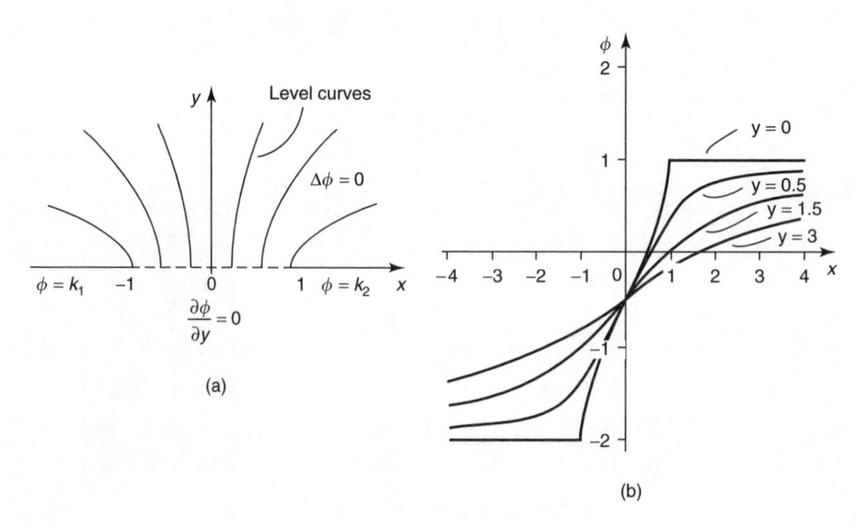

FIGURE 5.20
A mixed boundary value problem in the half plane with $k_1 = -2$ and $k_2 = 1$.

5.2.5 The Poisson Integral Formulas for the Half Plane and the Circle

The two important Poisson integral formulas were derived at the end of Section 2.5 in Chapter 2. For the sake of completeness, the formulas are repeated here because they extend to the half plane and to the circle the solution of a Dirichlet boundary value problem with an arbitrary Dirichlet condition. However, for all but the simplest functions f the integrals need to be computed numerically.

THEOREM 5.2.1 *Poisson's Integral Formula for a Half Plane*
Let $f(x)$ be a real valued function which is bounded and may be either continuous or piecewise continuous for $-\infty < x < \infty$. Then, when the integral exists, the function $\phi(x, y)$ defined as

$$\phi(x, y) = \frac{y}{\pi} \int_{-\infty}^{\infty} \frac{f(s)ds}{(x-s)^2 + y^2}$$

is harmonic in the half plane $y > 0$, and on the x-axis assumes the boundary condition $\phi(x, 0) = f(x)$ everywhere $f(x)$ is continuous. ◆

THEOREM 5.2.2 *Poisson's Integral Formula for a Disk*
Let $f(\theta)$ be a bounded piecewise continuous function defined for $-\pi < \theta \leqslant \pi$. Then the function $\phi(r, \theta)$ defined as

$$\phi(r, \theta) = \frac{1}{2\pi} \int_{0}^{2\pi} \frac{(r_0^2 - r^2) f(\psi) d\psi}{r_0^2 - 2r_0 r \cos(\psi - \theta) + r^2}$$

is harmonic in the disk $0 \leqslant r < r_0$, and on its boundary $r = r_0$ assumes the boundary condition $\phi(r_0, \theta) = f(\theta)$ everywhere $f(\theta)$ is continuous. ◆

Example 5.2.6 A Constant Dirichlet Condition on Each Half Axis
Use the Poison formula in Theorem 5.2.1 to find the function $\phi(x, y)$ that is harmonic in the upper half plane $y > 0$, and subject to the constant Dirichlet conditions $\phi(x, 0) = -1$ for $x < 0$ and $\phi(x, 0) = 2$ for $x > 0$.

SOLUTION
The problem is illustrated in Figure 5.21. Substituting the boundary conditions into Theorem 5.2.1 gives

$$\phi(x, y) = \frac{-1}{\pi} \int_{-\infty}^{0} \frac{y\,ds}{(x-s)^2 + y^2} + \frac{2}{\pi} \int_{0}^{\infty} \frac{y\,ds}{(x-s)^2 + y^2},$$

and hence

$$\phi(x, y) = \frac{1}{2\pi}(\pi + 6\text{Arctan}(x/y)) = 2 - (3/\pi)\textbf{Arctan}(y/x), \quad \text{with } y > 0.$$

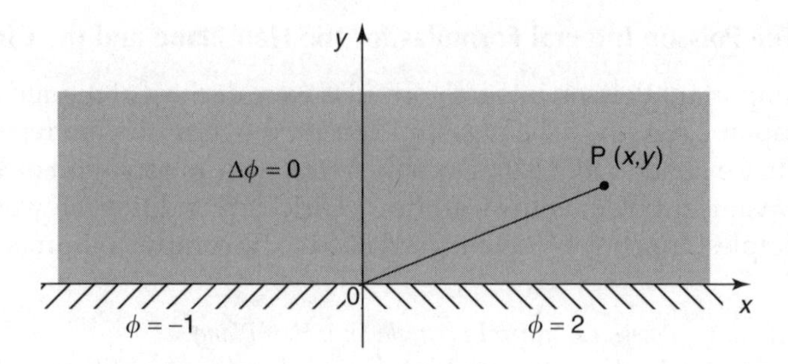

FIGURE 5.21
Constant Dirichlet conditions on each half axis.

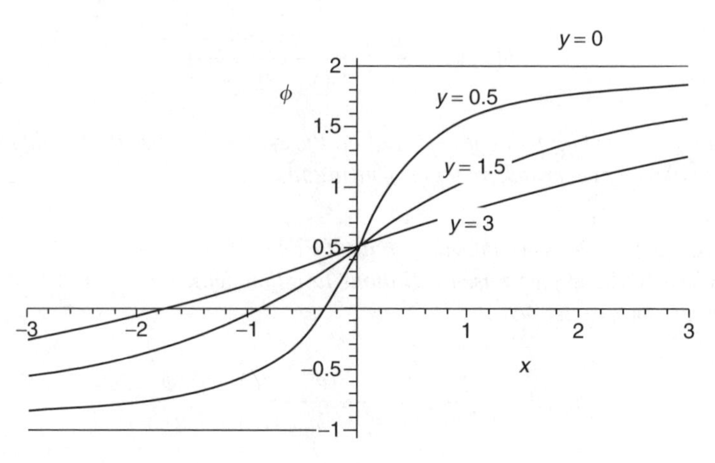

FIGURE 5.22
The variation of $\phi(x, y_0)$ for different values of y_0.

This is, of course, the result that would have been obtained had the solution been obtained from Equation (5.23). Figure 5.22 shows a plot of the solution for $y = 0$, 0.5, 1.5, and 3, where the plot for $y = 0$ is simply the Dirichlet boundary condition.

These results illustrate three important properties of the Laplace equation. The first is that even a localized boundary condition influences the solution for all x and y, the second is that a discontinuity in the Dirichlet conditions is smoothed out immediately, while the third is that at a discontinuity of the boundary condition, the Laplace equation assigns a value there equal to the mean of the values on the adjacent sides of the discontinuity. This last feature can be seen by inspection of Figure 5.22 because when $x = 0$ the curves for $y = 0.5$, 1.5, and 3 all pass through the value $y = \frac{1}{2}(-1 + 2) = 0.5$. ◇

The last two examples illustrate the results obtained by numerical integration when the function f is not constant and the integrals cannot be evaluated analytically.

Example 5.2.7

Use the Poison formula in Theorem 5.2.1 to find the function $\phi(x, y)$ that is harmonic in the upper half plane $y > 0$, and subject to the Dirichlet condition $\phi(x, 0) = \exp(-x^2)\tanh x$ on the x-axis.

SOLUTION

Substituting the Dirichlet condition into Theorem 5.2.1 gives

$$\phi(x, y) = \frac{y}{\pi}\int_{-\infty}^{\infty} \frac{\exp(-s^2)\tanh s}{(x-s)^2 + y^2}\,ds.$$

This integral is complicated, so instead of attempting to find an analytical solution, a numerical solution is obtained. Normally the fact that the interval of integration extends over the entire x-axis would lead to complications when using numerical integration. However, in this case the factor $\exp(-s^2)$ in the numerator of the integrand decays to zero sufficiently rapidly as $|s|$ increases for the integral to be approximated by an integral over a finite interval. As $e^{-9} = 0.00012$, we will take as our interval of integration $-3 \leqslant s \leqslant 3$, when we obtain the approximation

$$\phi(x, y) \approx \frac{y}{\pi}\int_{-3}^{3} \frac{\exp(-s^2)\tanh s}{(x-s)^2 + y^2}\,ds \quad \text{for } y > 0.$$

Setting $y = y_0 > 0$, where y_0 is a specific numerical value, this becomes

$$\phi(x, y_0) \approx \frac{y}{\pi}\int_{-4}^{4} \frac{\exp(-s^2)\tanh s}{(x-s)^2 + y_0^2}\,ds \quad \text{for } y_0 > 0,$$

where now $\phi(x, y_0)$ is a function only of x. A numerical routine used to compute and plot $\phi(x, y_0)$ as a function of x for the values $y_0 = 0, 0.3, 0.75, 1.5$, and 2.5 produced the results shown in Figure 5.23, where the result for $y_0 = 0$ is the Dirichlet boundary condition. These numerical results show that the solution tends to zero fairly rapidly as y increases, as might be expected along the line $y = y_0$ as y_0 increases.

If required, this same numerical routine will produce the numerical value for $\phi(x, y)$ at a specific point (x_0, y_0), where, for example, at the point $(0.5, 0.3)$ it gives the value $\phi(0.5, 0.3) = 0.1902$. ◇

Example 5.2.8 The Solution in a Disk

Use the Poison formula in Theorem 5.2.2 to find the function $\phi(r, \theta)$ that is harmonic in the disk $r = R$ that is centered on the origin, and on its boundary is subject to the Dirichlet condition $\phi(R, \theta) = \sin\theta$.

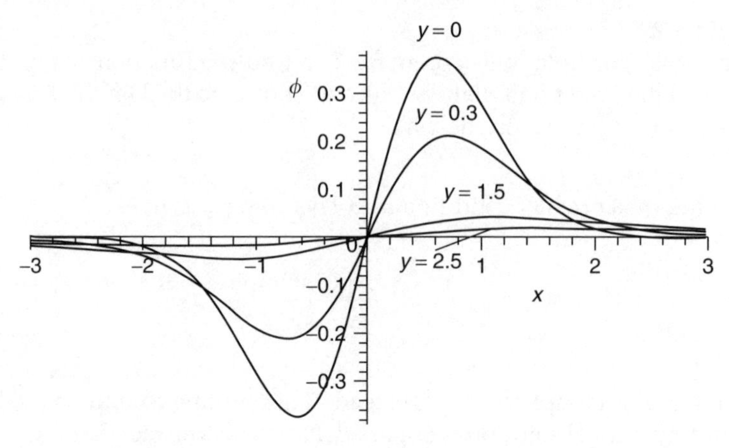

FIGURE 5.23
The solution in a half-plane subject to the Dirichlet condition $f(x) = \exp(-x^2)\tanh x$.

SOLUTION
A solution that depends on the ratio r/R, rather than on r and R separately, is more useful as it is nondimensional. Such a solution can be obtained from Theorem 5.2.2 by setting $r_0 = R$, dividing the numerator and denominator of the integrand by R^2, and then setting $\rho = r/R$ to obtain

$$\phi(\rho, \theta) = \frac{1}{2\pi}\int_0^{2\pi} \frac{(1-\rho^2)f(\psi)d\psi}{1-2\rho\cos(\psi-\theta)+\rho^2} \qquad \text{for } 0 \leqslant \rho < 1.$$

Substituting the Dirichlet condition $f(\theta) = \sin\theta$ into the integral gives

$$\phi(\rho, \theta) = \frac{1}{2\pi}\int_0^{2\pi} \frac{(1-\rho^2)\sin(\psi)d\psi}{1-2\rho\cos(\psi-\theta)+\rho^2}.$$

Setting $\rho = \rho_0$, where ρ_0 is a specific numerical value, this becomes

$$\phi(\rho_0, \theta) = \frac{1}{2\pi}\int_0^{2\pi} \frac{(1-{\rho_0}^2)\sin(\psi)d\psi}{1-2\rho_0\cos(\psi-\theta)+{\rho_0}^2} \qquad \text{for } 0 \leqslant \rho_0 < 1,$$

where now $\phi(\rho_0, \theta)$ is a function only of θ. A numerical routine used to compute and plot $\phi(\rho_0, \theta)$ as a function of θ for the values $\rho_0 = 0, 0.15, 0.45$, and 0.8 produced Figure 5.24, where the result for $\rho = 0$ is the Dirichlet boundary condition. ◇

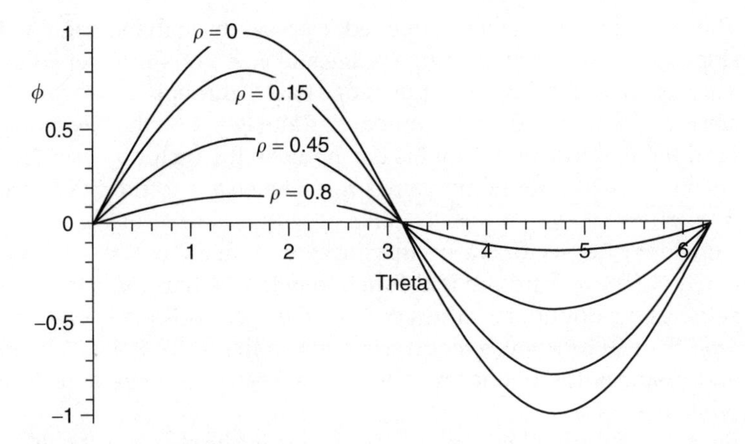

FIGURE 5.24
The solution $\phi(\rho, \theta)$ for $\rho = 0, 0.15, 0.45$, and 0.8, subject to the Dirichlet boundary condition $\phi(1, \theta) = \sin\theta$.

A simple but useful result concerning two-dimensional Dirichlet problems for harmonic functions is that if the boundary condition in Theorem 5.2.1 is replaced by

$$\tilde{\phi}(x, 0) = M + mf(x), \tag{5.49}$$

with m and M arbitrary constants, then the corresponding solution $\tilde{\phi}(x, y)$ is given by

$$\tilde{\phi}(x, y) = M + \frac{my}{\pi}\int_{-\infty}^{\infty}\frac{f(s)ds}{(x-s)^2 + y^2}. \tag{5.50}$$

This result can be deduced directly from the result of Theorem 5.2.2, although this is left as an exercise. Results from Equations (5.49) and (5.50) are almost self evident. Notice first that the linearity of the Laplace equation means that scaling a boundary condition by a constant m scales the solution by m. Secondly, adding a constant M to the boundary condition will not affect the harmonic nature of the solution, because $\Delta M = 0$, showing that $M =$ const. is an harmonic function. A corresponding result is true for Theorem 5.2.2.

Exercises 5.2

In Exercises 1 through 4, find the real function $\phi(x, y)$ that is harmonic in the given strip subject to the stated boundary conditions on its edges.

1. The strip $2 < x < 5$, with $\phi(2, y) = 1$ and $\phi(5, y) = 4$.
2. The strip $0 < y < 1$, with $\phi(x, 0) = 2$ and $\phi(x, \pi) = 3$.

3. The strip of width 1 whose upper edge passes through the origin with slope 2, with $\phi = 3$ on the upper edge and $\phi = 5$ on the lower edge.
4. The strip of width 2 whose upper edge passes through the origin with slope -1, with $\phi = 0$ on the upper edge and $\phi = 4$ on the lower edge.
5. Find the real function ϕ that is harmonic in the sector $0 < \theta < 2\pi/3$ and assumes the boundary values $\phi = 3$ when $\theta = 0$ and $\phi = 7$ when $\theta = 2\pi/3$.
6. Consider the sector D occupying the domain $\alpha \leqslant \theta \leqslant \beta$, with $0 \leqslant \alpha \leqslant \beta \leqslant \pi$. Find the real function ϕ that is harmonic in D and satisfies the boundary values $\phi = k_1$ on $\theta = \beta$, and $\phi = k_2$ on $\theta = \alpha$.
7. Find the real function ϕ that is harmonic in the sector $\pi/4 < \theta < 3\pi/4$ and assumes the boundary values $\phi = 5$ when $\theta = \pi/4$ and $\phi = 9$ when $\theta = 3\pi/4$.
8. Find the real function ϕ that is harmonic in the sector $\pi/2 < \theta < \pi$ and assumes the boundary values $\phi = -3$ when $\theta = \pi/2$ and $\phi = 6$ when $\theta = \pi$.

In Exercises 9 through 12 find the real function $\phi(x, y)$ that is harmonic in the half plane $y > 0$ and on the boundary assumes the indicated boundary condition. Examine the solution to see if it is either an even or an odd function and, if so, explain why this is to be expected.

9. $\phi(x, 0) = -3$ for $x < 0$ and $\phi(x, 0) = 4$ for $x < 0$.
10. $\phi(x, 0) = 1$ for $x < -1$, $\phi(x, 0) = 0$ for $-1 < x < 1$ and $\phi(x, 0) = -1$ for $x > 1$.
11. $\phi(x, 0) = 2$ for $x < -2$ and $\phi(x, 0) = 3$ for $-2 < x < 1$, $\phi(x, 0) = 1$ for $1 < x < 4$ and $\phi(x, 0) = 0$ for $x > 4$.
12. $\phi(x, 0) = 0$ for $|x| < 2$ and $\phi(x, 0) = 3$ for $-2 < x < -1$, $\phi(x, 0) = 1$ for $-1 < x < 1$ and $\phi(x, 0) = 3$ for $1 < x < 2$.
13. Show that the function $w = i(1 - z)/(1 + z)$ maps the unit disk $|z| = 1$ conformally onto the upper half of the w-plane. Use the function to find $\phi(x, y)$ that is harmonic inside the disk and on its boundary $|z| = 1$ assumes the values $\phi = 3$ for $\text{Im}\{z\} > 0$ and $\phi = -1$ for $\text{Im}\{z\} < 0$.
14. Show that the function $w = i(4 - z)/z$ maps the disk $|z - 2| < 2$ conformally onto the upper half of the w-plane. Use the mapping to find the function $\phi(x, y)$ that is harmonic inside the disk and on its boundary $|z - 2| = 2$ assumes the value $\phi = 1$ for $\text{Im}\{z\} > 0$ and $\phi = -2$ for $\text{Im}\{z\} < 0$.
15. Show that the function $w = 2z/[2 + (1 + i)z]$ maps the interior of the disk $|z - i| < 1$ conformally onto the upper half of the w-plane. Use the mapping to find the function $\phi(x, y)$ that is harmonic inside the disk and on its boundary $|z - i| = 1$ assumes the values $\phi = 1$ for $\text{Re}\{z\} > 0$ and $\phi = -1$ for $\text{Re}\{z\} < 0$.
16. Use the conformal mapping in Exercise 15 to obtain a different form of solution for the boundary value problem discussed in Example 5.2.2. Compute the values $\phi(0, 1.1)$ and $\phi(-0.2, 1.5)$ using both the solution

in Example 5.2.2 and the solution to this exercise, and show they are the same. This illustrates the fact that although different conformal transformations may be used to map the circle onto the upper half plane, each giving a different form of the solution, the solutions are equivalent.

17. Show that $w = \sin z$ maps the semi-infinite strip $0 < x < \pi/2$, $y > 0$ conformally onto the first quadrant in the w-plane. Use the mapping to find the function $\phi(x, y)$ that is harmonic inside the semi-infinite strip, and on its boundary satisfies the boundary conditions $\phi = 1$ on $x = 0, y > 0, \phi = 3$ on $0 < x < \pi/2, y = 0$, and $\phi = 3$ on $x = \pi/2, y > 0$.
18. Show that $w = z + 1/z$ maps the upper half of the z-plane from which has been deleted the disk $|z| < 1$, $\text{Im}\{z\} > 0$ conformally onto the upper half of the z-plane. Use the mapping to find the function $\phi(x, y)$ that is harmonic in this domain and satisfies the boundary conditions $\phi = 3$ on $-\infty < x < 1, \phi = 0$ on the semicircle, and $\phi = 1$ on $1 < x < \infty$.
19. Formulate and solve the mixed boundary value problem which under the mapping $w = z + 1/z$ is equivalent to a problem in which ϕ is harmonic in the quarter annulus $2 < r < 3$ for $0 < \theta < \pi/2$, and such that on its boundary
    (i) ϕ satisfies the Dirichlet condition $\phi = 3$ on $r = 2$ and $\phi = 5$ on $r = 3$ for $0 < \theta < \pi/2$, and
    (ii) ϕ satisfies the homogeneous Neumann condition $\partial\phi/\partial n = 0$ on $\theta = 0$ and $\theta = \pi/2$.

The following exercises use Theorems 5.2.1 and 5.2.2 to establish some fundamental properties of real variable solutions of the two-dimensional Laplace equation, so they do not involve conformal mapping.

20. Use Theorem 5.2.1 to prove that if $\phi_1(x, y)$ and $\phi_2(x, y)$ are each harmonic in the half plane $y > 0$, and both satisfy the same boundary conditions $\phi_i(x, 0) = f(x)$ $(i = 1, 2)$ on the x-axis, then they are identical. This uses Theorem 5.2.1 to establish the *uniqueness* of the solution of a Dirichlet problem for the Laplace equation in the half plane.
21. Let the functions $\phi_1(x, y)$ and $\phi_2(x, y)$ be harmonic in the half plane $y > 0$, and satisfy the boundary conditions $\phi_1(x, 0) = f_1(x)$ and $\phi_2(x, 0) = f_2(x)$ on the x-axis. If $|f_1(x) - f_2(x)| < \varepsilon$ for all x and $\varepsilon > 0$, use Theorem 5.2.1 to show that $|\phi_1(x, y) - \phi_2(x, y)| < \varepsilon$. This proves that a function that is harmonic in the half plane $y > 0$ that satisfies a Dirichlet condition on the x-axis depends *continuously* on the boundary data. That is, a small change in the boundary data from $f_1(x)$ to $f_2(x)$ produces a correspondingly small change in the solution. (Hint: you will find the integral inequality $|\int_a^b h(x)dx| < \int_a^b |h(x)|\, dx$ useful, where $h(x)$ is a real function.)
22. Use Theorem 5.2.2 to prove that if $\phi_1(r, \theta)$ and $\phi_2(r, \theta)$ are each harmonic in the disk $r < r_0$, and both satisfy the same boundary conditions $\phi_i(r_0, \theta) = f(\theta)$ $(i = 1, 2)$ on the boundary $r = r_0$, then they are

identical. This uses Theorem 5.2.2 to establish the *uniqueness* of the solution of a Dirichlet problem for the Laplace equation in a disk. This result could have been deduced from Exercise 20 because a disk can always be transformed into a half plane by a conformal transformation.

23. Let the functions $\phi_1(r, \theta)$ and $\phi_2(r, \theta)$ are each harmonic in the disk $r < r_0$, and satisfy the boundary conditions $\phi_1(r_0, \theta) = f_1(\theta)$ and $\phi_2(r_0, \theta) = f_2(\theta)$ for $0 < \theta < 2\pi$. If $|f_1(\theta) - f_2(\theta)| < \varepsilon$ for all θ and ε. Use Theorem 5.2.2 to show that $|\phi_1(r, \theta) - \phi_2(r, \theta)| < \varepsilon$ at any point in the disk. This proves that a function that is harmonic in the disk and that satisfies a Dirichlet condition on its boundary depends *continuously* on the boundary data. This result could have been deduced from Exercise 21 because a disk can always be transformed into a half plane by a conformal transformation.
24. Use Theorem 5.2.2 to prove that the value of a harmonic function $\phi(r, \theta)$ at the center of a disk C of radius r_0 centered on the origin is equal to the average value of ϕ on the boundary of C. Deduce that the result remains true for any disk C^* of radius ρ with its center at an arbitrary point z^*. This is the *Gauss mean value theorem for two-dimensional harmonic functions.*
25. Let $u(r, \theta)$, $v(r, \theta)$ be solutions of the Dirichlet problem for the Laplace equation in a disk $0 \leq r \leq r_0$ corresponding, respectively, to the boundary conditions $u(r_0, \theta) = f(\theta)$ and $v(r_0, \theta) = g(\theta)$. Let $f(\theta)$ only differ from $g(\theta)$ on the interval $\alpha \leq \theta \leq \beta$, with $0 < \alpha < \beta < 2\pi$. If $(\tilde{r}, \tilde{\theta})$ is any point in the disk, prove by using Theorem 5.2.2 that $u(\tilde{r}, \tilde{\theta}) \neq v(\tilde{r}, \tilde{\theta})$, and show that an upper bound for the difference $|u(r, \theta) - v(r, \theta)|$ is given by

$$\frac{1}{2\pi}\frac{(r_0 + \tilde{r})}{(r_0 - \tilde{r})}\int_\alpha^\beta |f(x) - g(x)|\, dx.$$

This shows that a change in the Dirichlet conditions, on however small a segment of the boundary of the disk, will influence the solution throughout the disk, and that the effect on the solution need not necessarily be small.

5.3 Steady-State Temperature Distribution

To show the relevance of Laplace's equation to heat conduction problems, we begin by deriving the partial differential equation that governs unsteady heat flow in a homogeneous solid heat conducting material that occupies some domain R in which no heat source exists. The method adopted is to study the heat balance in an arbitrary volume V within R. This will involve equating the rate at which heat flows into (or out of) V across its bounding surface S, to the rate at which heat is stored (or lost) in V.

The fundamental law governing heat flow under such conditions was formulated by Joseph Fourier, and it is known as *Fourier's law of heat conduction*. The law states that heat flows in the direction of decreasing temperature and that the rate of flow of heat h across a plane area, called the *heat flux* across the area and measured in joules (for example) per unit time, is proportional to the area and to the temperature gradient normal to the area. Let the temperature at a point with position vector $\mathbf{x}$ at time t be $T(\mathbf{x}, t)$. Then if $\hat{\mathbf{n}}$ is the unit normal to the plane area pointed in the direction of increasing temperature, the temperature gradient in this direction is given by the *directional derivative*

$$\frac{\partial T}{\partial n} = \hat{\mathbf{n}} \cdot \operatorname{grad} T. \tag{5.51}$$

The relationship between the heat flux h, the area A and the temperature gradient $\partial T/\partial n$ may be expressed quantitatively by introducing a constant of proportionality k and writing

$$h = -kA\hat{\mathbf{n}} \cdot \operatorname{grad} T. \tag{5.52}$$

The constant k is called the *thermal conductivity* of the material and we assume it to be an absolute constant whose value is determined solely by the nature of the material in which the heat conduction is taking place. The negative sign is necessary in Equation (5.52) because $\hat{\mathbf{n}}$ points in the direction of increasing temperature, while heat flows in the reverse direction.

If $\hat{\mathbf{n}}$ is taken to be the outward drawn normal to S, the heat flux dh entering V through an element of surface area $d\Sigma$ of S is

$$dh = k\frac{\partial T}{\partial n}\, d\Sigma. \tag{5.53}$$

This will be positive when $\partial T/\partial n > 0$, because then the temperature outside S will exceed the temperature inside it, causing heat to enter V. The total heat flux h entering V through S is found by integrating in Equation (5.53) over S to obtain

$$h = \int_S k\frac{\partial T}{\partial n}\, d\Sigma. \tag{5.54}$$

Now let the specific heat of the material be σ and its density be ρ (a constant), then the heat dQ stored in a volume element dv is

$$dQ = \rho c T\, dv. \tag{5.55}$$

Provided heat is neither created nor removed from the material the heat stored in volume V is found by integrating Equation (5.55) over V to obtain

$$Q = \int_V \rho c T\, dv. \tag{5.56}$$

Thus the rate at which heat is stored in V is

$$\frac{dQ}{dt} = \frac{d}{dt}\int_V \rho c T\, dv = \int_V \rho c \frac{\partial T}{\partial t}\, dv. \tag{5.57}$$

Since $h = dQ/dt$, by equating Equations (5.54) and (5.57) we find that

$$\int_S k \frac{\partial T}{\partial n}\, d\Sigma = \int_V \rho c \frac{\partial T}{\partial t}\, dv. \tag{5.58}$$

Using Equation (5.51) in the integral on the left of Equation (5.58), and applying the Gauss divergence theorem brings us to the result

$$\int_V k\,\mathrm{div}(\mathrm{grad}\, T)dv = \int_V \rho c \frac{\partial T}{\partial t} dv. \tag{5.59}$$

Combining terms and using the fact that div(grad T) $= \Delta T$ this becomes

$$\int_V \left(\frac{\partial T}{\partial t} - \kappa^2 \Delta T\right) dv = 0, \tag{5.60}$$

where $\kappa^2 = k/\rho c$ is a constant called the *coefficient of diffusivity* that depends only on the material. Because the volume V was arbitrary, this can only be true if the integrand vanishes, so we arrive at the equation for the temperature

$$\frac{\partial T}{\partial t} = \kappa^2 \Delta T. \tag{5.61}$$

This is the partial differential equation that governs the temperature distribution in the solid heat conducting material. Once T is known, Equation (5.52) determines the unsteady heat flux h in the direction $\hat{\mathbf{n}}$ at any time t and position $\mathbf{x}$ in the solid material. This partial differential equation is called the *heat equation*, and also the *diffusion equation* because it also describes an unsteady diffusion process.

The specification of the temperature on the boundary of a domain involves a *Dirichlet problem* for the heat equation governing the temperature distribution in the domain, while the specification of the heat flux on the boundary involves a *Neumann problem* for the heat equation. When the heat flux across a surface is zero it follows from Equations (5.51) and (5.52) that $\partial T/\partial n = 0$, so a homogeneous Neumann problem is involved, and this is the condition to be applied to a surface that is *thermally insulated*.

If it is possible for a steady-state temperature distribution to develop after a suitable laps of time, the temperature T is then no longer dependent on the time t so that $\partial T/\partial t = 0$, with the result that the heat Equation (5.61) reduces to the Laplace equation, and

$$\Delta T = 0. \tag{5.62}$$

Depending on the geometry of the problem, this will involve either the two- or a three-dimensional Laplace equation.

Should heat be generated throughout the medium at a rate $W(\mathbf{x}, t) > 0$ joules per unit volume per unit time by, for instance, electrical means or an exothermic chemical reaction, or removed by some cooling process or endothermic chemical reaction ($W < 0$), a volume integral of $W(\mathbf{x}, t)$ must be added to the left side of Equation (5.59). Combining the terms under a single integral sign and reasoning as before shows that the heat equation then becomes

$$c\rho\frac{\partial T}{\partial t} = k\Delta T + W(\mathbf{x}, t). \tag{5.63}$$

When a steady-state solution is possible for this equation, when $W(\mathbf{x}, t)$ becomes $W_0(\mathbf{x})$, the temperature distribution satisfies the partial differential equation

$$\Delta T = -\frac{1}{k}W_0(\mathbf{x}). \tag{5.64}$$

This equation is called the *Poisson equation* to which it is closely related.

Inspection of Equation (5.52) reveals it to be the scalar product of a vector element of area $\mathbf{A}$ with magnitude A, and a unit normal $\hat{\mathbf{n}}$. This leads to the definition of the *heat flux vector*

$$\mathbf{H} = -k \operatorname{grad} T, \tag{5.65}$$

that can be found once the temperature distribution T is known. This vector has the property that its direction at a point P is the direction of the heat flow at P, while its magnitude is equal to the heat flux through a unit area with normal $\hat{\mathbf{n}}$ located at P. The result from Equation (5.65) shows the important physical result that the heat flux is greatest in the direction of grad T.

It follows directly from Equation (5.65) that there can be no heat flux along an *isothermal* (a line or surface of constant temperature), so the heat flux vector at a point P can have no component along the isothermal through P. Thus an isothermal through P and the heat vector at that point must be orthogonal.

A line that is everywhere tangent to the heat flux vector $\mathbf{H}$ is a line along which heat flows, and it is called a *heat flow line*. In the two-space dimensions that are of concern to us here, lines belonging to the family of isothermals and to the family of heat flow lines form orthogonal trajectories in two-space dimensions, with corresponding results in three-space dimensions.

The equation for the heat flow lines is easily found once the temperature distribution has been found. Along a two-dimensional isothermal $T = \text{const.}$,

$$\frac{dT}{dx} = 0 = \frac{\partial T}{\partial x} + \frac{\partial T}{\partial y}\frac{dy}{dx}.$$

So the gradient $(dy/dx)_{T(\mathrm{P})}$ of the isothermal through an arbitrary point P is

$$\left(\frac{dy}{dx}\right)_{T(\mathrm{P})} = -\left(\frac{\partial T}{\partial x}\right)_{\mathrm{P}} \Big/ \left(\frac{\partial T}{\partial y}\right)_{\mathrm{P}}. \tag{5.66}$$

Because the isothermal and heat flow lines through P are orthogonal, the product of their slopes at P must equal -1. So from this and Equation (5.66) the heat flow line through P must have the slope

$$\left(\frac{\partial T}{\partial y}\right)_{\mathrm{P}} \Big/ \left(\frac{\partial T}{\partial x}\right)_{\mathrm{P}}.$$

Dropping the suffix P, since P was arbitrary, we conclude that the slope of dy/dx of the heat flow lines is everywhere given in terms of T by

$$\frac{dy}{dx} = \left(\frac{\partial T}{\partial y}\right) \Big/ \left(\frac{\partial T}{\partial x}\right). \tag{5.67}$$

Integration of Equation (5.67) then yields the equation of the heat flow lines.

Although the next results are not needed for this chapter, we nevertheless take this opportunity to use Equation (5.65) to derive two frequently used boundary conditions. The first is the condition to be applied at an interface separating two heat-conducting solids with different thermal conductivities.

Let the temperature distributions be T_1 and T_2 in the two materials, and let their conductivities be k_1 and k_2. Take $\mathbf{A} = A\hat{\mathbf{n}}$ be a vector element of area with magnitude A and unit normal $\hat{\mathbf{n}}$ located at the interface between the two materials. Then the heat flux leaving material 1 and entering A must equal the heat flux leaving the other side of A and entering material 2. Equating these heat fluxes gives

$$A\hat{\mathbf{n}} \cdot (-k_1 \operatorname{grad} T_1) = A\hat{\mathbf{n}} \cdot (-k_2 \operatorname{grad} T_2),$$

which is equivalent to

$$k_1 \frac{\partial T_1}{\partial n} = k_2 \frac{\partial T_2}{\partial n}, \tag{5.68}$$

where $\partial/\partial n$ is differentiation normal to the interface. Thus the condition in Equation (5.68) is the boundary condition to be applied at the interface and clearly when the two materials are identical, no interface exists and $k_1 = k_2$ so that the result in Equation (5.68) reduces to the continuity of the temperature gradient. The *thermal insulation condition* $\partial T/\partial n = 0$ follows from Equation (5.68) when one of the materials has a thermal conductivity of zero (it is a thermal insulating material).

Finally, we use Equation (5.65) to derive a boundary condition that is appropriate for certain exterior boundaries. Suppose heat is lost from an exterior boundary at a rate proportional to the difference between the surface temperature T and the ambient temperature T_0 outside the body. In such a case the heat flux leaving the surface through a vector element $\mathbf{A} = A\hat{\mathbf{n}}$ must equal $\alpha A(T - T_0)$, where α is a constant of proportionality. Equating heat fluxes across A gives

$$A\hat{\mathbf{n}} \cdot (-k \operatorname{grad} T) = A\alpha(T - T_0),$$

leading to the nonhomogeneous boundary condition

$$k\frac{\partial T}{\partial n} + \alpha T = \alpha T_0. \tag{5.69}$$

This boundary condition is called *Newton's law of cooling*, and it is often used to model a low temperature cooling process at the surface of a body. However, because our subsequent concern is with the use of conformal mapping, it must be remembered that only Dirichlet conditions and homogeneous Neumann conditions map directly from the z-plane to the w-plane and that conditions like Equation (5.69) are altered under such a mapping and so will not be considered here.

Leaving aside these more general considerations concerning steady-state temperature distributions, we return to the main purpose of this section which is to illustrate how conformal mapping may be used to find two-dimensional steady-state temperature distributions. Before doing this let us first view the relationship between isothermals and heat flow lines from a somewhat different standpoint. As the temperature distribution T is harmonic, if it is regarded as the real part of an analytic function $w = f(z)$, it will always have associated with it a function ψ that is its harmonic conjugate. Making the identifications $u = T$ and $v = \psi$, the function ψ follows directly from T using the arguments in Section 1.4. It thus follows that the level curves $\psi =$ const. are none other than the heat flow lines.

This means that if we consider a general analytic function

$$f(z) = T(x, y) + i\psi(x, y), \tag{5.70}$$

the family of lines $T(x, y) =$ const. represent isothermals and the family of lines $\psi(x, y) =$ const. represent the corresponding heat flow lines. The choice of the analytic function $f(z)$ determines the shape of the isothermals. By selecting a suitable form for $f(z)$, it is possible to match the isothermals to the boundary shape and boundary conditions is a domain representing some shape of interest. The complete temperature distribution is then given by $T(x, y) = \text{Re}\{f(z)\}$, while the heat flow lines are given by $\text{Im}\{f(z)\} = \psi(x, y) = \text{const}$. The change of emphasis here, starting as it does from an analytic function $f(z)$, rather than a temperature distribution T, is the idea underlying the indirect method mentioned earlier. This idea will be used again when considering fluid flows and electrostatic problems.

In view of the result from Equation (5.62), all of the standard boundary value problems and specific examples considered in Section 5.2 may be interpreted as two-dimensional steady-state temperature distributions, and these can be extended in simple ways by use of result from Equation (5.49). By way of illustration let us apply result Equation (5.49) to Example 5.2.1 with $\phi = T$, the scale factor $m = 100$, and the additive constant $M = 300$. Then the boundary conditions on $v = 0$ become $T(u, 0) = 600$ for $-\infty < u < -2$, $T(u, 0) = 300$ for $-2 < u < 1$, and $T(u, 0) = 100$ for $1 < u < \infty$, and the solution becomes

$$T(u, v) = 100 + \frac{1}{\pi}\left[300\,\mathbf{Arctan}\left(\frac{v}{u+2}\right) + 200\,\mathbf{Arctan}\left(\frac{v}{u-1}\right)\right], \quad v > 0. \quad (5.71)$$

This, then, is the two-dimensional steady-state temperature distribution in a semi-infinite heat-conduction slab of metal occupying the space $v > 0$, on the plane boundary $v = 0$ of which the temperature is $T = 600$ for $-\infty < u < -2$, $T = 300$ for $-2 < u < 1$, and $T = 100$ for $1 < u < \infty$.

Example 5.3.1 A Half Space with an Insulated Strip

Find the temperature distribution $T(x, y)$ in the half space $y > 0$ occupied by a heat-conducting material when $T = 0$ for $x < -4$, $y = 0$, $T = 200$ for $x > 4$, $y = 0$, and the strip $-4 < x < 4$, $y = 0$ is thermally insulated.

SOLUTION

The problem is illustrated on the left of Figure 5.25, where the thermal insulation on the strip $-4 < x < 4$, $y = 0$ corresponds to the imposition of the homogeneous Neumann condition $\partial T/\partial y = 0$ on the strip.

This problem is of the type illustrated in Figure 5.18 in connection with the mapping $w = \text{Arcsin}\, z$ in Example 5.2.5, though here the insulation extends over the strip $-4 < x < 4$, $y = 0$, instead of the strip $-1 < x < 1$, $y = 0$. The appropriate mapping in this case is $w = \text{Arcsin}\, z/4$, which maps the upper half of the z-plane, cut along the x-axis from $-\infty < x < -4$ and from $4 < x < \infty$ onto the semi-infinite strip $-\pi/2 < u < \pi/2$, $v > 0$ as shown on the right of Figure 5.25. Let $T(x, y)$ become $\overline{T}(u, v)$ under the mapping. Then the

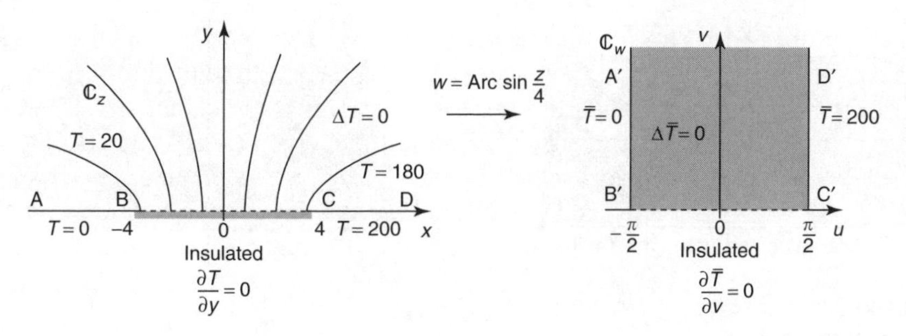

FIGURE 5.25
A half space with a thermally insulated strip.

solution in the w-plane follows from (5.46) with $k_1 = 0$, $k_2 = 200$, and $\phi = \overline{T}$. The result of the substitution is

$$\overline{T}(uv) = 100 + \frac{200}{\pi}u, \tag{5.72}$$

where now $u = \text{Re}\{w\} = \text{Re}\{\text{Arcsin}\, z/4\}$. Setting $a = 4$ in Equation (4.64) of Section 4.4 shows the required solution to be

$$T(x, y) = 100 + \frac{200}{\pi}\text{Arcsin}\left(\frac{\sqrt{(x+4)^2 + y^2} - \sqrt{(x-4)^2 + y^2}}{8}\right), \tag{5.73}$$

where the range of the function Arcsin in Equation (5.73) lies within the interval $(-\pi/2, \pi/2)$. Some representative isothermals are shown on the left of Figure 5.25.

Notice that in this case the simple form of the solution in Equation (5.72) could have been deduced without appeal to Example 5.2.5. This is because constant Dirichlet conditions exist on each of the semi-infinite boundaries, while the plane face $-\pi/2 < u < \pi/2, v = 0$ is thermally insulated so heat can only flow parallel to the u-axis.

Consequently the solution of the Laplace equation in the w-plane can only depend on u so it reduces to $d^2\overline{T}/du^2 = 0$, subject to the boundary conditions $\overline{T} = 0$ when $u = -\pi/2$ and $\overline{T} = 200$ when $u = \pi/2$, which yields the solution from Equation (5.72).

In fact, for the same reasons, Equation (5.72) is also the solution in any finite rectangle $-\pi/2 < u < \pi/2, v_1 < v < v_2$ with the Dirichlet conditions shown in Figure 5.25, provided a homogeneous Neumann condition is applied to the plane faces $-\pi/2 < u < \pi/2, v = v_1$ and $-\pi/2 < u < \pi/2, v = v_2$. ◇

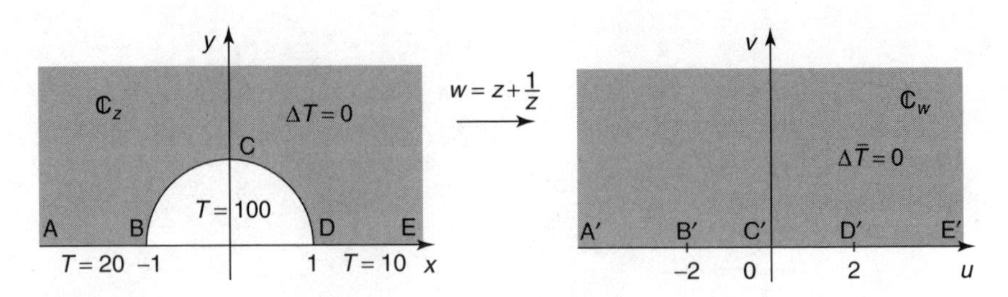

FIGURE 5.26
The indented slab.

Example 5.3.2 The Temperature Distribution within an Indented Slab

Find the temperature distribution within a semi-infinite indented slab of heat conducting material, when the semicircular indentation and boundary conditions are as shown on the left in Figure 5.26.

SOLUTION

The solution in the w-plane is given by Equation (5.27), with $k_1 = 20$, $k_2 = 100$, $k_3 = 10$, $u_1 = -2$, and $u_2 = 2$, from which it is found to be

$$\bar{T}(u, v) = 10 + \frac{1}{\pi}\left[-80\,\mathbf{Arctan}\left(\frac{v}{u+2}\right) + 90\,\mathbf{Arctan}\left(\frac{v}{u-2}\right)\right],$$

where $T(x, y)$ has become $\bar{T}(u, v)$ under the mapping. However,

$$u(x, y) = \mathrm{Re}\{z + 1/z\} = \frac{x(x^2 + y^2 + 1)}{x^2 + y^2}$$

and

$$v(x, y) = \mathrm{Im}\{z + 1/z\} = \frac{y(x^2 + y^2 - 1)}{x^2 + y^2}.$$

So the temperature distribution in the (x, y)-plane follows by replacing $\bar{T}(u, v)$ by $T(x, y)$ in the previous solution, and using these expressions for u and v to obtain:

$$T(x, y) = 10 + \frac{1}{\pi}\left[-80\,\mathbf{Arctan}\left(\frac{v(x, y)}{u(x, y)+2}\right) + 90\,\mathbf{Arctan}\left(\frac{v(x, y)}{u(x, y)-2}\right)\right].$$

◇

Example 5.3.3 The Temperature Distribution in a Semi-Infinite Slab Heated Internally along a Slit

Find the temperature distribution in a semi-infinite slab of heat-conducting material when heated internally along a semi-infinite slit, as shown in Figure 5.27.

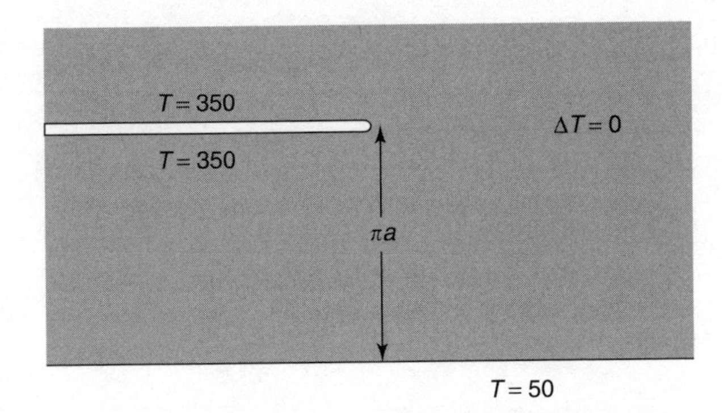

FIGURE 5.27
An internally heated slab of heat-conducting material.

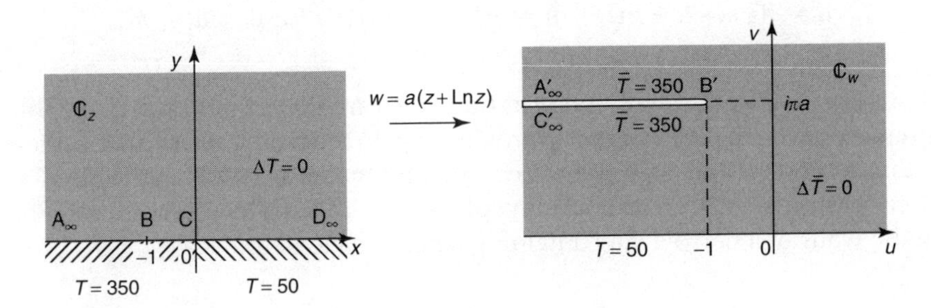

FIGURE 5.28
The mapping $w = a(z + \mathrm{Ln}\, z)$.

SOLUTION
This problem may be solved by appeal to Example 4.5.4. The mapping $w = a(z + \mathrm{Ln}\, z)$ maps the upper half of the z-plane shown on the left of Figure 5.28 onto the cut upper half of the w-plane shown on the right. We identify the problem shown in Figure 5.27 with the diagram on the right of Figure 5.28 by using a suitable choice of axes and origin, and let $T(x, y)$ become $\overline{T}(x, y)$ under the mapping.

The required temperature distribution in the z-plane then follows from Equation (5.22) as

$$T(\theta) = 50 + \frac{300\theta}{\pi}, \qquad \text{for } 0 \leqslant \theta \leqslant \pi.$$

To find the temperature distribution in the w-plane we first set $z = re^{i\theta}$ in $w = a(z + \mathrm{Ln}\, z)$ to obtain

$$w = are^{i\theta} + \ln_e r + ia\theta.$$

The isothermals in the w-plane are the lines $\theta = \alpha$ (const.), so these correspond to the lines

$$w = are^{i\alpha} + a\ln_e r + ia\alpha,$$

where now r is to be regarded as a parameter, and on these lines

$$t = 50 + \frac{300\alpha}{\pi}, \qquad 0 \leqslant \alpha \leqslant \pi.$$

So, for fixed α and $r = \xi$, a parameter, the parametric representation of the isothermals $\overline{T} = 50 + (300\alpha)/\pi$ within the cut half space is

$$u = a(\xi\cos\alpha + \ln_e \xi),\ v = a(\xi\sin\alpha + \alpha) \qquad \text{for } 0 < \alpha < \pi.$$

Representative isothermals and heat flow lines are shown in Figure 5.29. This situation can be regarded as an approximation to the temperature distribution within a semi-infinite slab of heat conducting material that is maintained at a temperature $T = 50$ on its bounding plane face, and at a temperature $T = 350$ on the walls of a narrow slit containing a heating element. ◇

Example 5.3.4 The Temperature Distribution within a Nonsimply Connected Slab

Find the temperature distribution within a semi-infinite slab of heat-conducting material occupying the half plane $y > 0$, and penetrated by a cylindrical hole of unit radius with its axis normal to the plane and passing through the point $x = 0$, $y = 3$, when the boundary conditions are as shown on the left of Figure 5.30.

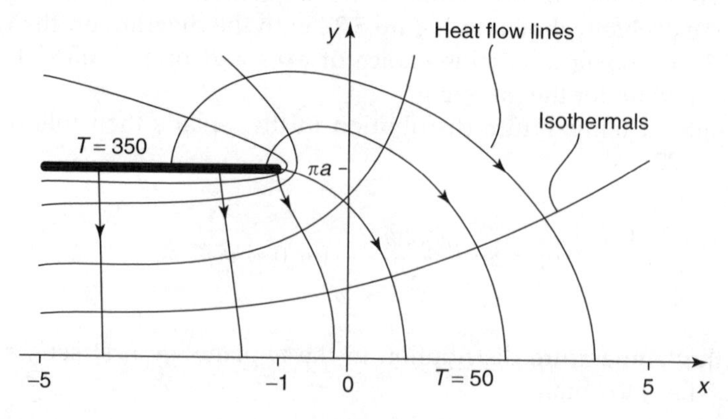

FIGURE 5.29
Representative isothermals and heat flow lines.

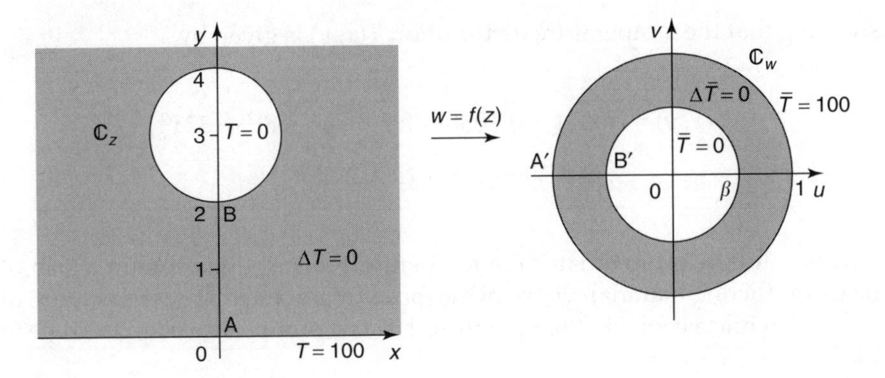

FIGURE 5.30
The mapping $w = (z - i2\sqrt{2})/(z + i2\sqrt{2})$.

SOLUTION
The domain in which the heat flows is not simply connected because the boundary of the hole represents an internal boundary. To solve this problem we must try to map the region onto a simpler one which also has only one internal boundary. The mapping in Example 4.3.5 is appropriate because it maps the domain in the z-plane onto the annulus on the right of Figure 5.30, and both domains have one internal boundary. Setting $\alpha = 3$, $\rho = 1$ in that mapping shows that the required function is

$$f(z) = \frac{z - i2\sqrt{2}}{z + i2\sqrt{2}}.$$

The bounding surface $y = 0$ in the z-plane of the half-space maps to the unit circle $|w| = 1$ in the w-plane, and the boundary of the circular hole to the circle $|w| = \beta$, where $\beta = (3 - 2\sqrt{2})/17$. Let $T(x, y)$ becomes $\bar{T}(u, v)$ under the mapping.

Then the temperature distribution in the annulus is given by Equation (5.40) as

$$\bar{T}(R) = 100\left(1 - \frac{\ln_e R}{\ln_e \beta}\right), \quad \beta < R < 1.$$

After substituting for β this becomes

$$\bar{T}(R) = 100(1 + 0.2176 \ln_e R), \quad \beta < R < 1.$$

However, $R = |w| = |f(z)|$, so

$$R = \frac{[(x^2 + y^2 - 8)^2 + 32x^2]^{1/2}}{x^2 + (y + 2\sqrt{2})^2},$$

showing that the temperature distribution $T(x, y)$ is given by

$$T(x,y) = 100\Big(1 + 0.1088\ln_e[(x^2 + y^2 - 8)^2 + 32x^2] \\ - 0.2176\ln_e[x^2 + (y + 2\sqrt{2})^2]\Big).$$

This solution approximates the temperature distribution within a slab of heat-conducting material whose plane face is maintained at a temperature of $T = 100$, while a coolant at temperature $T = 0$ is pumped through the hole in the slab.

This example illustrates how, when boundary value problems in nonsimply connected domains are to be solved by conformal mapping, the domain must be mapped onto a similar nonsimply connected domain; that is, the number of internal closed boundaries in each region must be the same. In general, the solution of such problems is difficult because although one of the boundaries can be mapped onto another boundary of convenient shape, it does not necessarily follow that this will be true of the other boundaries. The method was successful in this case because each boundary in the z-plane mapped onto one of the same family of constant coordinate lines (R = const.) in the w-plane. ◇

Exercises 5.3

1. Use Equation (5.49) together with the solution of Example 5.2.1 to find the temperature distribution in the half plane $v > 0$, when on the boundary $v = 0$, $T(u, 0) = 800$ for $-\infty < u < -2$, $T(u, 0) = 350$ for $-2 < u < 1$, and $T(u, 0) = 50$ for $1 < u < \infty$. Check the result by direct calculation.
2. Use Theorem 5.1.1 with the solution of Example 5.2.2 of Section 5.2 to find the temperature distribution in the unit disk shown in Figure 5.31 subject to the given boundary conditions.
3. Find the temperature distribution in the infinite strip shown in Figure 5.32 subject to the given boundary conditions.
4. Find the temperature distribution in the infinite strip shown in Figure 5.33 subject to the given boundary conditions.
5. Find the temperature distribution in a slab of heat-conducting material occupying the first quadrant subject to the boundary conditions $T(x, 0) = 35$ for $x > 0$ and $T(0, y) = 100$ for $y > 0$.
6. Use the transformation $w = z^2$ to find the temperature distribution in the domain shown in Figure 5.34 subject to the given boundary conditions.
7. Use the transformation $w = z^2$ to find the temperature distribution in the domain shown in Figure 5.35 subject to the given boundary conditions.

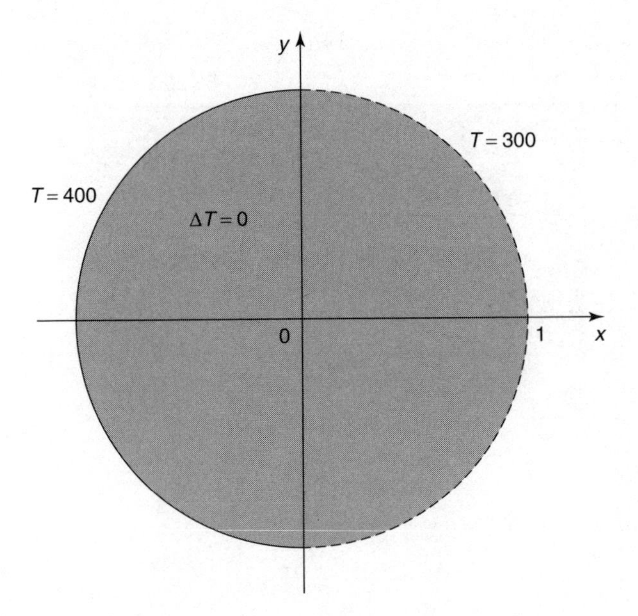

FIGURE 5.31

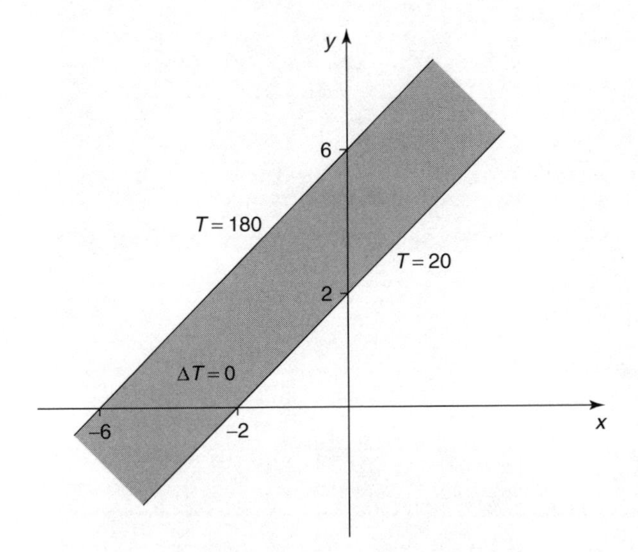

FIGURE 5.32

8. Use the method of Example 5.2.3 to find the temperature distribution in the domain shown in Figure 5.36 subject to the given boundary conditions.
9. By means of the transformation $w = [(1 + z)/(1 - z)]^2$, find the temperature distribution in the domain shown in Figure 5.37 subject to the given boundary conditions.

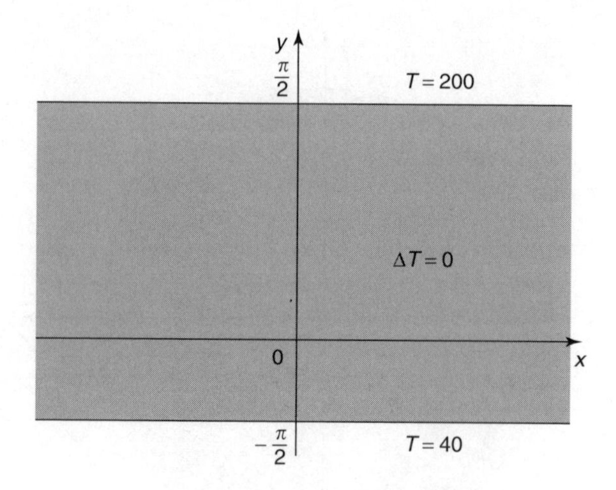

FIGURE 5.33

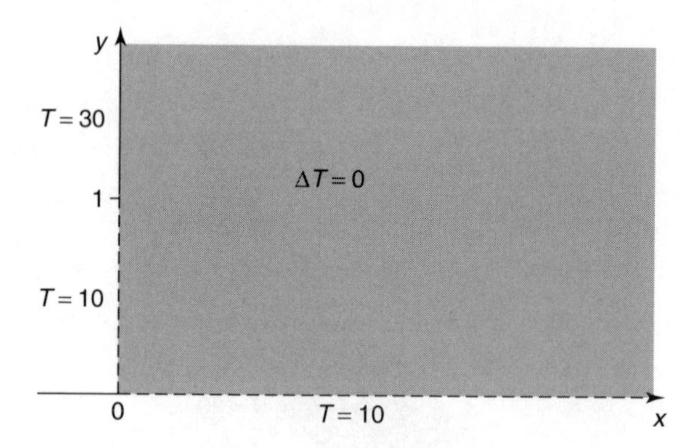

FIGURE 5.34

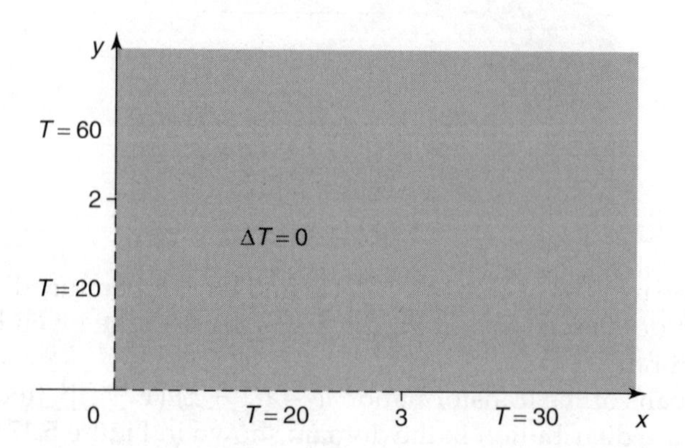

FIGURE 5.35

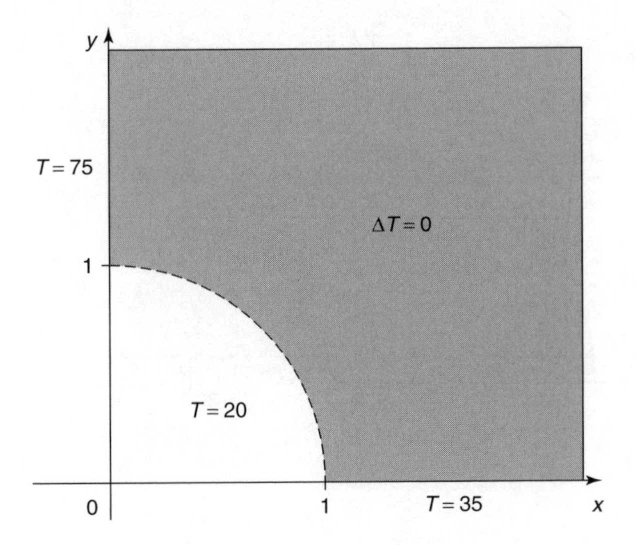

FIGURE 5.36

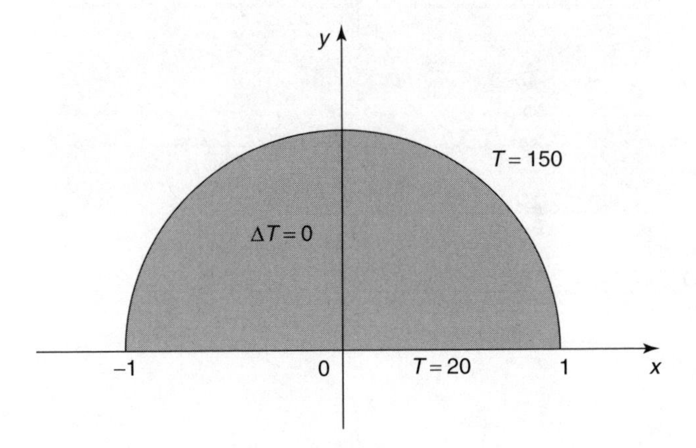

FIGURE 5.37

10. Use the mapping $w = i[(1 - z)/(1 + z)]$ to find the temperature distribution in the half-disk shown in Figure 5.38 subject to the given boundary conditions.
11. Use the mapping $w = \sqrt{z^2 + 1}$ to find the temperature distribution in the upper half-plane $y > 0$ when the temperature on the plane boundary and along the cut $x = 0, 0 < y < 1$ is as shown in Figure 5.39.
12. Use the mapping $w = i[(1 + z)/(1 - z)]$ to find the temperature distribution in the disk shown in Figure 5.40 subject to the given boundary conditions.

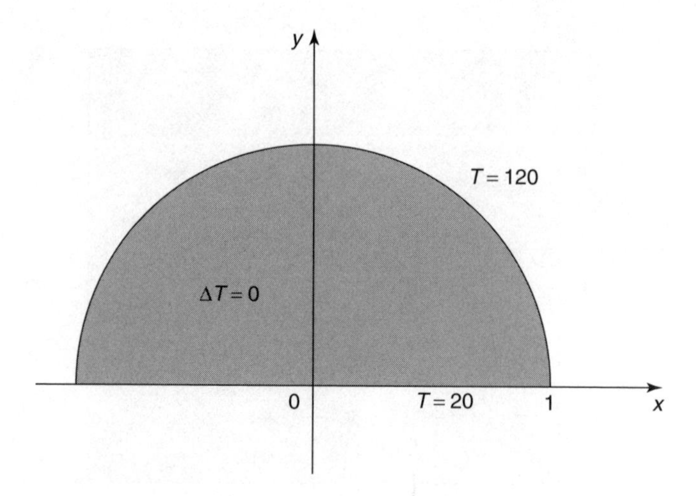

FIGURE 5.38

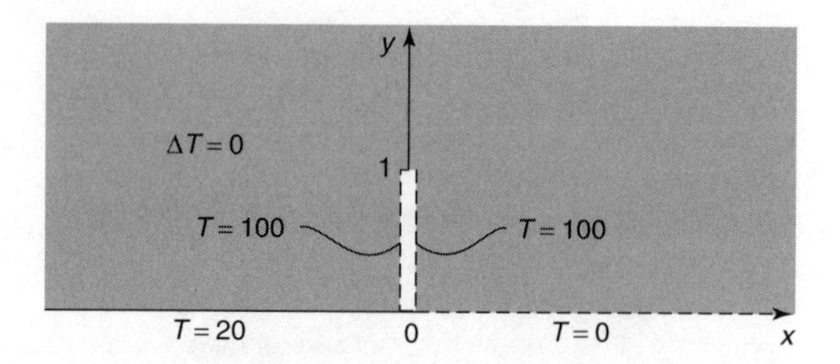

FIGURE 5.39

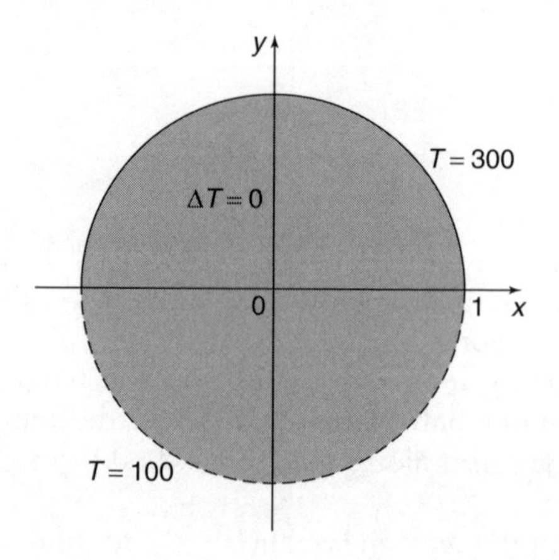

FIGURE 5.40

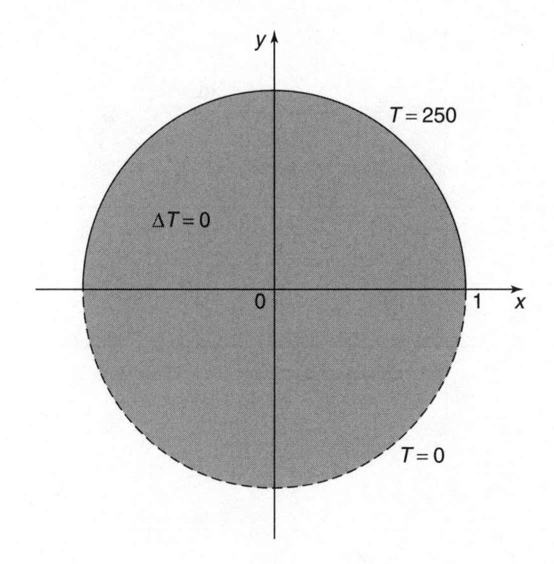

FIGURE 5.41

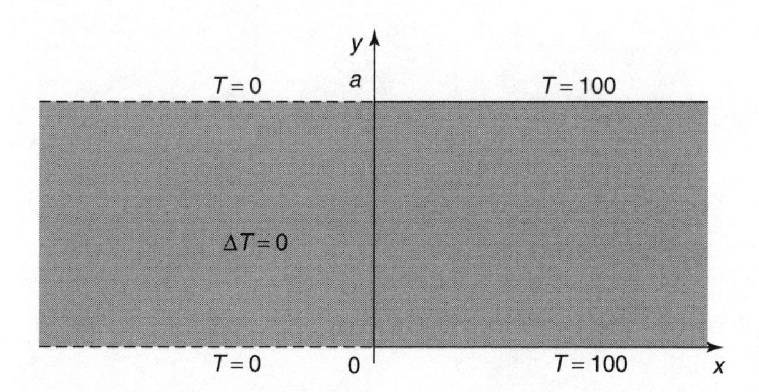

FIGURE 5.42

13. Use the mapping $w = [(1 + z)/(1 - z)]$ to find the temperature distribution in the disk shown in Figure 5.41 subject to the given boundary conditions.
14. Use the mapping $w = e^{z/a}$ to find the temperature distribution in the infinite strip $0 < y < a$ shown in Figure 5.42 subject to the given boundary conditions. What is the form of the solution (a) when $x \gg 0$ and (b) when $x \ll 0$?
15. Use the mapping $w = \cosh(\pi z/a)$ to find the temperature distribution in the semi-infinite strip shown in Figure 5.43 subject to the given boundary conditions.

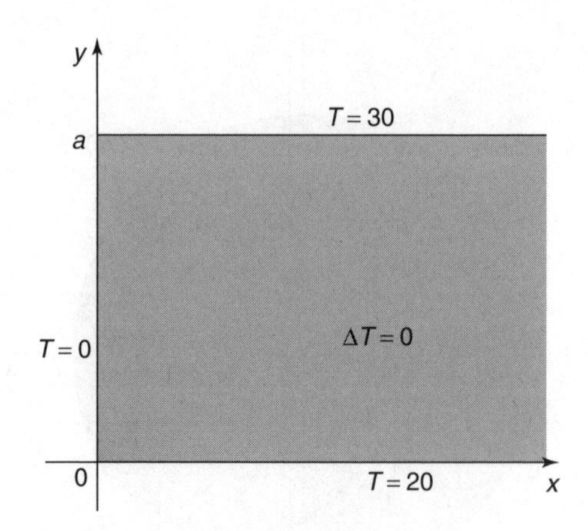

FIGURE 5.43

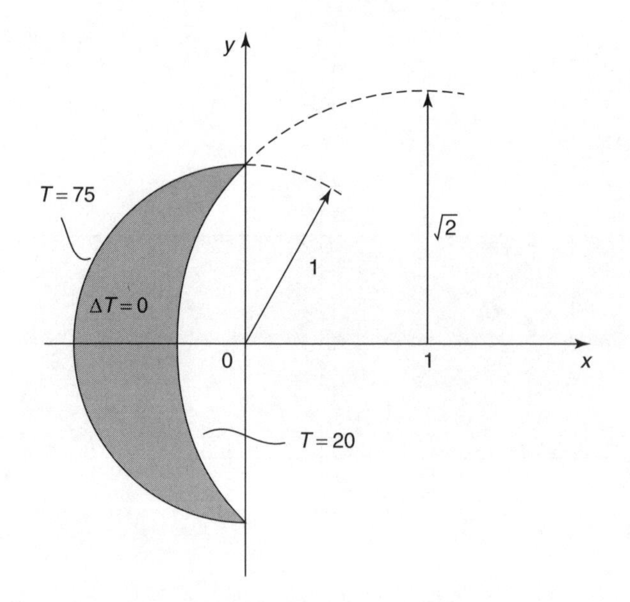

FIGURE 5.44

16. Find the temperature distribution in the crescent shaped domain shown in Figure 5.44 subject to the given boundary conditions. What is the form taken by the isothermals? (Hint: Use Example 4.3.4.)
17. Find the temperature distribution in the rectangular domain shown in Figure 5.45 subject to the given boundary conditions.
18. Use the transformation $w = e^z$ to find the temperature distribution in the infinite strip $-\infty < x < \infty$, $0 < y < \pi/2$ shown in Figure 5.46 subject to the given boundary conditions.

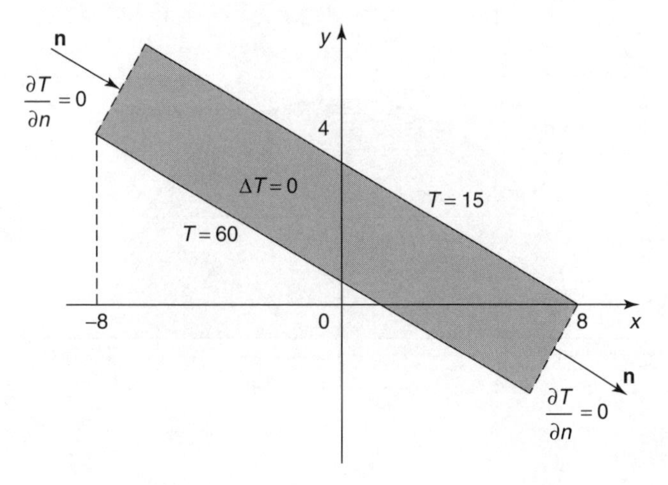

FIGURE 5.45

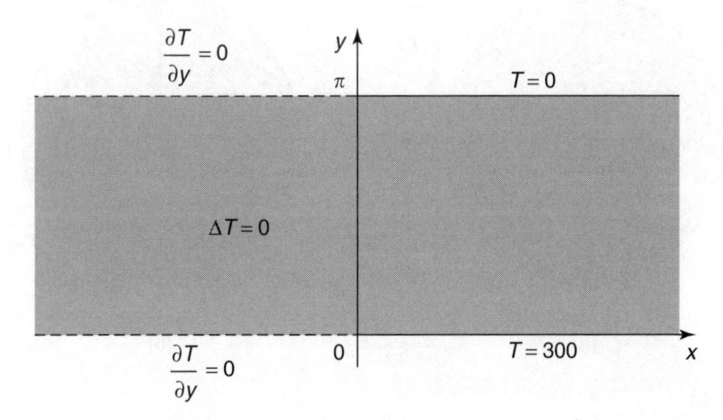

FIGURE 5.46

19. Use the transformation $w = i[(1 - z)/(1 + z)]$ to find the temperature distribution in the half disk shown in Figure 5.47 subject to the given boundary conditions. (Hint: Consider the temperature distribution in a unit disk with suitable boundary conditions.)
20. Use Theorem 5.2.2 to prove that temperature distribution in a unit disk subject to the boundary conditions given in Figure 5.48(a) is given by

$$T(x, y) = \frac{T_1}{\pi}\left\{\textbf{Arctan}\left[\left\{\frac{(1+r)^2}{1-r^2}\tan\left(\tfrac{1}{2}(\phi - \theta)\right)\right\}\right]\right.$$
$$\left. + \textbf{Arctan}\left[\left(\frac{(1+r)^2}{1-r^2}\right)\tan\tfrac{1}{2}\theta\right]\right\},$$

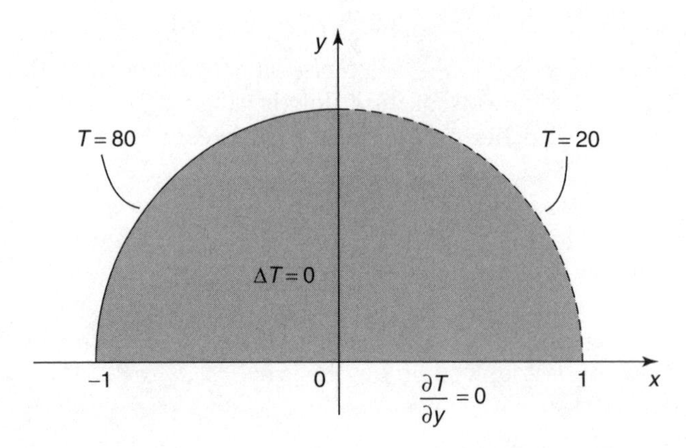

FIGURE 5.47

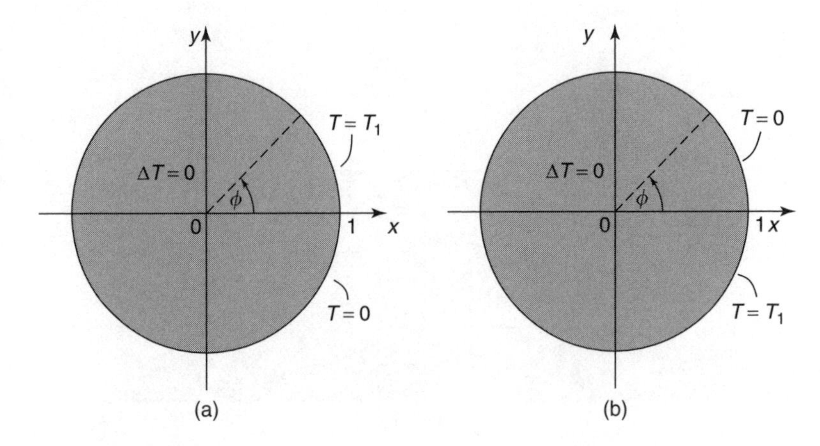

FIGURE 5.48

for any point (r, θ) in the disk, with $0 < \phi < \pi$. Use this result to deduce the solution corresponding to the boundary conditions given in Figure 5.48(b).

21. Use Theorem 5.2.1 to show that the temperature distribution in the half plane $y > 0$, when the boundary $y = 0$ is maintained at the steady temperature $T(x, 0) = a\cos mx + b\sin nx$ is given by

$$T(x, y) = ae^{-mx}\cos mx + be^{-nx}\sin nx,$$

for (x, y) any point in the half plane $y > 0$.

5.4 Steady Two-Dimensional Fluid Flow

A fluid is a continuous medium that offers very little resistance to internal movement, flows to take up the shape of any container, and is such that its volume is hardly affected by pressure. The internal resistance to the relative motion between layers of fluid is called *viscosity,* the motion of a fluid is called its *flow*, and the general study of fluid flow is called *hydrodynamics.* A flow is best visualized by considering the paths followed by particles of the fluid, that are called *streamlines*, and by virtue of its definition, the tangent at each point of a streamline is directed along the fluid velocity vector at that point.

An important idealization of a real fluid is the model that treats a fluid as *inviscid* (nonviscous), and *incompressible* (its volume and hence its density are unchanged by pressure). This model is often used to describe slow flows in liquids such as water that have a low viscosity, and even in a gas such as air when the speed of flow is well below the speed of sound in the gas because only then may the effect of compressibility in the gas be neglected.

The hydrodynamic equations governing the flow of such a fluid are:
The equation of conservation of mass

$$\frac{\partial \rho}{\partial t} + \operatorname{div}(\rho \mathbf{q}) = 0, \tag{5.74}$$

The equation of motion

$$\frac{\partial \mathbf{q}}{\partial t} + (\mathbf{q} \cdot \operatorname{grad})\mathbf{q} + \frac{1}{\rho} \operatorname{grad} p = \mathbf{F}, \tag{5.75}$$

The equation of state

$$\rho = f(p), \tag{5.76}$$

where t is the time, ρ is the fluid density, $\mathbf{q}$ is the fluid velocity vector, p is the fluid pressure, and $\mathbf{F}$ is the external force (gravity, etc.) acting on the fluid. The function $f(\rho)$, sometimes called the *constitutive law* for the fluid, is prescribed and models the particular form of the fluid under consideration.

To simplify the class of flows (solutions) described by Equations (5.74) through (5.76) to the point where analytic functions can be used, it is necessary to impose some restrictions on the flow which are the fluid is required to be steady (independent of time), two-dimensional, and irrotational. By a *steady flow*, we mean a flow that may have different velocities at different points in the plane, but which is such that the velocity at any fixed point is independent of the time. To understand the meaning of *irrotational flow,* we must appeal to the result established in hydrodynamics that the vector quantity $\boldsymbol{\zeta} = \operatorname{curl} \mathbf{q}$ called the *vorticity* of the flow, which is obtained by taking a line integral around any closed path in the fluid, is an invariant of the flow. The vorticity measures the circulation or internal rotation of the fluid around the closed path. As $\boldsymbol{\zeta}$ is an invariant for any particular flow, a flow that starts with zero vorticity ($\boldsymbol{\zeta} = \mathbf{0}$), called an *irrotational flow,* will continue to be irrotational for all time (Kelvin's circulation theorem). For a reason that will be come apparent later, irrotational flows are called *potential flows*.

Let us work with two-dimensional rectangular Cartesian coordinates, with the unit vector $\mathbf{i}$ in the positive x-direction and the unit vector $\mathbf{j}$ in the positive y-direction. Setting $\mathbf{q} = q_1\mathbf{i} + q_2\mathbf{j}$, and considering steady inviscid, incompressible ($\rho =$ const.), and irrotational flow without body forces ($\mathbf{F} = \mathbf{0}$), reduces Equations (5.74), (5.75), and $\boldsymbol{\zeta} = \mathbf{0}$ to

Conservation of mass

$$\frac{\partial q_1}{\partial x} + \frac{\partial q_2}{\partial y} = 0, \tag{5.77}$$

Conservation of momentum

$$q_1\frac{\partial q_1}{\partial x} + q_2\frac{\partial q_1}{\partial y} + \frac{1}{\rho}\frac{\partial p}{\partial x} = 0, \tag{5.78}$$

$$q_1\frac{\partial q_2}{\partial x} + q_2\frac{\partial q_2}{\partial y} + \frac{1}{\rho}\frac{\partial p}{\partial y} = 0, \tag{5.79}$$

Irrotational condition

$$\frac{\partial q_2}{\partial x} - \frac{\partial q_1}{\partial y} = 0. \tag{5.80}$$

Taken together, Equations (5.77) and (5.80) are seen to be the Cauchy–Riemann equations for the analytic function

$$f(z) = q_1 - iq_2, \tag{5.81}$$

so that the components q_1 and q_2 of the velocity $\mathbf{q}$ are harmonic functions, with the functions q_1 and $-q_2$ harmonic conjugates.

To discover how the pressure p is related to $\mathbf{q}$ we make use of the vector identity

$$(\mathbf{q}\cdot \text{grad})\mathbf{q}^2 = \tfrac{1}{2}\,\text{grad}\,\mathbf{q}^2 - \mathbf{q}\times \text{curl}\,\mathbf{q}$$

in the momentum Equation (5.75). Then for steady incompressible flow without body forces in Equation (5.75) becomes

$$\text{grad}\left(\tfrac{1}{2}\rho\mathbf{q}^2 + p\right) = \rho\mathbf{q}\times \text{curl}\,\mathbf{q}. \tag{5.82}$$

Forming the scalar product of this equation with $\mathbf{q}$, and using the properties of the triple scalar product, simplifies Equation (5.82) to

$$\mathbf{q} \cdot \operatorname{grad}\left(\tfrac{1}{2}\rho\mathbf{q}^2 + p\right) = 0. \tag{5.83}$$

However, the operator $\mathbf{q} \cdot \operatorname{grad}$ is a directional derivative in the direction of the flow along a streamline, so Equation (5.83) shows that

$$\tfrac{1}{2}\rho\mathbf{q}^2 + p = \text{const.} \tag{5.84}$$

along a streamline. The result from Equation (5.84) is called *Bernoulli's equation*, and for two-dimensional flows it takes the simpler form

$$\tfrac{1}{2}\rho(q_1^2 + q_2^2) + p = \text{const.} \tag{5.85}$$

along a streamline in the flow. In particular, this result shows that the pressure $p(x, y)$ is greatest at the point where the speed of flow $q = (q_1^2 + q_2^2)^{1/2}$ is the least, and vice versa. Notice that the derivation of Bernoulli's equation did not make use of the irrotational condition, so Equation (5.85) is true along streamlines in fluids with or without circulation. Points in a flow where the velocity vanishes are called *stagnation points*, and Bernoulli's Equation (5.85) shows them to be points of greatest pressure. In an irrotational flow $\operatorname{curl}\mathbf{q} = \mathbf{0}$, so from vector analysis $\mathbf{q}$ must have the form

$$\mathbf{q} = \operatorname{grad}\phi, \tag{5.86}$$

where the scalar function $\phi(x, y)$ is called the *velocity potential for the flow*, although for historical reasons some books use the definition $\mathbf{q} = -\operatorname{grad}\phi$. Notice that the velocity $\mathbf{q}$ is not altered if an arbitrary constant ϕ_0 is added to the velocity potential ϕ, because $\operatorname{grad}(\phi + \phi_0) = \operatorname{grad}\phi = \mathbf{q}$. In hydrodynamics this fact is used to remove unnecessary additive constants from the solution by equating them to zero. It will be seen later that in electrostatics this same fact is used to adjust the reference potential to a convenient value.

Setting $\mathbf{q} = q_1\mathbf{i} + q_2\mathbf{j}$ in Equation (5.86) and equating components gives the two important results that

$$q_1 = \frac{\partial\phi}{\partial x}, \quad q_2 = \frac{\partial\phi}{\partial y}. \tag{5.87}$$

Substituting Equation (5.87) into Equation (5.77) then gives

$$\frac{\partial^2\phi}{\partial x^2} + \frac{\partial^2\phi}{\partial y^2} = 0,$$

showing that the velocity potential ϕ is harmonic.

If we associate with the velocity potential $\phi(x, y)$ the conjugate harmonic function $\psi(x, y)$, called the *stream function* for the flow, it follows that the complex function

$$w(z) = \phi + i\psi \tag{5.88}$$

is an analytic function of $z = x + iy$. The function $w(z)$ is called the *complex potential* for the flow. The stream function ψ may always be found from a known velocity potential ϕ by integrating the Cauchy–Riemann equations, that in this notation become

$$\frac{\partial \phi}{\partial x} = \frac{\partial \psi}{\partial y} \quad \text{and} \quad \frac{\partial \psi}{\partial x} = -\frac{\partial \phi}{\partial y} \tag{5.89}$$

Conversely, given a complex potential $w(z)$, the velocity potential may always be recovered from it by taking its real part

$$\phi(x, y) = \text{Re}\{w(z)\}. \tag{5.90}$$

We know from our examination of the Cauchy–Riemann equations that

$$w'(z) = \frac{\partial \phi}{\partial x} + i\frac{\partial \psi}{\partial x},$$

so after use of Equation (5.89) this becomes

$$w'(z) = \frac{\partial \phi}{\partial x} - i\frac{\partial \phi}{\partial y},$$

and so because of Equation (5.87),

$$w'(z) = q_1 - iq_2. \tag{5.91}$$

In terms of the complex potential $w(z)$ this shows that

$$q_1 = \text{Re}\{w'(z)\} \quad \text{and} \quad q_2 = -\text{Im}\{w'(z)\}, \tag{5.92}$$

while the speed

$$q = |w'(z)| = (q_1^2 + q_2^2)^{1/2} = \left[\left(\frac{\partial \phi}{\partial x}\right)^2 + \left(\frac{\partial \phi}{\partial y}\right)^2\right]^{1/2}. \tag{5.93}$$

When the complex quantity is interpreted as a two-dimensional vector describing the fluid velocity $\mathbf{q}$, it follows from Equation (5.90) that

$$\mathbf{q} = q_1 + iq_2 = \overline{w'(z)}. \tag{5.94}$$

The curves $\phi = \phi_0 =$ const., which are the level curves of ϕ, are the equipotentials of the flow and there can be no flow across an equipotential because $\mathbf{q} = \text{grad}\,\phi_0 = \mathbf{0}$.

At any point P in the flow the vector $\mathbf{q} = \text{grad}\,\phi$ is normal to the equipotential through P, showing that $\mathbf{q}$ must be tangent to the level curve $\psi(x, y) =$ const. through P. Because P was an arbitrary point this shows that the level curves $\psi(x, y) =$ const. must coincide with the streamlines of the flow. It is for this reason that the function ψ is called the *stream function.*

Clearly, any analytic function can be regarded as a complex potential. The basic idea behind the use of complex potentials is that instead of considering two-dimensional real problems in the (x, y)-plane, we set $z = x + iy$ and consider analytic functions in the z-plane. This changes the problem from one of calculating a real potential $\phi(x, y)$, into what is often a simpler problem of calculating a complex potential $w(z)$. The real potential ϕ can then be recovered from $w(z)$ through the result

$$\phi(x, y) = \text{Re}\{w(z)\}. \tag{5.95}$$

The solution of boundary value problems for the two-dimensional Laplace equation can then be found by using a complex potential, rather than by finding the real potential as in Section 5.3 (the direct method). This is the indirect method mentioned in Section 5.1.

A powerful technique that is often used when seeking solutions of the Laplace equation is the principle of linear superposition of simpler solutions and this has an immediate parallel when working with complex potentials. To understand this important idea let ϕ_1 and ϕ_2 be two harmonic functions. Then, setting $\phi = \phi_1 + \phi_2$, it follows at once that

$$\Delta\phi = \Delta\phi_1 + \Delta\phi_2 = 0, \tag{5.96}$$

so the new function ϕ is also harmonic. Thus if ϕ_1 and ϕ_2 are both solutions describing situations of physical interest, the solution $\phi = \phi_1 + \phi_2$ is a new and more complicated solution that combines the features of ϕ_1 and ϕ_2. This additivity of solutions is called the *linear superposition of solutions.*

Now let ϕ_1 and ϕ_2 be velocity potentials and ψ_1 and ψ_2 be their corresponding stream functions, so that $w_1 = \phi_1 + i\psi_1$ and $w_2 = \phi_2 + i\psi_2$ are two complex potentials. Then the linear superposition principle that allows the combination of ϕ_1 and ϕ_2 taken with the Cauchy–Riemann equations shows that

$w = w_1 + w_2$ is also a complex potential. Thus the addition of complex potentials yields another and more complicated complex potential that describes a more complicated flow. This property of complex potentials will be used later when considering specific examples.

To see how complex potentials, and thus flows, transform, suppose that

$$w = f(z) = u(x, y) + iv(x, y) \tag{5.97}$$

gives a one-one conformal mapping of a domain D in the z-plane onto an image domain D^* in the w-plane. Now consider a complex potential in D^* of the form

$$F(w) = \phi(u, v) + i\psi(u, v), \tag{5.98}$$

and transform it by means of Equation (5.97) to obtain

$$G(z) = F[f(z)] = \phi[u(x, y), v(x, y)] + i\psi[u(x, y), v(x, y)], \tag{5.99}$$

in domain D.

As the functions F and G are analytic, so also is the composite function $G(z)$ in Equation (5.99), so that $G(z)$ must be the complex potential of the flow in domain D. It follows from Equation (5.99) that the velocity potential in the z-plane is

$$\phi(x, y) = \phi[u(x, y), v(x, y)], \tag{5.100}$$

and the stream function is

$$\psi(x, y) = \psi[u(x, y), v(x, y)]. \tag{5.101}$$

If $\mathbf{q}$ has components (Q_1, Q_2) in the w-plane and (q_1, q_2) in the z-plane it follows directly from Equation (5.92) that

$$Q_1 = \text{Re}\{F'(w)\}, \quad Q_2 = -\text{Im}\{F'(w)\}, \tag{5.102}$$

and

$$q_1 = \text{Re}\{G'(z)\}, \quad q_2 = -\text{Im}\{G'(z)\}. \tag{5.103}$$

These results are, of course, related through the chain rule for the differentiation of the composite function because [see Equation (5.99)]

$$\frac{dG}{dz} = \frac{dF}{dw}\frac{dw}{dz}. \tag{5.104}$$

Before proceeding further, we again draw attention to two simple but useful conformal transformations that have already been considered.
These are the linear transformation

$$w = z - z_0, \tag{5.105}$$

that simply translates the origin to $z = z_0$ without change of scale or rotation, and

$$w = e^{-i\alpha} z, \tag{5.106}$$

that rotates the axes about the origin through an angle α without changing the scale.
In order to solve some typical fluid flow problems, it is first necessary to discuss the most common types of boundary conditions that arise. In what follows, every boundary γ is assumed to be rigid and impenetrable, so as fluid neither enters nor leaves a boundary the component of fluid velocity normal to a boundary must be zero. Thus, if $\mathbf{n}$ is normal to γ, the flow must satisfy the boundary condition

$$\mathbf{n} \cdot \mathbf{q} = 0 \quad \text{on } \gamma. \tag{5.107}$$

Because $\mathbf{q}$ is related to the velocity potential ϕ by Equation (5.86), the above condition becomes the homogeneous Neumann condition

$$\frac{\partial \phi}{\partial n} = 0 \quad \text{on } \gamma. \tag{5.108}$$

However, as an ideal fluid is not viscous, it is possible for flow to take place tangential to a boundary, so that a boundary must coincide with a streamline $\psi = \text{const}$. As no flow takes place across a streamline we see from this that any existing free surface (the interface between a fluid and air) must also coincide with a streamline. It thus follows directly that *any streamline may be replaced by a solid boundary or, where appropriate, by a free surface.*
Another important type of boundary condition that arises when studying fluid flow is the specification of the *flow at infinity*. Thus usually amounts to requiring the fluid to be at rest at infinity, or to be in uniform motion in some specified direction. Specific cases involving these conditions will be found in the examples that follow, and also in the exercises.
Two mathematical idealizations of simple physical processes that are of considerable importance in fluid flow must now be mentioned. The first is the line source that is used to introduce new fluid into the flow and the second is the vortex, that is used to introduce radially symmetric rotational motion into an irrotational flow.

A *line source* may be considered to be an idealization of a perforated tube of small diameter, with fluid from within the tube entering the fluid outside the tube at a constant rate through the perforations. A *vortex* may be considered to be an idealization of whirlpool-like behavior in a fluid. The core of the vortex is a tube of fluid of radius R with its axis normal to the plane of the flow, and inside the tube the vorticity of the fluid flow is $\boldsymbol{\zeta} = \boldsymbol{\omega}$ (const.) with the vector $\boldsymbol{\omega}$ directed along the axis of the vortex tube, while outside the tube the vorticity of the fluid flow is $\boldsymbol{\zeta} = \mathbf{0}$. Vortices in fluids are often found in wakes behind obstacles, while on an atmospheric scale a vortex may be taken as a crude model for a hurricane or tornado. By convention, the vorticity will be considered to be positive when the fluid rotation in the (x, y)-plane is counterclockwise, and negative when it is clockwise.

5.4.1 The Line Source and Line Sink

A *line source* is the idealized mathematical model of the process by which fluid is introduced into a two-dimensional steady flow by means of the outflow of fluid at a constant rate along a line of infinite length normal to the plane of the flow. It is assumed that the flow occurs steadily and uniformly along the line, and in a radially symmetric manner in any plane normal to the line source. The volume m of fluid flowing out from a line source per unit length of the line per unit time is called the *strength* of the line source. A negative line source is called a *line sink*, and it extracts fluid by means of an inflow of fluid into the line. For a source the strength $m > 0$, while for a sink $m < 0$.

Let us now find the complex potential for a line source of strength m located at the origin, and to accomplish this we must first find the velocity potential of the flow. In terms of the plane, cylindrical polar coordinates (r, θ), the velocity potential must satisfy the polar form of the Laplace equation

$$\frac{\partial^2 \phi}{\partial r^2} + \frac{1}{r}\frac{\partial \phi}{\partial r} + \frac{1}{r^2}\frac{\partial^2 \phi}{\partial \theta^2} = 0. \tag{5.109}$$

As the outflow is radially symmetric ϕ must be independent of θ, and so it must satisfy the ordinary differential equation

$$\frac{d^2 \phi}{dr^2} + \frac{1}{r}\frac{d\phi}{dr} = 0 \quad \text{or} \quad \frac{d}{dr}\left(r\frac{d\phi}{dr}\right) = 0.$$

Integration of this result shows the general solution to be

$$\phi = A \ln_e r, \tag{5.110}$$

where without loss of generality we have set the arbitrary additive integration constant equal to zero.

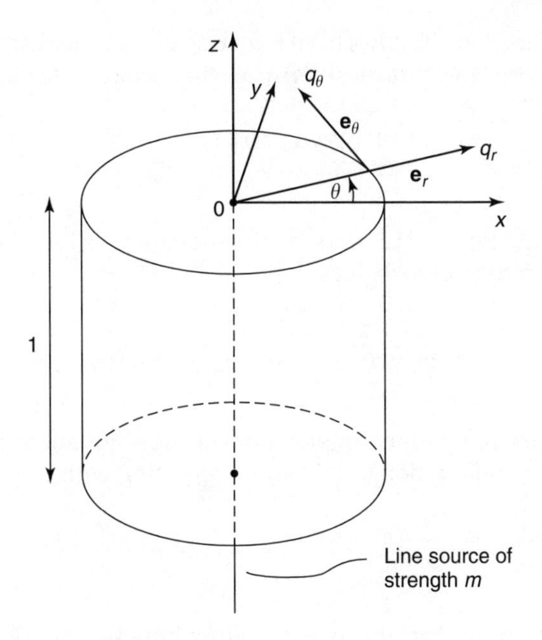

FIGURE 5.49
Flow out of a unit length of cylinder containing a line source.

If the radial component of the fluid velocity due to the line source at the origin is q_r and the transverse component is q_θ, then in terms of the radial unit vector $\mathbf{e}_r$ and the transverse unit vector $\mathbf{e}_\theta$ shown in Figure 5.49, we have $\mathbf{q} = q_r\mathbf{e}_r + q_\theta\mathbf{e}_\theta$. Now in plane cylindrical polar coordinates

$$\operatorname{grad}\phi = \frac{\partial\phi}{\partial r}\mathbf{e}_r + \frac{1}{r}\frac{\partial\phi}{\partial\theta}\mathbf{e}_\theta,$$

so making use of Equation (5.110) we find that

$$q_r = \frac{A}{r} \quad \text{and} \quad q_\theta = 0. \tag{5.111}$$

To determine A, consider the flow out of a cylindrical surface of radius r, of unit length in unit time, when a line source of strength m coincides with the axis of the cylinder. The outflow is simply $q_r \times$ (the area of the cylindrical wall) $= 2\pi r q_r = 2\pi A$. Equating this result to m gives $A = m/2\pi$, and so the velocity potential in Equation (5.110) becomes

$$\phi = \frac{m}{2\pi}\ln_e r. \tag{5.112}$$

The stream function ψ follows from Equation (5.112) after using the Cauchy–Riemann equations in cylindrical form [see Equations (1.65) in Section 1.4]

$$\frac{\partial\phi}{\partial r} = \frac{1}{r}\frac{\partial\psi}{\partial\theta}, \qquad \frac{\partial\psi}{\partial r} = -\frac{1}{r}\frac{\partial\phi}{\partial\theta}. \tag{5.113}$$

Substituting Equation (5.112) into the first equation of Equation (5.113) and integrating with respect to θ gives

$$\frac{\partial\psi}{\partial\theta} = \frac{m}{2\pi}, \qquad \text{so } \psi = \frac{m\theta}{2\pi} + f(r), \tag{5.114}$$

where f is an arbitrary function of r. Substituting Equation (5.112) into the second equation in Equation (5.113) and integrating with respect to r gives

$$\frac{\partial\psi}{\partial r} = 0, \qquad \text{so } \psi = h(\theta), \tag{5.115}$$

where h is an arbitrary function of θ. Equating Equation (5.114) and Equation (5.115), which must be identical, shows that $f(r) = 0$ and $h(\theta) = m\theta/2\pi$, so the stream function is

$$\psi = \frac{m\theta}{2\pi}. \tag{5.116}$$

Thus the complex potential $w = \phi + i\psi$ for a line source is thus

$$w = \frac{m}{2\pi}\ln_e r + i\frac{m\theta}{2\pi}. \tag{5.117}$$

This result can be written more concisely as

$$w = \frac{m}{2\pi}\operatorname{Ln} z. \tag{5.118}$$

The result from Equation (5.118) is the form in which it is most convenient to express the complex potential of a line source of strength m located at the origin $z = 0$. It follows from this that the equipotentials are concentric circles $r =$ const. centered on the origin $z = 0$, while the streamlines are the radial lines $\theta =$ const. radiating out from the origin.

The complex potential for a line source of strength m located at $z = z_0$ is simply

$$w = \frac{m}{2\pi}\operatorname{Ln}(z - z_0). \tag{5.119}$$

Away from singular points, a one-one conformal transformation maps a line source of strength m into another line source of identical strength. This follows because any simple closed curve surrounding a line source is mapped onto a similar closed curve, so the flows across each of the cylinders of unit length in a unit time remains the same.

5.4.2 The Vortex

Another complex potential that must be derived is for a vortex located at the origin. The diagram on the left of Figure 5.50 shows an element of a vortex tube of radius R, inside which $\boldsymbol{\zeta} = \boldsymbol{\omega}$, while outside $\boldsymbol{\zeta} = \mathbf{0}$, so the flow outside is irrotational.

We first find the velocity potential of a vortex tube of radius R. The flow in the vortex is assumed to be cylindrically symmetric about its axis which is normal to the plane of the flow. Two regions must be considered: one inside the vortex tube where $\boldsymbol{\zeta} = \boldsymbol{\omega}$, and the other outside it where $\boldsymbol{\zeta} = \mathbf{0}$. In the diagram on the right of Figure 5.50, C_1 is a circle of radius $r < R$ inside the vortex tube. The fluid speed on C_1 is everywhere tangential to C_1 with magnitude Q_θ. By definition, the circulation around C_1 is

$$\int_{C_1} \mathbf{q} \cdot d\mathbf{s} = \int_0^{2\pi} Q_\theta r d\theta = 2\pi Q_\theta, \quad (0 < r < R) \tag{5.120}$$

where $d\mathbf{s}$ is the line element on C_1.

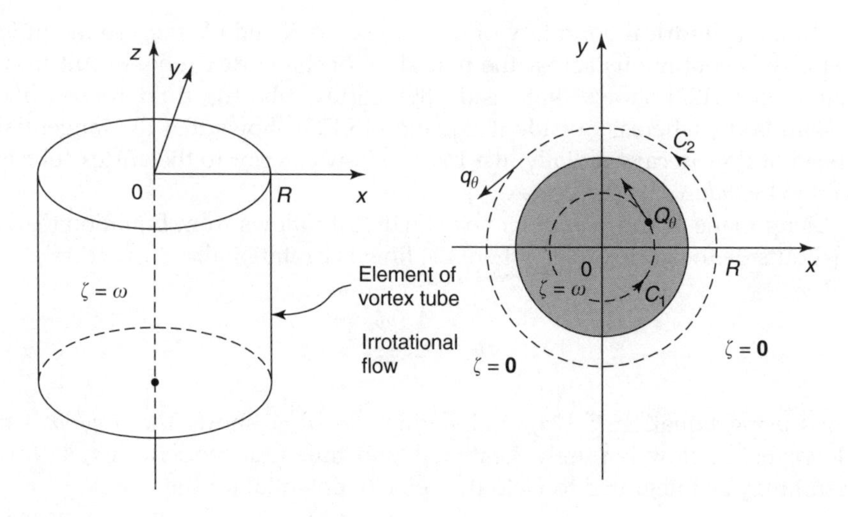

FIGURE 5.50
A cylindrical vortex tube.

Let S_1 be the region inside C_1 and take $d\boldsymbol{\sigma}$ to be the vector element of S_1 pointing in the sense of increasing z. Then by Stokes' theorem

$$\int_{C_1} \mathbf{q} \cdot d\mathbf{s} = \int_{S_1} \operatorname{curl} \mathbf{q} \cdot d\boldsymbol{\sigma} = \int_0^r 2\pi\omega\rho\, d\rho = \pi r^2 \omega, \quad (r < R). \tag{5.121}$$

Equating Equations (5.120) and (5.121) we find that inside the vortex tube

$$Q_\theta = \tfrac{1}{2} r\omega, \quad (0 < r < R) \tag{5.122}$$

Now let C_2 be a circle with interior S_2 that contains the vortex tube. Then, if the fluid speed tangential to C_2 is q_θ, an argument similar to the previous one shows that

$$\int_{C_2} \mathbf{q} \cdot d\mathbf{s} = \int_0^{2\pi} q_\theta r\, d\theta = 2\pi r q_\theta, \quad (r > R) \tag{5.123}$$

and hence by analogy with Equation (5.121)

$$\int_{C_2} \mathbf{q} \cdot d\mathbf{s} = \int_{S_2} \operatorname{curl} \mathbf{q} \cdot d\boldsymbol{\sigma} = \int_0^R 2\pi\omega\rho\, d\rho = \pi\omega R^2, \quad (r > R). \tag{5.124}$$

Equating Equations (5.123) and (5.124) shows that outside the vortex tube

$$q_\theta = \frac{R^2\omega}{2r}, \quad (r > R). \tag{5.125}$$

On the cylindrical boundary of the vortex $r = R$ and $Q_\theta = q_\theta$, so the fluid velocity is continuous across the boundary of the vortex tube. Result from Equation (5.122) shows that inside the vortex tube the fluid rotates like a solid body, whereas outside it Equation (5.125) shows that the tangential speed of flow decays radially like $1/r$. The flow exterior to the vortex tube is said to be *induced* by the vortex.

Using plane cylindrical polar coordinates, it follows from Equation (5.63) that outside the vortex tube, where the flow is irrotational,

$$q_\theta = \frac{1}{r}\frac{\partial \phi}{\partial \theta}.$$

Thus using Equation (5.125), and setting $\kappa = \frac{1}{2}R^2\omega$, shows that $\partial\phi/\partial\theta = \kappa$. However the flow is purely rotational, and thus independent of r, so this result may be integrated to yield the velocity potential for the vortex

$$\phi = \kappa\theta, \tag{5.126}$$

where once again, without loss of generality, the additive arbitrary constant has been set equal to zero.

Using the Cauchy–Riemann Equations (5.113), and reasoning as for the line source, shows that the stream function is

$$\psi = -\kappa \ln_e r. \tag{5.127}$$

It follows from Equation (5.126) that the equipotentials of a vortex of radius R located at the origin are the radial lines $\theta = \text{const.}$, $r > R$, so the streamlines must be concentric circles centered on the origin with $r > R$, so the complex potential for the vortex becomes

$$w = \kappa\omega - ik \ln_e r.$$

Setting $z = re^{i\theta}$ we see by inspection that this can be written more concisely as

$$w = -i\kappa \mathrm{Ln}\, z, \quad (r > R). \tag{5.128}$$

It is usual to call κ the *strength* of the vortex because from Equation (5.124) the circulation around any circle outside the vortex is seen to be $2\kappa\pi$. Combining Equations (5.105) and (5.128) shows that the *complex potential of a vortex* of strength κ located at the $z = z_0$ is

$$w = -i\kappa \mathrm{Ln}(z - z_0). \tag{5.129}$$

The radius of the vortex tube is arbitrary and so it may be chosen to suit the problem at hand. It may, for example, be shrunk to zero to give a line vortex by letting $R \to 0$ and $\omega \to \infty$ in such a way that $\kappa = \frac{1}{2}R^2\omega$ remains constant. Alternatively, if a circulation takes place about a cylindrical boundary of radius a, since any streamline may always be replaced by a solid boundary, the vortex type may be taken to coincide with the cylinder itself. The resulting flow is irrotational everywhere outside the cylinder, though due to the vortex the region where the flow takes place is no longer simply connected. This is reflected by the fact that the velocity potential ϕ in Equation (5.126) is no longer single valued, but is periodic with period 2π. Comparison of Equations (5.118) and (5.126) shows that a line vortex may be regarded as a line source with purely imaginary strength $m = -2\pi i\kappa$.

Taken in conjunction with the Bernoulli equation, the radially symmetric velocity distribution induced by a vortex started in a fluid initially at rest shows that the pressure distribution must also be radially symmetric. Thus, although inducing a flow, the position of a vortex will remain stationary relative to fixed axes. It follows from this that a vortex will only move by effects exterior to the vortex itself, that is, by the combined effects of any other vortices, line sources, or sinks that are present, or by fixed boundaries, and by any superimposed flow from infinity.

To determine the motion of a vortex in a flow, let there be a flow with complex potential $w(z)$ in which there is a vortex of strength κ at z_0. Then the complex potential of the vortex must be separated from the other effects causing the flow, so that

$$w(z) = -i\kappa \mathrm{Ln}(z - z_0) + w_1(z), \tag{5.130}$$

The motion of the vortex at z_0 is due only to the complex potential $w_1(z)$, so the components of its velocity are given by

$$\frac{dw_1}{dz} = q_1 - iq_2. \tag{5.131}$$

5.4.3 Uniform Parallel Flow

The simplest flow of all in which the velocity is everywhere finite and nonzero is one in which the streamlines are all parallel straight lines inclined to the x-axis at an angle α, and such that the speed of flow is $q = U$ (const.). Such a flow is called *uniform parallel flow* and it is illustrated in Figure 5.51.

If a and b are real and $\alpha = \tan(b/a)$, the velocity potential for the flow in which the Cartesian components q_1 and q_2 of the velocity vector $\mathbf{q}$ are $q_1 = a$ and $q_2 = b$ is easily seen to be

$$\phi = ax + by. \tag{5.132}$$

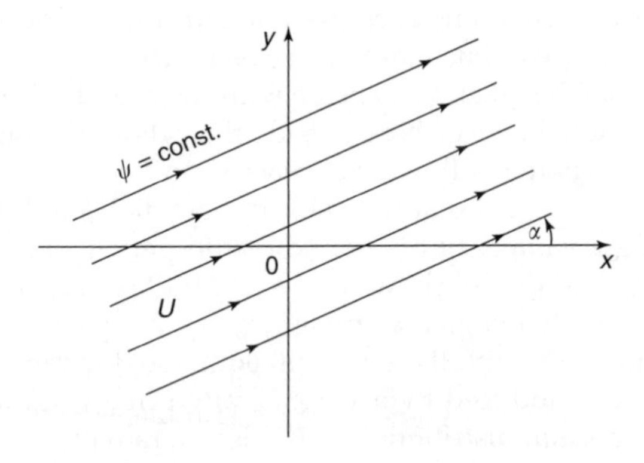

FIGURE 5.51
Uniform parallel flow inclined to the x-axis.

The speed of the flow is

$$U = (a^2 + b^2)^{1/2}. \tag{5.133}$$

The stream function ψ found from Equation (5.132) by means of the Cauchy–Riemann equations is

$$\psi = -bx + ay, \tag{5.134}$$

so that the complex potential $w = \phi + i\psi$ becomes

$$w = ax + by + i(-bx + ay).$$

Regrouping terms, this is expressible in terms of z as

$$w = (a - ib)z = Uze^{-i\alpha}. \tag{5.135}$$

This shows that flow at speed U parallel to the x-axis in the direction of increasing x is

$$w = Uz. \tag{5.136}$$

Thus the flow will be in the direction of increasing x when $U > 0$, and in the opposite direction when $U < 0$.

5.4.4 The Study of Steady Flows Using Complex Potentials

The visualization of a flow pattern involves finding the streamlines ψ = const. for a flow and these follow from its complex potential $w(z)$ from the result

$$\psi = \mathrm{Im}\{w(z)\}. \tag{5.137}$$

As any analytic function may be regarded as a complex potential $w(z)$, the flow it represents may be studied by means of result from Equation (5.137). Furthermore, because a streamline may always be replaced by a solid boundary, whenever such replacements are made the same complex potential then describes the flow past obstacles and boundaries. It is for this reason that the use of complex potentials is sometimes referred to as the *indirect method of solution*, for it can lead to the solution of flow problems without first finding the velocity potential by direct calculation.

When fluid flow occurs due to any combination of a uniform parallel flow, line sources and sinks and no rigid boundaries are present, the complex potential for the flow is the sum of the complex potentials of each of the effects giving rise to the flow. To understand why this is, notice that the determination of the complex potentials for a uniform flow, line sources and line sinks, and vortices involve no boundary conditions. Thus these individual solutions of the Laplace equation may be added to give a new solution which is the velocity potential for the flow resulting from their combined effects. Because of the Cauchy–Riemann equations, the same is true for the stream function for the combined flow. The determination of specific complex potentials when boundaries are present is more complicated, because the presence of boundary conditions renders the addition of complex potentials invalid.

To illustrate the construction of a complex potential for a flow without boundaries, we need only consider the complex potential for a uniform flow at infinity moving with speed U in the positive x-direction, and a line source of strength m located at the origin. Adding Equations (5.116) and (5.136) shows that the required complex potential is

$$w = Uz + \frac{m}{2\pi}\,\mathrm{Ln}\,z. \tag{5.138}$$

An important and useful class of flows are those in which a localized obstacle is involved, so that the flows far upstream and downstream are uniform and parallel to one another. It then follows from Equation (5.135) that far from the obstacle the flow must tend asymptotically to

$$w_\infty(z) = (a - ib)z = Uze^{-i\alpha}, \tag{5.139}$$

where the fluid speed at infinity is given by $U = (a^2 + b^2)^{1/2}$, and the angle α of inclination of the flow there relative to the real axis is $\tan\alpha = b/a$.

Suppose that $\mathbf{q}_\infty$ is the velocity of this uniform parallel flow at infinity. Then, if we superimpose on the whole system an equal and opposite velocity $-\mathbf{q}_\infty$, the fluid at infinity is reduced to rest and the obstacle moves through the fluid with the velocity $-\mathbf{q}_\infty$. Expressed differently, this is equivalent to adopting a reference frame (coordinate system) that moves with the fluid at infinity. In this reference frame, the fluid motion is generated by the motion of the obstacle, and far from the obstacle the fluid is at rest. Thus it is immaterial whether the obstacle is considered to be fixed or to move steadily, since a simple transformation converts one situation into the other.

Example 5.4.1 Uniform Parallel Flow with a Line Source

Find the complex potential for a flow which is uniform at infinity, where it is inclined to the positive x-axis at an angle α and moves to the right with speed U, given that line source of strength $2\pi m$ is located at the origin. Examine the nature of the flow.

SOLUTION

The uniform flow has the complex potential $w_1 = Uze^{-i\alpha}$, and the line source has the complex potential $w_2 = m\,\mathrm{Ln}\,z$. As no boundaries are present the complex potential w for the flow is $w = w_1 + w_2$, so that

$$w = Uze^{-i\alpha} + m\mathrm{Ln}\, z. \tag{5.140}$$

Setting $z = re^{i\theta}$, the stream function becomes

$$\psi = \mathrm{Im}\{w(z)\} = Ur\sin(\theta - \alpha) + m\theta,$$

so the streamlines are

$$Ur\sin(\theta - \alpha) + m\theta = \text{const.} \tag{5.141}$$

Some typical streamlines are shown in Figure 5.52, in which the one marked D separates the flow into two distinct regions, I and II. No fluid from region I can enter region II, and conversely. The flow is symmetrical about the straight line drawn through the origin O at an angle α to the positive x-axis, with the line passing through point S.

A point in the fluid where the fluid speed is zero is called a **stagnation point**. Thus the fluid speed $q = |dw/dz|$, it follows that a stagnation point occurs when $dw/dz = 0$ (a critical point of $w(z)$). For the complex potential Equation (5.140) there is only one stagnation point at $z = -(m/U)e^{i\alpha}$, shown as point S in Figure 5.52.

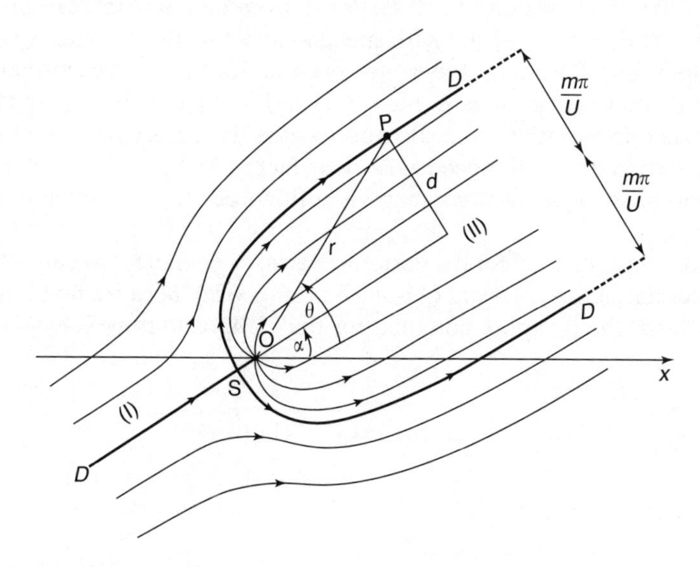

FIGURE 5.52
Uniform parallel flow at infinity with a line source at the origin.

This lies on the streamline D, and is the point where the dividing streamline from upstream in region I bifurcates to enclose region II. At S, $\theta = \pi + \alpha$, so the constant in Equation (5.141) becomes $m(\pi + \alpha)$. The equation for the dividing streamline D is thus

$$Ur\sin(\theta - \alpha) + m\theta = m(\pi + \alpha). \tag{5.142}$$

The distance $d = r\sin(\theta - \alpha)$ in Figure 5.52 is the perpendicular distance of point $P(r, \theta)$ on D from the axis of symmetry. Downstream $\theta \to \alpha$, so $d \to m\pi/U$, which is the asymptotic value of the half width of the enclosed region II. If region II is regarded as a solid obstacle, this flow models a possible flow around a long island of width $2m\pi/U$ when the end on which the flow is incident is rounded in the way shown in Figure 5.52. ◇

Example 5.4.2 A Vortex Pair

A vortex of strength κ instantaneously located at $z = ia$, and one of strength κ is instantaneously located at $z = -ia$. Examine the movement of the vortices, find the complex potential for the induced flow, and determine the nature of this flow.

SOLUTION

A combination of two vortices whose strengths have equal magnitudes but opposite senses is called a *vortex pair*. Before determining the streamlines for this vortex pair, we must find their effect on one another. Let P denote the point $z = ia$ and Q the point $z = -ia$. Result from Equation (5.125) shows that the vortex at P induces a flow with speed $\kappa/2a$, while the vortex at Q induces a corresponding flow with the same speed and in the same sense at P. The flows at P and Q are parallel to the x-axis and in the positive Ix-direction, as may be seen from Figure 5.53. Because of this the vortex pair moves in the positive x-direction with speed $\kappa/2a$. This fact is needed when determining the streamlines, because their shape is influenced by the movement of the vortices.

The complex potential for the vortex at P is $w_1 = -i\kappa\text{Ln}(z - ia)$, and the complex potential of the vortex at Q is $w_2 = i\kappa\text{Ln}(z + ia)$. Because no boundaries are involved, the complex potential for the composite flow is $w = w_1 + w_2$, giving

$$w = -i\kappa\text{Ln}(z - ia) + i\kappa\text{Ln}(z + ia),$$

or

$$w = -i\kappa\text{Ln}\left(\frac{z - ia}{z + ia}\right). \tag{5.143}$$

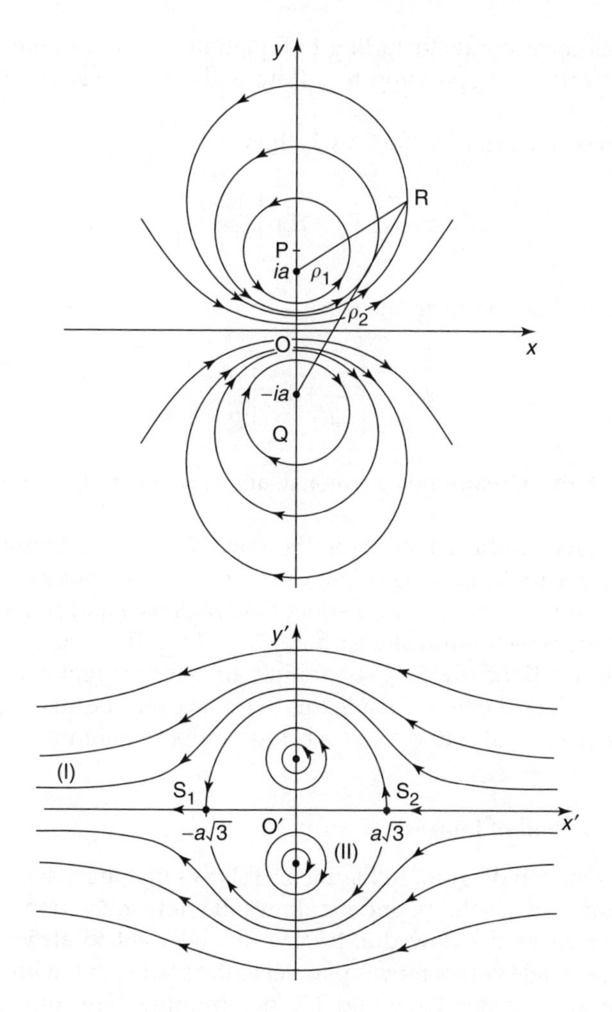

FIGURE 5.53
A vortex pair.

Setting $z - ia = \rho_1 \exp(i\theta_1)$ and $z + ia = \rho_2 \exp(i\theta_2)$ the stream function $\psi = \text{Im}\{w(z)\}$ is seen to be

$$\psi = -\kappa \ln_e \left| \frac{\rho_1}{\rho_2} \right|. \tag{5.144}$$

The instantaneous streamlines are thus the locus of all points R such that $\rho_1/\rho_2 =$ const. and so these streamlines must be coaxial circles with their limiting points at $z = \pm ia$, as shown in the lower diagram in Figure 5.53.

The relative streamlines follow by observing the flow from a reference frame that moves with the vortices, that is from a frame in which the vortices appear to be stationary. This is achieved by superimposing on the entire fluid flow a velocity equal and opposite to that of the vortices. In terms of a complex potential, this corresponds to adding to Equation (5.143) the complex potential $w = -(\kappa/2a)z$, corresponding to a flow with speed $\kappa/2a$ in the negative x-direction.

The complex potential for the flow is thus

$$w = -\kappa\left[\frac{z}{2a} + i\mathrm{Ln}\left(\frac{z - ia}{z + ia}\right)\right]. \tag{5.145}$$

The stream function in this frame is

$$\psi = \kappa\left[\frac{y}{2a} + \ln_e\left(\frac{\rho_1}{\rho_2}\right)\right], \tag{5.146}$$

and representative streamlines ψ = const. are shown in the lower diagram in Figure 5.53.

The flow is symmetrical about both the $x'$and y'-axes which move with the vortices. Relative to the moving origin O′, two stagnation points S_1 and S_2 are seen to occur at $x' = \pm a\sqrt{3}$. Two distinct flow regions, I and II, are separated by a streamline which bifurcates at both S_1 and S_2. The flow in region II is entirely enclosed. If the dividing streamline surrounding region II is replaced by a solid boundary, the flow outside this boundary may be taken to represent a flow past a fixed oval cylinder like a bridge support column. ◇

5.4.5 The Method of Images

Inspection of the top diagram in Figure 5.53 shows that the x-axis is a streamline. Replacing it by a solid boundary allows the flow in the upper half plane to be interpreted as the flow due to a vortex adjacent to an infinite plane boundary. The single vortex moves parallel to the plane $y = 0$ with speed $\kappa/2a$.

This suggests a general method for determining flow due to vortices located instantaneously at points $z_1, z_2, \ldots, z_n$ in the upper half of the z-plane in the presence of a solid boundary $y = 0$ (the real axis). The idea is that if the vortices at these points have strengths $\kappa_1, \kappa_2, \ldots, \kappa_n$, we associate with them a set of fictitious vortices with strengths of equal magnitude but opposite sign $-\kappa_1, -\kappa_2, \ldots, -\kappa_n$ located at the complex conjugate points $\bar{z}_1, \bar{z}_2, \ldots, \bar{z}_n$ in the lower half of the z-plane. Then by analogy with Example 5.4.2, the complex potential of the flow in the upper half of the z-plane is the sum of the complex potentials for this system of n real and n fictitious vortices. The points $\bar{z}_1, \bar{z}_2, \ldots, \bar{z}_n$ are called the *image points* of $z_1, z_2, \ldots, z_n$ with respect to the real axis, and the fictitious system of vortices with strengths $-\kappa_1, -\kappa_2, \ldots, -\kappa_n$ located at these points is called the *image system* corresponding to the vortex system in the upper half of the z-plane.

A strictly analogous argument shows that the image of a line source of strength m located at z_1 in the upper half of the z-plane, is a line source of the *same* strength m located at the image point $\bar{z}_1$ in the lower half of the z-plane.

Because a sink is a negative source, in the previous text we may replace the word "source" by "sink" provided we replace m by $-m$. Thus a vortex and its image have strengths with opposite signs, while the image of a source (sink) is a source (sink).

For example, using images, the complex potential for a source of strength m_1 at $-2 + i$, a sink of strength $-m_2$ ($m_2 > 0$) at $-1 + 2i$, a vortex of strength κ_1 at $1 + i$, and a vortex of strength $-\kappa_2$ at $2 + 3i$ in the presence of a plane solid boundary $y = 0$ is

$$\begin{aligned} w(z) &= \frac{m_1}{2\pi}\operatorname{Ln}(z+2-i) - \frac{m_2}{2\pi}\operatorname{Ln}(z+1-2i) - i\kappa_1\operatorname{Ln}(z-1-i) \\ &\quad + i\kappa_2\operatorname{Ln}(z-2-3i) + \frac{m_1}{2\pi}\operatorname{Ln}(z+2+i) - \frac{m_2}{2\pi}\operatorname{Ln}(z+1+2i) \\ &\quad + i\kappa_1\operatorname{Ln}(z-1+i) - i\kappa_2\operatorname{Ln}(z-2+3i). \end{aligned} \tag{5.147}$$

In this expression, the first four terms represent the complex potentials of the original system, and the last four represent the complex potentials of the image system.

The system has the x-axis as a streamline, so the flow in the original system is obtained from Equation (5.147) by confining z to the upper half of the z-plane $y > 0$.

5.4.6 Study of Steady Flows Using Conformal Mapping

The idea underlying the use of conformal mapping to study how uniform parallel steady flows in the z-plane are disturbed by the presence of arbitrary obstacles is based on the following simple approach. A transformation is sought that maps the streamlines in a uniform parallel flow in the w-plane onto the z-plane in such a manner that one of the streamlines maps onto the boundary of the obstacle. The stream function associated with the transformation will then describe all of the streamlines around the obstacle, provided the asymptotic form of the transformation for large z describes the undisturbed parallel flow at infinity.

Example 5.4.3 Flow Past an Inclined Flat Plate

Find how a uniform two-dimensional parallel flow in the (x, y)-plane moving with speed U in the positive x-direction is disturbed by the introduction of a circular cylinder of radius a at the origin with its axis normal to the plane of the flow. Determine the pressure distribution around the surface of the cylinder. Find how a uniform parallel flow with speed U is disturbed by the introduction of an infinite flat plate of width $4a$ at an angle α to the flow.

SOLUTION

The Joukowski transformation $w = z + 1/z$ has been seen to map the exterior of the unit circle $|z| = 1$ one-one and conformally onto the entire w-plane cut along the real axis from $w = -2$ to $w = 2$. The cut being necessary because the boundary of the unit circle maps onto the line segment removed by the cut, and the restriction on z must be imposed in order that the mapping is one-one because the interior of the unit circle also maps onto the cut w-plane.

With these ideas in mind, let us consider the following complex potential

$$w = \phi + i\psi = U\left(z + \frac{a^2}{z}\right), \tag{5.148}$$

where U and a are real and positive. The arguments used in Section 4.5 show that when considered as a transformation, Equation (5.148) maps the exterior of the circle $|z| = a$ onto the entire w-plane cut along the line $-2a < \phi < 2a$, $\psi = 0$.

For large z result Equation (5.148) has the asymptotic form $w \sim Uz$ corresponding to uniform parallel flow at infinity with speed U in the positive x-direction. Setting $z = x + iy$ in Equation (5.148) shows that the velocity potential ϕ and the stream function ψ are given by

$$\phi = Ux\left(1 + \frac{a^2}{x^2 + y^2}\right), \quad \psi = Uy\left(1 - \frac{a^2}{x^2 + y^2}\right). \tag{5.149}$$

The streamlines $\psi = 0$ correspond to the lines $|x| > a, y = 0$ and $x^2 + y^2 = a^2$, which together comprise the real axis exterior to the circle $|z| = a$ and the boundary of the circle itself. As this streamline has the required behavior at infinity, and it divides to pass around the surface of the cylinder, we conclude that the lines $\psi =$ const. describe all of the streamlines for the desired flow past the cylinder in the z-plane. The streamlines in the z-plane exterior to the circle $x^2 + y^2 = a^2$ are thus described by the equation

$$Uy\left(1 - \frac{a^2}{x^2 + y^2}\right) = \text{const.} \tag{5.150}$$

Some typical streamlines are shown on the left of Figure 5.54. It follows from Equation (5.148) that $dw/dz = 0$ when $z = \pm a$, so there are stagnation points S_1 and S_2 at these points.

To find the pressure distribution around the cylindrical surface we use Bernoulli's theorem given in Equation (5.85). The flow speed at infinity is U, so denoting the pressure at infinity by p_∞, we see that the constant in Equation (5.85) appropriate for this flow is $\frac{1}{2}\rho U^2 + p_\infty$. So, for flow past the cylinder Bernoulli's equation takes the form

$$p = p_\infty + \tfrac{1}{2}\rho U^2 - \tfrac{1}{2}\rho q^2, \tag{5.151}$$

where q is the speed of the flow at any point outside the cylinder.

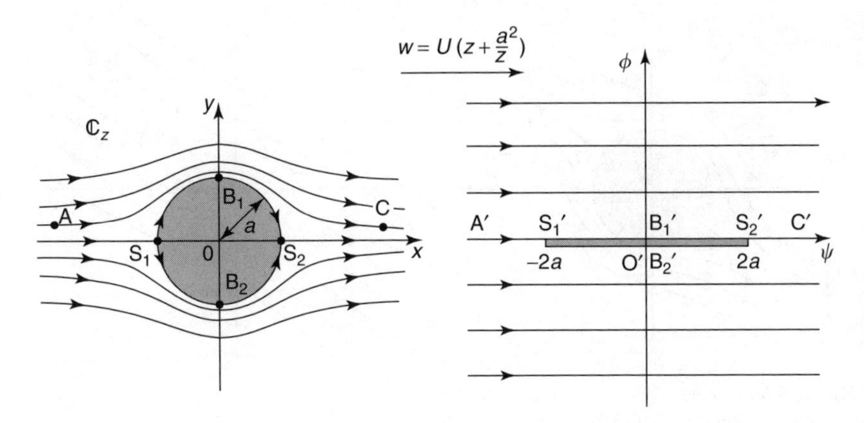

FIGURE 5.54
Flow past a cylinder.

Since $dw/dz = q_1 - iq_2$, it follows from Equation (5.148) that

$$U\left(1 - \frac{a^2}{z^2}\right) = q_1 - iq_2. \tag{5.152}$$

On the surface of the cylinder we may set $z = ae^{i\theta}$, when Equation (5.152) shows that on the cylinder at the polar angle θ,

$$q_1 = U(1 - \cos 2\theta), \quad q_2 = -U \sin 2\theta. \tag{5.153}$$

Using these results in Equation (5.151) with $q^2 = q_1^2 + q_2^2$ gives the pressure distribution on the cylinder as a function of the polar angle θ in the form

$$p = p_\infty + \tfrac{1}{2}\rho U^2 - \rho U^2(1 - \cos 2\theta) = p_\infty - \tfrac{1}{2}\rho U^2 + \rho U^2 \cos 2\theta. \tag{5.154}$$

This confirms that the maximum pressure occurs at the stagnation points S_1 when $\theta = \pi$ and at S_2 when $\theta = 0$. The minimum pressures occur at the points B_1 when $\theta = \pi/2$ and at B_2 when $\theta = -\pi/2$ shown on the left of Figure 5.54. Because the pressure distribution is symmetrical with respect to the x-axis the flow produces neither thrust nor lift on the cylinder. Thrust is defined here as a force in the x-direction, and lift as a force in the y-direction.

To find the flow past an infinitely long flat plate of width $4a$ inclined at an angle α to the x-axis it is necessary to consider the diagram on the right of Figure 5.54. In this diagram the cut from S_1' to S_2' is a barrier of length $4a$ and zero width aligned with the flow. This is the image of the circular boundary, and we shall regard it as the cross-section of an infinitely long plate of width $4a$,

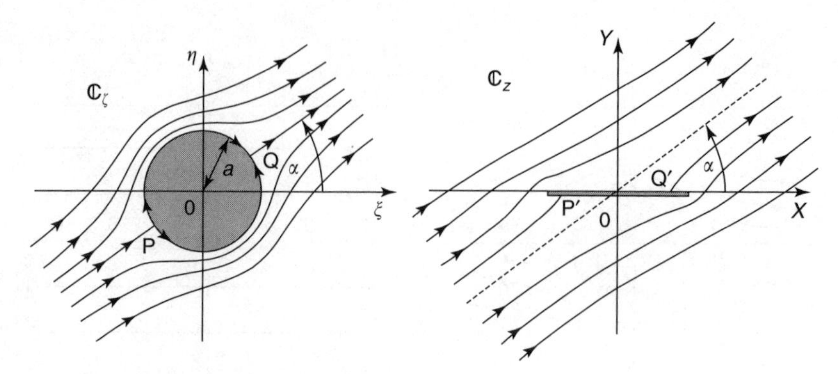

FIGURE 5.55
Flow past an inclined flat plate.

with its length normal to the plane of the flow. Positioned as it is along the streamline $\psi = 0$, the plate will not disturb the flow.

In the Joukowski transformation Equation (5.148) we now introduce a rotation through an angle α by setting $z = e^{-i\alpha}\zeta$, to obtain

$$w = U\left(e^{-i\alpha}\zeta + \frac{a^2 e^{i\alpha}}{\zeta}\right). \tag{5.155}$$

Result from Equation (5.155) now represents a uniform parallel flow, with speed U at infinity in the ζ-plane inclined at an angle α to the real ζ-axis, and incident on the cylinder $|\zeta| = a$. This is shown on the left of Figure 5.55 in which $\zeta = \xi + i\eta$, and P and Q are stagnation points.

Under the Joukowski transformation

$$Z = \zeta + \frac{a^2}{\zeta}, \quad \text{or} \quad \zeta = \frac{Z + (Z^2 - 4a^2)^{1/2}}{2}, \tag{5.156}$$

in which $Z = X + iY$, the cylinder $|\zeta| = a$ is mapped onto the same flat plate of width $4a$ as before, but it is now located at $-2a < X < 2a$, $Y = 0$ in the Z-plane. However, the direction of the uniform parallel flow at infinity in the ζ-plane is unaltered and maps onto a corresponding flow inclined at an angle α to the x-axis in the z-plane.

Substituting the expression for ζ in Equation (5.156) and simplifying the result shows that the complex potential for the flow illustrated on the right of Figure 5.55 is

$$w = U\left(Z\cos\alpha - i\sin\alpha\sqrt{Z^2 - 4a^2}\right). \tag{5.157}$$

It follows from this that $dw/dZ = 0$ at the points $Z = \pm 2a\cos\alpha$, which correspond to the two stagnation points in the flow. These are shown on the right of Figure 5.55 because the point P′ on the underside of the plate and the point Q′ on the top of the plate.

Because the stagnation points P′ and Q′ are points of maximum pressure, but are separated and on opposite sides of the plate, the stream exerts a couple on the plate which tends to turn it broadside to the flow. This is the reason why a drifting boat turns broadside to the prevailing current.

In conclusion, we need to draw attention to the fact that this solution is physically unrealistic at the ends of the plate because at these points the fluid speed becomes infinite. This is caused by the singularities of the Joukowski transformation at these two points. If, instead of the cylinder $|\zeta| = a$ the cylinder $|\zeta| = R$ is considered where $R > a$, this is mapped by the same transformation onto an elliptic cylinder in the Z-plane with its foci at $X = \pm 2a$, $Y = 0$, on which the flow is everywhere finite. The addition of a term $-i\kappa\,\mathrm{Ln}\,\zeta$ to Equation (5.155) corresponds to the incorporation of a circulation of strength κ about the cylinder. This produces lift and thus makes the flow more realistic for many applications. By locating the cylinder in a suitable manner, this same approach also enables flows with circulation about Joukowski airfoils to be studied. ◇

Example 5.4.4 Flow over a Submerged Log

Interpret the complex potential

$$w = \pi a U \coth\left(\frac{\pi a}{z}\right)$$

in terms of possible physical flow past an obstacle. Use it to give an example of a flow with a free surface.

SOLUTION

As w is a complex potential we shall set $w = \phi + i\psi$ and determine the streamlines $\psi = \text{const}$. First, however, setting $z = x + iy$ we have

$$w = \phi + i\psi = \pi a U \coth\left(\frac{\pi a}{z}\right) = \pi a U\left[\frac{\cosh(\pi a/z)}{\sinh(\pi a/z)}\right]$$

$$= \pi a U\left[\frac{\cosh\left(\frac{\pi a x}{x^2+y^2}\right)\cos\left(\frac{\pi a y}{x^2+y^2}\right) - i\sinh\left(\frac{\pi a x}{x^2+y^2}\right)\sin\left(\frac{\pi a y}{x^2+y^2}\right)}{\sinh\left(\frac{\pi a x}{x^2+y^2}\right)\cos\left(\frac{\pi a y}{x^2+y^2}\right) - i\cosh\left(\frac{\pi a x}{x^2+y^2}\right)\sin\left(\frac{\pi a y}{x^2+y^2}\right)}\right]. \tag{5.158}$$

This shows that when $|z|$ is large, w tends asymptotically to

$$w = \pi a U \frac{z}{\pi a} = Uz.$$

Thus, at infinity, there is a uniform flow with speed U moving parallel to the real axis in the positive x-direction.

Inspection of Equation (5.158) shows it to be real when $y = 0$, and also when $2ay \pm (2n + 1)(x^2 + y^2) = 0$ for $n = 0, 1, 2, \ldots$. The first of these conditions describes the real axis in the z-plane, while the second describes two families of circles. These circles are all tangent to the real axis at the origin and are contained within one or other of the two circles of radius a with their centers at $(0, \pm a)$. Therefore the real axis and the boundaries of the two largest circles represent the streamlines $\psi = 0$. Because the complex potential w is only single valued outside these two circles, the boundaries of each of which is a streamline, they may be regarded as solid obstacles in the fluid flow. These circles are shown as the disks in Figure 5.56.

The equation of the streamlines throughout the rest of the flow follows from the stream function ψ defined by Equation (5.158). Equating imaginary parts gives

$$\psi = \pi a U \left[\frac{\sin\left(\dfrac{\pi a y}{x^2 + y^2}\right)\cos\left(\dfrac{\pi a y}{x^2 + y^2}\right)}{\sinh^2\left(\dfrac{\pi a x}{x^2 + y^2}\right) + \sin^2\left(\dfrac{\pi a y}{x^2 + y^2}\right)} \right]. \tag{5.159}$$

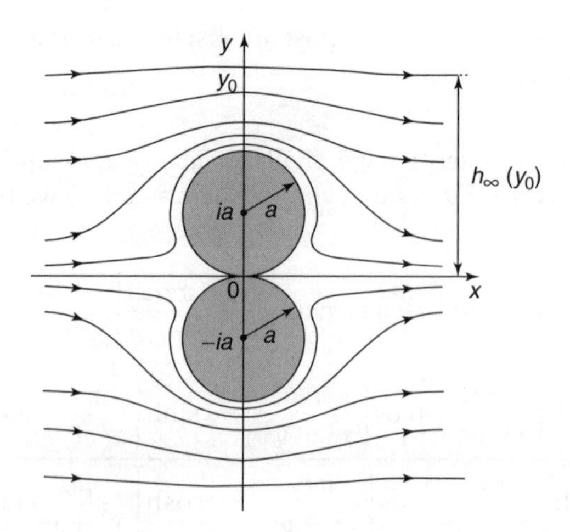

FIGURE 5.56
Flow past two touching cylinders.

This is an even function of x and an odd function of y, so the streamlines are symmetrical about both the x- and y-axes. Some typical streamlines are shown in Figure 5.56. In physical terms, the complex potential describes how uniform parallel flow at infinity is distorted by two infinitely long cylinders of radius a, broadside to the flow, and touching along one generator which passes through the origin and is normal to the plane of the flow.

From Equation (5.159), the streamline $\psi =$ const. passing through the point $(0, y_0)$ has the constant value

$$\psi = \psi_0 = \pi a U \cot\left(\frac{\pi a}{y_0}\right). \tag{5.160}$$

Far upstream and downstream ($x^2 + y^2 \gg 1$) the stream function becomes

$$\psi = Uy, \tag{5.161}$$

as would be expected from Equation (5.159). Equating Equations (5.160) and (5.161) gives for the asymptotic value $h_\infty(y_0)$ of y approached by this streamline at infinity

$$h_\infty(y_0) = \pi a \cot\left(\frac{\pi a}{y_0}\right). \tag{5.162}$$

As y_0 ($\geqslant 2a$) increases, so $h_\infty(y_0) \to y_0$, showing that far away from the cylinders there is very little disturbance to the incident uniform parallel flow.

Replacing the streamline along the real axis by a rigid boundary and considering y to be vertical permits an alternative interpretation of the flow as the flow of deep water over a submerged log. With this interpretation in mind, we may regard the streamline through $(0, y_0)$ in Figure 5.57 as the free

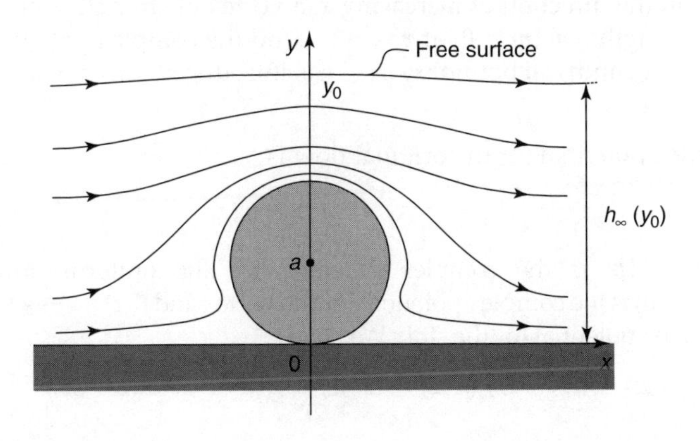

FIGURE 5.57
Flow over a submerged log.

surface of a flow over a submerged log. The elevation of the free surface above the log relative to the elevation $h_\infty(y_0)$ of the same surface at infinity is seen to be $y_0 - \pi a \cot(\pi a/y_0)$. ◇

We conclude this brief introduction to the study of fluid flow by offering a useful theorem.

THEOREM 5.4.1 The Milne–Thompson Circle Theorem
Let the complex potential $w_0 = f(z)$ describe a flow in which no boundaries exist and for which all the singularities of $f(z)$ lie outside the circle $|z| = a$. Then, if a cylindrical boundary with cross-section $|z| = a$ is introduced into the flow with its axis normal to the plane of the flow, the complex potential becomes

$$w = f(z) + \overline{f}(a^2/z)$$

PROOF

On the circle $|z| = a$ we have $z\bar{z} = a^2$ so that $(a^2/z) = \bar{z}$. Thus, on the circle, the given complex potential

$$w = f(z) + \overline{f}(a^2/z) = f(z) + \overline{f}(\bar{z}) = f(z) + \overline{f(z)}.$$

This shows that $w = \phi + i\psi$ is purely real, and so $\psi = 0$. Thus the boundary of the circle is a streamline.

The proof is completed by noticing that the singularities of w_0 and w are the same throughout the fluid. This follows because by hypothesis, as all of the singularities (vortices, line sources, and sinks) of $f(z)$ lie outside $|z| = a$, all of those of $f(a^2/z)$, and therefore of $\overline{f(a^2/z)}$ lie inside the cylinder, and so are excluded from the flow. ◆

Example 5.4.5 Use of the Circle Theorem

Fluid motion results from a uniform flow with speed U at infinity parallel to the x-axis in the direction of increasing x, a vortex of strength κ at $z = 3$ and a sink of strength $-m$ $(m > 0)$ at $z = -2a$. Find the complex potential for the flow when a cylindrical boundary $|z| = a$ is introduced into the flow.

SOLUTION
The complex potential for the original flow is

$$w = f_1(z) + f_2(z) + f_3(z),$$

where $f_1(z) = Uz$ is the complex potential for the uniform flow, $f_2(z) = -i\kappa \operatorname{Ln}(z - 3ia)$ is the complex potential for the vortex and $f_3(z) = -\frac{m}{2\pi} \operatorname{Ln}(z + 2a)$ is the complex potential for the sink. We have

$$\overline{f}_1(z) = Uz, \quad \text{so } \overline{f}_1(a^2/z) = Ua^2/z, \qquad \overline{f}_2(z) = i\kappa \operatorname{Ln}(z + 3ia),$$

so

$$\overline{f}_2(a^2/z) = i\kappa \operatorname{Ln}(a^2/z + 3ia) \quad \text{and} \quad \overline{f}_3(z) = -(m/2\pi) \operatorname{Ln}(z + 2a),$$

so that

$$\bar{f}_3(a^2/z) = -(m/2\pi)\,\text{Ln}(a^2/z + 2a).$$

Thus by the circle theorem, the complex potential for the modified flow caused by the introduction of the cylinder $|z| = a$ is

$$w = f_1(z) + f_2(z) + f_3(z) + \bar{f}_1(a^2/z) + \bar{f}_2(a^2/z) + \bar{f}_3(a^2/z),$$

and so

$$w = U\left(z + \frac{a^2}{z}\right) - i\kappa\,\text{Ln}(z - 3ia) + i\kappa\,\text{Ln}\left(\frac{a^2}{z} + 3ia\right) - \frac{m}{2\pi}\,\text{Ln}(z + 2a)$$
$$- \frac{m}{2\pi}\,\text{Ln}\left(\frac{a^2}{z} + 2a\right). \qquad \diamond$$

Exercises 5.4

1. Find the complex potential for a line source of strength m located at $z = 1$ and a line sink of equal strength located at $z = -1$. Determine the form of the streamlines.
2. Find the streamlines of the flow resulting from the complex potential $w = \mu e^{i\alpha}/z$, in which μ and α are real.
3. Find the complex potential for a flow that is uniform at infinity, where it is inclined at an angle α to the x-axis and moves to the right with speed U, when a line sink of strength $-2m\pi$ $(m > 0)$ is located at the origin. Note any features of the flow that are of special interest.
4. Find the complex potential for a line source of strength $2m\pi$ $(m > 0)$ at $z = a$ and a line sink of equal strength at $z = -a$ (a real) in the presence of uniform parallel flow at infinity moving with speed U parallel to the x-axis in the direction of negative x. Find the stream function and the stagnation points. Show the streamline $\psi = 0$ encircles an oval shaped region in which the flow is trapped.
5. Find the complex potential for two line sources of equal strength $2m\pi$ $(m > 0)$ located at $z = a$ and $z = -a$ (a real). Locate any stagnation points, find the stream function and deduce the flow caused by a line source of strength $2m\pi$ parallel to a plane boundary and distant a from it.
6. By using the method of images find the complex potential for flow in the first quadrant caused by a source of strength $m_1(>0)$ at $z = 1$ and a source of strength $m_2(>0)$ at $z = i$. Find the stream function, and deduce the complex potential when the source of strength m_2 at $z = i$ is replaced by a sink of the same strength. Find the stream function.
7. Find the instantaneous complex potential for two vortices of strength κ located at $z = 1 + i$ and $z = -1 - i$, and two vortices of strength $-\kappa$ located at $z = 1 - i$ and $z = -1 + i$. Locate the stagnation points, find the stream function. Deduce the instantaneous streamlines due to a

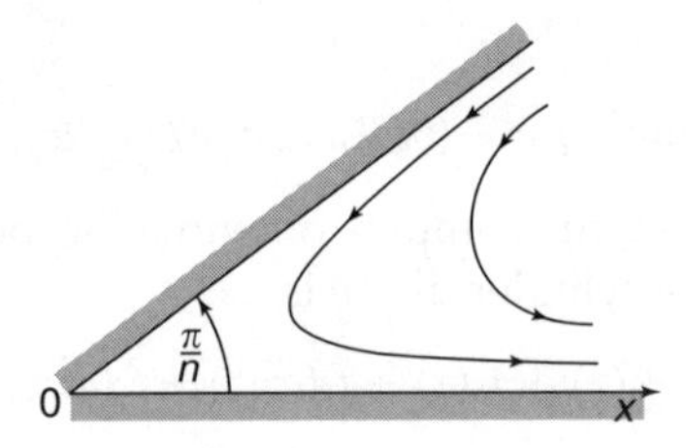

FIGURE 5.58
Flow inside a wedge.

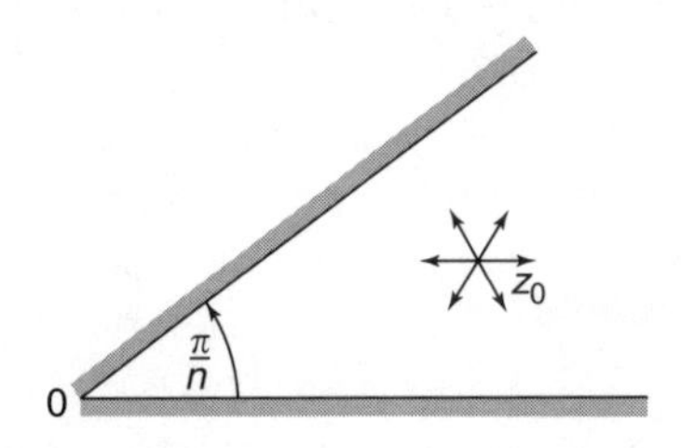

FIGURE 5.59
Flow inside a wedge caused by a source.

vortex of strength κ located at $z = 1 + i$ in the presence of the plane boundaries $y = 0, x > 0$ and $x = 0, y > 0$. What is the instantaneous velocity with which the vortex at $z = 1 + i$ will move?

8. Fluid flows inside the wedge $0 < \text{Arg}\{z\} < \pi/n$ as shown in Figure 5.58. Find the stream function using the transformation $\zeta = z^n$, and then considering the complex potential $w = A\zeta$ in the ζ-plane. Find the streamlines when $n = 2$, and the components of the velocity at points on the line $y = \alpha x$ ($\alpha > 0$) in the first quadrant.
9. Using the transformation $\zeta = z^n$, find the complex potential for the flow produced inside the wedge $0 < \text{Arg}\{z\} < \pi/n$ due to a line source of strength m located parallel to the walls of the wedge at the point z_0 as shown in Figure 5.59. Show the fluid speed tends to zero far from the line source. (Hint: Introduce an appropriate image source in the ζ-plane.)
10. Show that

$$w = A\left(z^4 + \frac{a^8}{z^4}\right)$$

with A real and positive, is the complex potential for the flow illustrated in Figure 5.60. Prove that P and Q are stagnation points for the flow, and find the pressure distribution around the circular surface PQ of radius $r = a$ with $0 < \theta < \pi/4$ in terms of the pressure p_S at a stagnation point.
11. Show that the complex potential

$$w = U(z^2 + 1)^{1/2}$$

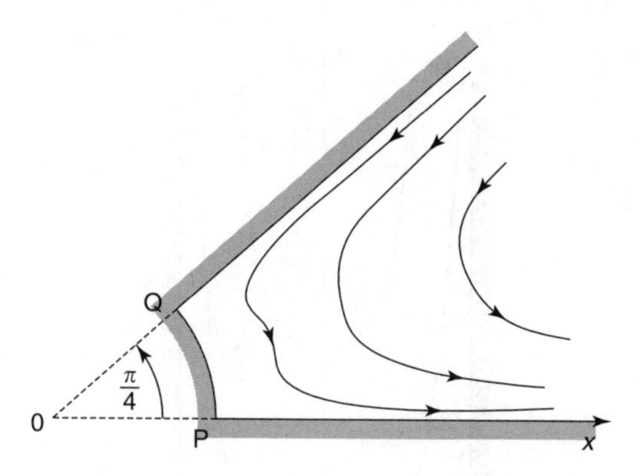

FIGURE 5.60

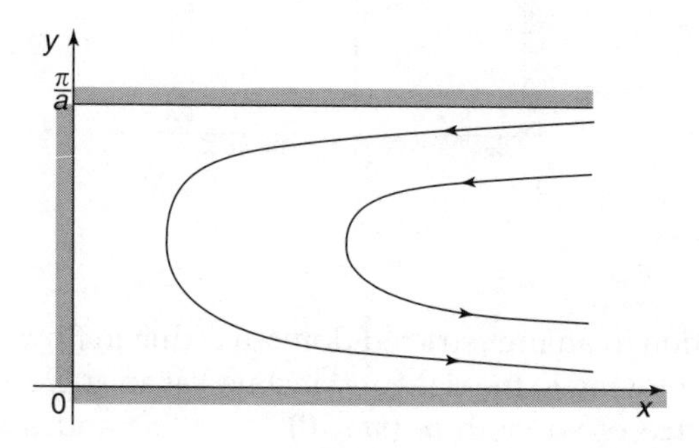

FIGURE 5.61

describes flow along the x-axis and over a vertical barrier $x = 0$, $0 < y < 1$ in the z-plane when at infinity the flow moves with speed U in the positive x-direction. Find the equation of the streamlines $\psi = \alpha U$ with $\alpha \geqslant 0$ a parameter.

12. Show that the complex potential $w = U \cosh az$ describes the flow illustrated in Figure 5.61. Use it to derive the equation of the streamlines and find the fluid speed on the streamline $\psi = cU$ as a function of y and the parameter c.
13. Find the complex potential that describes the flow illustrated in Figure 5.62. Use it to derive the equation of the streamlines and find the location of the maximum pressure on the streamline $\psi = cU$.
14. Fluid motion in an unrestricted domain is due to an instantaneous complex potential

$$w = i\kappa \mathrm{Ln}(z + 4ia) + \frac{\mu e^{i\alpha}}{z - 2ia},$$

with μ, α and a real. Find the instantaneous complex potential when a circular cylinder with cross-section $|z| = a$ is introduced into the flow with its axis at the origin and normal to the z-plane.

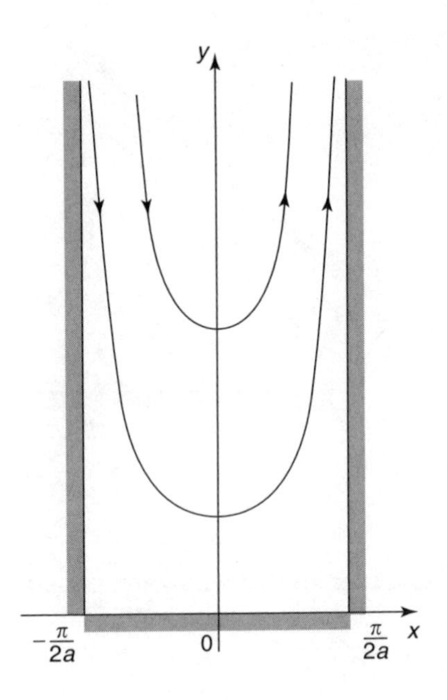

FIGURE 5.62

15. Fluid motion in an unrestricted domain is due to flow with speed U at infinity moving to the right and inclined at an angle α to the x-axis, a line source of strength m ($m > 0$) at $z = 2ia$ and a sink of equal strength at $z = -3a$. Find the complex potential when a cylinder of cross-section $|z| = a$ is introduced into the flow with its axis at the origin and normal to the z-plane.

Exercises of Greater Difficulty

16. Write down the complex potential of a set of $2n + 1$ vortices of strength κ located at the points $z = 0, z = \pm a, \pm 2a, \ldots, \pm na$ on the x-axis, where $a > 0$. By letting $n \to \infty$, using the infinite product

$$\sin\frac{\pi z}{a} = \frac{\pi z}{a}\left(1 - \frac{z^2}{a^2}\right)\left(1 - \frac{z^2}{2^2 a^2}\right)\left(1 - \frac{z^2}{3^2 a^2}\right)\cdots,$$

and discarding an arbitrary constant, show that the complex potential of this infinite row of vortices is

$$w = -i\kappa \mathrm{Ln}\left(\sin\frac{\pi z}{a}\right).$$

Find the equation of the streamlines, and by showing that the vortex at $z = 0$ remains at rest, prove that the infinite row of vortices remains at rest.

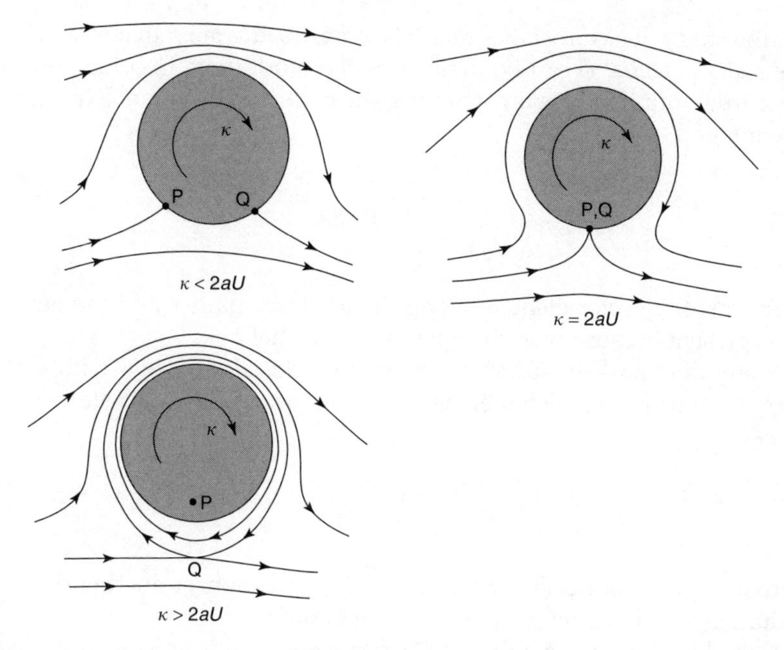

FIGURE 5.63

17. Explain why

$$w = U\left(z + \frac{a^2}{z}\right) + i\kappa \operatorname{Ln} z$$

is the complex potential for uniform flow with speed U in the positive x-direction past a cylinder of radius a centered on the origin about which there is a circulation of strength κ. Find the two stagnation points P and Q for the flow and show that (a) they are distinct and both lie on the surface of the cylinder if $\kappa < 2aU$ (b) they coalesce and lie on the surface of the cylinder is $\kappa = 2aU$ and (c) they are distinct, with one inside the cylinder (physically unreal) and one outside if $\kappa > 2aU$. Find the stream function and verify the main features of Figure 5.63. Find the speed around the surface of the cylinder as a function of θ where $z = ae^{i\theta}$, and use this in case (a) to show the pressure variation around the surface of the cylinder is asymmetric and so generates an upward thrust (lift).

5.5 Two-Dimensional Electrostatics

The *electrostatic field* (electric field) **E** at a point P in space is the force that would be experienced by a unit charge were it to be placed at P. It is known from the study of electrostatics that **E** is a *conservative force*: that is, an *electrostatic scalar potential* ϕ can be defined as the work done in bringing a unit charge from infinity along any path γ to the point P in question. Expressed as a line integral

$$\phi = -\int_{\infty}^{P} \mathbf{E} \cdot d\mathbf{s}, \tag{5.163}$$

where $d\mathbf{s}$ is the vector element along the arbitrary path γ and the negative sign is present because work is done *against* the field.

For any domain V in space, bounded by a closed surface S with surface vector element $d\mathbf{S}$ in which there is a total charge ρ, Gauss' law states that

$$\int_S \mathbf{E} \cdot d\mathbf{S} = \frac{4\pi\rho}{\varepsilon}, \tag{5.164}$$

where ε is a constant called the *permittivity* of the nonconducting dielectric medium in which the field and charge are located.

We shall be concerned only with two-dimensional problems in the sense already defined. Accordingly, we set $\mathbf{E} = E_1\mathbf{i} + E_2\mathbf{j}$ where $\mathbf{i}$ and $\mathbf{j}$ are unit vectors in the x- and y-directions or, when it is convenient to use the vector interpretation of complex numbers, $\mathbf{E} = E_1 + iE_2$. Result from Equation (5.163) then becomes

$$\phi = -\int_{\infty}^{P} E_1\,dx + E_2\,dy, \tag{5.165}$$

where the integral is now taken from the point at infinity in the (x, y)-plane along any path γ to the point P, and dx and dy are the components of $d\mathbf{s}$. Partial differentiation of Equation (5.165) with respect to x and y gives

$$E_1 = -\frac{\partial \phi}{\partial x}, \quad E_2 = -\frac{\partial \phi}{\partial y}, \tag{5.166}$$

so that

$$\mathbf{E} = -\operatorname{grad} \phi. \tag{5.167}$$

Notice that here, unlike the situation in fluid flow because work is involved when charges are moved in an electric field, we have adopted the convention from mechanics that a negative sign is introduced in Equation (5.167) when defining the field **E** in terms of the gradient of an electrostatic potential ϕ (see Equation (5.85)).

To interpret Gauss' law in two dimensions we take the domain V to be a cylinder of arbitrary height normal to the (x, y)-plane with an arbitrary curve of cross-section C. Applying the divergence theorem to the result from Equation (5.164) for such a domain that is free from charge gives

$$\int_S \mathbf{E} \cdot d\mathbf{S} = \int_V \operatorname{div} \mathbf{E} \, dV = 0. \qquad (5.168)$$

Because C, and thus V, are arbitrary, Equation (5.168) is only possible of $\operatorname{div} \mathbf{E} = 0$, so that $\mathbf{E}$ must satisfy the equation

$$\frac{\partial E_1}{\partial x} + \frac{\partial E_2}{\partial y} = 0. \qquad (5.169)$$

Combining Equations (5.166) and (5.169) we find that

$$\frac{\partial^2 \phi}{\partial x^2} + \frac{\partial^2 \phi}{\partial y^2} = 0,$$

which shows that the electrostatic potential ϕ is a harmonic function.

Lines of force are the lines that are everywhere tangent to the electrostatic field vector $\mathbf{E}$. Thus a line of force through P is the line along which a free electric charge placed at P would move.

As $\operatorname{grad} \phi$ is normal to the equipotential $\phi = \text{const.}$, and $\mathbf{E} = -\operatorname{grad} \phi$, it follows that lines of force and equipotentials are mutually orthogonal trajectories. Thus, in two dimensions, the lines of force may be identified with the curves $\psi = \text{const.}$, where ψ is the conjugate harmonic function associated with ϕ by the Cauchy–Riemann equations. By analogy with fluid flow, ψ is again called the *stream function*, and the analytic function

$$w(z) = \phi(x, y) + i\psi(x, y) \qquad (5.170)$$

is called the *complex potential* for the electrostatic field.

Differentiating Equation (5.170) with respect to z and using the Cauchy–Riemann equations gives

$$w'(z) = \frac{\partial \phi}{\partial x} + i\frac{\partial \psi}{\partial x} = \frac{\partial \phi}{\partial x} - i\frac{\partial \phi}{\partial y}.$$

From Equation (5.166) we now find that in electrostatics

$$w'(z) = -E_1 + iE_2, \qquad (5.171)$$

so using the vector interpretation of complex numbers we have

$$\mathbf{E} = -\overline{w'(z)}. \tag{5.172}$$

Result from Equation (5.171) should be compared with the corresponding result for fluid flow given in Equation (5.91). It will be seen that the only difference is a reversal of sign due to the definition $\mathbf{q} = \text{grad}\,\phi$ used in Equation (5.86) and the definition $\mathbf{E} = -\text{grad}\,\phi$ used in Equation (5.167). The intensity E (strength) of the field $\mathbf{E}$ is given by

$$E = |w'(z)| = (E_1^2 + E_2^2)^{1/2} = \left[\left(\frac{\partial \phi}{\partial x}\right)^2 + \left(\frac{\partial \phi}{\partial y}\right)^2\right]^{1/2}. \tag{5.173}$$

Points where $E = 0$ are called *equilibrium points* or *neutral points* in the electrostatic field, because a free charge placed at such a point would experience no force, and so would be in equilibrium. It follows from Equation (5.171) that such neutral points occur wherever $dw/dz = 0$, and they are the analog of stagnation points in fluid flow.

At this point it is useful to compare the situations in heat conduction and electrostatics, each of which involves a potential. An immediate correspondence occurs because the isothermals in heat flow correspond to electrostatic equipotentials, the heat flow lines correspond to electrostatic lines of force, and the boundary conditions in these two types of problem are equivalent, in the sense that both a temperature and an electrostatic potential are specified on a boundary.

The correspondence between electrostatic problems and related fluid flows is different. In a fluid flow the vector field describes the fluid velocity vector, while in electrostatics it describes the electric field vector, though the boundary conditions to be applied to each of these vectors can be different. Whereas a streamline adjacent to a boundary must be tangential to it (except where a stagnation point occurs on the boundary), in a domain with electrically conducting boundaries, lines of force can originate from and terminate on such boundaries. Thus, care must be exercised when relating these two types of problem. An example of this form of relationship will be considered later when discussing an electrically conducting cylinder in a uniform electrostatic field.

We now derive the complex potential for a uniform electrostatic field of intensity E_0, inclined at an angle α to the x-axis and directed in the sense of increasing x. Because of this we have

$$E_1 = E_0 \cos\alpha, \quad E_2 = E_0 \sin\alpha,$$

so substituting these results into Equation (5.166) gives

$$\frac{\partial \phi}{\partial x} = -E_0 \cos \alpha, \quad \frac{\partial \phi}{\partial y} = -E_0 \sin \alpha,$$

where ϕ is the electrostatic potential. Integrating the first result with respect to x gives

$$\phi = -E_0 x \cos \alpha + f(y) + \phi_0,$$

where $f(y)$ is an arbitrary function of y and ϕ_0 is an arbitrary integration constant. Similarly, integrating the second result with respect to y gives

$$\phi = -E_0 y \sin \alpha + g(x) + \phi_1,$$

where $g(x)$ is an arbitrary function of x and ϕ_1 is another arbitrary integration constant.

Because these two expressions must be identical, it follows that

$$f(y) = -E_0 y \sin \alpha \quad \text{and} \quad g(x) = -E_0 \cos \alpha, \text{ while } \phi_0 = \phi_1$$

so the required electrostatic potential is

$$\phi = -E_0(x \cos \alpha + y \sin \alpha) + \phi_0. \tag{5.174}$$

Because the choice of reference potential is arbitrary, hereafter we set $\phi_0 = 0$.

Using the Cauchy–Riemann equations to determine the harmonic conjugate ψ of ϕ, we find that

$$\psi = -E_0(y \cos \alpha - x \sin \alpha). \tag{5.175}$$

Thus, in terms of x and y, the required complex potential is

$$w = -E_0(x \cos \alpha + y \sin \alpha) - iE_0(y \cos \alpha - x \sin \alpha),$$

which is easily shown to be

$$w = -E_0 z e^{-i\alpha}. \tag{5.176}$$

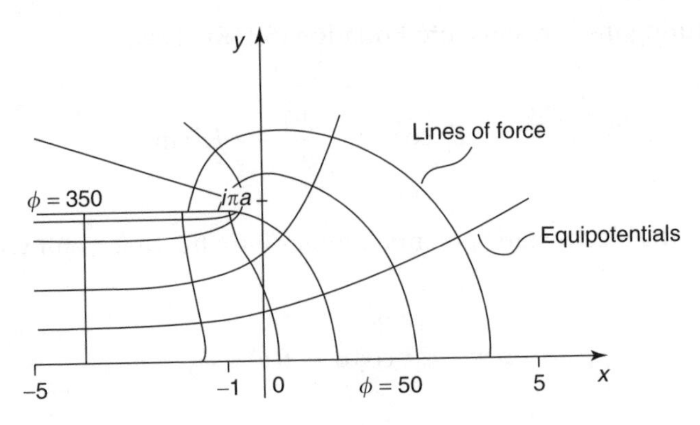

FIGURE 5.64
Representative equipotentials and lines of force in a parallel plate capacitor.

Then, this is the complex potential for a parallel electric field of intensity E_0, inclined to the x-axis at an angle α and directed in the sense of increasing x. The analog in heat conduction is a uniform heat flux in this same direction, and in fluid flow a uniform flow moving from left to right and inclined at an angle α to the x-axis.

Dirichlet problems arise naturally in electrostatics because they determine the electrostatic potential distribution throughout a region that often takes the form of a cavity filled with a dielectric, when the potential on the walls is given. The specification of the Neumann condition $\partial\phi/\partial n = 0$ on part or all of the boundary is seen from Equation (5.167) to be equivalent to requiring the normal component of **E** to vanish on the boundary.

The equivalence in mathematical terms of the determination of a temperature distribution involving Dirichlet conditions and the electrostatic potential distribution has already been mentioned. We now illustrate this in the case of the heat conduction problem in Example 5.3.3 in Section 5.3. A comparison of Figures 5.27 and 5.64 shows how the pattern of isothermals and heat flow lines that occur in the heat flow case relate to the equipotentials and lines of force in the electrostatic case.

Similarly, Example 5.3.4 can be interpreted as the electrostatic potential distribution due to a conducting cylinder of unit radius at the potential $\phi = 0$ parallel and at a distance of 2 length units from an infinite conducting plane at the potential $\phi = 100$. A shift and scaling of solutions is always possible as in Equation (5.49) thereby providing solutions for a range of related problems. Such a shift of the entire solution by the addition of a constant corresponds to the choice of a reference potential which is, in any case, arbitrary. Thus it is the potential *differences* that are important, rather than the potentials themselves.

Example 5.5.1 Potential Distribution in a Hollow Cylinder with a Semicircular Cross-Section

Find the equipotentials within a hollow conducting cylinder whose curve of cross-section is the semicircle shown in Figure 5.65, given that the circular

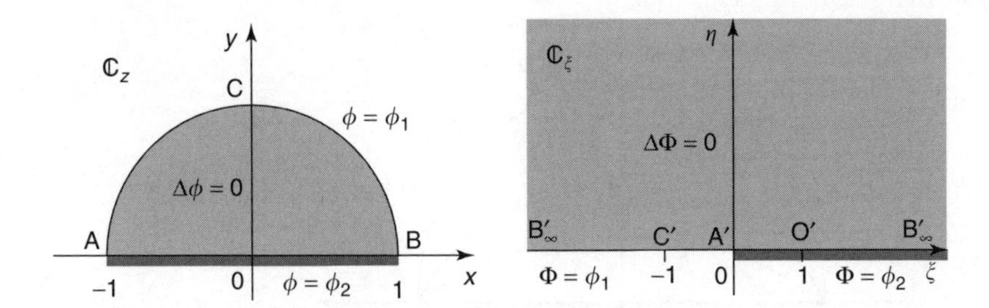

FIGURE 5.65
Electrostatic potential problem for a hollow semicircular cylinder.

boundary is maintained at a potential $\phi = \phi_1$, and the plane boundary at the potential $\phi = \phi_2$. Find the field on the plane bounding surface.

SOLUTION
A routine calculation establishes that the transformation

$$\zeta = \left(\frac{1+z}{1-z}\right)^2 \tag{5.177}$$

maps the half-disk in the z-plane shown on the left of Figure 5.65 one-one and conformally onto the upper half of the ζ-plane shown on the right. Under this transformation let the potential $\phi(x, y)$ in the z-plane become the potential $\Phi(\xi, \eta)$ in the ζ-plane. Then from Equation (5.23) the solution of the potential problem in the ζ-plane is

$$\Phi = \phi_2 + \left(\frac{\phi_1 - \phi_2}{\pi}\right)\text{Arg}\,\zeta.$$

Using the result

$$\text{Arg}\,\zeta = 2\text{Arg}\left(\frac{1+z}{1-z}\right) = 2\,\mathbf{Arctan}\left(\frac{2y}{1-x^2-y^2}\right),$$

the potential in the z-plane is found to be

$$\phi(x, y) = \phi_2 + \frac{2}{\pi}(\phi_1 - \phi_2)\,\mathbf{Arctan}\left(\frac{2x}{1-x^2-y^2}\right). \tag{5.178}$$

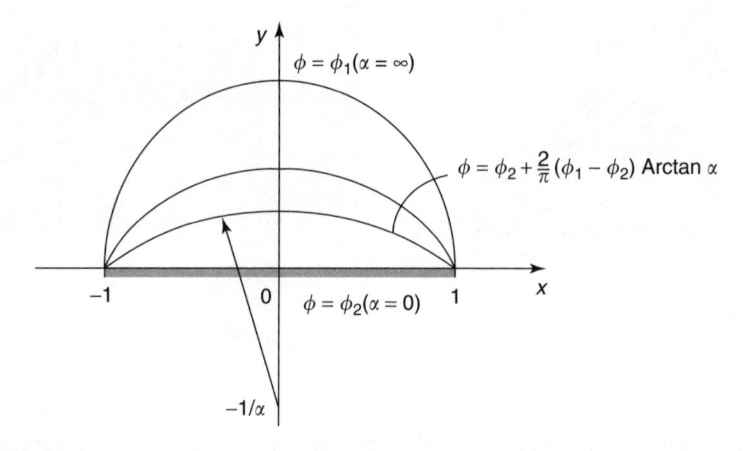

FIGURE 5.66
Equipotentials within a semicircular cylinder.

The equipotentials ϕ = const. are the curves

$$\frac{2y}{1 - x^2 - y^2} = \alpha,$$

where α is a parameter such that $0 \leqslant \alpha < \infty$. These form the family of circles with centers at $(0, -1/\alpha)$ and radius $(1 + (1/\alpha^2))^{1/2}$, as shown in Figure 5.66. Using Equation (5.166) with [Equation (5.178)] and $\mathbf{E} = -\text{grad}\,\phi$ shows that on the plane surface $-1 < x < 1, y = 0$, the electrostatic field **E** has the components

$$E_1 = -\left.\frac{\partial \phi}{\partial x}\right|_{y=0} = 0 \quad \text{and} \quad E_2 = -\left.\frac{\partial \phi}{\partial y}\right|_{y=0} = \frac{4(\phi_2 - \phi_1)}{\pi(1 - x^2)}.$$

So on the plane surface the vector **E** is in the y-direction. ◇

Example 5.5.2 The Potential Distribution between Eccentric Cylinders

The circles $|z| = 1$ and $|z - \alpha| = \rho$, with $0 < \alpha < 1$ and $\rho < 1 - \alpha$, are the boundaries of conducting cylinders with their axes normal to the z-plane. Find the potential distribution between the cylinders if the inner cylinder is maintained at the potential $\phi = \phi_1$ and the outer cylinder at the potential $\phi = \phi_2$.

SOLUTION

The transformation needed to map the geometrical configuration in the z-plane shown on the left of Figure 5.67 one-one and conformally onto the annulus in the ζ-plane shown on the right was discussed in Example 4.3.6. Under this mapping, let the potential $\phi(x, y)$ become the potential $\Phi(\xi, \eta)$ in the ζ-plane.

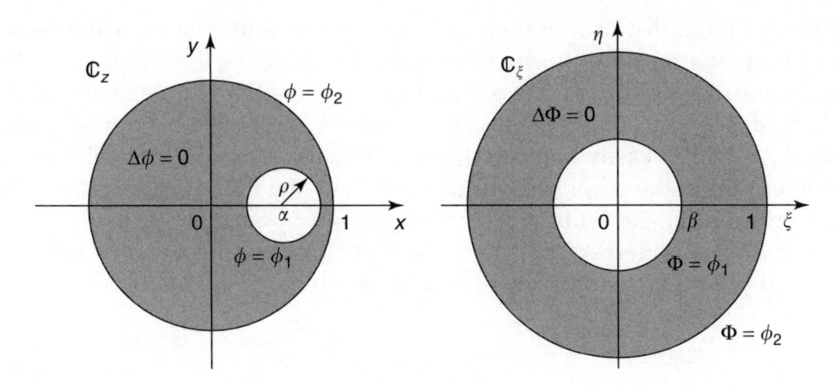

FIGURE 5.67
Conducting eccentric cylinders.

In the notation of that example we replace w by $\zeta = \xi + i\eta$ to obtain the required transformation

$$\zeta = \frac{z - c}{cz - 1}, \tag{5.179}$$

where c is the real root of

$$\alpha c^2 - (1 + \alpha^2 - \rho^2)c + \alpha = 0, \tag{5.180}$$

which lies within the unit circle $|z| = 1$. Transforming Equation (5.179) maps the circle $|z| = 1$ and the associated boundary condition $\phi = \phi_2$ onto the circle $|\zeta| = 1$ with the boundary condition $\Phi = \phi_2$, and the circle $|z - \alpha| = \rho$ onto the circle $|\zeta| = \beta$ with the associated boundary condition $\Phi = \phi_1$, where

$$\beta = \left|\frac{\alpha - \rho - c}{c(\alpha - \rho) - 1}\right|. \tag{5.181}$$

The solution of the electrostatic potential problem in the ζ-plane is

$$\Phi(\zeta) = \phi_2 + \left(\frac{\phi_1 - \phi_2}{\ln_e \beta}\right) \ln_e|\zeta|. \tag{5.182}$$

Combining Equations (5.179) and (5.182) to obtain the solution in the z-plane gives

$$\phi(x, y) = \phi_2 + \left(\frac{\phi_1 - \phi_2}{\ln_e \beta}\right) \ln\left|\frac{z - c}{cz - 1}\right|,$$

where $z = x + iy$ is any point between the circles on the left of Figure 5.67.

This potential distribution may be used to determine the capacitance per unit length of a capacitor formed by these two cylinders; that is the ratio C of the magnitude $|\rho|$ of the charge ρ stored per unit length of the capacitor to the magnitude $|\phi_1 - \phi_2|$ of the potential difference between the cylinders. Such a quantity, which is easily measured, provides a simple experimental method for determining the displacement of the axis of the inner cylinder from the axis of the outer cylinder because with ρ, ϕ_1 and ϕ_2 fixed, $C = C(\alpha)$.

We shall not undertake the calculation of $C(\alpha)$. It will suffice to mention that in physics texts the charge ρ per unit length of the cylinders is

$$\rho = \int_{|z|=1} \sigma(x, y)ds, \tag{5.183}$$

where ds is the line element around $|z| = 1$, and σ is the surface charge per unit area. This is given in terms of the electric field by

$$\sigma(x, y) = \pm\frac{\varepsilon}{4\pi}E_n, \tag{5.184}$$

where E_n is the component of $\mathbf{E}$ normal to $|z| = 1$, and ε is the permittivity of the dielectric material between the cylinders (often air). ◇

5.5.1 Standard Problems Involving the Complex Electrostatic Potential

Although working with the complex potential is not necessary in electrostatic problems, its use often simplifies computations. The following standard problems are useful in this context.

A Simple Dirichlet Problem in the Half Plane

Starting from the standard solution established in Equation (5.23), with $\alpha = \pi$, and using the polar form of the Cauchy–Riemann equations a routine calculation shows that

$$w(\zeta) = \phi_2 + i\left(\frac{\phi_2 - \phi_1}{\pi}\right)\mathrm{Ln}\,\zeta \tag{5.185}$$

is the complex potential for the electrostatic problem involving the Dirichlet conditions shown in Figure 5.68. Using the notation $w = \Phi + i\Psi$, the electrostatic potential becomes

$$\Phi = \phi_2 + \left(\frac{\phi_1 - \phi_2}{\pi}\right)\mathrm{Arg}\,\zeta \tag{5.186}$$

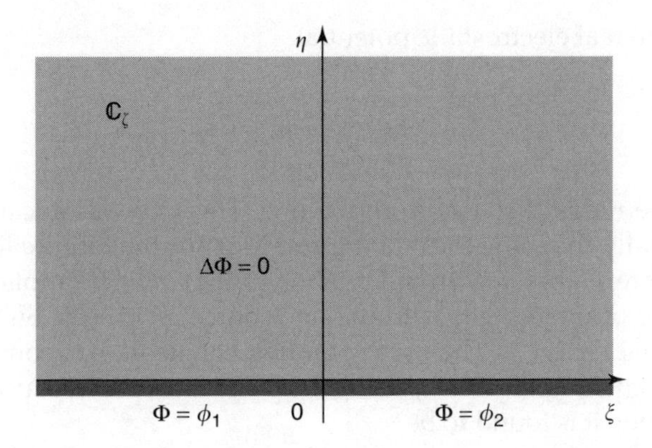

FIGURE 5.68
A simple Dirichlet problem in the half plane.

and the stream function becomes

$$\Psi = \left(\frac{\phi_2 - \phi_1}{\pi}\right)\ln_e|\zeta|. \tag{5.187}$$

As an example of the use of the complex potential, and of Equation (5.185) in particular, we return to the last part of Example 5.5.1.

The complex potential in the ζ-plane for the problem shown in Figure 5.65 was given in Equation (5.185), so using Equation (5.177) this becomes

$$w(z) = \phi_2 + i\left(\frac{\phi_2 - \phi_1}{\pi}\right)\text{Ln}\left[\left(\frac{1+z}{1-z}\right)^2\right].$$

To find the field on $-1 < x < 1, y = 0$ we use the fact that $dw/dz = -E_1 + iE_2$. Then either by direct calculation or, more simply from Equations (5.185) and (5.177) using the result $dw/dz = (dw/d\zeta)(d\zeta/dz)$, and setting $y = 0$ this reduces to

$$\left.\frac{dw}{dz}\right|_{y=0} = i\frac{4(\phi_2 - \phi_1)}{\pi(1 - x^2)},$$

so $E_1 = 0$ and $E_2 = [4(\phi_2 - \phi_1)/\pi(1 - x^2)]$, in agreement with the previous result.

5.5.2 The Infinite Line Charge

An infinitely long line charge of strength e per unit length normal to the (x, y)-plane in a dielectric medium of permittivity ε, that passes through the origin, has the real electrostatic potential

$$\phi = -\frac{2e}{\varepsilon}\ln_e r \tag{5.188}$$

where r is the radial distance from the line. This expression can be derived using essentially the same form of argument as for the infinite line source in fluid flow. Here, Gauss' law from Equation (5.164) must be applied to a cylindrical volume of unit length with the line source as its axis. Such a cylinder contains a total charge e. The corresponding conjugate harmonic function ψ follows from Equation (5.188) by using the polar form of the Cauchy–Riemann equations, when it is found to be

$$\psi = -\frac{2e\theta}{\varepsilon}. \tag{5.189}$$

Thus the complex potential $w = \phi + i\psi$ becomes

$$w = -\frac{2e}{\varepsilon}(\ln_e r + i\theta) = -\frac{2e}{\varepsilon}\,\mathrm{Ln}\,z.$$

Correspondingly, the complex potential for such a line charge through the point z_0 is

$$w = -\frac{2e}{\varepsilon}\,\mathrm{Ln}(z - z_0). \tag{5.190}$$

Figure 5.69 shows the equipotentials (concentric circles) and lines of force (rays) through z_0 due to a line charge at z_0. As conformal mapping maps a cross-section of a cylinder surrounding a line charge one-one onto a similar

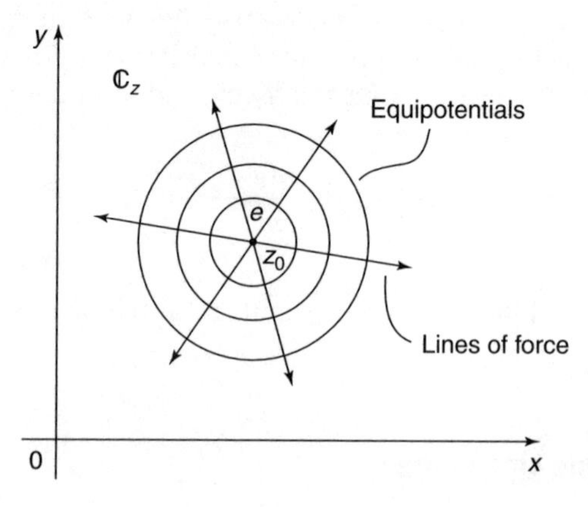

FIGURE 5.69
Equipotentials and lines of force of an infinite line charge at z_0.

cylinder with the same charge, so a transformation maps a line charge at z_0 into one of the same strength.

5.5.3 The Image of a Line Charge in an Infinite Conducting Plane at Zero Potential

Consider the complex potential

$$w(z) = -2e\mathrm{Ln}(z - z_0) + 2e\mathrm{Ln}(z - \bar{z}_0), \tag{5.191}$$

corresponding to a line charge of strength e per unit length, and one of strength $-e$ per unit length at $\bar{z}_0$ in a medium with permittivity $\varepsilon = 1$. The electrostatic potential $\phi = \mathrm{Re}\{w(z)\}$ is

$$\phi = -e\ln_e[(x - x_0)^2 + (y - y_0)^2] + e\ln_e[(x - x_0)^2 + (y + y_0)^2], \tag{5.192}$$

where $z_0 = x_0 + iy_0$. As $\phi = 0$ on the real axis, this axis may be regarded as a conducting plane at zero potential. Thus, either Equation (5.191) represents the complex potential of two infinite line charges of equal and opposite signs in the entire complex plane or as the complex potential in the upper half of the z-plane due to a line charge of strength e at z_0 in the presence of a conducting plane $y = 0$ at zero potential. In this interpretation, the line charge at $\bar{z}_0$ is called the *image* in the z-plane of the line charge at z_0. Conversely, the line charge at $\bar{z}_0$ has for its image the line charge at z_0. The arrangement of line charges is shown in Figure 5.70.

Notice that in electrostatics a line charge and its image have strengths of equal but *opposite* sign, whereas in a fluid flow the image of a line source has the same strength m as its image.

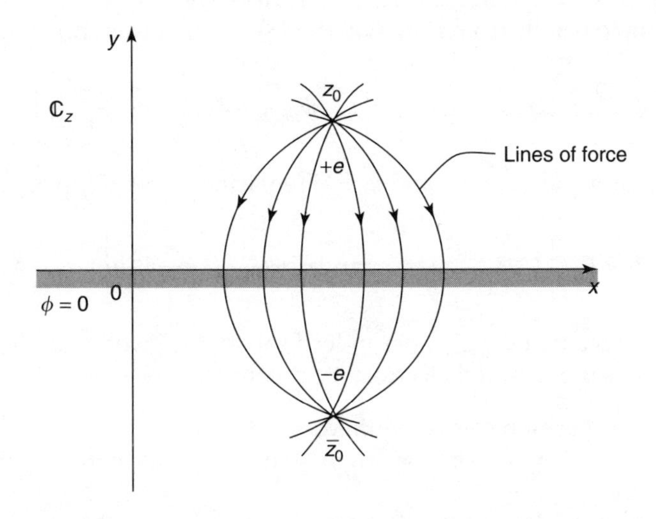

FIGURE 5.70
The image at $\bar{z}_0$ of a line charge at z_0 and some lines of force.

5.5.4 A Conducting Cylinder at the Potential ϕ in a Uniform Field

Let $|z| = a$ be the curve of cross-section of a conducting cylinder at the constant potential $\phi = \phi_0$, with its axis normal to the z-plane. We now find the complex potential when this cylinder is placed in a uniform electric field of intensity E_0 at infinity, with the field parallel to the x-axis at infinity and directed in the sense of positive x.

It is usual to deduce this complex potential from the real electrostatic potential ϕ subject to the appropriate boundary conditions, but here we will adopt a different approach and find it by comparison with a fluid flow problem solved in Section 5.4.

Consider the flow of fluid past a cylinder of radius a which is inserted into the flow with its axis normal to the direction of flow, as considered in the left of Figure 5.54. This is shown again in Figure 5.71(a). The streamlines are then the lines $\psi_{\text{fluid}} = \text{const.}$, and their orthogonal trajectories, the lines of equipotentials, are the lines $\phi_{\text{fluid}} = \text{const.}$

In the analogous electrostatic case, the lines of force that are parallel at infinity and suitably close to the y-axis at infinity must terminate on the cylinder. This suggests that the same pattern of orthogonal trajectories should be found in the electrostatic case, although there the lines of force must be the lines shown in Figure 5.71(b) that terminate on the cylinder, while the equipotentials are the other lines shown there. If this is so, a comparison of Figure 5.71(a) and (b) shows that the roles of the streamlines and the equipotentials in the fluid flow are reversed in the electrostatic case. Thus ϕ_{electric} can be deduced from ψ_{fluid}, and ψ_{electric} can be deduced from ϕ_{fluid}. Now Figure 5.71(b) is obtained from Figure 5.71(a) by a clockwise rotation through $\pi/2$, but in the process $y \to x$ and $x \to -y$. The potential functions in fluid flow and electrostatics have opposite signs, however, to compensate the direction of the y-axis must be reversed in the electrostatic case when $U \to -E_0$.

Appealing to results from Equation (5.148) where it was shown that

$$\phi_{\text{fluid}} = Ux\left(1 + \frac{a^2}{x^2 + y^2}\right), \quad \psi_{\text{fluid}} = Uy\left(1 - \frac{a^2}{x^2 + y^2}\right),$$

interchanging ϕ and ψ, and x and y, and replacing U by $-E_0$ gives

$$\phi_{\text{electric}} = -E_0 x\left(1 - \frac{a^2}{x^2 + y^2}\right) \quad \text{and} \quad \psi_{\text{electric}} = -E_0 y\left(1 + \frac{a^2}{x^2 + y^2}\right).$$

We must check that ϕ_{electric} does indeed satisfy the given boundary conditions, and so is the required electrostatic potential.

1. ϕ_{electric} is harmonic as is required;
2. $\phi_{\text{electric}} \to -E_0 x$ far from the origin, so it satisfies the correct condition at infinity;
3. On the boundary of the circle $x^2 + y^2 = a^2$ we see that $\phi_{\text{electric}} = 0$, whereas it should be ϕ_0.

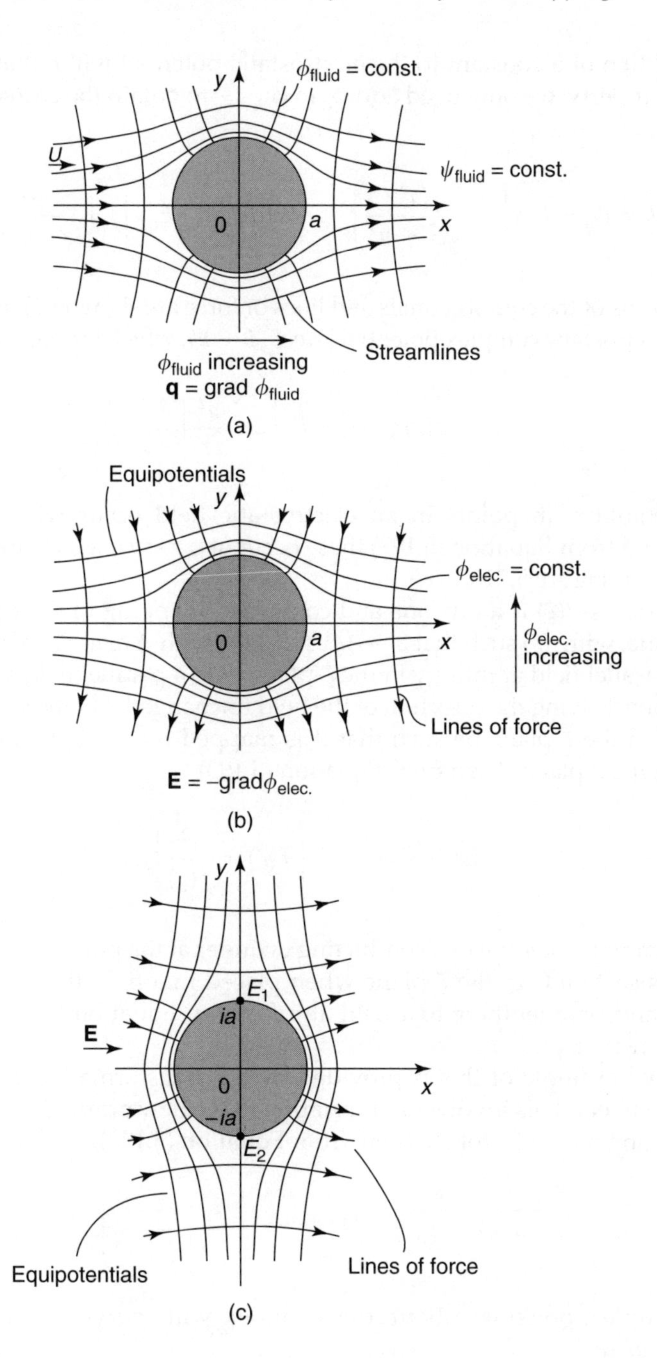

FIGURE 5.71
A comparison of fluid flow past a cylinder and a conducting cylinder in a uniform electric field.

The addition of a constant to the electrostatic potential will not alter its harmonic, property, we only need add ϕ_0 to ϕ_{electric} to obtain the correct potential function

$$\phi = \phi_0 - E_0 x\left(1 - \frac{a^2}{x^2 + y^2}\right), \qquad \text{with } \psi = -E_0 y\left(1 + \frac{a^2}{x^2 + y^2}\right).$$

The patterns of the equipotentials and lines of force are shown in Figure 5.71(c).

The electrostatic complex potential is $w = \phi + i\psi$, which is easily shown to be

$$w(z) = \phi_0 - E_0\left(z - \frac{a^2}{z}\right).$$

The equilibrium points in an electrostatic field occur when $dw/dz = 0$ ($\mathbf{E} = \mathbf{0}$) and from Equation (5.193) these occur at $z = \pm ia$, which are the points E_1 and E_2 in Figure 5.71(c).

Suppose $z = f(\zeta)$ is a one-one and conformal mapping of the ζ-plane onto the z-plane, with $f(\zeta)$ such that $z \to A\zeta$ as $|\zeta| \to \infty$, with A real. Then $w = f(\zeta)$ will map a parallel field at infinity in the ζ-plane onto a parallel field at infinity in the z-plane, leaving the direction of the field unchanged. Let the simple closed curve C in the ζ-plane be such that it is mapped by $z = f(\zeta)$ onto the circle $|z| = a$ in the z-plane. Then from Equation (5.193)

$$w(\zeta) = \phi_0 - E_0\left(f(\zeta) - \frac{a^2}{f(\zeta)}\right), \tag{5.193}$$

is the complex potential of a conducting cylinder at the potential $\phi = \phi_0$ with the cross-section C in the ζ-plane when placed in a field that is uniform at infinity and parallel there to a field at infinity in Equation (5.193), but now with intensity $E_0 A$.

A trivial example of this is provided by the transformation $z = A(\zeta - c)$, with $A > 0$ real. This involves a translation to the new origin $\zeta = c$ and a uniform scaling by the factor A. Then, from Equation (5.193),

$$w(\zeta) = \phi_0 - E_0 A\left((\zeta - c) - \frac{a^2}{A^2(\zeta - c)}\right)$$

is a complex potential about the cylinder with curve of cross-section $|\zeta - c| = a/A$.

Exercises 5.5

Find the electrostatic potential in the domains shown in the following diagrams subject to the given boundary conditions

1.

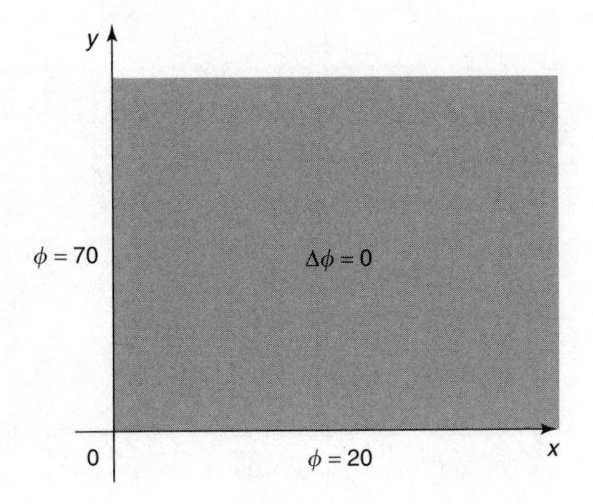

FIGURE 5.72
$\phi = 20$ on $y = 0, x > 0, \phi = 70$ on $x = 0, y > 0$.

2.

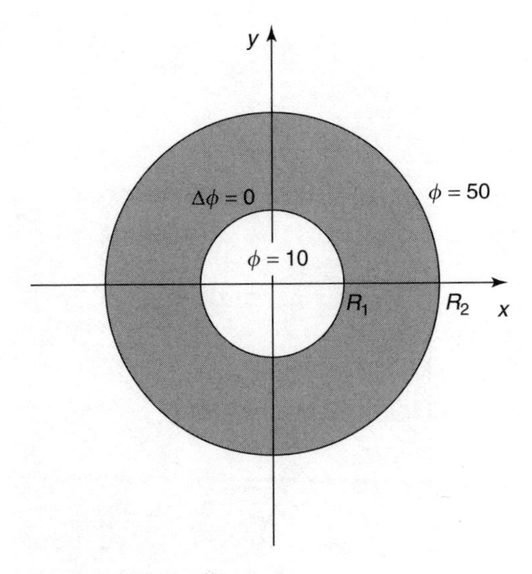

FIGURE 5.73
$\phi = 10$ on $r = R_1, \phi = 50$ on $r = R_2$.

3.

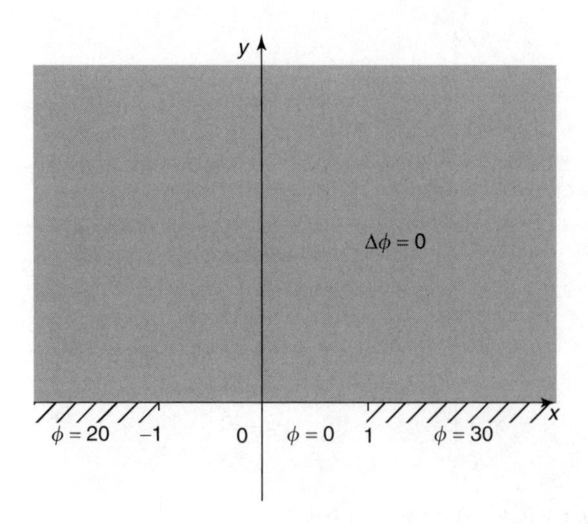

FIGURE 5.74
$\phi = 20$ on $y = 0, x < -1, \phi = 0$ on $-1 < x < 1, y = 0, \phi = 30$ on $x > 1, y = 0$.

4.

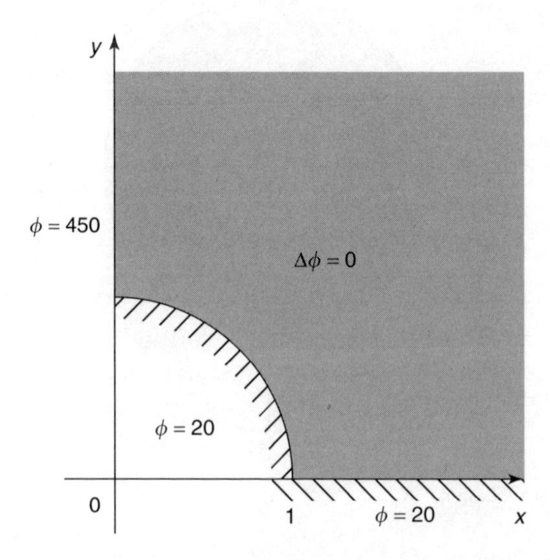

FIGURE 5.75
$\phi = 450$ on $x = 0, y > 1, \phi = 20$ on the circular arc in the first quadrant, $\phi = 20$ on $x > 1, y = 0$.

5.

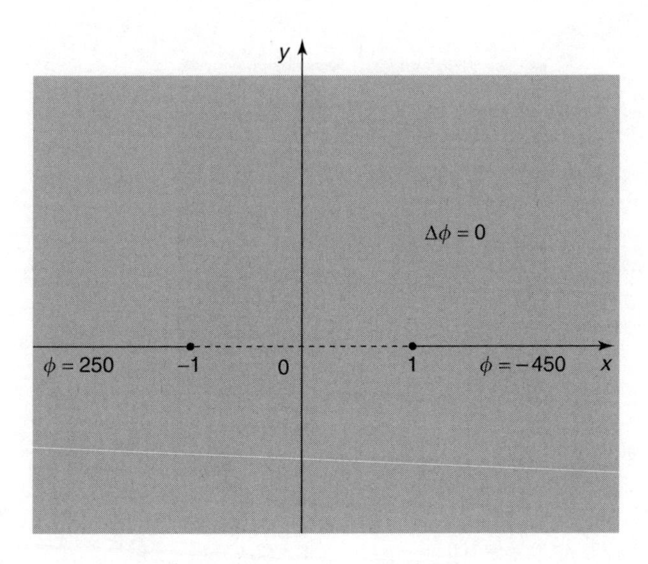

FIGURE 5.76
$\phi = 250$ on $x < -1, y = 0$, $\phi = -450$ on $x > 1, y = 0$, $\partial\phi/\partial n = 0$ on $-1 < x < 1, y = 0$.

6.

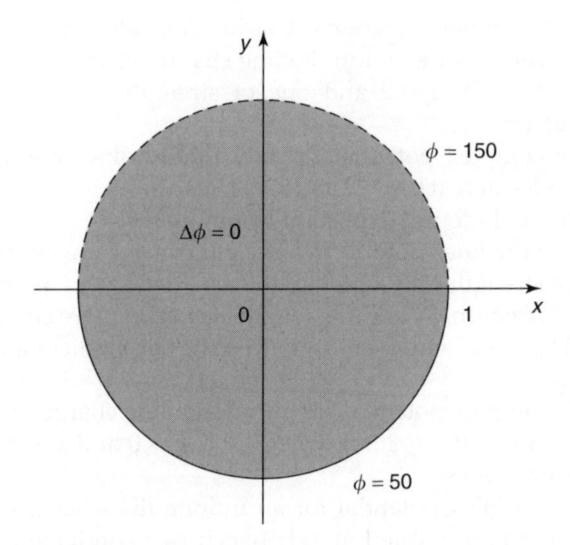

FIGURE 5.77
$\phi = 150$ on the semicircle with $y > 0$, $\phi = 50$ on the semicircle with $y < 0$.

7.

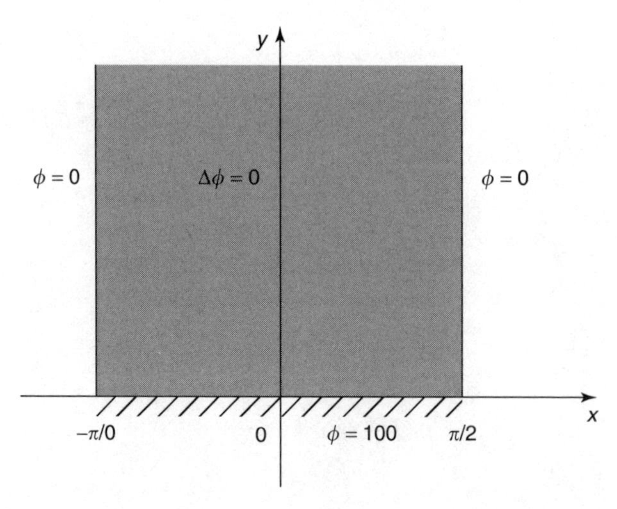

FIGURE 5.78
$\phi = 0$ on $x = \pm\pi/2, y > 0, \phi = 100$ on $-\pi/2 < x < \pi/2, y = 0$.

8. Find the component of the electrostatic field in Example 5.5.1 along the ray
$$\operatorname{Arg} z = \alpha \quad \text{for } 0 < \alpha < \pi, 0 < r < 1$$
9. Find the components of the electrostatic field along the line $y = x$ in a system comprising an infinite line charge of strength $2e$ per unit length located at $z = 2$ and one of strength $-e$ per unit length located at $z = -1$.
10. Find the complex potential for two infinite line charges each of strength e located at $z = -2$ and $z = 2$, respectively. Sketch the equipotentials and identify any equilibrium points.
11. Prove that the lines of force in a system comprising an infinite line charge of strength e per unit length located at z_1 and an infinite line charge of strength $-e$ per unit length located at z_2 are circles passing through z_1 and z_2 with their centers on the perpendicular bisector of z_1 and z_2.
12. Find the complex potential for an infinite line charge of strength e per unit length at $z = 1 + i$ when $x > 0, y = 0$ and $x = 0, y > 0$ are conducting planes.
13. Find the complex potential for an infinite line charge of strength e per unit length located at z_0 between two conducting planes at zero potential, as shown in Figure 5.79.
14. Show that the linear fractional transformation $w = u + iv$, with $w = \frac{1}{i}[(z-1)/(z+1)]$, maps the unit circle $|z| = 1$ onto the upper half plane, and find the images of the points $= -i, z = 1, z = i$ and $z = -1$ on the real w-axis $v = 0$. A conducting circular cylinder $|z| = 1$ has its

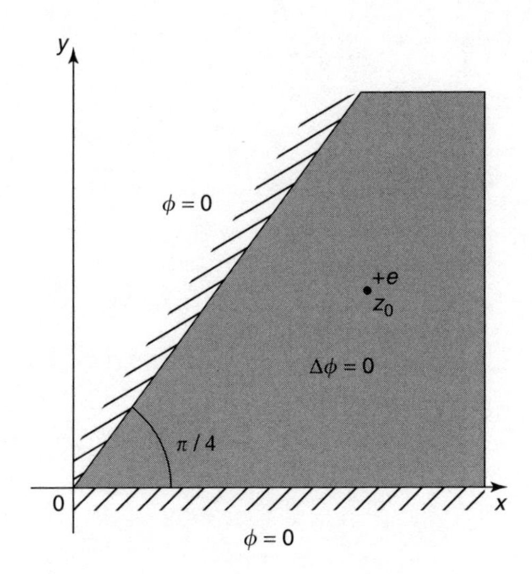

FIGURE 5.79
A line charge at z_0 between two conducting plates forming a wedge.

surface $0 < \theta < \pi/2$ in the first quadrant at a constant potential $\phi = \phi_0$, and the remainder of its surface $\pi/2 < \theta < 2\pi$ at the constant potential $\phi = \phi_2$. Find the potential $\phi(x, y)$ at an arbitrary point (x, y) inside the cylinder. Use a simple geometrical argument to show that the equipotential through a point $Q(x_0, y_0)$ inside the cylinder is a circular arc drawn through the point (x_0, y_0) and the points $z = 1$ and $z = i$ on which the potential is $\phi\ (x_0, y_0)$.

Solutions to Selected Odd-Numbered Exercises

Solutions 1.1

1. (a) $4 + 5i$ (b) $2 - 25i$ (c) $1 + 6i$ (d) $6 + 20i$
3. (a) $2\sqrt{2}$ (b) $\frac{1}{4}(1 + i)$ (c) $2\sqrt{2}$
5. $-\frac{1}{5}(6 + 2i)$
7. $z = -\frac{1}{3}(5 + i)$
9. No solution necessary.
11. If $z = x + iy$ then $z^2 = (\bar{z})^2$ implies that $2ixy = 0$. Thus either $x = 0$, in which case z is purely imaginary, or $y = 0$, in which case z is purely real.
13. Either if $a = 0$, when $a + ib$ is purely imaginary or if $b = 0$, when $a + ib$ is purely real because $|(a + ib)z| = |a + ib||z|$.
15. $|w| = 2\sqrt{2}, \operatorname{Arg}(w) = \tan^{-1}\left(\dfrac{\sqrt{3}-1}{\sqrt{3}+1}\right) - \pi.$
17. $\sin 5\theta = 5\cos^4\theta \sin\theta - 10\cos^2\theta \sin^3\theta + \sin^5\theta.$
19. $\sin^4\theta = \frac{1}{8}(3 - 4\cos 2\theta + \cos 4\theta).$
21. Yes, substitute for z and use the periodicity properties of sine and cosine.
23. Either verify the identity by cross-multiplication, or by using the fact that the expression on the left can be summed because it is a geometrical series with common ratio z. The Lagrange identity then follows as outlined, where multiplication of both the numerator and denominator of the expression on the right by $z^{-1/2}$ simplifies the manipulation by introducing the required denominator $2\sin(\frac{1}{2}\theta)$ from the outset.
25. The fourth roots are:

$$w_k = \frac{1}{2^{1/8}}\left(\cos\left(\frac{(8k+3)\pi}{16}\right) + i\sin\left(\frac{(8k+3)\pi}{16}\right)\right),$$

with $k = 0, 1, 2, 3.$

27. Setting $w = -i^{3/4}$ and raising both sides of the equation to the power 4 gives $w^4 = (-i^{3/4})^4 = (-1)^4(i^{3/4})^4 = i^3 = -i$, so the roots are the fourth roots of $-i$. These are: $w_k = \cos(\frac{1}{8}(4k-1)\pi) + i\sin(\frac{1}{8}(4k-1)\pi)$ with $k = 0, 1, 2, 3, 4$.
29. Set $w = u + iv$ and $z = x + iy$ in $w^2 = z$, and solving for u and v in terms of x and y gives $x = u^2 - v^2$ and $y = 2uv$, while $u^2 + v^2 = (x^2 + y^2)^{1/2} = |z|$. We also have $u^2 = \frac{1}{2}(|z| + x)$ and $v^2 = \frac{1}{2}(|z| - x)$. The result now follows by using the fact that when $y > 0$, either $u > 0$ and $v > 0$, or $u < 0$ and $v < 0$, while when $y < 0$, either $u > 0$ and $v < 0$, or $u < 0$ and $v > 0$, and then choosing u and v so that their product has the same sign as y; that is the sign of $\operatorname{sgn}(y)$.
31. The square roots found using Exercise 29 are: $\pm(1 + i)/\sqrt{2}$.
33. Theorem 1.1.1 applies: $z_1 = -1, z_2 = -2 - i, z_3 = -2 + i$.
35. Theorem 1.1.2 applies: $z_1 = -3$, $z_2 = -1 - i\sqrt{3}$, $z_3 = -1 + i\sqrt{3}$. The moduli $|z_1|$, $|z_2|$, and $|z_3|$ are all greater than 1, confirming that there are no roots inside the unit circle.

Solutions 1.2

1. Exterior and boundary points of the unit circle centered on $z = 3$.
3. Points inside the annulus $R_1 \leqslant r < R_2$, above the real axis with points on the inner boundary included.
5. Points inside the wedge $\alpha < \theta < \beta$ in the first quadrant such that $a < y < b$.
7. Points inside and on the rectangle $-2 \leqslant x \leqslant 3$, $-1 \leqslant y \leqslant 2$ from which have been removed the points inside but not on a circle of unit radius centered on the origin.
9. Points inside an ellipse centered on the origin with a semimajor axis a and a semiminor axis b.
11. Points on and below the part of the hyperbola with asymptote $y = bx/a$ lying in the first quadrant.
13. The perpendicular bisector of the line joining $z = 1 + i$ and $z = 1 - i$: the line $y = -x$.
15. A circle of radius $\sqrt{1/(\alpha - 1)}$ centered on the origin.
17. $-1 \leqslant a \leqslant 3$.
19. $\frac{1}{4}(x-2)^2 + \frac{1}{9}(y+1)^2 > 1$. A domain.

Solutions 1.3

1. (a) $-3 + 2i$ (b) -6
3. (a) $\sqrt{5}(7 - i)$ (b) $\sqrt{10}(3 + i)$
5. Cartesian form: $u = (x^2 - y^2)/(x^2 + y^2)^2$, $v = -2xy/(x^2 + y^2)^2$;
   Polar form: $u = (1/r^2)\cos 2\theta$, $v = -(1/r^2)\sin 2\theta$.

7. If $w = f(z)$, the region is the strip $1 \leqslant \text{Re}\{w\} \leqslant 5$.
9. If $w = f(z)$, the region is the interior of the disk $|w| \leqslant 1$, and the interior of the annulus $2 \leqslant |w| \leqslant 4$.
11. $3 - 2i$, continuous.
13. No limit, not continuous.
15. A routine calculation starting from the definition of a derivative in Equation (1.57). The condition $z \neq -a/b$ is necessary if the derivative is to remain finite.
17. (a) Yes (b) Yes (c) yes (d) Yes, if $g(z)$ never vanishes (e) Yes (f) Yes (g) Yes, if $g(z) + f(z)$ never vanishes.
19. $-10i$
21. 9

Solutions 1.4

1. Analytic for all z.
3. Analytic for $z \neq 0$.
5. Nondifferentiable.
7. Analytic for $\text{Im}\{z\} \geqslant 0$.
9. Nondifferentiable. Actually $f(z) = \bar{z}$.
11. $f(z) = e^{2z}, f'(z) = 2e^{2x}(\cos 2y + i \sin 2y), f'(z) = 2e^{2z}$.
13. $f(z) = \cosh z, f'(z) = \sinh x \cos y + i \cosh x \sin y, f'(z) = \sinh z$.
15. Use the Cauchy–Riemann equations to rewrite Equations (1.8) and (1.9).
17. Both functions satisfy the Laplacian equation $\Delta\phi = 0$, but they do *not* satisfy the Cauchy–Riemann equations.
19. $f(z) = \sinh z + c$ (c real).
21. $f(z) = z \sin z + c$ (c real).
23. $f(z) = \ln(1/z^3) + 4$.
25. The result follows directly from the Cauchy–Riemann equations. If, for example, $u = \text{Re}\{f(z)\} = \text{const.}$, then $u_x = u_y = 0$, and thus $v_x = v_y = 0$, showing $v = \text{const.}$ and hence that $f(z) = \text{const.}$
27. If $f(z)$ is analytic then $\Delta u = 0$ and $|f'(z)|^2 = u_x^2 + u_y^2$. Set $\phi = u^n$, then $\phi_x = nu^{n-1}u_x$ and $\phi_{xx} = n(n-1)u^{n-2}u_x^2 + nu^{n-1}u_{xx}$ and, similarly, $\phi_{yy} = n(n-1)u^{n-2}u_y^2 + nu^{n-1}u_{yy}$. Thus $\Delta\phi = \Delta\{[u(x, y)]^n\} = n(n-1)$ $[u(x, y)]^{n-2}|f'(z)|^2 + n[u(x, y)]^{n-1}\Delta u(x, y)$, but $\Delta u = 0$, so the last term vanishes and the result is proved.
29. If $\Delta\{|f(z)|\} = 0$, from Exercise 26 we have $\Delta\{|f(z)|\} = |f'(z)|^2/|f(z)| = 0$ so $|f'(z)|^2 = 0$, but $|f'(z)|^2 = u_x^2 + v_x^2 = 0$ which is only possible if $u_x = v_x = 0$. The Cauchy–Riemann equations then imply that

$u_y = v_y = 0$, so as all of the first-order partial derivatives of u and v are zero, $u =$ const. and $v =$ const. and hence $f(z) =$ const.

Solutions 1.5

1. $\dfrac{3}{4(z-1)} - \dfrac{3}{4(z+1)} + \dfrac{3}{2(z+1)^2}$

3. $z^2 - 2z + 3 - \dfrac{4}{z+1} - \dfrac{2}{(z+1)^2}$

5. (a) $-I$ (b) i (c) $\exp(-9\pi^2/4)$

7. (a) $e(\cos 2 + i\sin 2)$ (b) $e^{-1}(\cos 2 - i\sin 2)$ (c) $e^{-3}(\cos 4 + i\sin 4)$

9. $z = i$: $\ln(z) = \frac{1}{2}(4n+1)\pi i,\ n = 0, \pm 1, \pm 2, \ldots$; $\mathrm{Ln}(z) = \frac{1}{2}\pi i$

11. $z = e^{-3}$: $\ln(z) = -3 + 2n\pi i,\ n = 0, \pm 1, \pm 2, \ldots$; $\mathrm{Ln}(z) = -3$

13. $z = 4$: $\ln(z) = \ln 4 + 2n\pi i,\ n = 0, \pm 1, \pm 2, \ldots$; $\mathrm{Ln}(z) = \ln 4$

15. $e^{-(2n+1)\pi}[\cos(\ln 4) + i\sin(\ln 4)],\ n = 0,\ \pm 1,\ \pm 2, \ldots :\ e^{-\pi}[\cos(\ln 4) + i\sin(\ln 4)]$

17. $\cos(4n+1) + i\sin(4n+1),\ n = 0, \pm 1, \pm 2, \ldots : \cos 1 + i\sin 1$

19. $\exp[(4n-1)\pi],\ n = 0, \pm 1, \pm 2, \ldots : e^{-\pi}$

21. $3e^{2n\pi}[\cos(2n\pi - \ln 3) + i\sin(2n\pi - \ln 3)],\ n = 0, \pm 1, \pm 2, \ldots :$ $3[\cos(\ln 3) - i\sin(\ln 3)]$

23. $\frac{1}{2}(2n+1)\pi + i\,\mathrm{arccosh}\,5,\ n = 0, \pm 1, \pm 2, \ldots$

25. $\frac{1}{2}(4n+1)\pi + i\,\mathrm{arcsinh}\,3,\ n = 0, \pm 1, \pm 2, \ldots$

27. $\frac{1}{2}(4n+1)\pi + i\ln\sqrt{6},\ n = 0, \pm 1, \pm 2, \ldots$

29. $\frac{1}{6}(12n \pm 1)\pi i,\ n = 0, \pm 1, \pm 2, \ldots$

31.
$$\begin{aligned}
|\sin z|^2 &= \sin^2 x\cosh^2 y + \cos^2 x\sinh^2 y \\
&= \sin^2 x\cosh^2 y + (1 - \sin^2 x)\sinh^2 y \\
&= \sin^2 x(\cosh^2 y - \sinh^2 y) + \sinh^2 y \\
&= \sin^2 x + \sinh^2 y, \quad \text{so as } \sin^2 x \geqslant 0,
\end{aligned}$$

$\sinh^2 y \leqslant |\sin z|^2.$

As $\sinh y \leqslant \sinh|y|$, taking the square root gives, $\sinh|y| \leqslant |\sin z|$. Similarly,

$$\begin{aligned}
|\sin z| &= (1 - \cos^2 x)\cosh^2 y + \sin^2 x\sinh^2 y \\
&= -\cos^2 x(\cosh^2 y - \sinh^2 y) \\
&= -\cos^2 x + \cosh^2 y \leqslant \cosh^2 y.
\end{aligned}$$

However, $\cosh y \geqslant 1$, so taking the square root gives $|\sin z| \leqslant \cosh y$, and thus $\sinh|y| \leqslant |\sin z| \leqslant \cosh y$.

33. Set $\dfrac{P(z)}{Q(z)} = \dfrac{A_1}{z - z_1} + \dfrac{A_2}{z - z_2} + \cdots + \dfrac{A_n}{z - z_n}$,

 then

$$\frac{(z - z_i)P(z)}{Q(z)} = (z - z_i)\left(\frac{A_1}{z - z_1} + \cdots + \frac{A_{i-1}}{z - z_{i-1}} + \frac{A_{i+1}}{z - z_{i+1}} + \cdots + \frac{A_n}{z - z_n}\right) + A_i(*)$$

 As $Q(z_i) = 0$, the left side of this result can be written as

$$\frac{(z - z_i)P(z)}{Q(z)} = P(z)\Big/\left(\frac{Q(z) - Q(z_i)}{z - z_i}\right),$$

 and in the limit as $z \to z_i$ this becomes $P(z_i)/Q'(z_i)$. Combining this result with the limiting value of (*) as $z \to z_i$ we find that $A_i = P(z_i)/Q'(z_i)$, for $n = 1, 2, \ldots, n$.

$$\frac{z^2 + z - 3i}{(z - 2)(z - 3)(z + i)} = \frac{-\frac{1}{5}(9 - 12i)}{z - 2} + \frac{\frac{1}{10}(33 - 21i)}{z - 3} - \frac{\frac{1}{10}(5 + 3i)}{(z + i)}.$$

Solutions 2.1

1. Parameterization is not unique, but an obvious one is: AB: $x(s) = s$, $y(s) = (1 + s)/2$ for $1 \leqslant s \leqslant 3$, when $dz/ds = (1 + \frac{1}{2}i)$; BC: $x(s) = s$, $y(s) = (7 - s)/2$ for $3 \leqslant s \leqslant 5$, when $dz/ds = (1 - \frac{1}{2}i)$.

$$\int_{AB} f(z)dz = \tfrac{19}{2} + 16i, \int_{BC} f(z)dz = \tfrac{61}{2} - 4i, \int_{ABC} f(z)dz = 40 + 12i.$$

3. Parameterization is not unique, but an obvious one is: AB: $x(s) = s$, $y(s) = s$ for $0 \leqslant s \leqslant \pi$, when $dz/ds = 1 + i$; BC: $x(s) = s$, $y(s) = \pi$, when $dz/ds = 1$.

$$\int_{AB} f(z)dz = -i\sinh\pi, \int_{BC} f(z)dz = 2i\sinh\pi,$$
$$\int_{ABC} f(z)dz = i\sinh\pi.$$

5. 0
7. $-4\pi i$
9. Self checking.
11. The two branches of the square root are $w_0(z) = r^{1/2}e^{i\theta/2}$ and $w_1(z) = -r^{1/2}e^{i\theta/2}$. If $z = -1$, $r = |z| = 1$, and $\mathrm{Arg}(z) = \pi$, then $w_0(-1) = i$ and $w_1(-1) = -i$. Thus the branch to be used is $w_0(z)$, and on C, it becomes $w_0(z) = z^{1/2} = e^{i\theta/2}$, while on C, $dz/d\theta = ie^{i\theta}$.

Integration around C starts from $z = i$, corresponding to $\theta = \pi/2$, so it must terminate when $\theta = 5\pi/2$, yielding

$$\int_C z^{1/2} dz = \int_{\pi/2}^{5\pi/2} (e^{i\theta/2})(ie^{i\theta})d\theta = \tfrac{2\sqrt{2}}{3}(1 - i).$$

13. If $z = re^{i\theta}$, then $z^{\alpha} = e^{\alpha \ln z} = \exp\{\alpha[\ln_e r + (\theta + 2k\pi)i]\} = r^{\alpha}\exp[\alpha(\theta + 2k\pi)i]$. The branch used requires that $1^{\alpha} = 1$, so as $|z| = 1$, $\text{Arg}(z) = 0$, the integer k must be such that $1 = \exp[\alpha(\theta + 2k\pi)i]$, which is only possible if $k = 0$. Thus, for z^{α} we must use the branch corresponding to $k = 0$, that on $|z| = 1$ becomes $z^{\alpha} = e^{\alpha\theta i}$. The integration around C is to start from $z = 1$, corresponding to $\theta = 0$, so it must terminate when $\theta = 2\pi$. On C, $z = e^{i\theta}$, so $dz/d\theta = ie^{i\theta}$.
15. $\sinh z = \sinh x \cos y + \cosh x \sin y$, so $|\sinh z| = (\sinh^2 x + \sin^2 y)^{1/2} < (\sinh^2 x + 1)^{1/2} = \cosh x$. For z on C, $-1 \leqslant x \leqslant 1$, so $|\sinh x| \leqslant \cosh 1$.

As $|z - a||z + a| = |z^2 - a^2| \leqslant \left||z|^2 - |a|^2\right|$, and for z on $|z| = 1$,

$$\frac{|\sinh z|}{|z - a||z + a|} \leqslant \frac{\cosh 1}{\left|1 - |a|^2\right|}, \text{ but on } C,\ z = e^{i\vartheta},\ dz/d\theta = ie^{i\theta},\ \text{so } |dz| = d\theta.$$

$$\text{Thus } \int_C \frac{|\sinh z||dz|}{|z - a||z + a|} \leqslant \int_0^{2\pi} \frac{\cosh 1}{\left|1 - |a|^2\right|} d\theta = \frac{2\pi \cosh 1}{\left|1 - |a|^2\right|}.$$

17. $1.424 + 0.369i$
19. $-2.288 - 1.317i$

Solutions 2.2

1. $1, -1, i, -i$; isolated.
3. $2i, -2i$, and at $n\pi$ for $n = 0, \pm 1, \pm 2, \ldots$; isolated.
5. Removable singularity at the origin; singularities at $n\pi i$ for $n = \pm 1, \pm 2 \ldots$; isolated.
7. $\pm\frac{1}{2}(4n + 1)\pi + i \,\text{arccosh}\, 2$, for $n = 0, \pm 1, \pm 2, \ldots$; isolated.
9. 0 and $i/(n\pi)$, $n = \pm 1, \pm 2, \ldots$; all isolated with the exception of the origin which is a limit point.
11. Setting $f = u + iv$ and proceeding as in the proof of Theorem 2.2.2 gives

$$\int_C f(z)dz = -\int_D \left(\frac{\partial v}{\partial x} + \frac{\partial u}{\partial y}\right) dx\, dy + i\int_D \left(\frac{\partial u}{\partial x} - \frac{\partial v}{\partial y}\right) dx\, dy.$$

Setting $f(z) = x$ implies that $u = x$ and $v = 0$, so the result reduces to

$$\int_C f(z)dz = i\int_D dx\, dy = iD.$$

Setting $f(z) = y$ implies that $u = 0$ and $v = y$, causing the result to reduce to

$$\int_C f(z)dz = -\int_C dx\,dy = -D.$$

As $\bar{z} = x - iy$,

$$\int_C \bar{z}\,dz = \int_C (x - iy)dz = iD - i(-D) = 2iD.$$

13. (i) $4\pi i$. (ii) $-\pi i$.
15. (i) $3\pi i/2$. (ii) $-3\pi i/2$.
17. (i) Deform the circle $|z - \frac{5}{4}| = \frac{1}{2}$ into the circle $|z - 1| = \frac{1}{2}$, or any other circle centered on $z = 1$ and containing only the singularity at $z = 1$. Integrate to show the required result is $\pi(3 - i)$. (ii) Deform the circle $|z - \frac{13}{4}| = \frac{1}{2}$ into the circle $|z - 3| = \frac{1}{2}$, or any other circle centered on $z = 3$ and containing only the singularity at $z = 3$. Integrate to show the required result is $\pi(i - 9)$.
19. Only the singularity at $z = 1$ lies inside the cardioid, so deform C into any circle C^* centered on $z = 1$ containing only this one singularity. Then as

$$\frac{2z^2 + (3 - 2i)z - (1 + 6i)}{(z^2 - 1)(z - 2i)} = \frac{2}{z - 1} - \frac{1}{z + 1} + \frac{1}{z - 2i},$$

it follows in the usual way that

$$\int_{C^*} \frac{2}{z - 1} dz = 4\pi i.$$

Solutions 2.3

1. $-\frac{7}{4} + 6i$
3. $\frac{1}{3}(-22 + 4i)$
5. $\frac{1}{2}\cosh 2(\cos 2 - \cos 4) + \frac{1}{2}i \sinh 2(\sin 2 - \sin 4)$
7. $\cos 2 \cosh 3 - \cos 1 \cosh 2 - i\,(\sin 2 \sinh 3 + \sin 1 \sinh 2)$
9. Use partial fractions to write the integral as

$$3\int_{-1-i}^{1+i} \frac{dz}{z} - \int_{-1-i}^{1+i} \frac{1}{z - 2} dz,$$

and cut the z-plane along the negative real axis up to and including the origin, and take any path from $-1 - i$ to $1 + i$ that does not cross the cut. The integral equals $\frac{1}{2}\ln_e 5 + i(\frac{5}{4}\pi + \text{Arctan}\,\frac{1}{3})$.

11. $-6 + 6\sqrt{3}i$

13. Inspection shows that the function is continuous and bounded.

$$I_1 = \int_0^1 x\,dx + \int_1^2 1 \cdot dx + \int_0^2 1 \cdot i dy = \tfrac{3}{2} + 2i.$$

$$I_2 = \int_0^2 0 \cdot i dy + \int_0^1 x\,dx + \int_1^2 1 \cdot dx = \tfrac{3}{2}.$$

15. $\int_0^{1+i} z \cos z dz = \cosh 1(\sin 1 + \cos 1) - \cos 1 \sinh 1 - 1 + i[\sinh 1(\cos 1 - \sin 1) + \sin 1 \cosh 1].$

Solutions 2.4

1. $-16\pi i/e^3$
3. $i\pi^2$
5. $2\pi e i/5$
7. $v(x, y) = y \cos x \cosh y - x \sin x \sinh y$ (after using the condition that $v(0, 0) = 0$). Thus $f(z) = u + iv = z \cos z$, and so

$$\int_C \frac{f(z)}{z-1} dz = 2\pi i \cos 1.$$

9. Singularities of the denominator occur at $z = 0$ and $z = 1$. Possible choices for C are:

   (i) C contains neither singularity, so $I = 0$
   (ii) C contains only the singularity at $z = 0$, so $I = 1$
   (iii) C contains only the singularity at $z = 1$, so $I = e$
   (iv) C contains both singularities, so $I = 1 + e$.

11. $2\pi/\sqrt{3}$
13. $2\pi - 4\pi/\sqrt{3}$. Remember there are two singularities in C, so the contribution from each must be added.
15. 0. Remember there are two singularities in C, so the contribution from each must be added.

Solutions 2.5

1. $I = \dfrac{3(2\sqrt{3}\pi - 3)i}{2\pi}$
3. $I = 0$ when there are no singularities inside C; $I = 1$ when only the singularity at $z = 0$ is inside C; $I = -\frac{1}{2}\pi$ when only the singularity at $z = 1$ is inside C; $I = 1 - \frac{1}{2}i$ when the singularities at $z = 0$ and $z = -\frac{1}{2}\pi$ are inside C.
5. $I = \dfrac{2\pi(a^2 - 3)}{e(1 - a^2)^2} + \dfrac{\pi}{ae^a(1 - a)^2}.$

A negative sign has been introduced because integration is around C_- and not around C.

7. $I = -\dfrac{\pi(1+i)}{2a^3e^a}$

A negative sign has been introduced because integration is around C_- and not around C.

9. From the Cauchy integral theorem for derivatives we have

$$f^{(n)}(z_0) = \frac{n!}{2\pi i}\int_C \frac{f(z)}{(z-z_0)^{n+1}}\,dz$$

and from the same theorem replacing $f(z)$ in the integrand by $f^{(n-m)}(z)$ we have the equivalent expression for $f^{(n)}(z_0)$

$$f^{(n)}(z_0) = \frac{m!}{2\pi i}\int_C \frac{f^{(n-m)}(z)}{(z-z_0)^{m+1}}\,dz\,,$$

from which the result follows.

11. From the Schläfli formula, setting $z = z_0 + re^{i\theta}$ with C the prescribed simple closed contour containing z_0, we have

$$\begin{aligned}
P_n(z_0) &= \frac{1}{2\pi i}\int_0^{2\pi} \frac{[(z_0+re^{i\theta})^2-1]^n ire^{i\theta}}{2^n(re^{i\theta})^{n+1}}\,d\theta,\\
&= \frac{1}{2\pi}\int_0^{2\pi}\left(\frac{r^2e^{2i\phi}+2rz_0e^{i\theta}+r^2e^{i\theta}}{2re^{i\theta}}\right)^n d\theta\\
&= \frac{1}{2\pi}\int_0^{2\pi}\left(z_0+\tfrac{1}{2}re^{i\phi}[e^{-i(\theta-\phi)}+e^{i(\theta-\phi)}]\right)^n d\theta\\
&= \frac{1}{2\pi}\int_0^{2\pi}[z_0+(z_0^2-1)^{1/2}\cos u]^n\,du,
\end{aligned}$$

with $u = \theta - \phi$. However, as ϕ is a constant and the integrand $\cos u$ is even and periodic with period 2π, we may integrate over the interval $-\phi$ to $\pi - \phi$ and double the result to get

$$P_n(z) = \frac{1}{\pi}\int_{-\phi}^{\pi-\phi}\left[z_0+(z_0^2-1)^{1/2}\cos\theta\right]^n d\theta,$$

where the dummy variable u has been replaced by θ. The periodicity of the integrand now allows the limits of integration to be shifted to $0 \leqslant \theta \leqslant \pi$ giving the required result. Both this Laplace formula and the Rodrigues formula in Exercise 10 show that $P_3(x) = \frac{1}{2}(5x^3 - 3x)$.

Solutions 2.6

1. $\dfrac{1}{2\pi}\displaystyle\int_0^{2\pi}(3+5Re^{i\theta})d\theta = \dfrac{3}{2\pi}\displaystyle\int_0^{2\pi}d\theta + \dfrac{5R}{2\pi}\displaystyle\int_0^{2\pi}e^{i\theta}\,d\theta = 3 = f(0).$

3. $u(x, y)$ is harmonic so the maximum/minimum theorem applies. Maximum at $(\frac{1}{2}\pi, \frac{1}{2}\pi)$ where the value is $\cosh\pi/2$. Minima along $x = 0, 0 \leqslant y \leqslant \frac{1}{2}\pi$ where the value is zero.

5. Proceed as in Example 2.6.4.

7. Reason directly as in Example 2.6.1.

9. Consider $g(z) = f(z)/z$. Then $g(z)$ is analytic in the punctured disk $|z| < R$ because it is the quotient of two analytic functions where the denominator never vanishes. It is also differentiable at the origin because the fact that $f(0) = 0$ implies that $\lim_{z\to 0}(f(z)/z) = f'(0)$, and $f'(z)$ is analytic at $z = 0$. On $|z| = r$, with $0 < r < 1$ we have $|g(z)| = |f(z)|/|z| < 1/r$. Thus from the maximum modulus principle $|g(z)| < 1/r$ for $|z| \leqslant r$. Taking the limit as $r \to 1$ we have $|f(z)| \leqslant |z|$ for $|z| < 1$. $|f(z)| < (M/R)|z|$ for $|z| < R$. As $g(0) = f'(0)$, it follows that $|f'(0)| \leqslant 1$. If $g(z)$ attains its maximum at a point z_0 inside $|z| = 1$, it follows by the maximum modulus principle that $g(z) \equiv$ constant, and so $f(z) = e^{i\alpha}$ for some constant α, so that $f(z) = e^{i\alpha}z$.

11. This exercise provides a different proof of the Liouville theorem.

$$f(a) - f(b) = \frac{1}{2\pi i}\int_C \left(\frac{1}{z-a} - \frac{1}{z-b}\right)dz$$
$$= \frac{(a-b)}{2\pi i}\int_C \frac{f(z)}{(z-a)(z-b)}dz.$$

On C, $z = Re^{i\theta}$, so

$$f(b) - f(a) = \left(\frac{a-b}{2\pi i}\right)\int_0^{2\pi}\frac{f(Re^{i\theta})iRe^{i\theta}}{(Re^{i\theta}-a)(Re^{i\theta}-b)}d\theta.$$

If $f(z)$ is bounded for all z, then for some suitably large M we have $|f(z)| < M$, so that then

$$\left|\frac{f(Re^{i\theta})iRe^{i\theta}}{(Re^{i\theta}-a)(Re^{i\theta}-b)}\right| < \frac{MR}{(R-|a|)(R-|b|)},$$

and this vanishes as $R \to \infty$, so $f(b) - f(a) = 0$ for any a and b, and so we see that $f(z) =$ constant.

13. Because $f(z)$ is analytic everywhere, if z_0 is arbitrary then

$$f^{(m)}(z_0) = \frac{m!}{2\pi i}\int_{C_R}\frac{f(z)}{(z-z_0)^{m+1}}dz,$$

for any simple closed-curve C_R encircling z_0. Take C_R to be the circle $|z| = R$, so $z = Re^{i\theta}$ on C_R.
Then,

$$f^{(m)}(z_0) = \frac{m!}{2\pi i}\int_0^{2\pi} \frac{f(Re^{i\theta})iRe^{i\theta}}{(Re^{i\theta} - z_0)^{m+1}}\,d\theta,$$

and so

$$|f^{(m)}(z_0)| \leqslant \frac{m!}{2\pi}\int_0^{2\pi} \frac{|f(Re^{i\theta})|\,R}{|Re^{i\theta} - z_0|^{m+1}}\,d\theta \leqslant \frac{m!\,R}{2\pi(R - |z_0|)^{m+1}}\int_0^{2\pi} |f(Re^{i\theta})|d\theta,$$

but $|f(z)| < M|z|^n$ for suitably large $|z|$, so

$$|f^{(m)}(z_0)| \leqslant \frac{m!\,RM}{2\pi(R - |z_0|)^{m+1}}\int_0^{2\pi} R^n\,d\theta = \frac{m!\,R^{n+1}M}{2\pi(R - |z_0|)^{m+1}}.$$

Letting $R \to \infty$ we see that $|f^{(m)}(z_0)| = 0$ if $m > n$, so all orders of derivatives of $f(z)$ exceeding the nth vanish for arbitrary z_0, showing that $f(z)$ must be a polynomial whose degree does not exceed n.

Solutions 2.7

1. By definition

$$\begin{aligned}\text{P.V.}\int_2^8 \frac{dx}{x-4} &= \lim_{\varepsilon\to 0}\int_2^{4-\varepsilon} \frac{dx}{x-4} + \lim_{\varepsilon\to 0}\int_{4+\varepsilon}^{8} \frac{dx}{x-4} \\ &= \lim_{\varepsilon\to 0}\ln_e\varepsilon - \ln_e 2 + \ln_e 4 - \lim_{\varepsilon\to 0}\ln_e\varepsilon = \ln_e 2.\end{aligned}$$

The ordinary improper integral is given by

$$\begin{aligned}\int_2^8 \frac{dx}{x-4} &= \lim_{\varepsilon\to 0}\int_2^{4-\varepsilon} \frac{dx}{x-4} + \lim_{\delta\to 0}\int_{4+\delta}^{8} \frac{dx}{x-4} \\ &= \lim_{\varepsilon\to 0}\ln_e\varepsilon - \ln_e 2 + \ln_e 4 - \lim_{\varepsilon\to 0}\ln_e\delta = \ln_e 2 + \lim_{\substack{\varepsilon\to 0,\\ \delta\to 0}}\ln_e(\varepsilon/\delta),\end{aligned}$$

so as the last double limit is indeterminate the improper integral is divergent.

3. The integrand has no singularities inside Γ_R, so

$$\begin{aligned}\int_{\Gamma_R} z\exp(-z^2)dz = 0 = \int_{\text{OA}} z\exp(-z^2)dz + \int_{\text{AB}} z\exp(-z^2)dz \\ + \int_{\text{BO}} z\exp(-z^2)dz.\end{aligned}$$

$z = x$ on OA, $z = Re^{i\theta}$ on AB and $z = re^{i\alpha}$ on BO. The second integral on the right vanishes exponentially in the limit as $R \to \infty$, so in the limit

$$0 = \int_0^{\infty} x\exp(ix^2)dx + \int_{\infty}^{0} re^{i\alpha}\exp[-(re^{i\alpha})^2]ire^{i\alpha}dx.$$

After changing the dummy variable r to x, reversing the limits, simplifying the result, using trigonometric identities for double angles and recognizing that the value of the first integral on the right is 1/2, we obtain

$$\int_0^\infty x\exp(-x^2\cos 2\alpha)\cos[x^2\sin 2\alpha - 2\alpha]dx = \tfrac{1}{2}.$$

5. If $g(z)$ denotes the integrand, we see that it has a simple zero at $\frac{1}{2}\pi + i$, so by the Cauchy integral formula

$$\int_{\Gamma_R} g(z)dz = -\frac{\pi}{e} + i\frac{\pi}{e}\left(1 + \tfrac{1}{2}\pi\right).$$

However,

$$\int_{\Gamma_R} g(z)dz = \int_{\text{AB}} g(z)dz + \int_{\text{BCA}} g(z)dz,$$

with $z = x$ on AB and $z = Re^{i\theta}$ on BCA. As $g(z)$ satisfies Jordan's lemma the second integral on the left vanishes in the limit as $R \to \infty$, so

$$\int_{-\infty}^{\infty} \frac{(x+1)e^{ix}}{x^2 - \pi x + 1 + \frac{1}{4}\pi^2}dx = -\frac{\pi}{e} + \frac{i\pi}{e}(1 + \tfrac{1}{2}\pi),$$

from which I_1 and I_2 follow by equating real and imaginary parts.

7. The zeros of the denominator $1 + z^{2n}$ of $f(z)$ occur at $z_r = e^{i(2r+1)\pi/2n}$, for $r = 0, 1, \ldots, 2n-1$, so only the zero $z_0 = e^{i\pi/2n}$ lies inside OABO. Thus

$$\int_{\text{OABO}} f(z)dx = \int_{\text{OABO}} \frac{f_1(z)}{(z - z_0)}dz \quad \text{with}$$

$$f_1(z) = \frac{z^{2m}}{(z - z_1)(z - z_2)\cdots(z - z_{2n-1})}.$$

Then by the Cauchy integral theorem

$$\int_{\text{OABO}} f(z)dz = 2\pi i f_1(z_0) = -\frac{i\pi}{n}\exp[i(2m+1)\pi/2n].$$

This follows by writing $1 + z^{2n} = (z - z_0)(z - z_1)\cdots(z - z_{2n-1})$, differentiating with respect to z and setting $z = z_0$ in the result to show that $(z_0 - z_1)(z_0 - z_2)\cdots(z_0 - z_{2n-1}) = 2nz_0^{2n-1}$, and finally substituting for z_0. So

$$-\frac{i\pi}{n}\exp[i(2m+1)\pi/2n] = \int_{\text{OA}} f(z)dz + \int_{\text{AB}} f(z)dz + \int_{\text{BO}} f(z)dz.$$

We have $z = x$ on OA, $z = Re^{i\theta}$ on AB and $z = re^{i\pi/n}$ on OB. Letting $R \to \infty$ causes the integral around AB to vanish because the integrand vanishes like $1/R^{m/n}$, while after changing the dummy variable from r to x the integral along BO becomes

$$\int_{\text{BO}} f(z)dz = -\int_{\text{OB}} f(z)dz = -\exp[i(2m+1)\pi/n]\int_0^{\infty} \frac{x^{2m}}{1+x^{2n}}dx.$$

So combining results gives

$$(1-\exp[i(2m+1)\pi/n])\int_0^{\infty} \frac{x^{2m}}{1+x^{2n}}dx = -\frac{i\pi\exp[i(2m+1)\pi/2n]}{n},$$

from which the answer then follows.

9. The denominator of $f(z)$ has simple zeros at $z = ia$ and $z = ib$ inside Γ_R. the result then follows in the usual way after using the Cauchy integral theorem and letting $R \to \infty$, because the integral around the circular arc vanishes and the result reduces to

$$I = \int_{-\infty}^{\infty} \frac{\cos mx}{(x^2+a^2)(x^2+b^2)}dx = \frac{\pi}{(a^2-b^2)}\left(\frac{e^{-mb}}{b} - \frac{e^{-ma}}{a}\right).$$

11. Proceed as in Exercise 2.7.5. Contour (a) is necessary when $m > 0$ because then the integral around the semicircle vanishes as $R \to \infty$, after which the result follows in the usual way. It is obvious that the integral vanishes when $m = 0$, because in this case the integrand is identically zero. Contour (b) is necessary when $m < 0$ because only then will the integral around the semicircle vanish as $R \to \infty$. If contour (a) is used when $m < 0$, the integral around the semicircle will *diverge* as $R \to \infty$.

13. The denominator of $f(z)$ has a zero at the origin, about which the contour is indented, and a zero at $z = ia$ inside Γ_R. The contribution to the integral around Γ_R made by the indentation as $\rho \to 0$ is $i\pi/a^2$, while the contribution made by the singularity inside Γ_R is $-i\pi e^{-ma}/a^2$. So in the limit as $\rho \to 0$ and $R \to \infty$,

$$\int_{-\infty}^{\infty} \frac{\sin mx}{x(x^2+a^2)}dx = \frac{1}{i}\left(\frac{i\pi}{a^2} - \frac{i\pi}{a^2}e^{-ma}\right) = \frac{\pi}{a^2}(1-e^{-ma}).$$

As the integrand is even this simplifies to

$$\int_0^{\infty} \frac{\sin mx}{x(x^2+a^2)}dx = \frac{\pi}{2a^2}(1-e^{-ma}).$$

15. The reasoning proceeds in the usual manner after noticing that only the zero of the denominator at $z = 3i$ lies *inside* the contour, while that the zero at $z = 1$ lies *outside* in an indented semicircle around which integration is in the negative sense. The contribution to the integral at $z = 3i$ is found to be $-\dfrac{\pi}{10e^3}\left(\frac{1}{3}+i\right)$, while the contribution

made by integrating around the indentation in the negative sense is $-i\pi e^{i}/10$. The integral around the semicircle C_R vanishes in the limit, so combining results and proceeding to the limits gives

$$\text{P.V.}\int_{-\infty}^{\infty} \frac{\cos x + i\sin x}{(x+9)(x-1)}dx = \left(-\frac{\pi}{30e^3} + \frac{\pi}{10}\sin 1\right) + i\left(-\frac{\pi}{10e^3} - \frac{\pi}{10}\cos 1\right).$$

So equating the corresponding real and imaginary parts of this equation gives

$$I_1 = -\frac{\pi}{10}\left(\sin 1 + \frac{1}{3e^3}\right), \quad I_2 = \frac{\pi}{10}\left(\cos 1 - \frac{1}{e^3}\right).$$

The P.V. symbol is necessary because of the zero of $f(z)$ at $z = -1$.

17. The denominator of $f(z)$ has a quadruple zero at $z = i$ inside the contour ABCD, so from the Cauchy integral theorem

$$\int_{\Gamma_R} f(z)dx = \frac{5\pi}{16}.$$

The integral around the semicircle vanishes in the limit at $R \to \infty$, so

$$\int_{-\infty}^{\infty} \frac{dx}{(x^2+1)^4} = \frac{5\pi}{16},$$

but the integrand is even so that

$$\int_{0}^{\infty} \frac{dx}{(x^2+1)^4} = \frac{5\pi}{32}.$$

19. The denominator of $f(z)$ has a double zero at $z = i$ inside the contour ABCA, so from the Cauchy integral formula

$$\int_{\Gamma_R} \frac{z^2}{(z^2+1)^2}dz = \frac{\pi}{2}.$$

The integral around the semicircular arc vanishes in the limit as $R \to \infty$, so in the limit

$$\int_{-\infty}^{\infty} \frac{x^2}{(x^2+1)^2}dx = \frac{\pi}{2},$$

but the integrand is even, so

$$\int_{0}^{\infty} \frac{x^2}{(x^2+1)^2}dx = \frac{\pi}{4}.$$

21. Only the simple zero of the denominator of $f(z)$ at $z = i$ lies in the contour Γ_R, so setting $g(z) = z^\alpha/(z+i)$, we have $f(z) = g(z)/(z-i)$. It then follows from the Cauchy integral theorem that

$$\int_{\Gamma_R} f(z)dz = 2\pi i g(i) = 2\pi i(i^\alpha/(2i)) = \pi e^{i\alpha\pi/2}.$$

The integral around the semicircular indentation at the origin vanishes in the limit at $\rho \to 0$, as does the integral around the semicircle of radius R as $R \to \infty$, so combining terms and proceeding to the limits gives

$$\int_{AB} f(z)dz + \int_{CA} f(z)dz = \pi e^{i\alpha\pi/2}.$$

On AB, $z = re^{0i} = r$, while on CA, $z = re^{\pi i}$, so

$$\int_0^\infty \frac{r^\alpha}{1+r^2}dr + \int_\infty^0 \frac{(re^{\pi i})^\alpha e^{\pi i/2}}{1+(re^{\pi i})^2}dr = \pi e^{i\pi\alpha/2}$$

which simplifies to

$$(1+e^{i\alpha\pi})\int_0^\infty \frac{r^\alpha}{1+r^2}dr = \pi e^{i\alpha\pi/2}, \quad \text{and so} \int_0^\infty \frac{r^\alpha}{1+r^2}dr = \frac{\pi e^{i\alpha\pi/2}}{1+e^{i\alpha}}.$$

Replacing the dummy variable r by x and equating the real parts of this equation gives

$$\int_0^\infty \frac{x^\alpha}{1+x^2}dx = \frac{\pi}{2}\left(\frac{\cos\frac{1}{2}\alpha\pi[1+\cos\alpha\pi] + \sin\frac{1}{2}\alpha\pi\sin\alpha\pi}{1+\cos\alpha\pi}\right)$$
$$= \frac{\pi\cos\frac{1}{2}\alpha\pi}{(1+\cos\alpha\pi)}.$$

Using the trigonometric identity $2\cos^2\frac{1}{2}\theta = 1+\cos\theta$, and setting $= \alpha\pi/2$, the integral simplifies to

$$\int_0^\infty \frac{x^\alpha}{1+x^2}dx = \frac{\pi}{2\cos\left(\frac{1}{2}\alpha\pi\right)}.$$

23. $g(z)$ has no singularities in Γ_R so

$$\int_{\Gamma_R} g(z)dz = 0 = \int_\rho^R r^{\alpha-1}e^{-r}dr + \int_0^\beta R^{\alpha-1}e^{(\alpha-1)i\theta}e^{-Re^{i\theta}}iRe^{i\theta}\,d\theta$$
$$+ \int_R^\rho r^{\alpha-1}e^{i(\alpha-1)\beta}\exp(-re^{i\beta})e^{i\beta}\,dr$$
$$+ \int_\beta^0 \rho^{\alpha-1}e^{i(\alpha-1)\theta}\exp(-\rho e^{i\theta})i\rho e^{i\theta}\,d\theta.$$

The second integral on the right vanishes as $R \to \infty$ because of Jordan's lemma, and the last integral vanishes as $\rho \to 0$ because of the factor ρ^α. The results follow by proceeding to the limits and equating the respective real and imaginary parts on each side of this equation, having made use of the fact that the first integral on the right is the gamma function $\Gamma(\alpha)$.

25. The denominator of $f(z)$ has a double zero at $z = i$ inside the contour Γ_R, and the branch point of the logarithmic function at the origin is excluded by a semicircular indentation. We can write

$$f(z) = \frac{\ln z}{(z-i)^2(z+i)^2}, \text{ so if } g(z) = \frac{\ln z}{(z+i)^2}, \text{ then } f(z) = \frac{g(z)}{(z-i)^2}.$$

It then follows from the Cauchy integral theorem that

$$\int_{\Gamma_R} f(z)dz = \int_{\Gamma_R} \frac{g(z)}{(z-i)^2} dz = 2\pi i g'(z)\big|_{z=i} = 2\pi i\left(\tfrac{1}{4}i - \tfrac{1}{4}i \ln i\right).$$

Using the principal branch of the logarithmic function we have $\ln i = \frac{1}{2}\pi i$, so substituting this into the above result gives

$$\int_{\Gamma_R} f(z)dx = -\tfrac{1}{2}i + \tfrac{1}{4}\pi^2 i.$$

The integral around the semicircle of radius ρ vanishes in the limit $\rho \to 0$, and the integral around the semicircle of radius R vanishes in the limit at $R \to \infty$, so proceeding to the limits and using the results that on AB $z = re^{0i} = r$ while on CD $z = re^{i\pi} = -r$, gives

$$\int_{AB} f(z)dz + \int_{CD} f(z)dz = -\tfrac{1}{2}\pi + \tfrac{1}{4}\pi^2 i,$$

where AB is the integral of $f(z)$ along the positive real axis from 0 to infinity, and CD is the integral of $f(z)$ along the top of the cut above the negative real axis from $-\infty$ to the origin. Substituting for z this becomes

$$\int_0^\infty \frac{\ln_e r}{(r^2+1)} dr + \int_0^\infty \frac{\ln_e r + i\pi}{(r^2+1)^2} dr = -\tfrac{1}{2}\pi + \tfrac{1}{4}\pi^2 i.$$

The result follows directly from this by replacing the dummy variable r by x and equating the real parts on either side of the equation to obtain

$$\int_0^\infty \frac{\ln_e x}{(x^2+1)^2} dx = -\tfrac{1}{4}\pi.$$

Solutions 3.1

1. $z_1 = \frac{1}{2}(1+i), z_2 = \frac{1}{2}i, z_3 = \frac{1}{4}(-1+i), z_4 = -\frac{1}{4}, z_5 = -\frac{1}{8}(1+i), z_6 = -\frac{1}{8}i$; $|z_n| = 2^{-n/2}$, so $|z_n|$ and hence z_n tend to zero as $n \to \infty$.

3. $z_1 = \frac{3}{2}(1-i)$, $z_2 = -\frac{9}{2}i$, $z_3 = -\frac{27}{4}(1+i)$, $z_4 = -\frac{81}{4}$, $z_5 = -\frac{243}{8}(-1+i)$, $z_6 = \frac{729}{8}i$; $|z_n| = (3/\sqrt{2})^n$, so $|z_n|$ and hence z_n increase without bound as $n \to \infty$.
5. $z_1 = \cosh 1$, $z_2 = \cosh 2$, $z_3 = \cosh 3$, $z_4 = \cosh 4$, $z_5 = \cosh 5$, $z_6 = \cosh 6$; all terms z_n are real and they tend to infinity as $n \to \infty$.
7. Divergent, but with cluster points at 1 and $3i$.
9. Divergent. No cluster points.
11. $z_n = \text{Ln}\, k + n\pi i/2$, so the sequence is divergent.
13. Use the fact that $e^k = \lim_{n\to\infty}(1 + k/n)^n$. Thus $\lim_{n\to\infty}(1 + 2/n)^n = e^2$ and $\lim_{n\to\infty}(1 - 3/n)^n = e^{-3}$, so $\lim_{n\to\infty} z_n = e^2(2 - 3i)$ and $\lim_{n\to\infty} Z_n = e^{-3}(1 + 2i)$. Thus $\lim_{n\to\infty}(z_n Z_n) = e^{-1}(8 + i)$, $\lim_{n\to\infty}(z_n/Z_n) = -\frac{1}{5}e^5(4 + 7i)$.
15. Use the sum of a geometric series to show that $\sum_{n=1}^{\infty} z_n = \frac{3}{2} + i$.
17. Absolutely convergent by ratio test.
19. Divergent by comparison with the harmonic series $\sum_{1}^{\infty} 1/n$ which is divergent.
21. $$\frac{i}{(n+i)(n+1+i)} = i\left(\frac{1}{n+i} - \frac{1}{n+1+i}\right).$$

    So,

    $$\sum_{n=0}^{\infty}\left(\frac{i}{(n+i)(n+1+i)}\right) = i\left[\left(\frac{1}{i} - \frac{1}{1+i}\right) + \left(\frac{1}{1+i} - \frac{1}{2+i}\right) + \cdots\right].$$

    Removing brackets and canceling terms this reduces to $i(1/i) = 1$. The terms in the rewritten series *telescope*.

Solutions 3.3

1. $R = 1$; $|z| < 1$.
3. $R = \infty$; all z.
5. $R = 0$; only converges as the single point $z = 0$.
7. $1/4$; $|z - 1| < \frac{1}{4}$.
9. $R = 4$; $|z - i| < 4$.
11. $R = 1$; $|z + 2i| < 1$.
13. $\limsup |a_n|^{1/n} = 3$, so $R = 1/3$ and $|z - 2i| < \frac{1}{3}$.
15. Differentiate the geometric series and set $m = n - 1$ in the result to obtain $1/(1-z)^2 = \sum_{m=0}^{\infty}(m+1)z^m$. Replacing z by $3z$ and changing m to n gives the required series $1/(1-3z)^2 = \sum_{n=0}^{\infty}(n+1)(3z)^n$ for $|z| < \frac{1}{3}$.
17. Uniformly convergent by the M-test. Term by term differentiation gives

    $$\cos z = \sum_{n=0}^{\infty}(-1)^n \frac{z^{2n}}{(2n)!}.$$

19. $$\begin{aligned}\operatorname{Arctan} z &= \int_0^z \frac{dt}{1+t^2}\\ &= \int_0^z (1-t^2+t^4-t^6+\cdots)dt\\ &= z-\frac{z^3}{3}+\frac{z^5}{5}-\cdots \quad \text{for } |z|<1.\end{aligned}$$

This is justified because the series for $1/(1+t^2)$ is uniformly convergent for $0 \leqslant t^2 < 1$.

Solutions 3.4

1. The ratio test shows the radius of convergence to be $R = \infty$, so $\sin z$ is an entire function.

3. $\dfrac{1}{2z-1} = \sum\limits_{n=0}^{\infty} 2^n z^n; R = \frac{1}{2}.$

5. $\sum\limits_{n=0}^{\infty}(-1)^n z^{2n+2}/(n+1); R = 1.$

7. $\dfrac{z^2}{(1+z)^2} = \frac{1}{4} + \sum\limits_{n=0}^{\infty}(-1)^{n+1}\dfrac{(n-2)(z-1)^{n+1}}{2^{n+3}}; R = 2.$

9. $\operatorname{Ln}(i) = \ln_e|i| + i\operatorname{Arg}(i) = i\pi/2,$

so

$$\operatorname{Ln} z = \tfrac{1}{2}\pi i + \sum_{n=1}^{\infty}(-1)^{n-1}(z-i)^n/(ni^n); R = 1.$$

11. Set $f(z) = g(z) = 1/(1-z) = \sum\limits_{n=0}^{\infty} z^n$ for $|z| < 1$. Then $a_n = b_n = 1$ for $n = 0, 1, \ldots$. Thus

$$\frac{1}{(1-z)^2} = \sum_{n=0}^{\infty}(1+n)z^n \quad \text{for } |z|<1.$$

13. $\tan z = z + \frac{1}{3}z^3 + \frac{2}{15}z^5 + \frac{17}{315}z^7 + \cdots.$

15. $\tanh z = z - \frac{1}{3}z^3 + \frac{2}{15}z^5 - \frac{17}{315}z^7 + \cdots.$

17. $z \operatorname{ctn} z = 1 - \frac{1}{3}z^2 - \frac{1}{45}z^4 - \frac{2}{945}z^6 + \cdots$. $\operatorname{ctn} z = z^{-1} - \frac{1}{3}z - \frac{1}{45}z^3 - \frac{2}{945}z^5 + \cdots.$

19. Substituting the Maclaurin series for $f(z)$ into the equation to be satisfied by $f(z)$ gives

$$\sum_{n=0}^{\infty} a_n z^n = 1 + \sum_{n=0}^{\infty} a_n z^{n+1} + \sum_{n=0}^{\infty} a_n z^{n+2}.$$

Separating out terms and shifting the summation index so each remaining summation contains the power z^{n+2}, this becomes

$$a_0 + a_1 z + \sum_{n=0}^{\infty} a_{n+2} z^{n+2} \equiv 1 + a_0 z + \sum_{n=0}^{\infty} a_{n+1} z^{n+2} + \sum_{n=0}^{\infty} a_n z^{n+2}.$$

Equating the coefficients of corresponding powers of z on each side of this identity is now a simple matter because all summations contain the same power z^{n+2}, so as a result we find that: $a_0 = 1$, $a_1 = 1$ and the recurrence relation $a_{n+2} = a_n + a_{n+1}$ for $n = 0, 1, 2, \ldots$. Finding the quotient $f(z) = 1/(1 - z - z^2)$ up to and including the term in z^4 by the approach used in Exercises 13 through 18, gives $f(z) = 1 + z + 2z^2 + 3z^3 + 5z^4 + \cdots$, confirming that $a_0 = 1$, $a_1 = 1$, $a_2 = 2$, $a_3 = 3$ and $a_4 = 5$. Finding the radius of convergence of the series for $f(z)$ is not straightforward because from the recurrence relation $|a_n/a_{n+1}| = |a_{n+2}/a_{n+1} - 1|$, so to determine the limiting value as $n \to \infty$, it is necessary to know the general form of a_n. This can be found by solving the recurrence relation as a linear difference equation and when this is done it turns out that $R = \frac{1}{2}(\sqrt{5} - 1)$, although the details of this calculation are omitted.

Solutions 3.5

1. (a) $$f(z) = \frac{1}{z-2} + \frac{1}{z-3} = -\left[\frac{1}{2}\left(1 - \frac{z}{2}\right)^{-1} + \frac{1}{3}\left(1 - \frac{z}{3}\right)^{-1}\right]$$
for $0 \leqslant |z| < 2$.

Expanding by the binomial theorem gives

$$f(z) = -\left[\frac{1}{2}\sum_{n=0}^{\infty}\left(\frac{z}{2}\right)^n + \frac{1}{3}\left(\frac{z}{3}\right)^n\right]$$
$$= \frac{-5}{6} - \frac{13}{36}z - \frac{35}{216}z^2 - \frac{97}{1296}z^3 - \cdots, \quad 0 \leqslant |z| < 2.$$

(b) $$f(z) = \frac{1}{z}\left(1 - \frac{2}{z}\right)^{-1} - \frac{1}{3}\left(1 - \frac{z}{3}\right)^{-1} \quad \text{for } 2 < |z| < 3.$$

Expanding by the binomial theorem gives

$$f(z) = \sum_{n=1}^{\infty}\frac{2^{n-1}}{z^n} - \frac{1}{3}\sum_{n=0}^{\infty}\left(\frac{z}{3}\right)^n \quad \text{for } 2 < |z| < 3.$$

3. To expand about $z = i$ write

$$f(z) = \frac{1}{z-1} + \frac{1}{z+1} = \frac{1}{(z-i) - (1-i)} + \frac{1}{(z-i) + (1+i)}.$$

Set $Z = z - i$ to obtain for $0 \leqslant |Z| < \sqrt{2}$

$$\begin{aligned} f(z) &= \frac{1}{Z-(1-i)} + \frac{1}{Z+(1+i)} \\ &= -\left(\frac{1}{1-i}\right)\left(1-\frac{Z}{1-i}\right)^{-1} + \left(\frac{1}{1+i}\right)\left(1+\frac{Z}{1+i}\right)^{-1} \\ &= -\left(\frac{1+i}{2}\right)\sum_{n=0}^{\infty}\left(\frac{Z}{1-i}\right)^{n} + \left(\frac{1-i}{2}\right)\sum_{n=0}^{\infty}\left(\frac{Z}{1+i}\right)^{n} \\ &= -\left(\frac{1+i}{2}\right)\sum_{n=0}^{\infty}\left(\frac{1+i}{2}\right)^{n}(z-i)^n + \left(\frac{1-i}{2}\right)\sum_{n=0}^{\infty}(-1)^n\left(\frac{1-i}{2}\right)^{n}(z-i)^n, \end{aligned}$$

for $0 \leqslant |z-i| < \sqrt{2}$.

5. The expansion follows by writing

$$f(z) = \frac{1}{z-2} - \frac{2}{(z^2+1)}$$

and using either the binomial theorem or the Maclaurin theorem to expand each term in powers of z. The result is

$$f(z) = 2\sum_{n=0}^{\infty}\frac{(-1)^{n+1}}{z^{2n+2}} - \sum_{n=0}^{\infty}\frac{z^n}{2^{n+1}} \qquad \text{for } 1 < |z| < 2.$$

7. Use the hyperbolic identity $2\sinh A \cosh B = \sinh(A+B) + \sinh(A-B)$ with $A = z$ and $B = 5z$, and then sum the corresponding hyperbolic series. As the only singularity occurs at $z = 0$, and the hyperbolic functions are entire functions, the domain of convergence will be $|z| > 0$. The required Laurent series expansion is

$$f(z) = 2\sum_{n=0}^{\infty}\frac{(-1)^{n+1}}{z^{2n+2}} - \sum_{n=0}^{\infty}\frac{z^n}{2^{n+1}} \qquad \text{for } |z| > 0.$$

9. (a) The expansion near the origin can be found by replacing x in the Maclaurin series for $\exp(x)$ by $1/z$ and then multiplying the result by z^3 to obtain

$$f(z) = z^3 + z^2 + \frac{1}{2}z + \frac{1}{6} + \sum_{n=1}^{\infty}\frac{1}{(n+3)!\,z^n} \qquad \text{for } |z| > 0.$$

(b) The expansion in the neighborhood of the point at infinity follows in the usual way by setting $z = 1/\zeta$ to find a function $f(1/\zeta) = F(\zeta)$, expanding $F(\zeta)$ using its Maclaurin series and then setting $\zeta = 1/z$ to obtain

$$f(z) = \frac{1}{z^3} + \frac{1}{z^4} + \frac{1}{2!\,z^5} + \frac{1}{3!\,z^6} + \frac{1}{4!\,z^7} + \cdots, \qquad \text{for } |z| > 0.$$

11. Because $|z| > \frac{1}{2}$ we write

$$\begin{aligned} f(z) &= \mathrm{Ln}\left(\frac{z^2}{1-4z^2}\right) \\ &= \mathrm{Ln}\left(\frac{1}{1/z^2-4}\right) \\ &= \left[\left(\frac{1}{4}\right)\left(\frac{-1}{1-1/(4z^2)}\right)\right] \\ &\mathrm{Ln}\,\tfrac{1}{4} + \mathrm{Ln}(-1) - \mathrm{Ln}[1-1/(4z^2)]. \end{aligned}$$

The principal branch of the logarithmic function is involved, so writing $-1 = e^{i\pi}$, $\mathrm{Ln}\frac{1}{4} = \ln\frac{1}{4} - \ln 4$ and replacing z by $1/(4z^2)$ in the series expansion of $\mathrm{Ln}(1-z)$, we find that

$$f(z) = -\ln 4 + i\pi + \sum_{n=1}^{\infty} \frac{1}{n4^n z^{2n}} \qquad \text{for } |z| > \tfrac{1}{2}.$$

Solutions 3.6

1. Zero of order 5 at $z = 0$, and simple zeros at $z = 2n\pi i$, for $n = \pm 1, \pm 2, \dots$.
3. Zero of order 2 at $z = 0$ and simple zeros at $z = n\pi/3$, for $n = \pm 1, \pm 2, \dots$.
5. Zero of order 2 at $z = 0$ and simple zeros at $z = n\pi i$, for $n = \pm 1, \pm 2, \dots$ (zeros of $\sinh z$) and simple zeros at $z = n\pi$ for $n = \pm 1, \pm 2, \dots$ (zeros of $\sin z$).
7. Simple poles at $\frac{1}{\sqrt{2}}(1 \pm i)$, $\frac{1}{\sqrt{2}}(\pm 1 + i)$, the fourth roots of -1, while $z = \infty$ is a regular point.
9. $z = 3$ is an essential singularity and $z = \infty$ is a regular point.
11. Simple poles at $z = 2n\pi i$, for $n = 0, \pm 1, \pm 2, \dots$, while $z = \infty$ is a limit point of poles.
13. $z = 0$ is a pole of order 3 and $z = n\pi$ for $n = \pm 1, \pm 2, \dots$, are simple poles, while $z = \infty$ is a limit point of poles.

Solutions 3.7

1. $\mathrm{Res}[f(z), 0] = 1$, $\mathrm{Res}[f(z), \pm 1] = -\frac{1}{2}$.
3. $\mathrm{Res}[f(z), -1] = -28$.
5. $\mathrm{Res}[f(z), -ia] = \frac{ie^a}{2a}$, $\mathrm{Res}[f(z), ia] = -\frac{i}{2ae^a}$.
7. $\mathrm{Res}[f(z), \infty] = 0$.
9. $\mathrm{Res}[f(z), n\pi] = (-1)^n/(n\pi)$ for $n = \pm 1, \pm 2, \dots$. $z = 0$ is a pole of order 2.

11. $\text{Res}[f(z), \exp(i\pi/2n)] = -\exp(i\pi/2n)/2n$.
13. $\text{Res}[f(z), -1] = \exp[(\alpha - 1)\text{Ln}(-1)] = \exp\{(\alpha - 1)[\ln 1 + \pi i]\} = e^{\alpha\pi i}$.

Solutions 3.8

1. Simple poles inside the contour at $z_1 = e^{\pi i/4}$ and $z_2 = e^{3\pi i/4}$.

$$\text{Res}[f(z), z_1] = \text{Res}[f(z), z_2] = -\frac{i}{2\sqrt{2}}, \text{ so}$$

$$\int_C f(z)dz = 2\pi i\left(-\frac{i}{2\sqrt{2}} - \frac{i}{2\sqrt{2}}\right) = \pi/\sqrt{2},$$

where C is the total contour. Hence the result because as $R \to \infty$ $\lim_{|z|\to\infty} |f(z)| = 0$, causing the integral around C_R to vanish.

3. Only the simple pole at $z = i$ is inside the contour, with $\text{Res}[f(z), i] = -\frac{1}{2}i/e^n$. The integral around the semicircular contour vanishes by Jordan's lemma, so

$$\int_{-\infty}^{\infty} \frac{e^{inx}}{x^2+1} dx = 2\pi i\left(-\tfrac{1}{2} i/e^n\right) = \pi/e^n,$$

so equating the real parts gives

$$\int_{-\infty}^{\infty} \frac{\cos nx}{x^2+1} dx = \pi/e^n.$$

The integrand is an even function of x, and so

$$\int_0^{\infty} \frac{\cos nx}{x^2+1} dx = \frac{\pi}{2e^n}.$$

Had imaginary parts been equated the result would have been

$$\int_0^{\infty} \frac{\sin nx}{x^2+1} dx = 0,$$

which is to be expected because this integrand is an odd function.

5. The zeros of $\cosh z$ are at $z_n = (1 \pm 2n)\pi i/2$, so the only simple pole in the strip is at $z_0 = \pi i/2$. $\text{Res}[f(z), z_0] = \lim_{z\to\pi i/2}[(z - \pi i/2)\cosh\frac{1}{3}z/\cosh z] = -i\sqrt{3}/2$. This is an indeterminate form and the result follows by an application of L'Hospital's rule. The integrals over the ends of the strip vanish as $R, S \to \infty$, so

$$\int_{-\infty}^{\infty} \frac{\cosh\frac{1}{3}x}{\cosh x} dx + \int_{\infty}^{-\infty} \frac{\cosh\frac{1}{3}(x+\pi i)}{\cos(x+\pi i)} dx$$
$$= 2\pi i \,\text{Res}[f(z), \pi i/2] = \pi\sqrt{3}.$$

Thus, as $\cosh\frac{1}{3}(x+\pi i) = \frac{1}{2}\cosh\frac{1}{3}x + i\frac{\sqrt{3}}{2}\sinh\frac{1}{3}x$ and $\cosh(x+\pi i) = -\cosh x$,

$$\int_{-\infty}^{\infty} \frac{\cosh\frac{1}{3}x}{\cosh x}dx + \frac{1}{2}\int_{0}^{\infty} \frac{\cosh\frac{1}{3}x}{\cosh x}dx + i\frac{\sqrt{3}}{2}\int_{0}^{\infty} \frac{\sinh\frac{1}{3}x}{\cosh x}dx = \pi\sqrt{3}.$$

Equating the real parts gives

$$\int_{-\infty}^{\infty} (\cosh\tfrac{1}{3}x/\cosh x)dx = 2\pi/\sqrt{3},$$

but the integrand is even, so

$$\int_{0}^{\infty} (\cosh\tfrac{1}{3}x/\cosh x)dx = \pi/\sqrt{3}.$$

7. The poles $z_1 = 0$ and $z_2 = i$ are inside the contour. $\text{Res}[f(z), 0] = 1/(2\pi)$ and $\text{Res}[f(z), i] = e^{-\alpha}/(2\pi)$. The integral at the right-hand end of the strip vanishes in the limit as $R \to \infty$, so taking account of integrating around the two quarter circle indentations in the negative sense and letting their radii tend to zero, we are left with the result

$$\int_{0}^{\infty} \frac{e^{i\alpha x}}{e^{2\pi x}-1}dx + \int_{\infty}^{0} \frac{e^{i\alpha(x+iy)}}{e^{2\pi(x+i)}-1}dx + \int_{1}^{0} \frac{e^{i\alpha(iy)}}{e^{2\pi iy}} idy$$
$$- \frac{2\pi i}{4}\left(\frac{1}{2\pi} + \frac{e^{-\alpha}}{2\pi}\right) = 0,$$

and so

$$(1-e^{-\alpha})\int_{0}^{\infty} \frac{e^{i\alpha x}}{e^{2\pi x}-1}dx = \tfrac{1}{4}(1+e^{-\alpha}) + i\int_{0}^{1} \frac{e^{-\alpha y}}{e^{2\pi iy}-1}dy.$$

Simplifying the last shows that

$$i\int_{0}^{1} \frac{e^{-\alpha y}}{e^{2\pi iy}-1}dy = \tfrac{1}{2}\int_{0}^{1} e^{-\alpha y}\cot\pi y dy - \frac{i}{2a}(1-e^{-\alpha}),$$

and so

$$\int_{0}^{\infty} \frac{e^{i\alpha x}}{e^{2\pi x}-1}dx = \tfrac{1}{4}\left(\frac{1+e^{-\alpha}}{1-e^{-\alpha}}\right) + \frac{1}{2(1-e^{-\alpha})}\int_{0}^{1} e^{-\alpha y}\cot\pi y dy - \frac{i}{2\alpha},$$

from which the result follows by equating imaginary parts.

9. $f(z)$ has a simple pole at $z = 0$, where $\text{Res}[f(z), 0] = 1$. By Jordan's lemma the integral around the semicircle vanishes as $R \to \infty$, so that

$$\lim_{\varepsilon\to 0}\int_{-\infty}^{-\varepsilon} f(x)dx - \left(\tfrac{1}{2}\right) \times \text{Res}[f(z), 0] + \lim_{\varepsilon\to 0}\int_{\varepsilon}^{\infty} f(x)dx = 0,$$

giving

$$\int_{-\infty}^{\infty} \frac{\sin \alpha x}{x}\, dx = \pi,$$

but the integrand is an even function, so

$$\int_{0}^{\infty} \frac{\sin \alpha x}{x}\, dx = \tfrac{1}{2}\pi.$$

11. The only simple pole of the integrand inside the contour is as $z = i$, where

$$\mathrm{Res}[f(z), i] = -\frac{i(2n-2)!}{2^{2n-1}[(n-1)!]^2}$$

leading to the result

$$\int_{-\infty}^{\infty} \frac{dx}{(1+z^2)^n} = \frac{\pi(2n-2)!}{2^{2n-2}[(n-1)!]^2}.$$

13. Here $\alpha = 1/3$ and $R(z) = 1/(z^2 + z + 1)$, showing that as the degree of the numerator of $R(z)$ is less than that of the denominator, the result of Exercise 12 is applicable. There two simple poles, one at $z_+ = e^{2\pi i/3}$ and the other at $z_- = e^{4\pi i/3}$. Because the principal value of $z^{-1/3}$ is to be used,

$$\mathrm{Res}[z^{-1/3}R(z), z_+] = \frac{e^{2\pi i/9}}{e^{2\pi i/3} - e^{4\pi i/3}},$$

and

$$\mathrm{Res}[z^{-1/3}R(z), z_-] = \frac{e^{4\pi i/9}}{e^{4\pi i/3} - e^{2\pi i/3}}.$$

So,

$$\int_{0}^{\infty} \frac{z^{-1/3}}{z^2 + z + 1}\, dz = \left(\frac{2\pi i}{1 - e^{-2\pi i/3}}\right) \times \sum(\text{sum of residues}) = \frac{4\pi}{3}\sin\left(\tfrac{1}{9}\pi\right).$$

15. The only simple pole inside the contour occurs at $z = ia$, so as the principal branch of the logarithmic function is involved we must set $z = ae^{\pi i/2}$.

$$\mathrm{Res}[f(z), ae^{\pi i/2}] = \left(\frac{\ln a + \pi i/2}{2ia}\right)^2 = R \text{ (say)}.$$

The integrals around C_R and C_ε vanish for the usual reasons, so as $z = x$ on the positive real axis and $z = xe^{\pi i}$ on the negative real axis, in the limit as $R \to \infty$ and $\varepsilon \to 0$ we find that

$$\int_{0}^{\infty} \frac{(\ln x)^2}{x^2 + a^2}\, dx + \int_{\infty}^{0} \frac{(\ln a + \pi i/2)^2 e^{\pi i}}{x^2 + a^2}\, dx = 2\pi i R.$$

The required results follow by using the standard result

$$\int_0^\infty 1/(x^2 + a^2)dx = \pi/(2a),$$

and equating real and the imaginary parts.

17. Proceed as in Example 3.8.6.
19. Let $f(z) = 11$, and $g(z) = z^5 - 8z^2$. Now $|g(z)| \leq |z^5| + |-8z^2| = |z|^5 + 8|z|^2$, so on $|z| = 1$, we have $|g(z)| < 1 + 8 = 9 < 11 = |f(z)|$. Thus from Rouché's theorem $P(z) = f(z) + g(z)$ has the same number of zeros inside $|z| = 1$ as $f(z)$, namely zero.
21. The results are self checking.

Solutions 3.9

1. Self checking.
3. $f(t) = (t - a)^2 \exp(bt)$.
5. $f(t) = \dfrac{1}{2a^3}(\sin at - at \cos at)$
7. Using the contour in Figure 3.22 as in Example 3.9.1, and the fact that the principal branch of the square root function is to be used, we set $s = re^{i\pi}$ on BC and $s = re^{-i\pi}$ on DE. The residue is zero at $s = 0$, and the integrals around $C_R^{(\pm)}$ vanish as $R \to \infty$ by virtue of Lemma 3.9.1, so because no singularities are found inside the contour, in the limit as $R \to \infty$ and $\varepsilon \to 0$, we are left with

$$\frac{1}{2\pi i}\int_{c-i\infty}^{c+i\infty} e^{st}e^{-a\sqrt{s}}ds + \frac{1}{2\pi i}\left\{\int_0^\infty e^{-rt}e^{-ia\sqrt{r}}dr + \int_0^\infty e^{-rt}e^{ia\sqrt{r}dr}\right\} = 0,$$

and so

$$\begin{aligned} f(t) &= -\frac{1}{2\pi i}\left\{\int_0^\infty e^{-rt}e^{-ia\sqrt{r}}dr + \int_0^\infty e^{-rt}e^{ia\sqrt{r}dr}\right\} \\ &= \frac{1}{\pi}\int_0^\infty e^{-rt}\sin a\sqrt{r}dr. \end{aligned}$$

The change of variable $r = u^2$ gives the required inverse transform

$$\mathcal{L}^{-1}\left\{e^{-a\sqrt{s}}\right\} = \frac{2}{\pi}\int_0^\infty ue^{-u^2t}\sin(au)du,$$

in which a and t appear as parameters in the integrand. After some effort, this last integral can be evaluated to give

$$f(t) = \mathcal{L}^{-1}\left\{e^{-a\sqrt{s}}\right\} = a(4\pi t^3)^{-1/2}e^{-a^2/4t},$$

showing $f(t)$ to be another function with a Laplace transform which is *not* of exponential order.

Solutions 4.1

1. Maps to $v = \frac{1}{3}(1 - u)$
3. A similar parabola: unit magnification, rotation $-\pi/3$, translation -1.
5. $w = \frac{5}{2}(1 - i\sqrt{3})z - 4 + i(5\sqrt{3} - 2)$
7. (i) $w = \left(\dfrac{\delta}{\beta - \alpha}\right)(z - \alpha)$, maps points onto the strip for which $\text{Im}\{w\} > 0$.

   (ii) $w = \left(\dfrac{\delta}{\beta - \alpha}\right)(\beta - z)$, maps points onto the strip for which $\text{Im}\{w\} < 0$.
9. $w = \frac{1}{6}(1 - i\sqrt{3})z + \frac{1}{6}(5 - \sqrt{3}) + i(5 + \sqrt{3})$. Use the fact that a contraction factor of $1/3$ and a rotation of $-\pi/3$ is produced by the constant $a = \frac{1}{6}(1 - i\sqrt{3})$ in a linear transformation $w = az + b$. The fixed point $z = b/(1 - a)$, so as a is known and the fixed point $z = 1 + i$, the value of b follows from $b = (1 - a)(1 + i)$.

Solutions 4.2

1. Part of the parabola $y = 4x^2$ with the arrow directed away from the point $(-3, 16)$, that is away from $z = -3 + 16i$.
3. Rectangular hyperbola $y = 1/x$ in first quadrant, with arrow directed down the curve.
5. Single arc of cycloid $x = a(t - \sin t)$, $y = a(1 - \cos t)$ with arrow directed away from the origin.
7. arctan 8
9. $\arctan \dfrac{\sqrt{2}}{2 - \sqrt{2}}$
11. (a) $\rho(z) = 2|\cos 2z|$; $\rho(0) = 2$; (b) $\rho(z) = e^x$: $\rho(2 + i) = e^2$; (c) $\rho(z) = |1/z|$: $\rho(1 + i\sqrt{3}) = \frac{1}{2}$; (d) $\rho(z) = 4|\sinh 4z|$; $\rho(i) = 4\sin 4$
13. Maps to the region $\dfrac{1}{r_2} < \rho < \dfrac{1}{r_1}$ with $-\pi/3 \leqslant \varphi \leqslant -\pi/4$ in the w-plane.

Solutions 4.3

1. $w_1 = z + 1$ (translation), $w_2 = 1/w_1$ (inversion), $w_3 = w = 4iw_2 - 2i$ (magnification, a positive rotation by $\pi/2$, and a translation) [Hint: To find w_3, set $w_3 = Aw_2 + B = 2i(1 - z)/(1 + z)$ and solve for A and B.]
3. $w_1 = z$ (identity mapping), $w_2 = 1/w_1$ (an inversion), and $w_3 = w = 1 - w_2$ (no magnification, a rotation through π, and a translation).

5. (i) $z = 1$ (twice) (ii) $z = \frac{1}{\sqrt{2}}(1 \pm i)$ (iii) $z = \frac{1}{8}(1 \pm i\sqrt{15})$
7. Boundary of $|z| = 1$ maps to $|w - i| = 1$, with the interior of the unit circle mapping to the exterior of $|w - i| = 1$.
9. The circle $|z| = 1$ maps onto the real w-axis, and the circle $|z| = \frac{1}{2}$ maps onto the circle $u^2 + v^2 - \frac{5}{3}v + \frac{1}{4} = 0$ in the w-plane. [Hint: Solve for z and then find $z\bar{z} = x^2 + y^2$ when determining the images of the boundaries.]
11. $w = 3i(z + 1)/(3 - z)$.
13. $w = z/(2 - z)$.
15. $w = [2(1 - i)z + 8i]/z$: the strip $-4 < \text{Im}\{w\} < 0$.
17. $\alpha = 2\sqrt{2}$, when the cusp on the boundary will be located at $z = 1$, which maps to $w = \infty$; both map onto the wedge $0 < \text{Arg}\{w\} < \arctan(2\sqrt{2})$, although each in a different way.
19. $w = i(z - 2i)/(z + 2i)$.

Solutions 4.4

1. (i) Onto the upper half of the w-plane. (ii) Onto the lower half of the w-plane. Cut along the negative real axis in the w-plane if the strip $-\pi/a \leqslant y \leqslant \pi/a$ is to map one-one onto the w-plane.
3. The complete strip maps onto the first quadrant of the w-plane. Strip (i) maps onto the interior of $|w| = 1$ lying within that quadrant, with strip (ii) mapping onto the remainder of the quadrant.
5. The hyperbolic strip in the right half of the w-plane between the hyperbolas in Equation (4.62) with $a = -1$ and $a = 2$, and the ellipses in Equation (4.63) with $b = \pi/6$ and $b = \pi/3$.
13. (i) The w-plane cut from $w = -1$ to $w = \frac{1}{2}(a + 1/a)$ when $0 < a < 1$, and (ii) the w-plane with cuts from $w = -\infty$ to $w = \frac{1}{2}(a + 1/a)$ and from when $-1 < a < 0$.

Solutions 4.5

1. $B\ (x_1 = 0) \to B'\ (w_1 = 0, \theta_1 = \pi - \alpha)$; $C\ (x_2 = 1) \to C'\ (w_2 = 1, \theta_2 = 0)$

$$w = f(z) = A\int z^{\alpha/\pi - 1} \cdot 1\,dz + B, \qquad \text{so } w = \frac{A\pi}{\alpha} z^{\alpha/\pi} + B.$$

$z = x_1 = 0$ maps to $w_1 = 0$, so $B = 0$. $z = x_2 = 1$ maps to $w_2 = 1$, so $A = \alpha/\pi$, and so $w = z^{\alpha/\pi}$.

3. $B\ (x_1 = -1) \to B'\ (w_1 = ia, \theta_1 = -\pi/2)$; $C\ (x_2 = 1) \to C'\ (w_2 = ia, \theta_2 = \pi/2)$

$$w = f(z) = A\int (z + 1)^{1/2}(z - 1)^{-1/2} dz + B,$$

so that

$$w = A\left[\sqrt{z^2 - 1} + \text{Ln}\left(z + \sqrt{z^2 - 1}\right)\right] + B.$$

$z = x_1 = -1$ maps to $w_1 = ia$, so $ia = A\,\mathrm{Ln}(-1) + B = Ai\pi + B$. $z = x_2 = 1$ maps to $w_2 = 0$, so $0 = a\ln_e(1) + B$ showing that $B = 0$. Hence

$$w = \frac{a}{\pi}\left[\sqrt{z^2-1} + \mathrm{Ln}\left(z + \sqrt{z^2-1}\right)\right].$$

5. $B\,(x_1 = -1) \to B'\,(w_1 = 0, \theta_1 = \pi/2); C\,(x_2 = 0) \to C'\,(w_2 = ia, \theta_2 = -\pi)$; $D\,(x_3 = 1) \to D'\,(w_3 = 0,\ \theta_3 = \pi/2)$. Thus $w = f(z) = A\int(z+1)^{-1/2} z(z-1)^{-1/2}dz + B$ and so $w = A\sqrt{z^2-1} + B$. $z = x_1 = -1$ maps to $w_1 = 0$, showing that $B = 0$. The conditions at D and D' are consistent and provide no more information because they also show that $B = 0$. $z = x_2 = 0$ maps to $w_2 = ia$ showing that $ia = Ai$, when $A = 1$ and thus

$$w = a\sqrt{z^2-1}.$$

7. $B\ (x_1 = -1) \to B'\ (w_1 = i\pi,\ \theta_1 = -\pi/2)$; $C\ (x_2 = 0) \to C'\ (w_2 = \infty, \theta_2 = \pi)$

$$w = A\int (z+1)^{1/2} z^{-1}\,dz + B,$$

and so

$$w = A\left[2\sqrt{z+1} + \mathrm{Ln}\left(\frac{\sqrt{z+1}-1}{\sqrt{z+1}+1}\right)\right] + B.$$

$z = x_1 = -1$ maps to $w_1 = ia$, showing that $i\pi = A\,\mathrm{Ln}(-1) + B = Ai\pi + B$. Setting $w = u + iv$ and writing $A = A_1 + iA_2$ and $B = B_1 + iB_2$ gives

$$w = (A_1 + iA_2)\left[2\sqrt{(z+1)} + \mathrm{Ln}\left(\frac{\sqrt{z+1}-1}{\sqrt{z+1}+1}\right)\right] + B_1 + iB_2.$$

On DE_∞ $v = 0$ for $x > 0$, so after finding v and equating the result to zero gives

$$0 = A_2\left[2\sqrt{x+1} + \mathrm{Ln}\left(\frac{\sqrt{x+1}-1}{\sqrt{x+1}+1}\right)\right] + B_2.$$

However, this must be true for all $x > 0$, and this is only possible if $A_2 = B_2 = 0$. Using these results in the expression for w gives

$$w = 2\sqrt{z+1} + \mathrm{Ln}\left(\frac{\sqrt{z+1}-1}{\sqrt{z+1}+1}\right).$$

Solutions 5.2

1. $\phi = 1 + (x - 2)$.
3. $\phi = 3 + \sqrt{2}(x - y)$.
5. $\phi = 3 + 6\theta/\pi$.
7. $\phi = 3 + 8\theta/\pi$.
9. $\phi = 4 + \frac{7}{\pi}\textbf{Arctan}(y/x)$. Neither even nor odd.

11. $$\phi = \frac{1}{\pi}\left[-\textbf{Arctan}\left(\frac{y}{x+2}\right) + 2\textbf{Arctan}\left(\frac{y}{x-1}\right) + \textbf{Arctan}\left(\frac{y}{x-4}\right)\right].$$

    The solution is neither even nor odd.

13. $$\phi = 3 - \frac{4}{\pi}\textbf{Arctan}\left(\frac{1 - x^2 - y^2}{2y}\right).$$

15. $$\phi = -1 + \frac{2}{\pi}\left[\textbf{Arctan}\left(\frac{v}{u-2}\right) - \textbf{Arctan}\left(\frac{v}{u}\right)\right], \text{ with}$$

    $$u = \frac{2x + x^2 + y^2}{x^2 + y^2 + 2x - 2y + 2}, \quad v = \frac{2y - x^2 - y^2}{x^2 + y^2 + 2x - 2y + 2}.$$

17. $$\phi = 3 - \frac{4}{\pi}\textbf{Arctan}(\cot x \tanh y).$$

19. The quarter circular annulus maps onto an elliptical annulus in the first quadrant of the w-plane between ellipses

    $$\frac{u^2}{(5/2)^2} + \frac{v^2}{(3/2)^2} = 1 \quad \text{and} \quad \frac{u^2}{(10/3)^2} + \frac{v^2}{(8/3)^2} = 1.$$

    The Dirichlet conditions are $\phi = 3$ on the inner boundary and $\phi = 5$ on the outer one. Homogeneous Neumann conditions are found on the segments of the x- and y-axes between the ellipses. Within the quarter elliptical domain the level curves of the solution curves follow from Equation (5.44) for $2 < R < 3$, on which the solution is

    $$\phi = 3 - \frac{2\ln_e 2}{\ln_e(3/2)} + \frac{2}{\ln_e(3/2)}\ln_e R.$$

21. $$|\phi_1 - \phi_2| = \frac{y}{\pi}\left|\int_{-\infty}^{\infty}\frac{[f_1(s) - f_2(s)]}{(x-s)^2 + y^2}ds\right| \leqslant \frac{y}{\pi}\int_{-\infty}^{\infty}\left|\frac{f_1(s) - f_2(s)}{(x-s)^2 + y^2}\right|ds.$$

    However, $|f_1(s) - f_2(s)| < \varepsilon$, so using this result and evaluating the integral gives

    $$|\phi_1 - \phi_2| < \frac{y\varepsilon}{\pi}\int_{-\infty}^{\infty}\frac{ds}{(x-s)^2 + y^2} = \frac{y\varepsilon}{\pi}\frac{\pi}{y} = \varepsilon.$$

Thus if the absolute value of the difference between the two boundary conditions is less than ε, so also is the absolute value of the difference between the two solutions. This establishes the *continuous dependence* of the solution on the boundary condition.

23. Reason as in Solution 21.
25. Set $w(r, \theta) = u(r, \theta) - v(r, \theta)$. Then $w(r, \theta)$ is harmonic in the disk $0 < r < r_0$ and $w(r_0, \theta) = f(\theta) - g(\theta)$ for $\alpha \leqslant \theta \leqslant \beta$, and zero on the rest of the boundary. From Theorem 5.2.2

$$|w(\tilde{r}, \tilde{\theta})| = \frac{1}{2\pi}\left|\int_\alpha^\beta \frac{(r_0^2 - \tilde{r}^2)[f(\chi) - g(\chi)]}{r_0^2 - 2\tilde{r}r_0\cos(\chi - \theta) + \tilde{r}^2}\,d\chi\right|$$
$$\leqslant \frac{1}{2\pi}\left(\frac{r_0 + \tilde{r}}{r_0 - \tilde{r}}\right)\int_\alpha^\beta |f(\chi) - g(\chi)|\,d\chi.$$

Thus the difference between the boundary conditions over the arc $r = r_0$ with $\alpha \leqslant \theta \leqslant \beta$ is seen to influence the solution throughout the disk, however small the arc may be. This result also shows that the influence on the solution of the change of the boundary condition on the arc is not necessarily small.

Solutions 5.3

1. Set $m = 150$, $M = 350$ to get

$$T(u, v) = 50 + \frac{1}{\pi}\left[\textbf{Arctan}\left(\frac{v}{u+2}\right) + 300\textbf{Arctan}\left(\frac{v}{u-1}\right)\right].$$

3. $T(x, y) = 180 - 40(x - y + 6)$, for (x, y) in the strip.

5. $T(x, y) = 35 + \frac{130}{\pi}\textbf{Arctan}\left(\frac{y}{x}\right)$, for (x, y) in the strip.

7. $T(x, y) = 30 + \frac{1}{\pi}\left[\textbf{Arctan}\left(\frac{2xy}{x^2 - y^2 + 4}\right) - 10\textbf{Arctan}\left(\frac{2xy}{x^2 - y^2 - 9}\right)\right]$,

for (x, y) in the first quadrant.

9. $T(x, y) = 20 + \frac{260}{\pi}\textbf{Arctan}\left(\frac{2y}{1 - x^2 - y^2}\right)$ for (x, y) in the half disk.

11. $T(x, y) = \frac{1}{\pi}\left[-80\textbf{Arctan}\left(\frac{v}{u+1}\right) + 100\textbf{Arctan}\left(\frac{v}{u-1}\right)\right]$, with

$$u = \frac{\operatorname{sgn}(x)}{\sqrt{2}}\left[x^2 - y^2 + 1 + \sqrt{(x^2 - y^2 + 1)^2 + 4x^2y^2}\right]^{1/2}, \quad v = \frac{xy}{u}$$

for (x, y) in the cut half plane.

13. $T(x, y) = 125 + \frac{250}{\pi}\,\textbf{Arctan}\left(\frac{2y}{1-x^2-y^2}\right)$ for (x, y) in the upper half disk,

$$T(x, y) = -125 + \frac{250}{\pi}\,\textbf{Arctan}\left(\frac{2y}{1-x^2-y^2}\right) \text{ for } (x, y) \text{ in the lower half disk.}$$

15. $$T(x, y) = 20 + \frac{1}{\pi}\left[30\textbf{Arctan}\left(\frac{\sin y(\pi x/a)\sin(\pi y/a)}{\cosh(\pi x/a)\cos(\pi y/a)+1}\right)\right]$$

$$-\frac{20}{\pi}\,\textbf{Arctan}\left(\frac{\sinh(\pi x/a)\sin(\pi y/a)}{\cosh(\pi x/a)\cos(\pi y/a)-1}\right),$$

for (x, y) in the semi-infinite strip. $T(x, y) \approx 20 + 10y/a$ for $x \gg 0$, $0 \leqslant y \leqslant a$.

17. $T(x, y) = 60 - \frac{45}{4\sqrt{3}}(x + \sqrt{3}y)$ for (x, y) in the rectangle.

19. $$T(x, y) = 80 + \frac{60}{\pi}\left[\textbf{Arctan}\left(\frac{1-x^2-y^2}{2y+(1+x)^2+y^2}\right)\right]$$

$$-\frac{60}{\pi}\,\textbf{Arctan}\left[\frac{1-x^2-y^2}{2y+(1+x)^2-y^2}\right] \quad \text{for } (x, y) \text{ in the half disk.}$$

21. $T(x, y) = \frac{y}{\pi}\int_{-\infty}^{\infty}\frac{a\cos ms + b\sin ns}{(x-s)^2+y^2}ds$. Setting $u = x - s$ this becomes

$$T(x, y) = -\frac{y}{\pi}\int_{\infty}^{-\infty}\frac{a\cos m(x-u) + b\sin n(x-u)}{u^2+y^2}du$$
$$= \frac{y}{\pi}\int_{-\infty}^{\infty}\frac{a\cos m(x-u) + b\sin n(x-u)}{u^2+y^2}du.$$

Expanding $\cos m(x-u)$ and $\sin n(x-u)$, using the fact that $\sin mu$ and $\sin nu$ are odd functions so the integrals of $\sin mu/(u^2+y^2)$ and $\sin nu/(u^2+y^2)$ with respect to u over $(-\infty, \infty)$ must vanish gives

$$T(x, y) = \frac{ya\cos mx}{\pi}\int_{-\infty}^{\infty}\frac{\cos mu}{u^2+y^2}du + \frac{yb\sin nx}{\pi}\int_{-\infty}^{\infty}\frac{\cos nu}{u^2+y^2}du.$$

Because $\int_{-\infty}^{\infty}\frac{\cos \alpha u}{u^2+y^2}du = \frac{\pi e^{-\alpha y}}{y}$, this becomes $T(x, y) = ae^{-my}\cos mx + be^{-ny}\sin nx$.

Solutions 5.4

1. $\frac{m}{2\pi}\operatorname{Ln}(z-1) - \frac{m}{2\pi}\operatorname{Ln}(z+1) = \frac{m}{2\pi}\operatorname{Ln}\left(\frac{z-1}{z+1}\right).$

$$\psi = \frac{m}{2\pi}\textbf{Arctan}\left(\frac{y}{x-1}\right) - \frac{m}{2\pi}\textbf{Arctan}\left(\frac{y}{x+1}\right).$$

The streamlines are circles passing through $z = 1$ and $z = -1$, with their centers on the perpendicular bisector of $z = 1$ and $z = -1$.

3. $w = Uze^{-i\alpha} - m\operatorname{Ln} z$. $z = re^{i\theta}$ gives $\psi = Ur\sin(\theta - \alpha) - m\theta$. Stagnation point at $z = (m/U)e^{i\alpha}$.

5. $w = m\operatorname{Ln}(z-a) + m\operatorname{Ln}(z+a) = m\operatorname{Ln}(z^2 - a^2)$. One stagnation point at $z = 0$. Stream function

$$\psi = m\,\textbf{Arctan}\left(\frac{2xy}{x^2 - y^2 - a^2}\right).$$

these are rectangular hyperbolas about each line source as a focus.

7. $w = i\kappa \operatorname{Ln}\left(\frac{z^2 + 2i}{z^2 - 2i}\right)$. One stagnation point at $z = 0$. Instantaneous velocity of vortex at $x = 1 + i$ has components $q_1 = \kappa/4$ and $q_2 = -\kappa/4$. Symmetry shows that the x- and y-axes are streamlines. Flow due to a single vortex is the flow in the first quadrant.

9. $w = \frac{m}{2\pi}\operatorname{Ln}(z^n - z_0^n) + \frac{m}{2\pi}\operatorname{Ln}(z^n - z_0^{-n})$. $|dw/dz| \to 0$ as $z \to \infty$.

11. The streamlines are $y = \alpha U\left[\frac{1 + \alpha^2 + x^2}{\alpha^2 + x^2}\right]^{1/2}$.

13. $w = \sin az$. Streamlines $\psi = U\cos ax \sinh ay = \text{const}$. On the streamline $\psi = cU$, the speed

$$q = \left(\cos^2 ax + \frac{c^2}{\cos^2 ax}\right)^{1/2}.$$

By Bernoulli's theorem the maximum pressure occurs when the speed is minimum. The maximum pressure on this streamline occurs when $x = \pm(1/a)\text{Arccos}\sqrt{c}$ for $c < 1$ and when $x = 0$ for $c \geqslant 1$.

15. $w = Uze^{-i\alpha} + Ue^{i\alpha}\left(\frac{a^2}{z}\right) + \frac{m}{2\pi}\operatorname{Ln}(z - 2ia) + \frac{m}{2\pi}\operatorname{Ln}\left(\frac{a^2}{z} - 2ia\right)$

$$- \frac{m}{2\pi}\operatorname{Ln}(z + 3a) - \frac{m}{2\pi}\operatorname{Ln}\left(\frac{a^2}{z} - 3a\right).$$

17. The stagnation points are determined by the condition $dw/dz = 0$, which occur at the roots of the quadratic equation $z^2/a^2 + (z/a)(i\kappa/aU) - 1 = 0$. Thus the stagnation points $z_{\pm}$ are located at

$$z_{\pm} = a\left(-\frac{i\kappa}{2aU} \pm \left(1 - \frac{\kappa^2}{4a^2U^2}\right)^{1/2}\right).$$

The stagnation points are real and distinct if $\kappa < 2aU$, they coalesce if $\kappa = 2aU$, and they occur as complex conjugates if $\kappa > 2aU$. Setting $\sin\alpha = \kappa/2aU$ it follows that $z = a(\pm\cos\alpha - i\sin\alpha)$ lies on $|z| = a$. Thus when $\kappa < 2aU$ the stagnation points are distinct and lie on the surface of the cylinder, symmetrically spaced relative to the y-axis, when $\kappa = 2aU$ they coalesce on the surface of the cylinder on the y-axis, and when $\kappa > 2aU$ one stagnation point (nonphysical) lies inside the cylinder and the other outside it, with each located on the y-axis. Setting $z = ae^{i\theta}$, the speed of flow q on the cylinder is determined from the result $dw/dz = q_1 - iq_2$, so as $q^2 = q_1^2 + q_2^2$ we have $q^2 = (2U\sin\theta + \kappa/a)^2$, and from Bernoulli's theorem the pressure p is then given by

$$p = p_\infty + \tfrac{1}{2}\rho U^2 - \tfrac{1}{2}\left(4U^2\sin^2\theta + (4\kappa U/a)\sin\theta + \kappa^2/a^2\right).$$

An inspection of the pressure distribution around the cylinder suggests that the net horizontal force is zero, while there is a net upward force, showing there is lift on the cylinder. This can be established analytically by first integrating the horizontal components of the forces acting on a unit length of the cylinder to find the net horizontal force H, and then by integrating the vertical components of the forces acting on a unit length of the cylinder to find the net vertical force V acting on the cylinder. As a result it is found that $H = 0$ and $V = 2\kappa\rho U$, so when the circulation κ is positive the force on the cylinder is upward generating lift, but when it is negative the force on the cylinder acts vertically downward.

Solutions 5.5

1. $\phi = 20 + \left(\dfrac{100}{\pi}\right)\textbf{Arctan}\left(\dfrac{y}{x}\right)$, (x, y) in first quadrant.

3. $\phi = 30 + \dfrac{1}{\pi}\left[20\textbf{Arctan}\left(\dfrac{y}{x+1}\right) - 30\textbf{Arctan}\left(\dfrac{y}{x-1}\right)\right]$, (x, y) in upper half plane.

5. $\phi = -100 - \dfrac{700}{\pi}\text{Arcsin}\left(\dfrac{\sqrt{(x+1)^2 + y^2} - \sqrt{(x-1)^2 + y^2}}{2}\right)$.

7. $\phi = \dfrac{100}{\pi}\left[\textbf{Arctan}\left(\dfrac{\cos x \sinh y}{\sin x \cosh y - 1}\right) - \textbf{Arctan}\left(\dfrac{\cos x \sinh y}{\sin x \cosh y + 1}\right)\right]$, (x, y) in the semi-infinite strip.

9. $w = -\dfrac{4e}{\varepsilon}\,\mathrm{Ln}(z-2) + \dfrac{2e}{\varepsilon}\,\mathrm{Ln}(z+1)$.

11. The result follows directly from the complex potential $w = -2e\,\mathrm{Ln}(z - z_1) + 2e\,\mathrm{Ln}(z - z_2)$. The equipotentials are circles that pass through the points z_1 and z_2 with their centers on the perpendicular bisector of the line through z_1 and z_2.

13. Use the transformation $w = z^4$ to map the octant onto the half plane.

$$w = -\frac{2e}{\varepsilon}\,\mathrm{Ln}(z^4 - z_0^4) + \frac{2e}{\varepsilon}\,\mathrm{Ln}(z^4 - (\overline{z}_0)^4).$$

Bibliography and Suggested Reading List

Basic Complex Analysis

Churchill, R. V., Brown, J. W., and Verhey, R., *Complex Variables with Applications*, 5th ed., New York, McGraw-Hill, 1990.

Fulks, W., *Complex Variables*, New York, Marcel Dekker, 1993.

Levinson, N. and Redheffer, R. M., *Complex Variables*, San Francisco, Holden-Day, 1970.

Marsden, J. E., *Basic Complex Analysis*, San Francisco, W. H. Freeman, 1973.

Matthews, J. H. and Howell, R. W., *Complex Analysis for Mathematics and Engineering*, Sudbury, MA, Jones and Bartlett, 1997.

Paliouras, J. D. and Meadows, D. S., *Complex Variables for Scientists and Engineers*, 2nd ed., New York, Macmillan, 1990.

Pennisi, L. L., *Elements of Complex Variables*, 2nd ed., New York, Holt, Rinehart and Winston, 1976.

Rubenfeld, L. A. A., *First Course in Applied Complex Variables*, New York, McGraw-Hill, 1976.

Saff, E. B. and Snider, A. D., *Fundamentals of Complex Analysis for Mathematics, Science and Engineering*, 2nd ed., Upper Saddle River, NJ, Prentice Hall, 1993.

Shilov, G. E., *Applied and Complex Analysis*, New York, Dover Reprint, 1996.

Wunch, D. A., *Complex Variables with Applications*, Reading, MA, Addison-Wesley, 1994.

Zill, D. G. and Shanahan, P. D., *A First Course in Complex Analysis with Applications*, Sudbury, MA, Jones and Bartlett, 1997.

Advanced and Reference Books on Complex Analysis

Ahlfors, L., *Complex Analysis*, 2nd ed., New York, McGraw-Hill 1966.

Bieberbach, L., *Conformal Mapping*, 4th ed., New York, Chelsea Reprint, 1954.

Carathedory, C., *Theory of Functions*, Volumes I and II, New York, Chelsea Reprint, 1954.

Henrici, P., *Applied and Computational Complex Analysis*, Volume 1, New York, Wiley, 1974.

Hille, E., *Analytic Function Theory*, Volumes I and II, Boston, Chelsea Reprint, 1973.

Markushevich, A.L., *Theory of Functions of a Complex Variable*, Upper Saddle River, NJ, Prentice Hall, 1965.

Noguchi, J., *Introduction to Complex Analysis*, Providence, RI, American Mathematical Society, Translations of Mathematical Monographs Volume 168, 1998.

Rudin, W., *Real and Complex Analysis*, 2nd ed., New York, McGraw-Hill, 1974.

Tutschke, W. and Vasudeva, H. L., *An Introduction to Complex Analysis: Classical and Modern Approaches*, Boca Raton, FL, Chapman and Hall/CRC Press, 2005.

Standard Reference Works on Applications of Complex Analysis to Physical Problems

Duffy, D. G., *Transform Methods for Solving Partial Differential Equations*, 2nd ed., Boca Raton, FL, Chapman and Hall/CRC Press, 2004.

Carslaw, H. S. and Jaeger, J. C., *Conduction of Heat in Solids*, London, Oxford University Press, 1948.

Muskhelishvili, N. I., *Some Basic Problems in the Mathematical Theory of Elasticity*, Groningen, Wolters-Noordhoff, 1963.

Sneddon, I. N., *The Use of Integral Transforms*, New York, McGraw-Hill, 1972.

Index

Absolute convergence, complex series, 213–214
  scaling and addition, 215
Addition, complex series, 215
Air foil-like shapes, conformal mapping, Joukowski transformation, 389–390
Algebraic functions, Laurent series, at infinity point, 274–276
Analytic functions. *See also* specific operations, e.g. Complex integration
  approach-based limits, 41–42
  Cartesian and polar form, 35–36
  Cauchy integral formula, derivatives, 138–140
  Cauchy–Riemann equations, 47–50
    harmonic conjugate, 61
    harmonic function, 58
    Laplacian operator, 58–59
    necessity of, 54–56
    orthogonal trajectories, 59–60
    polar form, 56–61
    proof and consequences, 53–63
    sufficiency conditions, 55–56
  complex functions
    identities and properties, 85–86
    limits, 41
    origin differentiability, 46–47
  complex numbers, 1–27
    complex conjugate, 2
  continuity, defined, 42–43
  curves, domains, and regions, 27–34
  derivatives, 81–86
    differentiation rules, 43–45, 82
    elementary functions, 82–83
    harmonic functions, 82
    hyperbolic functions, 84
    trigonometric functions, 83
  differentiation
    derivatives, 43–45, 82
    rules, 50
  disk-defined function, 36–37
  elementary functions, 63–86
    complex sine functions, roots of, 75
    derivatives, 82–83
    entire functions, 45–46
    exponential function, 66–67
    hyperbolic functions, 75–81
      cosine function roots, 77
      defined, 84–85
      inverse, 77–80
      inverse sine and derivatives, 80–81
    inverse trigonometric/hyperbolic functions, 77–80, 84–85
    logarithmic function, 67–73
    partial fraction expansion, repeated zeros, 65–66
    polynomials, 64
    rational functions, 64–65
    trigonometric functions, 73–75
      defined, 83
      inverse, 77–80, 84–85
  geometrical aspects, 333–334
  L'Hospital's rule, 50–53
  limit definition, 38–39
  limits and continuity, 37–38
  Taylor's theorem, series integration, 246–247
Annular domain of convergence
  boundary value problems
    Dirichlet condition, 433–435
    mixed boundary value problem transformation, 435–437

Annular domain of convergence (*Contd.*)
Laurent series, 257–258
Cauchy inequality, 278
expansion formula, 264–266
nonsingular point, 268–270
nonsingular point expansion, 270
origin expansion, 267–268
point at infinity, 275–276
singularities, 271–273
uniqueness, 263–264
Laurent's theorem, 260–263
Annulus region
conformal mapping
eccentric circles onto, 375–377
straight line and circle onto, 372–374
Laurent series, 257–258
Anomalous mapping, critical points, 355–356
Antiderivatives
definite integrals, 120–122
branched function, 125–126
logarithmic function, 126–127
simple solution, 124–125
indefinite integrals, 122–124
Arc
contour integrals, vanishing conditions, 181–182
defined, 27
Arcsin *z/a,* conformal mapping, 383–385
Arctan
boundary value problems
Dirichlet piecewise constant
general, in half plane, 425–427
indented quadrant, 432–433
simple, in half-plane, 427–429
electrostatic potential distribution, 505–506
steady-state temperature distribution, 452
steady-state temperature distribution, indented slab, 454
Area mapping, conformal mapping, critical points, 357–358
Area scale factor, 353
Argument, of complex numbers, 10–11
Arg(z)
Cauchy–Riemann equations, polar form, 57–61
complex numbers, 10–12
Asymptotic representation, contour integrals around arc, 182
Bernoulli's equation, steady two-dimensional flow, 469–474
inclined flat plate, 488–491
vortex, 479–480
Bessel functions, Laplace transform inversion, 330
Bilinear transformation. *See* Linear fractional transformation
Binomial theorem
Laurent series, 262–263
expansion formula, 265–266
nonsingular point expansion, 270
origin expansion, 268
point at infinity, 275–276
Maclaurin series, 251
multiplication, 252
point at infinity residue, 295
Taylor series, 245–247
Boundary point, curve theorem and, 31–34
Boundary value problems
Laplace equation
conformal mapping, 409–420
Dirichlet problems, 411–412
harmonic functions, 411
Neumann problems, 412–420
uniqueness theorem, 420
Dirichlet condition
annulus region, 433–435
constant condition, half axis, 439–441
infinite strip, 421–423
mixed boundary value problem, plane area, 437–438
piecewise constant
disk, 429–431
half plane, 424–429
indented quadrant, 431–433
sector problems, 423–424
solutions, 421–443
disk solution, 441–443
mixed boundary value problem, 435–437
Poisson integral formulas, 140–142
half plane and circle, 439

steady-state temperature distribution, 446–458
half space with insulated strip, 452–454
indented slab, 454
nonsimply connected slab, 456–458
semi-infinite slab, internally heated along slit, 454–456
steady two-dimensional flow, 466–495
complex potentials, 481–482
conformal mapping, 487
flow over submerged log, 491–494
images method, 486–487
inclined flat plate, 487–491
line source/line sink, 474–477
Milne–Thompson circle theorem, 494–495
uniform parallel flow, 480–481
line source, 482–484
vortex, 477–480
vortex pair, 484–486
two-dimensional electrostatics, 499–514
complex electrostatic potential, 508–509
conducting cylinder, uniform field potential, 512–514
infinite conducting plane at zero potential, infinite line charge, 511
infinite line charge, 509–511
potential distribution
eccentric cylinders, 506–508
hollow cylinder with semicircular cross section, 504–506
Bounded sequences
behavior categories, 204–207
defined, 203
Branched function
antiderivatives, definite integrals, 125–126
contour integrals, 180–182
logarithmic contour, 185–187
definite integral evaluation, 309–311
Laplace transform inversion, 327–328
logarithmic, 70–72
residue of, 295–296
Schwarz–Christoffel transformation, 396–405
Branch points, conformal mapping, 339–341
Bromwich contour, Laplace transform inversion, branched point, 328–329

Cartesian representation
Cauchy–Riemann equations, 47–50
polar form, 59–61
closed rectangular contours, 98–99
complex numbers, 1–4
functions, 35
Cauchy convergence principle
complex sequences, 210–211
uniform convergence, 222–223
Weierstrass *M*-test, 223–225
Cauchy–Goursat theorem
antiderivatives
definite integrals, 120–122
branched function, 125–126
logarithmic function, 126–127
simple solution, 124–125
indefinite integrals, 122–124
Cauchy integral formula, 130
contour deformation, 115–117
elementary integrals, 138–140
improper definite integrals, evaluation by contour integration, 159
Dirichlet integral, 175–177
Fresnel integrals, 166–167
rectangular integration, 172–175
indented contours, 178–179
Laurent series expansion, 265–266
residues, 296–298
many-valued functions, 184–185
Morera's theorem as converse of, 151
proof, 198–201
residues and, indented rectangle, 300–301
singularities, 113–115
theoretical background, 107
uniform convergence, complex functions, term-by-term differentiation, 227–229
Cauchy–Hadamard formula, power series
applications, 234
radius of convergence, 233
Cauchy inequality
complex integration, 154–156
Laurent series coefficients, 277–278

Cauchy integral formula, 128–134
  Cauchy inequality, 153–156
  complicated application, 131
  derivatives, 135–142
  elementary integral analysis, 138–139
  fundamental theorem of algebra, 152–153
  harmonic function, maximum/minimum principle, 148–149
  Laurent series, 261–263
  Liouville's theorem, 151–152
  maximum/minimum principle, 150–151
  maximum modulus theorem, 146–148
  mean value theorem, 145–146
  Morera's theorem, 151
  π-radius circle, cosine z, 150
  Poisson formula, half-plane and circular disk, 140–142
  repeated singularity integrals, 170–171
  results, 145–157
  simple application, 131
  singularity of improper integrals, 169–170
  Taylor series, 245–247
  trigonometric functions, 132–134
  typical integral applications, 139–140
  unit circle cosine application, 149–150
Cauchy integral theorem, 107–120
  Cauchy–Goursat proof, 198–200
  contour deformation, 112–115
    applications, 116–117
  definite integral evaluation, indented contours, 307–308
  isolated singularities, functions with, 107
  Laurent series expansion, 265–266
  many-valued functions, 183–184
  nonisolated singularities, functions with, 108–112
  uniform convergence, complex functions, term-by-term differentiation, 228–229
Cauchy principal value
  Dirichlet integral, 176–177
  fundamental definite integral, 164–165
Cauchy–Riemann equation
  analytic functions, 47–50
    derivatives, 81–86
    harmonic conjugate, 61
    Laplacian operator, 58–59
    necessity of, 54–56
    orthogonal trajectories, 59–60
    polar form, 56–61
    proof and consequences, 53–63
    sufficiency conditions, 55–56
  boundary value problems, Laplace equation and conformal mapping, 410
  conformal mapping, 354
  elementary functions, 64–86
  exponential function, 66–67
  Maximum Modulus theorem, 148
  steady two-dimensional flow, 468–474
    complex potentials, 481–482
    line source and line sink, 476–477
    uniform parallel flow, 481
    vortex, 479–480
  two-dimensional electrostatics, 501–504
    Dirichlet problem, half plane, 508–509
    infinite line charge, 510–511
Cauchy test for convergence, complex sequences (Cauchy sequences), 211–213
Chain rule
  boundary value problems
    Laplace equation and conformal mapping, 409–420
    steady two-dimensional flow, 472–474
  conformal mapping, 351
Circle of convergence
  Laurent series, 260–263
  Maclaurin series, 249–250
  power series
    continuity and repeated differentiability, 236–237
    qualitative behavior, 231–232
    radius calculations, 232–233
  Taylor series, 244–247
Circles
  boundary value problems, Poisson integral formula, 439
  conformal mapping
    eccentric circles onto annulus, 375–377
    exterior contact onto strip, 367–368

interior circle onto exterior circle, 364–365
interior circle onto half plane, 365–366
Joukowski transformation, 386–390
linear fractional transformation, 362–364
point of contact and crescents, 366–367
straight line onto annulus, 372–374
symmetry in complex plane, 370–371
contour, integrals around, 99–101
Milne–Thompson circle theorem, 494–495
Circular disk, Poisson integral formulas, 140–142
Closed curve, defined, 27
Closed disk, curve theorem and, 30–34
Closed rectangular contours, integrals, 97–99
Cluster point, complex sequences, 207
Coefficient
complex numbers, quadratic equations, 21–22
Laurent series, 256
Cauchy inequality, 277–278
of diffusivity, steady-state temperature distribution, 448–452
of polynomial, 22–23
power series, uniform convergence, 220
Taylor series, 243–244
undetermined, rational functions, 65
Comparison test, complex series scaling and addition, 215
Complementary error function, Laplace transform inversion, 330
Complex functions
Cauchy–Riemann theorem, 47–50
conjugate, 58
continuity, 42–43
differentiability at origin, 46–47
exponential function, 66–67
harmonic conjugate, 58
identities and properties, 85–86
limits, 41
natural logarithm, 67–72
trigonometric functions, 72–75
uniform convergence, 219–222
Cauchy convergence principle, 222–223
differentiation, 226–229
infinite series, continuous function, 225
term-by-term integration, 225–226
Weierstrass *M*-test, 223–225
zeros of, 287–288
Complex integration
antiderivatives
definite integrals, 120–122
branched function, 125–126
logarithmic function, 126–127
simple solution, 124–125
indefinite integrals, 122–124
Cauchy–Goursat theorem, 115–116
proof, 198–201
Cauchy integral formula, 128–134
Cauchy inequality, 153–156
complicated application, 131
derivatives, 135–142
elementary integral analysis, 138–139
fundamental theorem of algebra, 152–153
harmonic function, maximum/minimum principle, 148–149
Liouville's theorem, 151–152
maximum/minimum principle, 150–151
maximum modulus theorem, 146–148
mean value theorem, 145–146
Morera's theorem, 151
Poisson formula, half-plane and circular disk, 140–142
results, 145–157
simple application, 131
trigonometric functions, 132–134
typical integral applications, 139–140
unit circle cosine application, 149–150
Cauchy integral theorem, 107–120
contour deformation, 112–115
applications, 116–117
isolated singularities, functions with, 107
nonisolated singularities, functions with, 108–112

Complex integration (*Contd.*)
contours, 89–106
circular contour integrals, 99–101
doubly connected domain integrals, 101–102
four branch integration, 102–104
infinitely many branches, integration, 104–105
integral inequality, 94–96
linearity property, 93
partitioning, 93
rectangular contour integrals, 97–99
reversal and integration direction, 93
scaling property, 93
improper definite integrals, contour integration, 158–198
branch point functions, 180–182
Dirichlet integral, 175–177
Fresnel integrals, 165–167
fundamental definite integral, 159–165
indented rectangle, 177–178
Jordan's lemma, 162–165
logarithmic function, 185–187
many-valued functions, 182–185
polar indentation, 173–175
rectangular integration, 171–175
repeated singularity, 170–171
single singularity integral, 167–170
vanishing conditions, 181–182
Laplace inversion integral, 322–330
branched functions, 327–330
inversion of Laplace transform, 326–327
Laplace transform conditions, 323–326
Laurent series, 255–280
algebraic function, infinity neighborhood, 274–276
Cauchy inequalities, coefficients of, 277–278
expansion formula, 264–266
hyperbolic identity, 273–274
Laurent's theorem, 260–263
nonsingular point, 268–270
origin expansions, 266–268
polynomial function conditions, 278–279
singularities, 271–273
trigonometric identity and Maclaurin expansion, 276–278
uniqueness, 263–264
power series, 229–240
Cauchy–Hamard formula applications, 232–236
continuity and repeated differentiability, 236–237
properties of, 236–237
qualitative behavior, 229–232
radius of convergence, 232–233
repeated differentiation, 237–239
term-by-term integration, 239–240
residues and residue theorem, 288–301
applications, 297–298, 303–322
at infinity, 294–295
branched functions, 295–296, 309–310
circular deformation, 313
definite integral evaluation, 303–305
fundamental theorem of algebra, 322
higher pole order, 290, 292–293
integrands, 305–306
indented contours, 306–308
indented rectangle, 299–300
indented single pole, 298–299
Laurent series expansion, 294
meromorphic function residue, 290–293, 316–320
removable singularity, 289
Rouché's theorem, 320–322
series summation, 311–316
simple pole residue, 289–290, 292
integrands, 303–305
sine and cosine functions, 293–294
uniformly bounded cosines, 313–314
sequences, series, and convergence, 203–219
absolute convergence implies convergence theorem, 213–214
behavior categories, 204–207
Cauchy convergence principle, 210–211
Cauchy convergence test, 211–213
comparison test, 215
complex series sum, 214
convergence conditions, 214–215

nth root test, 216
ratio test, 215–216
real and imaginary limits, 208–209
scaling and addition, 215
series convergence testing, 216–218
sum, product, and quotient, 209–210
uniqueness of sequence limit, 207–208
singularities
classification, 281–282
essential, 285–287
poles, 283–284
order k, 284–285
removable, 282–283
Taylor series, 242–255
binomial expansion Maclaurin series, 251
linear combinations and products, 247–248, 251
Maclaurin series problems, 249–250
multiplication, 252
Taylor's theorem, 244–247
term-by-term integration, 251–252
uniqueness, 247
theory, 89
uniform convergence, 219–229
Cauchy convergence principle, 222–223
complex functional series, 221–222
differentiation term by term, 226–229
integration term-by-term, 225–226
geometric series, 222
infinite series, continuous functions, 225
Weierstrass M-test, 223–225
zeros of functions, 287–288
Complex numbers
analytic functions, 1–27
Cartesian representation, 1–2
complex conjugate, 2
curves, domains, and regions, 28–34
Complex plane
complex numbers, 2–3
conformal mapping composition, 336–337
circular symmetry, 370–371
Complex potential
steady two-dimensional flow, 470–474
streamline visualization, 481–482
submerged log, 491–494
vortex, 479–480
vortex pair, 484–486
two-dimensional electrostatics, 501–504
conducting cylinder, uniform field, 512–514
Dirichlet condition, half plane, 508–509
infinite conducting plane at zero potential, 511
Complex variables
analytic functions, 34–35
curves, domains, and regions, 28–34
limits and continuity, 37–38
quadratic equations, 22
Composition, conformal mapping, 336–337
construction by, 391
Conformal mapping
boundary value problems
Laplace equation, 409–420
Dirichlet problems, 411–412
harmonic functions, 411
Neumann problems, 412–420
uniqueness theorem, 420
mixed boundary value problem, transformation in, 435–437
steady two-dimensional flow, 487
curves, domains, and regions, 28–34
defined, 347–352
elementary functions, 378–391
Arcsin z/a, 383–385
composition, 390
cosine mapping, 385
crescent onto quadrant mapping, 390
exponential function, 379–380
half disk onto half plane, 390–391
hyperbolic functions, 385–386
Joukowski transformation, 386–390
logarithmic function, 380–382
sine-cosine functions, 381–383
geometrical aspects, 333–347
composition, 336–337
fixed points, 346–347
inversion mapping, 344–346
linear transformation, 334–336
many-valued functions, 338–341
nonuniform magnification, 341–343

Conformal mapping (*Contd.*)
geometrical aspects (*Contd.*)
principal branch mapping, 343–344
quadrant-onto-quadrant mapping, 337–338
linear fractional transformation, 360–377
annular domain, straight line and circle on, 372–374
circle exterior contact onto strip, 367–368
circle in complex plane symmetry, 370–371
circle interior onto circle exterior, 364–365
circle interior onto half plane, 365–366
circle point of contact and crescents, 366–367
crescent onto a wedge, 368–369
cross ratio formula, 364–369
disk onto disk mapping, 374–375
eccentric circles onto annulus, 375–377
straight lines and circles, 362–364
symmetry and, 369–370
symmetry preservation, 371
upper half-plane, mapping on disk, 371–372
Schwarz–Christoffel transformation, 392–405
finite slit, 401–403
infinite strip, 399–401
semi-infinite slit, 403–405
semi-infinite strip, 398–399
theorem
critical point mapping, 357–358
linear transformation, 352–354
local inverse conditions, 354–356
Riemann mapping, 358–359
Conformal transformations, complex numbers, 3
Conjugate harmonic functions, Cauchy–Riemann equations, polar form, 58–61
Conjugation, operation of, complex numbers, 2
Connected points, curves, 31–34
Conservation of mass, steady two-dimensional flow, 468
Conservation of momentum, steady two-dimensional flow, 468
Conservative force, electrostatics, 500
Constitutive law, steady two-dimensional flow, 467
Continuity
complex functions, 42–43
power series, 236–237
Continuous functions, uniform convergence, infinite series, 225
Contour, defined, 27
Contour deformation
Cauchy–Goursat theorem, 115–117
residue theorem, 313
Contour integrals
Cauchy integral theorem, nonisolated singularities, 113–115
complex integration, 89–106
circular contour integrals, 99–101
doubly connected domain integrals, 101–102
four branch integration, 102–104
infinitely many branches, integration, 104–105
integral inequality, 94–96
linearity property, 93
partitioning, 93
rectangular contour integrals, 97–99
reversal and integration direction, 93
scaling property, 93
improper definite integral evaluation, 158–187
arc vanishing conditions, 181–182
branched functions, 180–182
Dirichlet integral, 175–177
Fresnel integrals, 165–167
fundamental definite integral, 159–165
logarithmic contour, 185–187
many-valued functions, 182–185
rectangular integration, 171–175
indented rectangles, 177–179
polar exclusion, 173–175
repeated singularity, 170–171
single singularity, 167–170
Contraction
conformal mapping, 342–343
linear transformation, 335–336

Convergence. *See also* Annular domain of convergence; Uniform convergence
circle of
Laurent series, 260–263
Maclaurin series, 249–250
power series
continuity and repeated differentiability, 236–237
qualitative behavior, 231–232
radius calculations, 232–233
Taylor series, 244–247
complex series
absolute convergence and, 213–214
testing for, 216–218
Laurent series, 256–258, 260–263
necessary conditions for, 214
power series, qualitative behavior, 230–232
radius of
of Taylor's series, 248
power series
formulas for, 232–236
qualitative behavior, 231–232
Convergent sequences, defined, 207
Cosine
Cauchy integral formula, unit circle, 149–150
complex numbers, De Moivre's theorem, 15–16
conformal mapping, 385
hyperbolic functions, 75–77
residue of, 293–294
trigonometric functions, 72–75
Crescents, conformal mapping
circle point of contact and, 366–367
onto wedge, 368–369
Critical points, conformal mapping
analytic functions, 357–358
local inverse, 354–356
Cross ratio, conformal mapping
circle exterior contact onto strip, 367–368
crescent onto wedge mapping, 368–369
interior circle onto exterior circle, 364–365
straight lines and circles, 363–364
Cube roots of unity, complex numbers, 17–20
Curves, analytical functions, 27–34
Cut planes
contour integrals, branch point functions, 180–182
natural logarithm, 69–72
Cylindrical conduction
boundary value problems, Neumann conditions, conformal mapping, 415–420
electrostatic potential distribution
eccentric cylinders, 506–508
semicircular cross section, 504–506
uniform field, 512–514
steady two-dimensional flow
inclined flat plate, 487–491
line source and line sink, 475–477
submerged log, 492–494
vortex, 477–480

D'Alembert ratio test, complex series scaling and addition, 215–216
Definite integrals
antiderivatives, 120–122
branched function, 125–126
simple form, 124–125
improper, evaluation by contour integration, 158–187
arc vanishing conditions, 181–182
branched functions, 180–182
Dirichlet integral, 175–177
Fresnel integrals, 165–167
fundamental definite integral, 159–165
Ln z contour, 185–187
many-valued functions, 182–185
rectangular integration, 171–175
indented rectangles, 177–179
polar exclusion, 173–175
repeated singularity, 170–171
single singularity, 167–170
residue theorem evaluation, 303–305
Deformation, family of circles, residue theorem, 313
Degenerate polygons, Schwarz–Christoffel transformation, 392
Degenerate root, polynomial equations, 22
Deleted neighborhood, curve theorem and, 31–34

δ-neighborhood, curve theorem and, 31–34
De Moivre's theorem
  Cauchy–Riemann equations, polar form, 57–61
  complex numbers, polar form, 15–16
Derivatives
  analytic functions, 43–45, 81–86
    Cauchy–Riemann equations, 54–56
  Cauchy integral formula, 134–138
  inverse hyperbolic sine, 80–81
  uniform convergence, complex functions, term-by-term differentiation, 228–229
Differentiation. *See also* Chain rule
  analytic functions, 43–45, 50
    derivatives, 82
  boundary value problems
    Laplace equation and conformal mapping, 409–420
    steady two-dimensional flow, 472–474
  complex functions
    Cauchy–Riemann equations, 47–50
    origin, 46–47
  conformal mapping, 351–354
  inverse hyperbolic and trigonometric functions, 79–80
  power rules, repeated differentiation, 236–239
  power series, repeated characteristics, 236–237
  Taylor series, 243
  trigonometric functions, 73–75
  two-dimensional electrostatics, 500–501
  uniform convergence, complex functions, term-by-term differentiation, 226–229
Diffusion equation, steady-state temperature distribution, 448–452
Directional derivative, steady-state temperature distribution, 447
Direct methods, boundary value problems, Neumann conditions, conformal mapping, 416–420
Dirichlet condition
  boundary value problems
    annulus, 433–435
    conformal mapping, Laplace equation, 411–412
      electrostatics, 504
        half plane problems, 508–509
      infinite strip, Laplace solution, 421–423
      mixed boundary value problem transformation, 435–437
        in plane, 437–438
      Neumann problems and, 414–420
      piecewise constant
        in half plane, simple constant, 427–429
        in half plane general, 424–427
        indented quadrant, 431–433
        on disk, 429–431
      Poisson integral formulas
        disk solution, 441–442
        half axis, 439–440
        half plane and circle, 439
      sector problems, 423–424
      steady-state temperature distribution, 416–420
      uniqueness theorem, 420
    contour integration evaluation, 175–177
    steady-state temperature distribution, 448–252
      half space, insulated strip, 452–454
Disconnected points, curves, 31–34
Discontinuous functions, defined, 42–43
Disks
  analytic function and, 36–37
  boundary value problems
    piecewise constant Dirichlet problem, 429–431
    Poisson integral formula, 439
    Poisson integral formula solution, 441–442
  circular, Poisson integral formula, 140–142
  closed and opened disks, 30–34
  conformal mapping
    disk to disk mapping, 374–375
    half disk onto half plane, 390–391
    upper half plane onto, 371–372
  function definition, 36–37
Divergence theorem, electrostatics, 501
Divergent sequences, defined, 207
Divergent series, properties, 215
Domain of definition, conformal mapping, 333–334

Domains. *See also* Annular domain of convergence
analytical functions, 27–34
Cauchy–Riemann equations, sufficiency conditions, 55–56
circular contour integrals, 99–101
doubly connected domain integrals, 101–102
Doubly connected domain, integral round, 101–102

Eccentric circles, conformal mapping onto annulus, 375–377
Eccentric cylinders, electrostatic potential distribution, 506–508
Electrostatic field, basic properties, 499–500
Electrostatic scalar potential, defined, 500
Elementary functions, 63–86
complex sine functions, roots of, 75
conformal mapping, 378–391
Arcsin z/a, 383–385
composition, 390
cosine mapping, 385
crescent onto quadrant mapping, 390
exponential function, 379–380
half disk onto half plane, 390–391
hyperbolic functions, 385–386
Joukowski transformation, 386–390
logarithmic function, 380–382
sine-cosine functions, 381–383
derivatives, 82–83
entire functions, 45–46
exponential function, 66–67
hyperbolic functions, 75–81
cosine function roots, 77
defined, 84–85
inverse, 77–80
inverse sine and derivatives, 80–81
inverse trigonometric/hyperbolic functions, 77–80, 84–85
logarithmic function, 67–73
partial fraction expansion, repeated zeros, 65–66
polynomials, 64
rational functions, 64–65
trigonometric functions, 73–75
defined, 83
inverse, 77–80, 84–85
Ellipsis
defined, 203
mixed boundary value problem transformation, 436–437
Eneström–Kakeya theorem, 24–25
Entire function
defined, 44
elementary properties, 45–46
Equality
complex numbers, 11–12
of complex numbers, 2
Equation of motion, steady two-dimensional flow, 467
Equation of state, steady two-dimensional flow, 467
Equilibrium points, two-dimensional electrostatics, 502
Equipotential surface
boundary value problems, Laplace equation solutions, 421
electrostatic potential distribution, 506
infinite line charge, 510–511
Equivalent contours, Cauchy–Goursat theorem, 116–117
Error function
improper definite integrals, evaluation by contour integration, 159
Laplace transform inversion, 330
Maclaurin series, term-by-term integration, 251–252
Essential singularity
defined, 285–286
Laurent series, 282
Expansion formula, Laurent series, 264–266
Exponential function
conformal mapping, 379–380
defined, 66–67
Exponential order α, Laplace transformation, 323
Extended complex plane, defined, 29
External boundaries, circular contour integrals, 99–101
Extremum principle, Laplace equation, boundary value problems, 420

Family of circles, residue theorem, deformation, 313
Finite polygon, Schwarz–Christoffel transformation, 397–398

Finite series, Cauchy sequence, 212–213
Finite slit, Schwarz–Christoffel transformation, 401–403
First kind boundary value problems. *See* Dirichlet problems
Fixed points, conformal mapping, 346–347
  linear fractional transformation, lines and circles, 362–364
Four-branch function, integration, 102–104
Fourier's law of heat conduction, steady-state temperature distribution, 447
Fresnel integrals
  contour integration evaluation, 165–167
  Laplace transform inversion, 330
Functions
  antiderivatives, definite integrals, branched function, 125–126
  Cartesian and polar forms, 35–36
  Cauchy integral theorem, nonisolated singularities, 108–112
  Cauchy–Riemann equations, 53–61
  conjugate harmonic, 58
  disk definition, 36–37
  elementary, 63–86
    complex sine functions, roots of, 75
    derivatives, 82–83
    entire functions, 45–46
    exponential function, 66–67
    hyperbolic functions, 75–81
      cosine function roots, 77
      defined, 84–85
      inverse, 77–80
      inverse sine and derivatives, 80–81
    inverse trigonometric/hyperbolic functions, 77–80, 84–85
    logarithmic function, 67–73
    partial fraction expansion, repeated zeros, 65–66
    polynomials, 64
    rational functions, 64–65
    trigonometric functions, 73–75
      defined, 83
      inverse, 77–80, 84–85
  function of, in conformal mapping, 337
  harmonic, 58
  zeros of, 287–288
Fundamental strip, exponential function, 67
Fundamental theorem of algebra
  Cauchy integral formula, 152–154
  polynomial equations, 22
  zeros of function, 322

Gauss divergence theorem
  Neumann boundary value problems, conformal mapping, 413–420
  steady-state temperature distribution, 448–452
Gauss' law, electrostatics, 500–501
  infinite line charge, 510–511
Geometrical symmetry, Schwarz–Christoffel transformation, 392
Geometric properties, conformal mapping, 333–347
  composition, 336–337
  fixed points, 346–347
  inversion mapping, 344–346
  linear fractional transformation, 362
  linear transformation, 334–336
  many-valued functions, 338–341
  nonuniform magnification, 341–343
  principal branch mapping, 343–344
  quadrant-onto-quadrant mapping, 337–338
Geometric series
  repeated differentiation, 237–239
  scaling and addition, 216
  term-by-term integration, 239–240
  uniform convergence, 222
Global inverse, conformal mapping, 354–356
Green's theorem, nonisolated singularities, simple regions, 108–112

Half-annular domain, mixed boundary value problem transformation, 435–437
Half axis, boundary value problems, constant Dirichlet condition, 439–440
Half-ellipse, mixed boundary value problem transformation, 436–437
Half plane
  boundary value problems
    Dirichlet generalized piecewise constant, 424–427

Dirichlet problem, complex electrostatic potential, 508–509
Dirichlet simple piecewise constant, 427–429
mixed boundary value problem transformation, 437–438
Poisson integral formula, 439
conformal mapping
half disk onto, 390–391
interior circle onto, 365–366
onto disk, 371–372
Half-play structures, Poisson integral formulas, 140–142
Half space, insulated strip, steady-state temperature distribution, 452–454
Harmonic function
boundary value problems
Dirichlet condition
generalized piecewise constant, in half plane, 425–427
in annulus, 433–435
infinite strip, 421–423
sector-based, 423–424
two-dimensional problems, 442–443
Laplace equation and conformal mapping, 410–411
mixed boundary value problem transformation, 435–437
plane Dirichlet condition, 437–438
Neumann problems, conformal mapping, 413–420
steady two-dimensional flow, 468–474
Cauchy–Riemann equations
conjugate, 61
polar form, 58–61
derivatives, 82
electrostatic potential, 501–504
infinite line charge, 510–511
Maximum Modulus theorem, 148–149
Heat equation, steady-state temperature distribution, 448–452
Heat flow line, steady-state temperature distribution, 449–452
Heat flux, steady-state temperature distribution, 447
Heat flux vector, steady-state temperature distribution, 449–452
Higher transcendental functions, Laplace transform inversion, 330
Holomorphic function, defined, 287–288
Homogeneous Neumann condition
boundary value problems
conformal mapping, 412–420
Dirichlet in annulus, 434–435
mixed boundary value problem transformation, 438
steady two-dimensional flow, 473–474
steady-state temperature distribution, half space, insulated strip, 452–454
Homotropic contours, Cauchy-Goursat theorem, 116–117
Hydrodynamics, steady two-dimensional flow, 467
Hyperbolic functions
conformal mapping
elementary functions, 385–386
principal branch, 343–344
defined, 75–77, 84
inverse functions, 77–81
roots, 77
Hyperbolic identity, Laurent series, 273–274

Image points, steady two-dimensional flow, 486–487
Imaginary axis, complex numbers, 2–5
Imaginary numbers, complex numbers, 1–5
Imaginary variables, complex sequence limits, 208–209
Improper definite integral evaluation, contour integration, 158–187
arc vanishing conditions, 181–182
branched functions, 180–182
Dirichlet integral, 175–177
Fresnel integrals, 165–167
fundamental definite integral, 159–165
Ln z contour, 185–187
many-valued functions, 182–185
rectangular integration, 171–175
indented rectangles, 177–179
polar exclusion, 173–175
repeated singularity, 170–171
single singularity, 167–170
Inclined flat plate, steady two-dimensional flow, 487–491

Incompressible model, steady two-dimensional flow, 467
Indefinite integrals, antiderivatives, 121–124
Indented contours
 boundary value problems, indented quadrant, piecewise constant Dirichlet condition, 431–433
 definite integral evaluation, 306–308
 improper definite integrals, 158–159
 rectangular integration
  contour integral evaluation, 177–179
  polar exclusion, 173–175
 slab, steady-state temperature distribution, 454
Indented poles, residues at, 298–300
Indented rectangle, residues and residue theorem, 300–301
Indirect methods
 boundary value problems, Neumann conditions, conformal mapping, 420
 steady two-dimensional flow, complex potential solutions, 481–482
Inequality
 Cauchy inequality, 154–156
 complex numbers, 7–9
 contour integrals, 94
 Jordan inequality
  Fresnel integrals, 166–167
  fundamental definite integral, 161–165
  improper definite integrals, evaluation by contour integration, 159
Infinite complex sequence, defined, 203
Infinite line charge, two-dimensional electrostatics, 509–511
 infinite conducting plane at zero potential, 511
Infinitely many branches, integration, 104–105
Infinite series, Cauchy sequence, 212–213
Infinite strip, Schwarz–Christoffel transformation, 399–401
Infinity
 curves, domains, and regions, 28–34
 Laurent series in neighborhood of, 274–276
 steady two-dimensional flow, 473–474
  uniform parallel flow, 482–484
Insulated strips, boundary value problems
 isothermal line, 416
 steady-state temperature distribution, 452–454
Integrands, definite integral evaluation
 higher-order poles, 305–306
 simple poles, 303–305
Integration, term-by-term integration
 Maclaurin series, 251–252
 power series, 239–240
 uniformly convergent series, complex functions, 225–226
Internal boundaries, circular contour integrals, 99–101
Into mappings, defined, 334
Inverse functions
 hyperbolic and trigonometric, 77–81
 natural logarithm, 69–72
 trigonometric, 84
Inverse function theorem, conformal mapping, 353–354
Inverse Laplace transform, properties of, 324–326
Inverse tangent function, boundary value problems, generalized piecewise constant, Dirichlet condition in half plane, 425–427
Inversion mapping
 critical points, 357–358
 geometric functions, 344–346
Inviscid models, steady two-dimensional flow, 467
Inward spirals, complex sequences, 204–207
Irrotational flow, steady two-dimensional flow, 467–474
Isolated singularities
 Cauchy integral theorem, 107
 classification, 280–285
 defined, 44
 Laurent series, 255–256
 pole order of k, 282
  testing for, 284–285
Isometric curve, conformal mapping, exponential function, 380

Isothermal curves, boundary value problems, Laplace equation solutions, 421
Isothermal lines
  boundary value problems, Neumann conditions, conformal mapping, 416–420
  steady-state temperature distribution, 449–452
Isothermal surface, boundary value problems, Laplace equation solutions, 421

Jacobian matrices, conformal mapping, 353–354
Jordan curve theorem, 30
Jordan inequality
  fundamental definite integral, 161–165
  improper definite integrals, evaluation by contour integration, 159
    Fresnel integrals, 166–167
  Laplace transform inversion, 328–330
Jordan's lemma
  Dirichlet integral, 176–177
  fundamental definite integral, 162–165
  singularity of improper integrals, 169–170
Joukowski transformation
  conformal mapping, 386–390
  defined, 28
  steady two-dimensional flow, inclined flat plate, 487–491

Kelvin's circulation theorem, steady two-dimensional flow, 468

Laplace equation
  boundary value problems
    conformal mapping, 409–420
      Dirichlet problems, 411–412
      harmonic functions, 411
      Neumann problems, 412–420
      uniqueness theorem, 420
    solutions, 421–443
      annulus Dirichlet problem, 433–435
      constant Dirichlet condition, half axis, 439–441
      Dirichlet plane and mixed boundary value problem, 437–438
      disk solution, 441–443
      infinite strip, Dirichlet problem, 421–423
      mixed boundary value problem, 435–437
      piecewise constant Dirichlet problem
        disk, 429–431
        half plane, 424–429
        indented quadrant, 431–433
      sector Dirichlet problems, 423–424
    steady two-dimensional flow, 471–474
      line source and line sink, 474–477
  Cauchy–Riemann equations
    harmonic conjugate, 61
    polar form, 58–61
  complex numbers, 3
  conformal mapping, 353–354
  harmonic function derivatives, 82
  Poisson integral formulas, 140–142
  steady-state temperature distribution, 449–452
    half space, insulated strip, 452–454
Laplace inversion integral, complex integration, 322–330
  branched functions, 327–330
  inversion of Laplace transform, 326–327
  Laplace transform conditions, 323–326
Laplace transformation
  complex integration, 322–330
    conditions for existence, 323–326
  inversion of, 327–330
Laplace transform pairs, defined, 323–324
Laurent series
  basic principles, 203–204
  complex integration, 255–280
    algebraic function, infinity neighborhood, 274–276
    Cauchy inequalities, coefficients of, 277–278
    expansion formula, 264–266
    hyperbolic identity, 273–274
    Laurent's theorem, 260–263

Laurent series (*Contd.*)
complex integration (*Contd.*)
nonsingular point, 268–270
origin expansions, 266–268
polynomial function conditions, 278–279
singularities, 271–273
trigonometric identity and Maclaurin expansion, 276–278
uniqueness, 263–264
removable singularity, 283
residues and residue theorem, 289
deduction of, 294
indented poles, 299–300
meromorphic function, 290–292
principle of the argument application, 317–320
singularities classification, 281–285
Taylor series comparisons, 260
Laurent's theorem, convergent Laurent series, 260–263
Legendre polynomials
complex integration, 144–145
Laplace transform inversion, 330
Level curve, boundary value problems
Laplace equation solutions, 421
mixed boundary value problem transformation, 438
Level surface, boundary value problems, Laplace equation solutions, 421
L'Hospital's rule, analytic functions, 50–51
Limits
Cauchy–Riemann equations, 54–56
complex functions, 41
complex sequences, 206–207
absolute convergence and, 213–214
real and imaginary limits, 208–209
sum, product, and quotient, 209–210
uniqueness of, 207–208
continuity and, 37–38
definition of, 38–40
Limit superior notion, power series, qualitative behavior, 231–232
Limit theorems, 39–40
L'Hospital's rule, 51
Linear combinations
Maclaurin series, 251
Taylor's series, 247–248
Linear fractional transformation, conformal mapping, 360–362
circle exterior contact onto strip, 367–368
crescent onto wedge, 368–369
disk onto disk, 374–375
interior circle onto exterior circle, 364–365
interior circle onto half plane, 365–366
straight lines and circles, 362–364
symmetry and, 369–370
upper half-plane onto disk, 371–372
Linearity property
complex integrals, 93
Laplace transform pairs, 324
Linear scale factor, conformal mapping, 353
exponential function, 380
logarithmic function, 381–382
Linear superposition principle, steady two-dimensional flow, 471–474
Linear transformation
boundary value problems
Dirichlet condition, infinite strip, 422–423
mixed boundary value problem, 435–437
steady two-dimensional flow, 473–474
conformal mappings
fixed points, 346–347
geometrical aspects, 334–336
theorem, 352
quadrant to quadrant mapping, 337–338
Line sink, steady two-dimensional flow, 474–477
image points, 487
Line source, steady two-dimensional flow, 474–477
image points, 486–487
uniform parallel flow, 482–484
Liouville's theorem
Cauchy integral formula, 151–152
fundamental theorem of algebra, 153–154
Local inverse, conformal mapping, 354–356
Logarithmic function
conformal mapping, 380–382

contour integral evaluation, 185–187
defined, 67–72
definite integrals, 126–127
inverse hyperbolic and trigonometric functions, 78–81
Lunes, conformal mapping, circle point of contact and, 366–367

Maclaurin series
binomial expansion, 251
Laurent series
convergence, 258–260
hyperbolic identity, 273–274
point at infinity, 275–276
trigonometric identity, 276–277
linear combinations, 251
multiplication and expansion of, 252
pole order of k, 284
residue deduction, 294
simple expansions, 249–250
Taylor's theorem and, 246–247
term-by-term integration, 239–240
error function, 251–252
Magnification
conformal mapping, 339–341
linear transformation, 335–336
Many-valued properties
contour integrals
branched function, 180–182
evaluation techniques, 182–185
inverse hyperbolic and trigonometric functions, 78–81
logarithmic function, 68–72
Mappings
conformal mapping theorem, 352–354
curves, domains, and regions, 28–34
Maximum-minimum principle, Cauchy integral formula, 150–151
Maximum Modulus theorem
boundary value problems, 420
Cauchy integral formula, 146–148
Mean value theorem, Cauchy integral formula, 145–146
Meromorphic function
defined, 64–65, 288
residue of, 290–293
principle of, 316–320
summation of series, 311–315
Milne–Thompson circle theorem, steady two-dimensional flow, 494–495
Mixed boundary value problem, transformation, 435–437
Mobius transformation. *See* Linear fractional transformation
Modulus
complex numbers, 5
argument of, 10–11
polar form, 13–14
triangle inequality, 7–8
complex sequences, behavior categories, 205–207
contour integral evaluation, logarithmic contour, 186–187
Morera's theorem
Cauchy integral formula, 151
uniform convergence, complex functions, term-by-term differentiation, 227–229
Multiplication
complex numbers, 5–6
of series, Maclaurin series expansion, 252
Multiplicity *k*, zeros of functions, 287–288
Mutually orthogonal trajectories, Cauchy–Riemann equations, polar form, 58–61

Natural logarithm, of complex variable, 67–72
Neumann condition. *See also* Homogeneous Neumann condition
boundary value problems
conformal mapping, 412–420
electrostatics, 504
mixed boundary value problem transformation, 435–437
steady-state temperature distribution, 448–252
Neutral points, two-dimensional electrostatics, 502
Newton's law of cooling, steady-state temperature distribution, 451–452
Nonanalytic functions, Cauchy–Riemann equations, 49–50
Nonhomogeneous Neumann condition, boundary values, conformal mapping, 414–420
Nonisolated singularities, Cauchy integral theorem, 108–112

Nonsimply connected domain, conformal mapping, 358–359
Nonsimply connected slab, steady-state temperature distribution, 456–458
Nonsingular point, Laurent series expansion, 268–270
Nonuniform magnification, conformal mapping, 338–343
nth partial sum of the sequence, Cauchy sequences, 212–213
nth roots of unity, complex numbers, 17–19
nth root test
  complex series scaling and addition, 215–216
  conformal mapping, linear transformation, 339–341
  radius of convergence, power series, 233
nth term, defined, 203
Numerical solutions, boundary value problems, Dirichlet condition, Poisson integral formula, 441

One-one mappings, defined, 334
Onto mappings
  circles
    exterior contact onto strip, 367–368
    interior circle onto exterior circle, 364–365
    interior circle onto half plane, 365–366
    onto annulus, straight line mapping, 372–374
  crescent onto quadrant, 390
  crescent onto wedge, 368–369
  defined, 334
  disk onto disk mapping, 374–375
  eccentric circles onto annulus, 375–377
  half disk onto half plane, 390–391
  half plane onto disk, 371–372
  quadrant onto quadrant, 337–338
Open disk, curve theorem and, 30–34
Operation of conjugation, of complex numbers, 2
Oriented curves, conformal mapping, 349
Origins, Laurent series expansion, 266–268
Orthogonal trajectories, Cauchy–Riemann equations, polar form, 58–61
Outward spirals, complex sequences, 204–207

Parabolas, conformal mapping, 341–342
Parallelogram rule, complex numbers, 3–5
Parametric form
  complex integrals, 90–94
  curve theorem and, 31–34
Partial differential equation
  Cauchy–Riemann equations, polar form, 58–61
  steady-state temperature distribution, 446–447
Partial fraction expansion, rational functions, 65
Partitioning, contour integrals, 93
Permittivity constant, electrostatics, 500
Picard exceptional value, essential singularity, 286
Piecewise constant Dirichlet condition, boundary value problems
  disk, 429–431
  half plane
    generalized constant, 424–427
    simple constant, 427–429
  indented quadrant, 431–433
Piecewise smooth, properties, 31–34
Plane-related Dirichlet condition, mixed boundary value problem transformation, 437–438
Point at infinity
  conformal mapping, circle point of contact and crescents, 366–367
  essential singularity, 286
  Laurent series, 274–276
  residue at, 294–295
Poisson equation, steady-state temperature distribution, 449–452
Poisson integral formulas
  boundary value problems
    disk solution, 441–442
    half axis, 439–440
    half plane and circle, 439
  half-plane and circular disk, 140–142

Polar coordinates, fundamental definite integral, 160–165
Polar form
analytic functions, 35
Cauchy–Riemann equations, 56–61
closed rectangular contours, 98–99
complex numbers, 9–13
products and quotients, 13–14
improper definite integrals, evaluation by contour integration, rectangular integration, 173–175
Pole order of k
defined, 283–284
definite integral evaluation, 305–306
isolated singularities, 282
residue at, 290
testing for, 284–285
Poles
Laurent series, 255–256
basic principles, 203
rational functions, 64–66
residues at, 292–293
indented poles, 298–300
Polynomials
complex variables, 22
elementary functions, 64
Eneström–Kakeya theorem, 24–25
Laurent series, 278–279
rational functions, 64–66
real coefficients, 22–23
residues and residue theorem, principle of the argument application, 319–320
Rouché's theorem, 321–322
Positive sense, conformal mapping, 349
Positive square root, complex numbers, 5
Potential distribution, two-dimensional electrostatics
eccentric cylinders, 506–508
hollow cylinder with semicircular cross section, 504–506
Potential flow, steady two-dimensional flow, 468
Potential theory, boundary value problems, Laplace equations, 414–420
Power series
complex integration, 229–240
Cauchy–Hamard formula applications, 232–236
continuity and repeated differentiability, 236–237
properties of, 236–237
qualitative behavior, 229–232
radius of convergence, 232–233
repeated differentiation, 237–239
term-by-term integration, 239–240
Laurent series as, 258–259
Taylor's theorem and, 245–247
uniform convergence, 219–220
Preimage, conformal mapping, defined, 334
Principal argument, complex numbers, 10–11
Principal branch, conformal mapping, 340–344
Principle of the argument application, residue theorem, 316–320
Products
complex numbers, polar form, 13–14
of complex sequences, 209–210
of Taylor's series, 247–248
Proper subset, complex numbers, 2
Proportionality constant, steady-state temperature distribution, 451
Punctured disk, curve theorem and, 31–34
Purely imaginary numbers, limit properties, 47

Quadrant
boundary value problems, piecewise constant Dirichlet condition, indented quadrant, 431–433
conformal mapping
crescent onto, 390
onto quadrant, 337–338
Quadratic equations, complex numbers, roots of unity, 20–21
Qualitative behavior, power series, 229–232
Quotient definition
complex numbers, 6–7
polar form, 13–14
of complex sequences, 209–210

Radius of convergence
of Taylor's series, 248
power series
formulas for, 232–236
qualitative behavior, 231–232

Rational functions, applications, 64–66
Ratio test, complex series scaling and addition, 215–216
Ray, conformal mapping, 339–341
Real axis, complex numbers, 2–5
Real numbers
  complex numbers, 1–5
  natural logarithm, 67–72
Real variables
  boundary value problems, Dirichlet condition, infinite strip, 422–423
  limits of, 41–42
    complex sequences, 208–209
Rectangular integration
  conformal mapping, 342–343
  contour evaluation, 171–175
    indented contours, 177–179
  residues and residue theorem, 300–301
Regions, analytical functions, 27–34
Removable singularity
  classification, 281–285
  continuous functions, 43
  defined, 282–283
  residue at, 289
Repeated differentiation, power series, 236–239
Repeated singularity integrals, contour integration evaluation, 170–171
Repeated zeros, partial fraction expansion, 65–66
Residues and residue theorem
  complex integration, 288–301
    applications, 297–298, 303–322
    at infinity, 294–295
    branched functions, 295–296, 309–310
    circular deformation, 313
    definite integral evaluation, 303–305
    fundamental theorem of algebra, 322
    higher pole order, 290, 292–293
      integrands, 305–306
    indented contours, 306–308
    indented rectangle, 299–300
    indented single pole, 298–299
    Laurent series expansion, 294
    meromorphic function residue, 290–293, 316–320
    removable singularity, 289
    Rouché's theorem, 320–322
    series summation, 311–316
    simple pole residue, 289–290, 292
      integrands, 303–305
    sine and cosine functions, 293–294
    uniformly bounded cosines, 313–314
  contour integrals, 96
Riemann mapping theorem, conformal mapping, 358–359
Riemann sphere, stereographic coordinates, 29–30
Rodrigue's formula, complex integration, 144–145
Rotation, linear transformation, 335–336
Rouché's theorem, zeros of analytic function, 320–322

Scaling property
  complex integrals, 93
  complex series, 215
Schläfli integral formula, complex integration, 144–145
Schwarz–Christoffel transformation
  antiderivatives, indefinite integrals, 123–124
  conformal mapping, 392–405
    finite slit, 401–403
    infinite strip, 399–401
    semi-infinite slit, 403–405
    semi-infinite strip, 398–399
Sector problems, boundary value problems, Dirichlet conditions, 423–424
Semicircular mapping
  electrostatic potential distribution, 504–506
  mixed boundary value problem transformation, 436–437
Semi-infinite slab, internal heating, steady-state temperature distribution, 454–456
Semi-infinite slit, Schwarz–Christoffel transformation, 403–405
Semi-infinite strip, conformal mapping
  logarithmic function, 381–382
  Schwarz–Christoffel transformation, 398–399
Sequences, complex integration
  behavior categories, 204–207

defined, 203–204
real and imaginary limits, 208–209
sum, product, and quotient, 209–210
uniqueness of limits, 207–208
Series
Cauchy sequences, finite and infinite, 212–213
complex integration, 203–219
absolute convergence implies convergence theorem, 213–214
behavior categories, 204–207
Cauchy convergence principle, 210–211
Cauchy convergence test, 211–213
comparison test, 215
complex series sum, 214
convergence conditions, 214–215
*n*th root test, 216
ratio test, 215–216
real and imaginary limits, 208–209
scaling and addition, 215
series convergence testing, 216–218
sum, product, and quotient, 209–210
uniqueness of sequence limit, 207–208
divergent, properties, 215
geometric, uniform convergence, 222
power series
complex integration, 229–240
Cauchy–Hamard formula applications, 232–236
continuity and repeated differentiability, 236–237
properties of, 236–237
qualitative behavior, 229–232
radius of convergence, 232–233
repeated differentiation, 237–239
term-by-term integration, 239–240
uniform convergence, 220–221
Taylor series
basic principles, 203–204
complex integration, 242–255
binomial expansion Maclaurin series, 251
linear combinations and products, 247–248, 251
Maclaurin series problems, 249–250
multiplication, 252
Taylor's theorem, 244–247
term-by-term integration, 251–252
uniqueness, 247
Similarity of polygons, Schwarz–Christoffel transformation, 393–394
Simple arcs
complex integrals, 89–90
contour integrals, 94–96
defined, 27
Simple closed curve
Cauchy–Goursat proof, 198–200
complex integrals, 89–90
defined, 27
residues and residue theorem, principle of the argument application, 318–320
Simple pole
isolated singularities, 282
Laurent series, 255–256
residue at, 289–290, 292
indented poles, 298–300
integrand with, 303–305
Simple polygons, Schwarz–Christoffel transformation, 392
Simple regions, Cauchy integral theorem, nonisolated singularities, 108–112
Simply connected domain, conformal mapping, 358–359
Sine
complex numbers, De Moivre's theorem, 15–16
conformal mapping, logarithmic function, 381–382
hyperbolic functions, 75–77
inverse hyperbolic function and derivative, 80–81
trigonometric functions, 72–75
roots of, 73
Sine integral, Laplace transform inversion, 330
Singularity
complex integration series, 203–204
classification, 280–285
continuous functions, 43
essential, 285–286
improper definite integrals, contour integration evaluation, 167–170
repeated singularity, 170–171

Singularity (*Contd.*)
  Laurent series, 255–256
    expansions, 271–273
    point at infinity, 274–276
  removable, 281–285
Slab, steady-state temperature distribution, 454
  nonsimply connected slab, 456–458
Slit
  Schwarz–Christoffel transformation
    finite slit, 401–403
    semi-infinite slit, 403–405
  semi-infinite slab, internal heating along, steady-state temperature distribution, 454–456
Smooth curve
  contour integrals, 94–96
  properties, 31–34
Square root equations
  complex numbers, quadratic form, 20–21
  inverse hyperbolic and trigonometric functions, 78–81
Stagnation points, steady two-dimensional flow, 469–474
  inclined flat plate, 489–491
  uniform parallel flow, line source, 483–484
  vortex pair, 486
Steady flow measurements, boundary value problems, Neumann conditions, conformal mapping, 415–420
Steady-state solutions, boundary value problems, Neumann conformal mapping, 414–420
Steady-state temperature distribution, boundary value problems, 446–458
  half space with insulated strip, 452–454
  indented slab, 454
  Neumann conditions, conformal mapping, 415–420
  nonsimply connected slab, 456–458
  semi-infinite slab, internally heated along slit, 454–456
Steady two-dimensional flow, boundary value problems, 466–495
  complex potentials, 481–482
  conformal mapping, 487
  flow over submerged log, 491–494
  images method, 486–487
  inclined flat plate, 487–491
  line source/line sink, 474–477
  Milne–Thompson circle theorem, 494–495
  uniform parallel flow, 480–481
    line source, 482–484
  vortex, 477–480
  vortex pair, 484–486
Stereographic coordinates, defined, 29–30
Stokes' theorem, steady two-dimensional flow, vortex, 478–480
Straight lines, conformal mapping
  circle onto annulus, 372–374
  Joukowski transformation, 386–390
  linear fractional transformation, 362–364
Stream function
  steady two-dimensional flow, 470–474
    conformal mapping, 487
    submerged log, 492–494
    vortex, 479–480
  two-dimensional electrostatics, 501
    Dirichlet problem, half plane, 509
Streamlines, steady two-dimensional flow, 467
  image points, 487
  inclined flat plate, 488–491
  submerged log, 492–494
  vortex pair, 485–486
Strength of vortex, steady two-dimensional flow, 479–480
Strip, Schwarz–Christoffel transformation
  infinite strip, 399–401
  semi-infinite strip, 398–399
Submerged log, steady two-dimensional flow, 491–494
Sufficiency conditions, Cauchy–Riemann equations, 55–56
Sum function
  of complex sequences, 209–210
  of complex series, 214
  of series
    Cauchy sequences, 213
    repeated differentiation, 238–239
    Taylor series, 242–243
Summation of series
  Laurent series, 262–263

power series, repeated differentiation, 237–239
residue theorem, 311–315
closed form, 314–315
Symmetry
conformal mapping
eccentric circles onto annulus, 375–377
linear fractional transformation, 369–370
preservation of, 371
Schwarz–Christoffel transformation, 392

Taylor series
basic principles, 203–204
complex integration, 242–255
binomial expansion Maclaurin series, 251
linear combinations and products, 247–248, 251
Maclaurin series problems, 249–250
multiplication, 252
Taylor's theorem, 244–247
term-by-term integration, 251–252
uniqueness, 247
Laurent series comparisons, 260
Taylor's theorem
Cauchy–Goursat proof, 200–201
Cauchy–Riemann equations, sufficiency conditions, 55–56
series expansion, 244–247
Temperature distribution. *See* Steady-state temperature distribution
Term-by-term differentiation, uniform convergence, complex functions, 226–229
Term-by-term integration
Maclaurin series, 251–252
power series, 239–240
uniformly convergent series, complex functions, 225–226
Theory of residues, fundamental definite integral, Cauchy principal value, 165
Thermal conductivity, steady-state temperature distribution, 447–452
Thermal insulation, steady-state temperature distribution, 448–252
half space, insulated strip, 452–454
Thermal insulation condition, steady-state temperature distribution, 451
Time-dependent problems, boundary value problems, Neumann conformal mapping, 414–420
Topology, arcs and curves in, 30–34
Total boundaries, nonisolated singularities, simple regions, 113
Transformation, conformal mapping. *See also* Linear fractional transformation
defined, 334
Translation, linear transformation, 335–336
Triangle inequality
Cauchy–Goursat proof, 198–200
complex numbers, 7–8
limit definition, 39–40
Trigonometric functions
Cauchy integral formula, 132–134
complex numbers, De Moivre's theorem, 17
defined, 72–75, 83
inverse functions, 77–81
defined, 84
Trigonometric identity, Laurent series expansions, 276–277
Two-dimensional electrostatics, boundary value problems, 499–514
complex electrostatic potential, 508–509
conducting cylinder, uniform field potential, 512–514
infinite conducting plane at zero potential, infinite line charge, 511
infinite line charge, 509–511
potential distribution
eccentric cylinders, 506–508
hollow cylinder with semicircular cross section, 504–506
Two-dimensional problems, boundary value problems
Dirichlet conditions for harmonic functions, 442–443
Laplace equation
conformal mapping, 415–420
solutions, 421

Two-dimensional problems, boundary value problems (*Contd.*)
steady two-dimensional flow, 466–495
complex potentials, 481–482
conformal mapping, 487
flow over submerged log, 491–494
images method, 486–487
inclined flat plate, 487–491
line source/line sink, 474–477
Milne–Thompson circle theorem, 494–495
uniform parallel flow, 480–481
line source, 482–484
vortex, 477–480
vortex pair, 484–486

Unbounded sequences
behavior categories, 204–207
defined, 203
Laurent series, 256–257
Undetermined coefficients, rational functions, 65
Uniform boundedness, residue theorem, 313–314
Uniform convergence
complex functions
term-by-term differentiation, 226–229
term-by-term integration, 225–226
complex series, 219–229
Cauchy convergence principle, 222–223
complex functional series, 221–222
differentiation term by term, 226–229
integration term-by-term, 225–226
geometric series, 222
infinite series, continuous functions, 225
Weierstrass *M*-test, 223–225
infinite series, continuous functions, 225
Laurent series, 263
Taylor series, 245–247
Weierstrass *M*-test, 224–225
Uniform field, two-dimensional electrostatics, conducting cylinder, 512–514
Uniform parallel flow, steady two-dimensional flow, 480–481
inclined flat plate, 487–491
line source, 482–484
Uniqueness
boundary value problems, 420
conformal mapping, 353–354
critical points, 358
Laurent series, 263–264
Taylor's series, 247
Unit circle
Cauchy integral formula, 149–150
complex numbers, 17–19
inversion mapping, 345–346

Vanishing conditions, contour integrals around arc, 181–182
Vector representation
conformal mapping, 349–351
steady two-dimensional flow, 468–474
Velocity potential
Neumann boundary value problems, conformal mapping, 413–420
steady two-dimensional flow, 467–474
line source and line sink, 475–477
uniform parallel flow, 480–481
vortex, 477–480
Viscosity, steady two-dimensional flow, 467–474
Vortex, steady two-dimensional flow
cylindrical boundary, 477–480
image points, 486–487
pairing, 484–486
properties, 474

Weierstrass *M*-test
Cauchy convergence principle, 223–225
power series, qualitative behavior, 230–232
Winding number, residues and residue theorem, principle of the argument application, 318–320

Zero
polynomial equations, 22
rational functions, 64–66
repeated zeros, partial fraction expansion, 65–66

Zero potential, two-dimensional electrostatics, infinite conducting plane, line charge image, 511
Zeros of functions
classification, 287–288
fundamental theorem of algebra, 322
residues and residue theorem, principle of the argument application, 317–320
Rouché's theorem, 320–322

UNIVERSITY OF NOTTINGHAM
WITHDRAWN
FROM THE LIBRARY
10 0460463 1